MONTE CARLO METHODS IN CHEMICAL PHYSICS

ADVANCES IN CHEMICAL PHYSICS

VOLUME 105

MONTE CARLO METHODS IN CHEMICAL PHYSICS

Edited by

DAVID M. FERGUSON
Department of Medicinal Chemistry
University of Minnesota
Minneapolis, Minnesota

J. ILJA SIEPMANN
Department of Chemistry
University of Minnesota
Minneapolis, Minnesota

DONALD G. TRUHLAR
Department of Chemistry
University of Minnesota
Minneapolis, Minnesota

ADVANCES IN CHEMICAL PHYSICS
VOLUME 105

Series Editors

I. PRIGOGINE
Center for Studies in Statistical Mechanics
and Complex Systems
The University of Texas
Austin, Texas
and
International Solvay Institutes
Université Libre de Bruxelles
Brussels, Belgium

STUART A. RICE
Department of Chemistry
and
The James Franck Institute
The University of Chicago
Chicago, Illinois

An Interscience® Publication
JOHN WILEY & SONS, INC.
NEW YORK • CHICHESTER • WEINHEIM • BRISBANE • SINGAPORE • TORONTO

This book is printed on acid-free paper. ♾

Library of Congress Catalog Number 58-9935

ISBN: 0-471-19630-4

Printed in the United States of America.

10 9 8 7 6 5 4 3 2 1

CONTRIBUTORS TO VOLUME 105

GERARD T BARKEMA, ITP, Utrecht University, Utrecht, The Netherlands

DARIO BRESSANINI, Istituto di Scienze Matematiche Fisiche e Chimiche, Universitá di, Milano, sede di Como, Como, Italy

DAVID M. CEPERLEY, National Center for Supercomputing Applications and Department of Physics, University of Illinois at Urbana–Champaign, Urbana, Illinois

BRUCE W. CHURCH, Biophysics Program, Section of Biochemistry, Molecular and Cell Biology, Cornell University, Ithaca, New York

JUAN J. DE PABLO, Department of Chemical Engineering, University of Wisconsin—Madison, Madison, Wisconsin

FERNANDO A. ESCOBEDO, Department of Chemical Engineering, University of Wisconsin—Madison, Madison, Wisconsin

DAVID M. FERGUSON, Department of Medicinal Chemistry and Minnesota Supercomputer Institute, University of Minnesota, Minneapolis, Minnesota

DAVID G. GARRETT, Lockheed Martin TDS Eagan, St. Paul, Minnesota

MING-HONG HAO, Baker Laboratory of Chemistry, Cornell University, Ithaca, New York

WILLIAM L. HASE, Department of Chemistry, Wayne State University, Detroit, Michigan

J. KARL JOHNSON, Department of Chemical and Petroleum Engineering, University of Pittsburg, Pittsburg, Pennsylvania

DAVID A. KOFKE, Department of Chemical Engineering, State University of New York at Buffalo, Buffalo, New York

ANDRZEJ KOLINSKI, Department of Chemistry, University of Warsaw, Warsaw, Poland

MICHAEL MASCAGNI, Program in Scientific Computing and Department of Mathematics, University of Southern Mississippi, Hattiesburg, Mississippi

MARK E. J. NEWMAN, Santa Fe Institute, Santa Fe, New Mexico

M. P. NIGHTINGALE, Department of Physics, University of Rhode Island, Kingston, Rhode Island

GILLES H. PESLHERBE, Department of Chemistry and Biochemistry, University of Colorado at Boulder, Boulder, Colorado

PETER J. REYNOLDS, Physical Sciences Division, Office of Naval Research, Arlington, Virginia

HAROLD A. SCHERAGA, Baker Laboratory of Chemistry, Cornell University, Ithaca, New York

DAVID SHALLOWAY, Biophysics Program, Section of Biochemistry, Molecular and Cell Biology, Cornell University, Ithaca, New York

J. ILJA SIEPMANN, Department of Chemistry and Department of Chemical Engineering and Materials Science, University of Minnesota, Minneapolis, Minnesota

JEFFREY SKOLNICK, Department of Molecular Biology, The Scripps Research Institute, La Jolla, California

ASHOK SRINIVASAN, National Center for Supercomputing Applications, University of Illinois at Urbana–Champaign, Illinois.

ROBERT Q. TOPPER, Department of Chemistry, School of Engineering, The Cooper Union for the Advancement of Science and Art, New York, New York

ALEX ULITSKY, Biophysics Program, Section of Biochemistry, Molecular and Cell Biology, Cornell University, Ithaca, New York

CYRUS J. UMRIGAR Cornell Theory Center and Laboratory of Atomic and Solid State Physics, Cornell University, Ithaca, New York

JOHN P. VALLEAU, Chemical Physics Theory Group, Department of Chemistry, University of Toronto, Toronto, Canada

HAOBIN WANG, Department of Chemistry, University of California, Berkeley, Berkeley, California

PREFACE

The Monte Carlo method is pervasive in computational approaches to many-dimensional problems in chemical physics, but there have been few if any attempts to bring all the dominant themes together in a single volume. This volume attempts to do just that—at a state-of-art level focusing on the main application areas. More detailed general introductions to the Monte Carlo method can be found in the excellent textbooks by Hammersley and Handscomb (1964) and Kalos and Whitlock (1986).

The term *Monte Carlo*, which refers to the famous gambling city and arises from the role that random numbers play in the Monte Carlo method (see the second chapter in this volume, by Srinivasan et al.), was suggested by Metropolis, and the method itself was first used in 1949 (Metropolis and Ulam, 1949). Von Neumann and Ulam (1945) had earlier realized that it is possible to solve a deterministic mathematical problem, such as the evaluation of a multidimensional integral, with the help of a stochastic sampling experiment. In fact, it turns out that if the dimensionality of an integal is of order of 10 or larger, the Monte Carlo method becomes the preferred technique of numerical integration. Such multidimensional integrals play an important role in many branches of physics and chemistry, especially in statistical mechanics.

Several of the chapters in this volume are concerned with the calculation of thermodynamic ensemble averages for systems of many particles. An introduction to this key application area is presented in the first chapter (by Siepmann), and advanced work is discussed in the last six chapters in this volume (by de Pablo and Escobedo, Valleau, Kofke, Siepmann, Johnson, and Barkema and Newmann). There are a large number of monographs and edited volumes with a major emphasis on techniques like those described above and their application to a wide variety of molecular simulations. Allen and Tildesley (1987), Heerman (1990), Binder and Heerman (1992), Binder (1995), and Frenkel and Smit (1996) may be consulted as a core library in this area.

Although the simulation of ensembles of particles constitutes a huge literature, it is only one aspect of the use of Monte Carlo methods in chemical physics. Basically, whenever one has to calculate integrals over a high-dimensionality space, the Monte Carlo method is potentially the algorithm of choice. Thus, for example, in few-body simulations of chemical reactions, one must integrate over the initial conditions of the reactant molecules, and this application is covered in the sixth chapter in this volume (by Peslherbe et al.).

Feynman's path-integral method also involves multidimensional integrals, and the dimensionality of these integrals can be much higher than the number of actual physical degrees of freedom of the system. The application of Monte Carlo methods to this problem is covered in the fifth chapter in this volume (by Topper). The Monte Carlo method arises somewhat differently in applications of quantum mechanics based on the Schrödinger equation. There are two basic approaches. The first is based on a variational approach in which parameters in a trial function are varied, and variational expectation values over the trial function are evaluated by Monte Carlo quadratures. The second approach is different in that one actually solves the Schrödinger equation. The method is based on an analogy of the Schrödinger equation to a diffusion equation or stochastic process, which is quite natural to solve by a random walk. These approaches are discussed in the third and fourth chapters in this volume (by Bressanini and Reynolds and Nightingale and Umrigar).

In the simulation of ensembles of condensed-phase particles, one must often combine the techniques used to sample over the motions of the particles as a whole (these are the same techniques used to simulate a liquid of noble-gas atoms) with the techniques for sampling the internal motions (vibrations and rotations) of the molecules themselves. The latter problem becomes particularly significant for large molecules with internal rotations and other low-frequency, wide-amplitude motions. Often there are many local minima of the molecular potential-energy function, and one must develop specialized techniques to ensure that these are all sampled appropriately. This problem is especially covered in the seventh through eleventh chapters in this volume (by Skolnick and Kolinski, Scheraga and Hao, Church et al., Ferguson and Garrett, and de Pablo and Escobedo).

An interesting development of the last 10 years that has made Monte Carlo methods even more pervasive in the field of scientific computation is the prominence of parallel computation. Often, but not always, Monte Carlo algorithms are more amenable to parallelization than are competitive approaches. The quest for parallelism has led to a new appreciation for the power and advantages of uncorrelated samples, and also to many new research questions. For example, the ever-present search for methods of generating long strings of uncorrelated random numbers to drive the algorithms presents new challenges; this is discussed in the second chapter in this volume (by Srinivasan et al.).

The Monte Carlo method is so general, and its use is so pervasive in chemical physics, that there will surely be further development of these techniques in the future. Will the Monte Carlo approach to simulation be the single most widely used paradigm in computational science in the new millennium? Only time will tell, but the probability is high.

REFERENCES

M. P. Allen, and D. J. Tildesley, *Computer Simulation of Liquids*, Clarendon Press, Oxford, 1987.

K. Binder, *Monte Carlo and Molecular Dynamics Simulations in Polymer Sciences*, Oxford, New York.

K. Binder, and D. W. Heerman, *Monte Carlo Simulation in Statistical Physics: An Introduction*, 2nd ed., Springer, Berlin, 1992.

D. Frenkel, and B. Smit, *Understanding Molecular Simulation: From Algorithms to Applications*, Academic Press, New York, 1996.

J. M. Hammersley, and D. C. Handscomb, *Monte Carlo Methods*, Methuen, London, 1964.

D. W. Heerman, *Computer Simulation Methods in Theoretical Physics*, 2nd ed., Wiley, New York, 1990.

M. H. Kalos, and P. A. Whitlock, *Monte Carlo Methods*, Wiley, New York, 1986.

N. Metropolis, and S. Ulam, *J. Am. Stat. Assoc.* **44**, 335 (1949).

J. von Neumann, and S. Ulam, *Bull. Am. Math. Soc.*, **51**, 660 (1945).

DAVID M. FERGUSON
J. ILJA SIEPMANN
DONALD G. TRUHLAR

University of Minnesota

CONTENTS

MONTE CARLO METHODS IN CHEMICAL PHYSICS

ADVANCES IN CHEMICAL PHYSICS

VOLUME 105

AN INTRODUCTION TO THE MONTE CARLO METHOD FOR PARTICLE SIMULATIONS

J. ILJA SIEPMANN

Department of Chemistry and Department of Chemical Engineering and Materials Science
University of Minnesota, Minneapolis, MN 55455-0431

CONTENTS

This chapter makes a humble attempt to introduce some of the basic concepts of the Monte Carlo method for simulating ensembles of particles. Before we jump into the heart of the matter, we would like to direct the interested reader to a more general description of different Monte Carlo methods by Hammersley and Handscomb [1] and to the excellent textbooks on particle simulations by Allen and Tildesley [2], Binder and Heermann [3], and Frenkel and Smit [4].

In most cases, the Monte Carlo method is used not to compute the absolute value of an integral

$$I = \int f(x)\ dx \tag{1.1}$$

but rather to compute an average (or ratio)

$$\langle g \rangle = \frac{\int g(x) f(x)\ dx}{\int f(x)\ dx} \tag{1.2}$$

Advances in Chemical Physics, Volume 105, Monte Carlo Methods in Chemical Physics, edited by David M. Ferguson, J. Ilja Siepmann, and Donald G. Truhlar. Series Editors I. Prigogine and Stuart A. Rice.
ISBN 0-471-19630-4

which is a simpler problem. To understand the use of the Monte Carlo method, we consider a simple example, namely, the packing of N had disks in a defined area (X,Y). To construct a configuration of N disks, N combinations of two random numbers x and y with $0 \leq x \leq X$ and $0 \leq y \leq Y$ are generated. Any such combination defines the position of a hard-disk center. After all N hard-disk positions have been computed, a test for overlap of the disks is made. If an overlap occurs, the configuration is said to be forbidden. The procedure is then repeated and the configurational properties of the hard-disk system are calculated as unweighted averages over all allowed configurations (more generally, as weighted averages over all configurations, but in the special hard-disk case all forbidden configurations have a zero weight and all allowed configurations have the same weight). This method is referred to as "hit-and-miss integration" [2]. The disadvantage of the simple hit-and-miss method is that it is unlikely that any allowed configurations are generated at high densities. A very large number of configurations therefore have to be calculated in order to obtain a few that contribute to the sum over states.

While elegant in its simplicity, such an approach is extremely problematic when computing properties of physical systems. To appreciate this, we need to introduce some fundamental relationships of classical statistical mechanics. The volume of phase space that can be accessed by the N hard disks in the example above is called the *canonical partition function*, Q:

$$Q = \mathscr{C} \int \cdots \int \exp\left[-\frac{H(\mathbf{q}^N, \mathbf{p}^N)}{k_B T} \right] d\mathbf{p}^N \, d\mathbf{q}^N \tag{1.3}$$

where $\int \cdots \int$ symbolizes a multidimensional integral; for convenience we will use a single $\int$ regardless of the dimensionality of the integral for the remainder of this chapter. $\mathbf{q}^N$ and $\mathbf{p}^N$ stand for the coordinates and the conjugate momenta of all N particles, and $\mathscr{C}$ is a constant of proportionality relating the quantum-mechanical to the classical descriptions [5]. k_B is Boltzmann's constant, and T is the absolute temperature. The Hamiltonian, $H(\mathbf{q}^N, \mathbf{p}^N)$, expresses the total energy of an isolated system as a function of the coordinates and momenta of the constituent particles. H can be divided into a kinetic-energy part, K, and a potential-energy part, U. In the canonical ensemble (constant-NVT) the statistical average of an observable *mechanical* property X, a function of coordinates and/or momenta, can be expressed as

$$\langle X \rangle = \frac{1}{Q} \int X(\mathbf{q}^N, \mathbf{p}^N) \exp[-\beta H(\mathbf{q}^N, \mathbf{p}^N)] \, d\mathbf{p}^N \, d\mathbf{q}^N \tag{1.4}$$

where $\beta = (k_B T)^{-1}$. As K is a quadratic function of individual momenta, the integration over momenta can be carried out analytically. However, in computer simulation, we are interested mainly in properties that depend on the particle coordinates. Solving the multidimensional *configuration integral* over $\mathbf{q}^N$ is the main application of the Monte Carlo method in particle simulations.

II. THE METROPOLIS MONTE CARLO METHOD

How can the ratio of a multidimensional integral [such as in Eq. (1.4)] be calculated in a more efficient way than the purely random hit-and-miss integration? Here the Metropolis method [6] comes into action. The system follows now some "time" evolution instead of being generated from scratch every time. First, an initial configuration of the hard-disk system, in which no overlap occurs, is chosen arbitrarily. Then the trial configurations are generated step by step. One particle is picked at random and then displaced in the x and y directions by an amount that is calculated from two independent random numbers. If there is an overlap in the resulting trial configuration, the trial move is rejected, and the old configuration is counted again; if not, it is accepted and the next trial is based on the new configuration. After a sufficient number of trial steps, the configurational properties can be calculated as unweighted averages over all configurations that were visited during the course of the simulation. Here, we see clearly the essence of the Metropolis method: "instead of choosing configurations randomly, then weighting them with $\exp(-\beta U)$ (their Boltzmann factor), we choose configurations with a probability $\exp(-\beta U)$ and weight them evenly" [6]. Thus the Metropolis scheme is an *importance sampling* technique. Importance sampling ensures that the regions of phase space that make the largest contribution to the configurational averages are the regions that are sampled most frequently.

So far, we have only considered systems of hard disks in which all allowed configurations have the same potential energy and therefore make the same contribution to the ensemble average. A generalization is needed in order to treat systems in which states have different potential energies. The problem is to find a method of generating a sequence of random states in such a way that at the end of the simulation each state has occurred with the appropriate probability, that is, the phase space is sampled according to the equilibrium distribution appropriate to the ensemble. Here we run into a problem; to know the absolute probability of a configuration, a priori knowledge of the configurational partition function, Q_c (defined as $Q_c = \mathscr{C} \int \exp[-\beta H(\mathbf{q}^N)]$), is required, which in all except the simplest cases is unknown. The Metropolis method [6] provides a very simple and ingenious way to avoid the problem stated above. This is due to the fact that in the Metropolis scheme the probabilities of configurations appear only in ratios (see Eqs. 2.5 and 2.6).

Let us go back to generalities. We want to generate a *Markov chain* of trial configurations. A Markov chain is a sequence of trials in which the outcome of each trial is one of a set of states, the state space or configurational space, and depends only on the state that immediately precedes it [7]. Two states m and n are linked by a transition probability, π_{mn}, which is an element of a stochastic transition matrix, Π. Thus the probability distribution $\underline{\rho}(t)$ after t steps is given in terms of an arbitrary initial distribution $\underline{\rho}(0)$ by

$$\underline{\rho}(t) = \Pi^t \underline{\rho}(0) \tag{2.1}$$

The quantity $\underline{\rho}$ is a vector in which each element gives the probability for the corresponding state; thus the number of elements of $\underline{\rho}$ is equal to the number of possible states of the system. Because Π^t is also a stochastic matrix, there exists a limiting distribution, $\mathscr{P}$, which is given by

$$\mathscr{P} = \lim_{t \to \infty} \Pi^t \underline{\rho}(0) \tag{2.2}$$

This leads to an eigenvalue equation for the limiting distribution, with an eigenvalue unity

$$\Pi \mathscr{P} = \mathscr{P} \tag{2.3}$$

Once the limiting distribution is reached, it will persist; this result is called the *steady-state* condition. If there exists one eigenvalue of unity, there exists only one limiting distribution, which is independent of the initial distribution. It can be shown that a unique limiting distribution exists if any state of the system can be reached from any other state with a nonzero multistep transition probability. In this case the system (or Markov chain) is *ergodic*.

It is obvious that the stochastic matrix, Π, and its elements, π, must fulfill one condition, namely, that it does not destroy the limiting distribution once it is reached. In other words, in equilibrium the number of accepted trial moves that have the outcome that the system leaves a state i in a given time (number of Monte Carlo steps) must be exactly balanced by the number of moves from any other state that end in state i. To find the transition matrix that satisfies all the necessary conditions, the stronger condition of *microscopic reversibility* or *detailed balance* is conveniently used [2]:

$$\rho_m \pi_{mn} = \rho_n \pi_{nm} \tag{2.4}$$

where ρ_i is the equilibrium probability to find the system in state i. Suppose that the system is in state m. An attempt is then made to move to an adjacent state n, the trial configuration. Let α_{mn} be the symmetric probability matrix, often called the *underlying matrix* of the Markov chain, of choosing n as the trial state. The solution to Eq. (2.4) proposed by Metropolis et al. [6] is an asymmetric one in which

$$\begin{aligned} \pi_{mn} &= \alpha_{mn} \quad \text{if} \quad \rho_n \geq \rho_m \quad m \neq n \\ \pi_{mn} &= \alpha_{mn}\left(\frac{\rho_n}{\rho_m}\right) \quad \text{if} \quad \rho_n \geq \rho_m \quad m \neq n \end{aligned} \tag{2.5a}$$

and the overall probability that the system remains in state m is

$$\pi_{mm} = 1 - \sum_{m \neq n} \pi_{mn} \tag{2.5b}$$

When sampling from a (canonical) Boltzmann distribution, it follows from Eq. (1.4) that the limiting probability, ρ_m, is given by

$$\rho_m = \frac{\exp(-\beta U_m)}{Q_c} \tag{2.6}$$

As mentioned earlier, the limiting probabilities occur only in the ratio ρ_n/ρ_m and the value of the denominator, Q_c, is therefore not required. Hence, in the canonical ensemble, the acceptance of a trial move depends only on the Boltzmann factor of the energy difference, ΔU, between the states m and n. If the system looses energy, then the trial move is always accepted; if the move goes uphill, that is, if the system gains energy, then we have to play "a game of chance on $\exp(-\beta\,\Delta U)$" [6]. In practical terms, we have to calculate a random number, ξ uniformly distributed in the interval [0,1]. If $\xi < \exp(-\beta\,\Delta U)$, then we accept the move, and reject otherwise.

Another possibility for Π is the symmetric solution [8] given by

$$\begin{aligned} \pi_{mn} &= \frac{\alpha_{mn}\rho_n}{\rho_m + \rho_n} \qquad m \neq n \\ \pi_{mm} &= 1 - \sum_{m \neq n} \pi_{mn} \end{aligned} \tag{2.7}$$

Again the limiting probabilities occur only as a ratio and the value of Q_c is not required. In most computer simulations, the Metropolis acceptance criterion is chosen, because it appears to sample phase space more efficiently than the other schemes that have been proposed [9].

To emphasize how the Metropolis scheme works in practice, let us return once more to the hard-disk example. Here the matrix α_{mn} represents the probability of moving a particle to a new trial positions, which is achieved by the generation of two random numbers for displacements in the x and y directions. To make the underlying matrix symmetric, the displacements are calculated from $d_{max}(2\xi - 1)$, where d_{max} is a user-specified displacement that can be optimized to improve the sampling [2–4]. The elements of the vector $\underline{\rho}$ are positive and nonzero for allowed and zero for forbidden configurations. Thus it follows for a move to an allowed trial configuration

$$\pi_{mn} = \alpha_{mn} \qquad \rho_m = \rho_n > 0 \tag{2.8a}$$

and if the trial configuration results in an overlap

$$\pi_{mn} = 0 \qquad \rho_m > 0 \qquad \rho_n = 0 \tag{2.8b}$$

It is clear for this example that the condition of microscopic reversibility is fulfilled, because for the move from m to n and from n back to m the transition probability π_{mn} is the same if the underlying matrix is symmetric.

The presentation of the Monte Carlo method in terms of a Markov chain with a "time" parameter [see Eq. (2.1)] allows us to visualize the process of generating a succession of configurations as a dynamical one [10]. These dynamics are, of course, not the true physical dynamics. Nevertheless, in some cases the Markov time evolution resembles the kinetics of the system [11,12].

Finally, we should make some remarks about the choice of the underlying Markov matrix. There are two general types of trial moves: Glauber-like excitations [13], which change a single-site property; and Kawasaki-like excitations [14], which are caused by a two-site exchange of a single-site property. Combinations of different moves of the same class or of different classes in one simulation are also allowed, but care has to be taken that the choice of a move at a particular instance during the simulation is not a deterministic one because this might violate the condition of microscopic reversibility [9]. The same is true for the selection of a particle for which the move is attempted. As shown explicitly by Frenkel [9], it is not efficient to attempt a defined number, normally N, of Glauber-like moves at the same time. There is much freedom in the design of trial moves as long as the stochastic matrix Π yields the desired steady state. In particular, the underlying matrix α_{mn} need not be symmetric (which would require that the acceptance of a move is based also on the ratio α_{nm}/α_{mn} [2]), thus allowing rather "unconventional" and "unphysical" moves (referring to the fact that the moves do not correspond to the natural time evolution of the physical system) to enhance the speed with which configuration space is sampled, thereby making the Monte Carlo method "the most powerful and commonly used technique for analyzing complex problems" [15].

III. MONTE CARLO SIMULATIONS IN OTHER ENSEMBLES

In the last section we have assumed that we perform our simulation for a fixed number, N, of particles at constant temperature, T, and volume, V, the canonical ensemble. A major advantage of the Monte Carlo technique is that it can be easily adapted to the calculation of averages in other thermodynamic ensembles. Most "real" experiments are performed in the isobaric–isothermal (constant-NpT) ensemble, some in the grand–canonical (constant-μVT) ensemble, and even fewer in the canonical ensemble, the standard Monte Carlo ensemble, and near to none in the microcanonical (constant-NVE) ensemble, the standard ensemble for molecular-dynamics simulations.

Consider first the microcanonical ensemble. Obviously it is against the true spirit of the Monte Carlo method to keep the energy fixed (except for hard-particle simulations), since trial moves are either accepted or rejected according to the energy difference caused by them. We can circumvent this problem by introducing an additional degree of freedom, an energy *demon*, to the system, as suggested by Creutz [16]. Now we can fix the combined energy of the system and the demon. During the simulation, the demon's energy will decide in a very simple way if we have to accept or reject a trial move; specifically, if the energy of the trial configuration is smaller than the fixed combined energy, then we accept it and change the energy of the demon accordingly. Thus no random number is needed to decide on the acceptance of a trial move.

The isobaric–isothermal ensemble and its close relative, the isotension–isothermal ensemble, are often used in Monte Carlo simulations. Finite-size effects play a smaller role at constant pressure than at constant volume. Thus the system can change phase more freely and reduce its Gibbs free energy. The constant-NpT ensemble was introduced by Wood [17] with the limitation to hard particles and later extended by McDonald [18] to continuous potentials. In the isobaric–isothermal ensemble the volume of the simulation box is allowed to fluctuate. Therefore the coordinates $\mathbf{q}^N$ of the particles depend on the volume, and it is convenient to use scaled coordinates (assuming a cubic simulation box) such as $\mathbf{s} = V^{-1/3}\mathbf{q}$. Besides normal translational (and for molecules, also rotational and conformational moves), we can now attempt trial moves that change the volume (or the length) of the simulation box. In the Metropolis scheme such a random volume change is accepted if

$$\pi_{V \to V'} = \exp\left\{-\beta\left[U(\mathbf{s}^N, V') - U(\mathbf{s}^N, V) + p_{\text{ext}}(V' - V) - N\beta^{-1}\ln\left(\frac{V'}{V}\right)\right]\right\} > \xi \quad (3.1)$$

where p_{ext} is the specified pressure of the external pressure bath and ξ is again a uniform random number in the interval [0,1]. Care has to be taken for molecular systems. Here the internal length of the bonds may have to be kept fixed and only the center-of-mass coordinates can be scaled. In the isotension-isothermal ensemble [19,20] the simulation box is additionally allowed to change its shape. This is of particular importance in the simulation of solid systems.

In grand–canonical Monte Carlo the chemical potential is fixed whereas the number of particles fluctuates [21,22]. Here, besides the conventional trial displacements, moves are attempted that add or remove a particle and that are accepted with the following probabilities

$$\begin{aligned} \pi_{N \to N+1} &= \frac{V}{(N+1)\Lambda^3}\exp\{-\beta[U(\mathbf{q}^{N+1}) - U(\mathbf{q}^N) - \mu]\} > \xi \\ \pi_{N+1 \to N} &= \frac{(N+1)\Lambda^3}{V}\exp\{-\beta[U(\mathbf{q}^N) - U(\mathbf{q}^{N+1}) + \mu]\} > \xi \end{aligned} \quad (3.2)$$

where $\Lambda = \sqrt{h^2/(2\pi m k_B T)}$ is the thermal de Broglie wavelength. The advantage of the grand–canonical Monte Carlo technique is that it allows us to measure the difference in free energy, a thermal property, between a reference state of the system and the state in which the simulation is performed.

The "ideal" ensemble for simulations would be an ensemble where chemical potential, pressure, and temperature are constant following Gibbs' condition for coexisting phases. No such ensemble exists in reality, however, Panagiotopoulos [23] has suggested an ensemble that comes very close to the ideal, which he called the *Gibbs ensemble*. A simulation in the Gibbs ensemble is based on (at least) two subsystems (primed and unprimed), with N and N' particles and with volumes V and V', which are in thermodynamic contact. The difference of the chemical potential of the particles in the two simulation boxes is fixed by allowing particles to swap in some way through an imaginary interface from one box to the other (i.e., in equilibrium $\Delta\mu = 0$), but the absolute value of the chemical potential is not prespecified as an input parameter. The probability of acceptance for this move is given by

$$\pi_{i \in V' \to i \in V} = \frac{V(N')}{V'(N+1)} \exp\{-\beta[U(\mathbf{s}^{N+1}, V) - U(\mathbf{s}^{N}, V) + U(\mathbf{s}^{N'-1}, V') - U(\mathbf{s}^{N'}, V')]\} > \xi \quad (3.3)$$

To ensure that the two sub-systems are at the same pressure, the total volume of both simulations boxes is held constant, but volume can be traded between the systems (i.e., $V \to V + \Delta V$ and $V' \to V' - \Delta V$), as in the original work [23], or more generally (if more than one component is involved), the two boxes are kept in contact with a constant pressure bath [24]. In the former case, the probability of accepting the combined volume change is the product of two volume changes as in Eq. (3.1), but in the latter case it is analogous to Eq. (3.1). If the overall system is in the two-phase region, then the two subsystems will automatically find the densities and compositions suitable for phase coexistence. Thus, using simulations in the Gibbs ensemble it is straightforward to gain detailed information on phase coexistence lines.

IV. ASSESSING THE VALIDITY OF MONTE CARLO SIMULATIONS

The main advantage of the Monte Carlo method, namely, the great freedom in selecting a stochastic matrix Π that enhances the sampling of phase space, is also its main curse. Although a molecular dynamics program can be easily adapted to a variety of systems, a Monte Carlo program will always require tailoring of the underlying Markov matrix (e.g., the addition of new types of trial moves) to be efficient. Thus a general-purpose Monte Carlo program might be far from the optimal solution for a specific problem.

When designing a new Monte Carlo algorithm, it is imperative to show that the new algorithm samples the desired ensemble distribution. This proof is usually done by testing for the detailed balance criterion. It should also be kept in mind that in addition to the detailed balance criterion the stochastic matrix Π should connect all states with a nonzero (multistep) transition probability; that is, every state can be reached starting from any one state, leading to a Markov chain that is said to be ergodic or irreducible. However, in almost all cases there exists no simple mathematical proof of the ergodicity of a given Monte Carlo algorithm. Furthermore, it is always a good idea to test new Monte Carlo algorithms on simple problems for which an analytical solution exists or that are amenable to proven standard simulation algorithms.

If a Monte Carlo algorithm or program has passed all the tests mentioned above, then it can be applied to the problem of interest. However, the user has to remain vigilant to ensure the validity of the simulations. For very good discussions on this subject, see Refs. 4 and 25. A major problem for particle simulations is that the equilibrium state of the system is not known a priori; thus, the simulation has to be started from an initial configuration guessed by the user. This initial configuration may be far from any configuration that contributes substantially to the ensemble averages. To remove any bias stemming from the initial configuration, the Monte Carlo program has to complete the task of reaching the important regions of phase space and the steady-state or equilibrium distribution. But how can the user know at any point during the simulation whether the steady-state distribution has been reached? The obvious check is to monitor the values of various properties along the trajectory of the Monte Carlo simulation. Once equilibrium is reached, the average value of any property should remain steady (fluctuate around its mean value) throughout the remainder of the simulation. However, since different properties may approach an apparently steady state at different rates, it is advisable to judge for equilibration by considering a set of different properties. The other prudent safety measure is to start the simulation from two (or many) very different initial configuration and to ensure that the same steady-state distribution is reached irrespective of initial configuration. Along the same line, it is also worthwhile to test that the same steady state distribution is reached when a different *pseudo* random number sequence is used (see also the second chapter in this volume, by Srinivasan et al.).

Once the steady-state distribution has been reached, the production period of the simulation can be started. How long a production period is required to produce reliable results? Ideally, the length of the production period of the simulation over which the averages are calculated should be at least an order of magnitude longer than the number of Monte Carlo steps required to sample the slowest relaxation inherent to the system (which can be estimated from the corresponding correlation function). However, it might not always be obvious what this slowest relaxation is. A related problem is the estimation of the statistical uncertainties in the ensemble averages [see Eq. (1.4)]. Since a Markov chain is followed, successive configurations are most often strongly correlated. The statistical errors in the simulation are usually estimated by dividing the Markov chain into equal-length blocks. However, care has to be taken to ensure the statistical independence of the blocks (which of course is related to the slowest relaxation). The blocks can be considered independent (long enough) when the resulting apparent error (statistical error of the mean calculated from the individual blocks) becomes independent of their length [25]. Another good reliability test is to calculate the probability distributions for some properties. For example, if a fluid of chain molecules (with symmetric torsional potential) is simulated, then the distribution of dihedral angles should be symmetric.

Finally, there is the problem of finite-size errors, specifically, the results obtained for the (relatively) small, but periodic systems being simulated may not be those found in the thermodynamic limit. This problem is most evident when simulations are performed closed to critical points where fluctuations of macroscopic length scale can be observed [3,26,27]. However, problems can also occur close to phase transitions where the periodicity of the system might stabilize one phase over the other [4]. If possible, some test simulations for smaller and larger system sizes (number of particles) should therefore be carried out to evaluate the finite-size corrections. For the special case of the excess chemical potential, an expression for the leading [$\mathcal{O}(N^{-1})$] system-size dependence has been derived [28].

ACKNOWLEDGMENTS

Ian McDonald and Daan Frenkel are gratefully acknowledged for introducing me to and sharing with me their enthusiasm on the subject of this chapter.

REFERENCES

1. J. M. Hammersley and D. C. Handscomb, *Monte Carlo Methods* (Methuen, London, 1964).
2. M. P. Allen and D. J. Tildesley, *Computer Simulation of Liquids*, Clarendon Press, Oxford, 1987.
3. K. Binder and D. W. Hermann, *Monte Carlo Simulation in Statistical Physics: An Introduction*, 2nd ed., Springer, Berlin, 1992.
4. D. Frenkel and B. Smit, *Understanding Molecular Simulation: From Algorithms to Applications*, Academic Press, New York, 1996.

5. D. A. McQuarrie, *Statistical Mechanics*, Harper Collins, New York, 1976.
6. N. Metropolis, A. W. Rosenbluth, M. N. Rosenbluth, A. H. Teller and E. Teller, *J. Chem. Phys.* **21**, 1087 (1953).
7. W. Feller, *An Introduction to Probability Theory and Its Applications*, Vol. 1, Wiley, New York, 1950.
8. W. W. Wood and J. D. Jacobsen, "Monte Carlo Calculations in Statistical Mechanics," *Proceedings of the Western Joint Computer Conference*, San Francisco, 1959.
9. D. Frenkel, in *Computer Modelling of Fluids, Polymers and Solids*, C. R. A. Catlow, S. C. Parker and M. P. Allen, eds., Kluwer, Dordrecht, 1990.
10. O. G. Mouritsen, *Computer Studies of Phase Transitions and Critical Phenomena*, Springer, Berlin, 1984.
11. K. Binder and M. H. Kalos, in *Monte Carlo Methods in Statistical Physics*, 2nd ed., K. Binder, ed., Springer, Berlin, 1986.
12. K. Kremer, in *Computer Simulation in Chemical Physics*, M. P. Allen and D. J. Tildesley, eds., Kluwer, Dordrecht, 1993.
13. R. J. Glauber, *J. Math. Phys.* **4**, 294 (1963).
14. K. Kawasaki, in *Phase Transitions and Critical Phenomena*, C. Domb and M. S. Green, eds., Vol. 2, Academic Press, London, 1972.
15. R. Y. Rubinstein, *Simulation and Monte Carlo Methods*, Wiley, New York, 1981.
16. M. Creutz, *Phys. Rev. Lett.* **50**, 1411 (1983).
17. W. W. Wood, *J. Chem. Phys.* **48**, 415 (1968).
18. I. R. McDonald, *Chem. Phys. Lett.* **3**, 241 (1969); *Mol. Phys.* **23**, 41 (1972).
19. R. Najafabadi and S. Yip, *Scripta Metall.* **17**, 1199 (1983).
20. J. R. Ray and A. Rahman, *J. Chem. Phys.* **80**, 4423 (1984).
21. Z. W. Salsburg, J. D. Jacobson, W. Fickett, and W. W. Wood, *J. Chem. Phys.* **30**, 65 (1959).
22. G. E. Norman and V. S. Filinov, *High Temp. Res. USSR* **7**, 216 (1969).
23. A. Z. Panagiotopoulos, *Mol. Phys.* **61**, 813 (1987).
24. A. Z. Panagiotopoulos, N. Quirke, M. Stapleton, and D. J. Tildesley, *Mol. Phys.* **63**, 527 (1988).
25. J. P. Valleau, in *Computer Simulations in Materials Science*, M. Meyer and V. Pontikis, eds., Kluwer, Dordrecht, 1991.
26. A. Z. Panagiotopoulos, *Int. J. Thermophys.* **15**, 1057 (1994).
27. N. B. Wilding, *Phys. Rev. E* **52**, 602 (1995).
28. J. I. Siepmann, I. R. McDonald, and D. J. Frenkel, *J. Phys. Cond. Matt.* **4**, 679 (1992).

RANDOM NUMBER GENERATORS FOR PARALLEL APPLICATIONS

ASHOK SRINIVASAN

National Center for Supercomputing Applications
University of Illinois at Urbana–Champaign, Urbana, IL 61801

DAVID M. CEPERLEY

National Center for Supercomputing Applications and Department of Physics
University of Illinois at Urbana–Champaign, Urbana, IL 61801

MICHAEL MASCAGNI

Program in Scientific Computing and Department of Mathematics
University of Southern Mississippi, Hattiesburg, MS 39406

CONTENTS

Advances in Chemical Physics, Volume 105, Monte Carlo Methods in Chemical Physics, edited by David M. Ferguson, J. Ilja Siepmann, and Donald G. Truhlar. Series Editors I. Prigogine and Stuart A. Rice.
ISBN 0-471-19630-4

I. INTRODUCTION

Random numbers arise in computer applications in several different contexts, such as

1. In the Monte Carlo method to estimate a many-dimensional integral by sampling the integrand. Metropolis Monte Carlo or, more generally, Markov chain Monte Carlo (MCMC), to which this volume is mainly devoted, is a sophisticated version of this where one uses properties of random walks to solve problems in high-dimensional spaces, particularly those arising in statistical mechanics.
2. In modeling random processes in nature such as those arising in ecology or economics.
3. In cryptography, one uses randomness to hide information from others.
4. Random numbers may also be used in games, for example, during interaction with the user.

It is only the first class of applications to which this chapter is devoted, because these computations require the highest quality of random numbers. The ability to do a multidimensional integral relies on properties of uniformity of n-tuples of random numbers and/or the equivalent property that random numbers be uncorrelated. The quality aspect in the other uses is normally less important simply because the models are usually not all that precisely specified. The largest uncertainties are typically due more to approximations arising in the formulation of the model than those caused by lack of randomness in the random number generator.

In contrast, the first class of applications can require very precise solutions. Increasingly, computers are being used to solve very well defined but difficult mathematical problems. For example, as Dirac [1] observed in 1929, the physical laws necessary for the mathematical theory of a large part of physics and the whole of chemistry are completely known and it is only necessary to find precise methods for solving the equations for complex systems. In the intervening years fast computers and new computational methods have come into existence. In quantum chemistry, physical properties must be calculated to "chemical accuracy" (say, 0.001 Rydberg) to be relevant to physical properties. This often requires a relative accuracy of 10^5 or better. Monte Carlo methods are used to solve the "electronic structure problem" often to high levels of accuracy [2]. (See also chapters by Reynolds and Nightingale in this book). In these methods one can use from 10^7 to 10^{12} random numbers, and subtle correlations between these numbers could lead to significant errors.

Another example is from the numerical study of phase transitions. Renormalization theory has proved accurate for the basic scaling properties of simple transitions. The attention of the research community is now shifting to corrections to scaling, and to more complex models. Very long simulations (also of the MCMC type) are done to investigate this effect, and it has been discovered that the random number generator can influence the results [3–6]. As computers become more powerful, and Monte Carlo methods become more commonly used and more central to scientific progress, the quality of the random number sequence becomes more important.

Given that the quality (which we shall define in a moment) of random numbers is becoming more and more important, the unfortunate fact is that important aspects of quality are very difficult to prove mathematically. The best one can do today is test empirically. But an empirical test is always finite. We will report here tests on random number streams that are of record length (up to about 10^{12} numbers). However, they will have to be redone in a few years with even longer sequences. Also, important algorithms use random numbers in a way that is difficult to encapsulate in a test for which we know the answer, and so we must resort to general guidelines on safe ways to use random number generators in practice.

This chapter constitutes a brief review of recent developments in random number generation. There are several excellent reviews of the older literature. In particular, we recommend the reader seriously interested in random number generation read the lengthy introduction in Knuth [7] and the shorter introduction in the second edition of *Numerical Recipes* [8]. More information can also be found in Refs. 9–11.

We shall focus here on developments caused by widespread use of parallel computers to perform Monte Carlo calculations. Our impression is that individual users are porting random number generators to parallel computers in an ad hoc fashion, possibly unaware of some of the issues that come to the fore when massive calculations are performed. Parallel algorithms can probe other qualities of random number generators such as interprocess correlation. A recent review covers parallel random number generation in somewhat more depth [12]. The interested reader can also refer to Refs. 13–17 for work related to parallel random number generation and testing.

This chapter is structured as follows. First we discuss the desired properties that random number generators should have. Next we discuss several methods that have been used as generators, in particular on parallel computers. Then we discuss testing procedures and show some results of our extensive tests. Quasi-random numbers (QRNs) have recently been introduced as a way to achieve faster convergence than true random numbers. We briefly discuss these and give some guidelines concerning those applications for which they are likely to be most effective.

We have recently developed a library implementing several of the parallel random number generators and statistical tests of them on the most widely available multiprocessor computers. Documentation and software are available at

http://www.ncsa.uiuc.edu/Apps/SPRNG

II. DESIRED PROPERTIES OF RANDOM NUMBER GENERATORS

In this section we discuss some of the desired properties of good random number generators. We shall then explain specific implications of these for parallel random number generation.

First, let us define a random number sequence, $\{u_i\}$, where i is an integer. In this chapter we will be concerned exclusively with uniformly distributed numbers. Other distributions can be generated by standard techniques [18]. Uniform numbers can be either reals, by convention in (0,1) such as those returned by the FORTRAN **ranf** and C **drand48** functions, or integer, by convention in $[1,2^n)$ for some n close to the word size on the computer. We shall denote the reals by u_k and the integers by I_k. Clearly, for many purposes integer and real generators on a computer are virtually equivalent if n is large enough and if we define $u = I2^{-n}$.

A. Randomness

Let us consider the following experiment to verify the randomness of an infinite sequence of integers in $[1,d]$. Suppose we let you view as many numbers from the sequence as you wished to. You should then guess any other number in the sequence. If the likelihood of your guess being correct is greater than $1/d$, then the sequence is not random. In practice, to estimate the winning probabilities, we must play this guessing game several times.

This test has certain implications, the most important of which is the uniformity of the individual elements. If the sequence consists of integers in $[1,d]$, then the probability of getting any particular integer should be $1/d$. All good sequences are constructed to have this property. But applications in statistical mechanics as well as other real applications rely heavily on uniformity in higher dimensions, at least four dimensions, if not thousands.

We define uniformity in higher dimensions as follows. Suppose we define n-tuples $U_i^n = (u_{i+1}, \dots u_{i+n})$ and divide the n-dimensional unit hypercube into many equal subvolumes. A sequence is uniform if in the limit of an infinite sequence all the subvolumes have an equal number of occurrences of random n-tuples. For a random sequence this will be true for *all* values of n and all partitions into subvolumes, although in practice we test only for small values of n.

The following simple Metropolis Monte Carlo example demonstrates how correlations between successive pairs of random numbers can give incorrect results. Suppose we sample the movement of a particle along the x axis confined in a potential well that is symmetric about the origin: $V(x) = V(-x)$. The classic Metropolis algorithm is outlined in Figure 1. At each step in our calculations, the particle is moved to a trial position with the first random number (*sprng*( )) and then that step is accepted or rejected with the second random number.

Figure 2 shows the results of the particle density computed exactly (for a harmonic well) with a good random number sequence, and also with sequences that are deliberately chosen to be correlated or anticorrelated. In the latter case a high number is likely to be followed by a low number and a low number by a high number. The particle density is then not symmetric because movements to the right are more likely to be rejected than movements to the left, so that the final distribution is skewed. This occurs despite uniformity of the individual elements of the sequence.

Now let us imagine how this changes for N particles in three dimensions. The usual Metropolis MC algorithm for a simple classical fluid will use random numbers four at a time (three for the displacement and one for the acceptance test) so that the algorithm is potentially sensitive to correlations between successive quadruples of numbers. But it can also be sensitive to correlations of u_i with u_{i+4N} since one usually goes through the particles in order, causing $u_i, u_{i+4N}, \dots$ to be used for the same purpose on the same particle. Unfortunately, the usual linear congruential generator has correlations between numbers separated by distances that are powers of 2; so it

```
x = 0
repeat loop N times
   xtrial = x + δ * (sprng( ) − 0.5)
   if (exp(−β * (V(xtrial) − V(x))) > sprng( )) then
      x = xtrial
   endif
end loop
```

Figure 1. Algorithm for simulating movement of the particle.

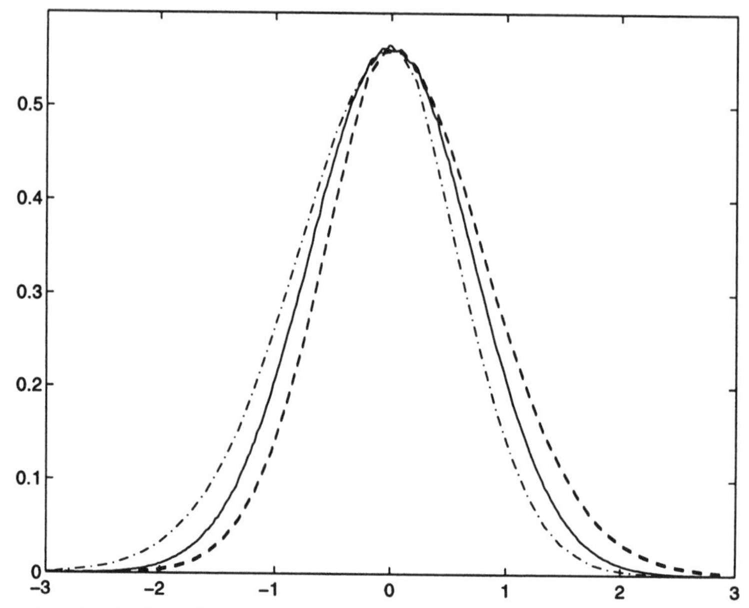

Figure 2. Distribution of particle positions in one-dimensional random-walk simulations. The solid line shows the results with an uncorrelated sequence; the bold dashed line, sequences with correlation coefficient = 0.2; and the dashed-dotted line for sequences with correlation coefficient = 0.2.

is not a good idea to simulate systems where N is a power of 2 with this generator. In general each Monte Carlo algorithm is sensitive to particular types of correlations, making it hard to define a universal test.

B. Reproducibility

In the early days of computers, it was suggested that one could make a special circuit element which would deliver truly random numbers. Computer vendors have not supplied such an element because it is not trivial to design a device to deliver high-quality numbers at a sufficient rate. Even more importantly, debugging codes would become much more difficult if each time the code was run a completely irreproducible sequence were to be generated. In Monte Carlo simulations, bugs may be manifest only at certain times, depending on the sequence of random numbers obtained. In order to detect the errors, it is necessary to repeat the calculations to find out how the errors occurred. The feature of reproducibility is also helpful while porting the program to a different machine. If we have a sample run from one machine available, then we can try an identical run on a different machine and verify that it ported correctly. Such reproducible random number generators are said to be *pseudo-random* (PRNG), and we shall call the numbers produced by such generators a PRNs.

There is a conflict between the requirements of reproducibility and randomness. On a finite memory computer, at any step k in the sequence the PRNG has an internal *state* specifiable conceptually by an integer S_k, where the size of this integer is not necessarily related to the word length of the computer. For each state S in the sequence, there is a mapping that gives a random number the user sees, $u_k = F(S_k)$. We also have an iteration process to determine the next state of the sequence from the current state, $S_{k+1} = T(S_k)$. All PRNGs can be classified by the internal state space, the mapping, and the iteration. The sequence is defined once we have specified the initial starting state S_0 known as the *seed*. Figure 3 illustrates the procedure for obtaining pseudo-random sequences described above.

Now let us return to the possible conflict between the properties of reproducibility and randomness. If it is reproducible, it cannot be perfectly random since knowing the sequence will make betting on the next number easy. How do we resolve the incompatibility between the two properties? In common sense terms, we mean that a good PRNG is one whose numbers are uncorrelated as long as you do not explicitly try to back out the mapping and iteration processes and use that to predict another member of the sequence. PRNGs have not been designed to be good cryptographic sequences.

C. Speed

It is, of course, desirable to generate the random numbers fast. While for some applications the generation of the random numbers is the limiting

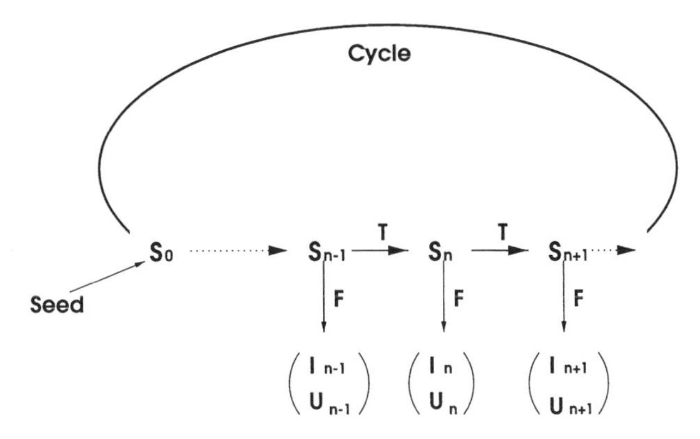

Figure 3. A pseudo-random sequence is defined by the internal state space, the mapping, the iteration, and the initial state.

factor, many generators take only a few clock cycles to deliver a new number so that usually the generation of random numbers is only a small fraction of the time required for the calculation. Hence speed is not a major concern unless, either the generator is extremely slow, or the remainder of the algorithm is extremely fast such as with a lattice spin model.

D. Large Cycle Length

A pseudo-random number generator is a finite-state machine with at most 2^p different states, where p is the number of bits that represent the state. One can easily see that the sequence must repeat after at most 2^p different numbers have been generated. The smallest number of steps after which the generator starts repeating itself is called the *period* or cycle length, L. Assuming that all cycles have the same length, the number of disjoint cycles (i.e. having no states in common) is then $2^p/L$.

A computer in 1997 might deliver 10^8 numbers per processor per second (or 2^{26}). Hence it will take 1 s to exhaust a generator with a 26-bit internal state and 1 year to exhaust one with 2^{51} internal states. This suggests that it could be dangerous to use the 32-bit generators developed for the microprocessors of the 1980s on today's computers. After the sequence is exhausted, the "true" error of a simple MC evaluation integral will no longer decrease, and one can be mislead into trusting an incorrect answer.

However, for many applications a small period will not in itself bias the results significantly. For example, in MCMC we can think of the "state" of the random walk as consisting both of the coordinates of the particles (say $3N$ position variables) and of the internal state of the PRNG. The walk will repeat itself only if all the coordinates are exactly the same. Hence, even if the random number sequence repeats, the particles will have moved on and have a different internal state. However, it is not a good idea to have repeating sequences, especially since it is easy to avoid.

E. Parallelization

We next mention the implications of correlation and cycle length on parallel pseudo-random number generators (PPRNGs).

In order to get higher speed, Monte Carlo applications make extensive use of parallel computers, since these calculations are particularly well suited to such architectures and often require very long runs. A common way to parallelize Monte Carlo is to put identical "clones" on the various processors; only the random number sequences are different. It is therefore important for the sequences on the different processors to be uncorrelated. That is, given an initial segment of the sequence on one process, and the random number sequences on other processes, we should not be able to predict the next element of the sequence on the first process. For example, it should not happen that if we obtain random numbers of large magnitude on one process, then we are more likely to obtain large numbers on another.

Consider the following extreme case to demonstrate the impact of correlations. Suppose we perform identical calculations on each process, expecting different results due to the presence of a different random number sequence. If, however, we use the same sequence on each process, then we will get an identical result on each process and the power of the parallel computer is wasted. Even worse, we may incorrectly believe that the errors are greatly reduced because all processes give identical results. Such cases routinely occur when users first port their MC codes to parallel computers without considering how the random number sequence is to be parallelized.

Even if the correlations across processes are not perfect, any correlation can affect the random walk. It is generally true that interprocessor correlation is less important that intraprocessor correlation, but that can depend on the application. The danger is that a particular parallel application will be sensitive to a particular correlation. New statistical tests have to be invented for correlation between processors.

The desire for *reproducibility*, when combined with *speed*, is also an important factor, and limits the feasible parallelization schemes. We shall next describe some common schemes for creating PPRNGs along with their merits.

1. *Central Server.* One can maintain one particular process that serves as a centralized random number generator for all the processes. Any process that requires a random number obtains it from that process by sending messages. Such a scheme reduces the speed greatly since interprocessor communication is very expensive and the process needing the PRN must have exclusive access to the server to ensure that there are no conflicts. It also hinders reproducibility because the different processes may request random numbers in different orders in different runs of the program, depending on the network traffic and implementation of the communication software.

2. *Cycle Division.* In one popular scheme the same iteration process is used on the different processes, but with widely separated seeds on each process. There are two related schemes: (a) the *leapfrog* method, where processor i gets u_i, u_{i+M}, ..., where M is the total number of processes (e.g., process 1 gets the first member of the sequence, process 2 the second, and so forth) and (b) in the *cycle splitting* method, where process $i + 1$ gets $u_{iL/M}$, $u_{iL/M+1}$, ..., where L is the cycle length and M is the number of processes. (That is, the first process will get the first L/M numbers, the second process the second L/M numbers, and so forth.)

Both methods require a fast way of advancing the PRNG a few steps; faster than iterating the sequence that number of steps. In the first method we need to be able to advance by M steps at each iteration. In the second method we need to be able to advance by L/M steps during initialization.

Statistical tests performed on the original sequence are not necessarily adequate for the divided sequence. For example, in the leapfrog method correlations M numbers apart in the original sequence become adjacent correlations of the split sequence.

Clearly, either method reduces the period of the original sequence by the number of processes. For example, with 512 nodes running at 100 Mflops (million floating-point operations per second), one will exhaust the sequence of a PRNG with a 46-bit internal state (the common real*8 or long rng) in only 23 minutes (min)! If the number of random numbers consumed is greater than expected, then the sequences on different processes could overlap.

3. *Cycle Parameterization.* Another scheme makes use of the fact that some PRNGs have more than one cycle. If we choose the seeds carefully, then we can ensure that each random sequence starts out in a different cycle, and so two sequences will not overlap. Thus the seeds are parameterized (i.e., sequence i gets a seed from cycle i, the sequence number being the parameter that determines its cycle). This is the case for the *lagged fibonacci generator* described in the next section.

4. *Parameterized Iteration.* Just as the disjoint cycles can be parameterized, so can many of the iteration functions for the internal state as with the *generalized feedback shift register* described in the next section. Here, sequence i gets iteration function T_i.

It is difficult to ensure reproducibility if the number of processors changes between one run and the next. This problem cannot be solved by the PPRNG in itself. In order to write a parallel MC application that gives identical results with a variable number of processors, it is necessary to write in terms of "virtual processors," each virtual processor having its own PRNG. Each physical processor would handle several virtual processors. It is unlikely that many programmers would go to this much trouble just to ensure that their code has this degree of portability unless it can be done automatically.

There are other consequences of the desire for reproducibility and speed. As an example, consider the simulation of a branching process. Suppose the generation of "neutron" paths is based on the outcome of various interactions between the neutrons and a medium. During a neutron flight, the neutron may collide with an atom and produce new neutrons (fission) or be absorbed (fusion). Efficient utilization of the processors requires good load balancing, and so one can move the computation of the statistics for new neutrons to different processors in order to keep the work evenly balanced. To ensure reproducibility in different runs, each neutron must be given a different random number sequence in a deterministic fashion. Thus, in a repetition of a run, even if the neutron is migrated to a different processor, the same random number sequence will be produced in determining its path. We also need to ensure that the random number sequence produced for each neutron is unique without incurring interprocessor communication. This can be accomplished by developing the ability to "spawn" unique sequences from an existing one.

We take the model that when a new process forks, a new sequence is generated for that process. Each sequence can be identified by a parameter if we parallelize the generators by any of the methods of parameterization described earlier. We ensure uniqueness of the new sequence by assigning each process a set $\mathscr{P}$ of parameters available for spawning. When a process forks, it partitions the elements of $\mathscr{P}$ among itself and its children, to create new sets of parameters available for spawning as shown in Figure 4. Since the sets available for spawning on each process are disjoint, different sequences are obtained on each process. There is, however, the risk that eventually all the parameters could be used up; so we should not spawn too often. In the next section we will discuss some generators with very large numbers of possible parameters, so that quite a lot of spawning could be done before we would be in danger of repeating parameters.

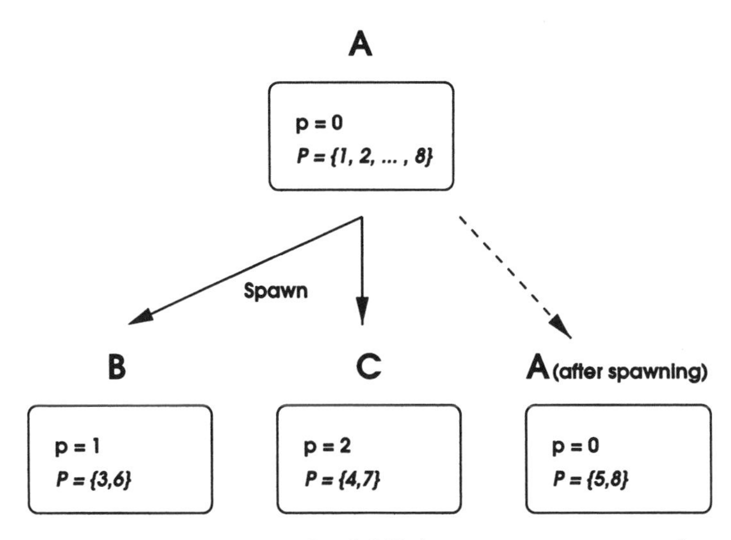

Figure 4. Process A spawns processes B and C. Each new process gets a new random-number sequence parameterized by p and a set of parameters $\mathscr{P}$ for spawning.

On today's parallel computers communication is very slow compared to floating-point performance. It is sometimes possible to use multiple random number sequences to reduce the communication costs in a Monte Carlo simulation. An example occurs in our parallel-path integral calculations [19]. The total configuration space of the imaginary time path is divided into subspaces based on the "imaginary time" coordinate so that a given processor has control over a segment of imaginary time. Most moves are made on variables local to each processor, but occasionally there are global moves to be made that move all values of imaginary time, where each processor has to evaluate the effect of the move on its variables. For these moves the number of messages can be cut in half by having available two different sequences of random numbers on each processor: (1) a local sequence which is different on each processor (and uncorrelated) and (2) a global sequence shared by all processors. The global sequence permits all the processors to "predict" without communication what the global move will be, evaluate its consequence on their variables, and then "vote" on whether that global move is acceptable. Thus half of the communication and synchronization cost is avoided compared to the scheme whereby a master processor would generate the global move, send it out to all the workers, and tally up the results from the various processors.

III. METHODS FOR RANDOM NUMBER GENERATION

In this section we describe some popular basic generators and their parallelization. We also mention about "combined generators," which are obtained from the basic ones.

A. Linear Congruential Generators

The most commonly used generator for pseudo-random numbers is the linear congruential generator (LCG) [20]:

$$x_n = ax_{n-1} + b(\text{mod } m) \tag{3.1}$$

where m is the modulus, a the multiplier, and c the additive constant or addend. The size of the modulus constrains the period, and it is usually chosen to be either prime or a power of 2.

This generator (with m a power of 2 and $c = 0$) is the de facto standard included with FORTRAN and C compilers. One of the biggest disadvantages to using a power-of-2 modulus is that the least significant bits of the integers produced by these LCGs have extremely short periods. For example, $\{x_n(\text{mod } 2^j)\}$ will have period 2^j [7]. In particular, this means that the least-significant bit of the LCG will alternate between 0 and 1. Since

PRNs are generated with this algorithm, some cautions to the reader are in order:

1. The PRN should not be split apart to make several random numbers since the higher order bits are much more random than the lower order bits.
2. One should avoid using the power-of-2 modulus in batches of powers of 2. (For example, if one has 1024 particles in three dimensions, one is using the PRNs 4096 at a time and the correlations between a PRN and one 4096 later may be large.)
3. Generators with large modulus are preferable to ones with small modulus. Not only is the period longer, but the correlations are much less. In particular, one should not use a 32-bit modulus for careful work. In spite of this known defect of power-of-2 LCGs, 48-bit multipliers (and higher) have passed many very stringent randomness tests.

Generally LCGs are best parallelized by parameterizing the iteration process through either the multiplier or the additive constant. Based on the modulus, different parameterizations have been tried.

1. *Power of 2 Modulus*

The parameterization chooses a set of additive constants $\{b_j\}$ that are pairwise relatively prime, $\gcd(b_i, b_j) = 1$, when $i \neq j$ so that the sequences are generated in different orders. The best choice is to let b_j be the jth prime less than $\sqrt{m/2}$ [14]. One important advantage of this parameterization is that there is an interstream correlation measure based on the spectral test that suggests that there will be good interstream independence.

2. *Prime Modulus*

When the modulus m is prime (usually very close to a power of 2 such as a Mersenne prime, $2^m - 1$, so that the operation of taking the modulus will be faster), a method based on using the multiplier, a, as the parameter to generate many sequences has been proposed. We start with a reference value of a and choose the multiplier for the jth stream as $a_j = a^{\ell_j} (\text{mod } m)$, where ℓ_j is the jth integer relatively prime to $m - 1$. This is closely related to the leapfrog method discussed earlier. Conditions on a and efficient algorithms for computing ℓ_j can be found in a recent work of one of the authors [16].

The scheme given above can be justified based on exponential sums, which is explained in Section IV.A. Two important open questions remain: (1) is it more efficient overall to choose m to be amenable to fast modular multiplication or fast calculation of the jth integer relatively prime to $m - 1$, and (2) does the good inter-stream correlation also ensure good intrastream independence via the spectral test?

B. Shift-Register Generators

Shift-register generators (SRGs) [21,22] are of the form

$$x_{n+k} = \sum_{i=0}^{k-1} a_i x_{n+i} (\text{mod } 2) \tag{3.2}$$

where the x_n and a_i values are either 0 or 1. The maximal period of $2^k - 1$ and can be achieved using as few as two nonzero values of a_i. This leads to a very fast random number generator.

There are two ways to make pseudo-random integers out of the bits produced by Eq. (3). The first, called the *digital multistep method*, takes n successive bits from Eq. (3) to form an integer of n bits. Then n more bits are generated to create the next integer, and so on. The second method, called the generalized feedback shift register, creates a new n-bit pseudo-random integer for every iteration of Eq. (3.2). This is done by constructing the n-bit word from the last bit generated, x_{n+k}, and $n - 1$ other bits from the k bits of SRG state. Thus a random number is generated for each new bit generated. While these two methods seem different, they are very related, and theoretical results for one always hold for the other. Serious correlations can result if k is small. Readers interested in more general information on SRGs should consult Refs. 21–23.

The shift-register sequences can be parameterized through the choice of a_i. One can systematically assign the values of a_i to the processors to produce distinct maximal period shift-register sequences [24]. This scheme can be justified based on exponential sum bounds, as in the case of the prime modulus LCG. This similarity is no accident, and is based on the fact that both generators are maximal period linear recursions over a finite field [25].

C. Lagged-Fibonacci Generators

The *additive lagged-fibonacci generator* (ALFG) is

$$x_n = x_{n-j} + x_{n-k} (\text{mod } 2^m), \qquad j < k \tag{3.3}$$

In recent years the ALFG has become a popular generator for serial as well as scalable parallel machines because it is easy to implement, is inexpensive to compute, and does well on standard statistical tests [11], especially when the lag k is sufficiently high (such as $k = 1279$). The maximal period of the ALFG is $(2^k - 1)2^{m-1}$ [26,27] and has $2^{(k-1)\times(m-1)}$ different full-period cycles [28]. Another advantage of the ALFG is that one can implement these generators directly in floating point to avoid the conversion from integer to floating point that accompanies the use of other generators. However, some care should be taken in the implementation to avoid floating-point roundoff errors [15].

In the previous sections we have discussed generators that can be parallelized by varying a parameter in the underlying recursion. Instead, the ALFG can be parameterized through its initial values because of the tremendous number of different cycles. We produce different streams by assigning each stream a different cycle. An elegant seeding algorithm that accomplishes this is described by Mascagni et al. [28].

An interesting cousin of the ALFG is the multiplicative lagged-Fibonacci generator (MLFG). It is defined by:

$$x_n = x_{n-j} \times x_{n-k} (\mathrm{mod}\ 2^m), \qquad j < k \tag{3.4}$$

Although this generator has a maximal period of $(2^k - 1)2^{m-3}$, which is a quarter the length of the corresponding ALFG [27], it has empirical properties considered to be superior to ALFGs [11]. Of interest for parallel computing is that a parameterization analogous to that of the ALFG exists for the MLFG [29].

D. Inversive Congruential Generators

An important new type of PRNG that, as yet, has not found any widely distributed implementation is the inversive congruential generator (ICG). This generator comes in two versions, the recursive ICG [30,31]

$$x_n = a\bar{x}_{n-1} + b (\mathrm{mod}\ m) \tag{3.5}$$

and the explicit ICG [32]

$$x_n = \overline{an + b} (\mathrm{mod}\ m) \tag{3.6}$$

In Eqs. (3.5) and (3.6) $\bar{c}$ denotes the multiplicative inverse modulo m in the sense that $c\bar{c} \equiv 1 (\mathrm{mod}\ m)$ when $c \neq 0$, and $\bar{0} = 0$.

An advantage of ICGs over LCGs are that tuples made from ICGs do not fall in hyperplanes [33,34]. Unfortunately, the cost of doing modular inversion is considerable: it is $O(\log_2 m)$ times the cost of multiplication.

E. Combination Generator

Better-quality sequences can often be obtained by combining the output of the basic generators to create a new random sequence as follows:

$$z_n = x_n \odot y_n \tag{3.7}$$

where $\odot$ is typically either the exclusive-or operator or addition modulo some integer m, and x and y are sequences from two independent generators. It is best if the cycle length of the two generators is relatively prime, for this implies that the cycle length of z will be the product of that of the basic generators. One can show that the statistical properties of z are no worse than those of x or y [11]. In fact, one expects it would be much superior but little has yet been proven to date.

Good combined generators have been developed by L'Ecuyer [35], based on the addition of linear congruential sequences.

IV. TESTING RANDOM NUMBER GENERATORS

We have so far discussed the desired features of random number generators and described some of the popular generators used in Monte Carlo applications. Now the question arises: as to how good the random number generators are. Whereas most generators are quite good for a variety of applications, popular generators have been found to be inadequate in a few applications [3–6].

Sophisticated Monte Carlo applications often spend only a small fraction of their time in the actual random number generation. Most of the time is taken up by other operations. It is therefore preferable to test the random number generator quickly, without actually using it in the application, to determine if it is good. For example, in the *runs test* we divide the random number sequence into blocks that are monotonically increasing, and determine the distribution of the lengths of each block. This is a good test of randomness because it is fast and looks at correlations between longer and longer groups of random numbers. (Using a sequence of length 10^{12} we expect to see runs as long as 15.)

However, a single statistical test is not adequate to verify the randomness of a sequence, because typical MCMC applications can be sensitive to various types of correlations. However, if the generator passes a wide variety of tests, then our confidence in its randomness increases. The tests suggested by Knuth [7] and those implemented in the DIEHARD package by Marsaglia [36] are a standard. Since there are many generators that pass these tests, there is no reason to consider one that is known to fail. (Of course, any generator will eventually fail most tests, so we always must state how many numbers were used in the test level of accuracy).

The second type of test is to run an application that uses random numbers in a similar manner to your applications, but for which the exact answer is known. For statistical mechanical applications, the two-dimensional Ising model (a simple lattice spin model) is often used since the exact answer is known and it has a phase transition so one expects sensitivity to long-range correlations. There are several different MCMC algorithms that can be used to simulate the Ising model and the random numbers enter quite differently in each algorithm. Thus this application is very popular in testing random number generators and has often detected defects in generators. We can test parallel generators on the Ising model application by assigning distinct random number sequences to subsets of lattice sites [37].

How good are the popular generators on the Ising model? Ferrenberg et al. [3] found that certain generators such as the particular shift-register generator they used (R250) failed with the Wolff algorithm, while the simple and much maligned 32-bit LCG called *CONG* did well. Similar defects in popular generators have also been observed by others [4–6]. However, with the Metropolis algorithm, CONG performed much worse than R250. Thus it appears that it is the combination of problem *and* generator that must be considered in choosing a suitable generator. Thus statistical tests are not adequate; we must also test the generators in our actual application.

Of course in almost all cases we do not know the exact answer. How then can we trust a given generator on our problem? The only general approach is to run our application with several different types of generators that are known to have good statistical properties. Hopefully, at least two will agree within the error bars and can be used on further similar runs with that algorithm. A word of caution: the test runs must be at least as long as the production runs because subtle correlations in the PRNG may not show up until long runs are made. This approach is quite computationally expensive and if done seriously, would increase the cost of all MC simulations by a factor of 2. (If you decide at the end that all generators are good, you can recover the time by averaging the results together.)

A. Parallel Tests

We now mention some tests of parallel random number generators.

Exponential sums. The exponential sum (Fourier transform of the density) of a sequence $u_0, \ldots, u_k$ is

$$C_g(k) = \sum_{j=0}^{k-1} e^{2\pi i x_j g}. \tag{4.1}$$

For a random sequence, $\langle |C_g(k)|^2 \rangle = k$ (for $g \neq 0$). This fact can be used to test correlation within, and between, random number sequences [13,16]. Consider two random sequences X and Y and define the exponential sum cross-correlation:

$$C_g(h, l, k) = \sum_{j=0}^{k-1} e^{2\pi i g(x_{h+j} - y_{l+j})} \tag{4.2}$$

In each term of this sum, we find the difference between an element of each sequence at a fixed offset apart. If this difference were uniformly distributed, then we should have $\langle | C(j, l, k)|^2 \rangle = k$.

Parallel Spectral Test. Percus and Kalos [14] have developed a version of the spectral test for parallel linear congruential generators.

Interleaved Tests. We create a new random sequence by interleaving several random sequences, and test this new sequence for randomness using standard sequential tests.

Fourier Transform Test. Generate a two-dimensional array of random numbers with each row in the array consisting of consecutive random numbers from one particular sequence. The two-dimensional Fourier transform can be performed. For a truly uncorrelated set of sequences, the coefficients (except the constant term) should be close to 0.

Blocking Test. In the blocking test we add random numbers from several streams as well as from within a stream. If the streams are independent, then the distribution of these sums will approach the normal distribution.

We now give some test results for the LCG with (parameterized) prime addend and a modified version of the LFG. Both of these generators performed acceptably in the sequential tests with 10^{11} random numbers. Preliminary results from other tests of PPRNG can also be found in the paper by Coddington [37].

We first tested the generators by interleaving about 1000 pairs of random sequences each containing about 10^8 PRNs. Both the generators passed this tests. Next we simulated parallelism on the 16×16 Ising model by using a different random sequence for each lattice site in the Metropolis algorithm. The LCG (with identical seeds) failed badly as can be seen from the dashed line in Figure 5. We then generalized the statistical tests, interleaving 256 sequences at a time. The LCG again failed, demonstrating the effectiveness of these tests in detecting nonrandom behaviour. The modified LFG passed these tests.

In the tests mentioned above, we had started each parameterized LCG with the same seed but used different additive constants. Even when we discarded the first million random numbers from each sequence, the sequences were still correlated. However, when we started the streams from

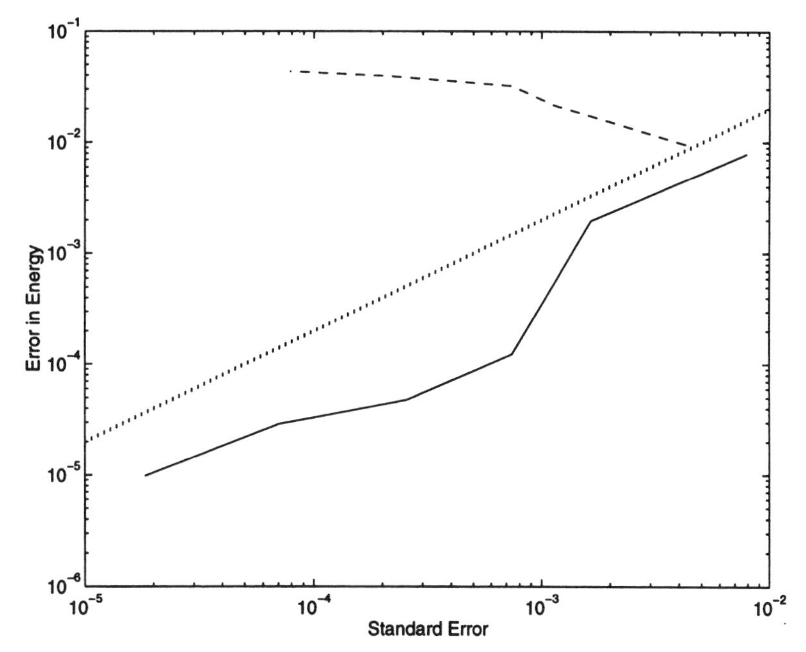

Figure 5. Plot of the actual error versus the internally estimated standard deviation of the energy error for Ising model simulations with the Metropolis algorithm on a 16×16 lattice with a different linear congruential sequence at each lattice site. The dashed line shows the results when all the linear congruential sequences were started with the same seeds but with different additive constants. The solid line shows the results when the sequences were started with different seeds. We expect around 95% of the points to be below the dotted line (which represents an error of two standard deviations) with a good generator.

different, systematically spaced seeds, the LCG passed all statistical tests (including parallel ones) with up to 10^{10}–10^{12} numbers. We finally repeated the parallel Metropolis algorithm simulation with 10^{11} random numbers, and both generators performed acceptably, as can be seen from the solid line in Figure 5.

Apart from verifying the quality of two particular generators, our results illustrate that it is commonplace for generators to pass some test and fail others. It is important to test parallel generators the way they will be seeded, since correlations in the seeding can lead to correlations in the resulting sequences.

V. QUASI-RANDOM NUMBERS

It is well known that the error estimate from any MC calculations (or in general from a simulation) converges as $N^{-1/2}$, where N is the number of random samples or equivalently the computer time expended. Recently there has been much research into whether nonrandom sequences could result in fast convergence, but still have the advantages of MC in being applicable to high-dimensional problems.

Consider the following integral:

$$I = \int_0^1 d^d x f(x) \tag{5.1}$$

The Monte Carlo estimate for this integral is

$$\hat{I}_n = \frac{1}{n} \sum_{i=1}^{n} f(x_i) \tag{5.2}$$

where the points x_i are to be uniformly distributed in the s-dimensional unit cube. An important, and very general, result that bounds the absolute error of the Monte Carlo estimate is the Koksma–Hlwaka inequality [25]:

$$|I - \hat{I}| \leq D_s(x_i) V(f) \tag{5.3}$$

where $D_s(x_i)$ is the discrepancy of the points x_i as defined by Eq. (5.4) and $V(f)$ is the total variation of f on $[0,1)^s$ in the sense of Hardy and Krause [38]. The total variation is roughly the average absolute value of the sth derivative of $f(x)$.

Note that Eq. (5.3) gives a deterministic error bound for integration because $V(f)$ depends only on the nature of the function. Similarly, the discrepancy of a point set is a purely geometric property of that point set. When given a numerical quadrature problem, we must cope with whatever function we are given, it is really only the points, x_i, that we control. Thus one approach to efficient integration is to seek point sets with small discrepancies. Such sets are necessarily not random but are instead referred to as *quasi-random numbers* (QRNs).

The s-dimensional discrepancy of the N points, x_i is defined as

$$D_s(x_i) = \sup_B \left| \frac{\#(B, x_i)}{N} - \lambda(B) \right| \tag{5.4}$$

Here B is any rectangular subvolume inside the s-dimensional unit cube with sides parallel to the axes and volume $\lambda(b)$. The function $\#(B, x_i)$ counts the number of points of x_i in B. Hence a set of points with a low discrepancy covers the unit cube more uniformly than one with a large discrepancy. A lattice has a very low discrepancy, however adding additional points to the lattice is slow in high dimensions. The discrepancy of a lattice does not decrease smoothly with increasing N.

A remarkable fact is that there is a lower (Roth) bound to its discrepancy [39]:

$$D_s^*(x_i) \geq C_s N^{-1}(\log N)^{(s-1)/2}, \qquad s > 3 \tag{5.5}$$

This gives us a target to aim at for the construction of low-discrepancy point sets (quasi-random numbers). For comparison, the estimated error with random sequences is

$$\langle (I - \hat{I})^2 \rangle = \sqrt{\frac{1}{N}\left[\int f^2 - \left(\int f\right)^2\right]} \tag{5.6}$$

The quantity in brackets, the variance, only depends on f. Hence the standard deviation of the Monte Carlo estimate is $O(N^{-1/2})$. This is much worse than the bounds of Eq. (5.5) as a function of the number of points. It is this fact that has motivated the search for quasi-random points.

What goes wrong with this argument as far as most high-dimensional problems in physical science is that the Koksma–Hlwaka bound is extremely poor for large s. Typical integrands become more and more singular the more one differentiates. The worse case is the Metropolis rejection step—the acceptance is itself discontinuous. Even assuming that $V(f)$ were finite, it is likely to be so large (for large s) that truly astronomical values of N would be required for the bound in Eq. (5.5) to be less than the MC convergence rate given in Eq. (5.6).

We will now present a very brief description of quasi-random sequences. Those interested in a much more detailed review of the subject are encouraged to consult the recent work of Niederreiter [25]. An example of a one-dimensional set of quasi-random numbers is the van der Corput sequence. First we choose a base, b, and write an integer n in base b as $n = \sum_{i=0}^{\log_b n} a_i b^i$. Then we define the van der Corput sequence as $x_n = \sum_{i=0}^{\log_b n} a_i b^{-i-1}$. For base $b = 3$, the first 12 terms of the van der Corput sequence are

$$\left\{0, \frac{1}{3}, \frac{2}{3}, \frac{1}{9}, \frac{4}{9}, \frac{7}{9}, \frac{2}{9}, \frac{5}{9}, \frac{8}{9}, \frac{1}{27}, \frac{10}{27}, \frac{19}{27}, \ldots\right\} \tag{5.7}$$

One sees intuitively how this sequence, while not behaving in a random fashion, fills in all the holes in a regular and low-discrepancy way.

There are many other low-discrepancy point sets and sequences. Some, like the Halton sequence [40], use the van der Corput sequence. There have been many others that are thought to be more general and have provably smaller discrepancy. Of particular note are the explicit constructions of Faure [41] and Niederreiter [42].

The use of quasi-random numbers in quadrature has not been widespread because the claims of superiority of quasi-random over pseudo-random for quadrature have not been shown empirically, especially for high-dimensional integrals and never in MCMC simulations [43]. QRNs work best in low-dimensional spaces where the spacing between the points can be made small with respect to the curvature of the integrand. QRNs work well for very smooth integrands. The Boltzmann distribution exp($-V(R)/k_B T$) is highly peaked for R, a many-dimensional point. Empirical studies have shown that the number of integration points needed for QRN error to be less than the MC error increases very rapidly with the dimensionality. Until the asymptotic region is reached, the error of QRN is not very different from simple MC. Also there are many MC tricks that can be used to reduce the variance such as importance sampling and antithetic sampling. Finally, MC has a very robust way of estimating errors; that is one of its main advantages as a method. Integrations with QRNs have to rely on empirical error estimates since the rigorous upper bound is exceedingly large. It may happen that the integrand in some way correlates with the QRN sequence so that these error estimates are completely wrong, just as sometimes happens with other deterministic integration methods.

There seem to be some advantages of quasi-random over pseudo-random for simple smooth integrands that are effectively low dimensional (say, less than 15 dimensions), but they are much smaller than one is lead to expect from the mathematical results.

ACKNOWLEDGMENT

We wish to acknowledge the support of DARPA through grant number DABT 63-95-C-0123 and the NCSA for computational resources.

REFERENCES

1. P. A. M. Dirac, *Proc. Roy. Soc. London Ser A* **123**, 734 (1929).
2. J. B. Anderson, in *Understanding Chemical Reactivity*, S. R. Langhoff, ed., Kluwer, Dordrecht, The Netherlands, 1995.
3. A. M. Ferrenberg, D. P. Landau, and Y. J. Wong, "Monte Carlo Simulations: Hidden Errors from 'Good' Random Number Generators," *Phys. Rev. Lett.* **69**, 3382–3384 (1992).
4. P. Grassberger, "On Correlations in 'Good' Random Number Generators," *Phys. Lett. A* **181**(1), 43–46 (1993).
5. W. Selke, A. L. Talapov, and L. N. Schur, "Cluster-Flipping Monte Carlo Algorithm and Correlations in 'Good' Random Number Generators," *JETP Lett.* **58**(8), 665–668 (1993).

6. F. Schmid and N. B. Wilding, "Errors in Monte Carlo Simulations Using Shift Register Random Number Generators," *Int. J. Mod. Phys. C* **6**(6), 781–787 (1995).
7. D. E. Knuth, *The Art of Computer Programming*, Vol. 2: *Seminumerical Algorithms*, 2nd ed., Addison-Wesley, Reading, MA, 1981.
8. W. H. Press, S. A. Teukolsky, W. T. Vetterling and B. P. Flannery, *Numerical Recipes in FORTRAN*, 2nd ed., Cambridge Univ. Press, New York, 1994.
9. S. K. Park and K. W. Miller, "Random Number Generators: Good Ones Are Hard to Find," *Commun. ACM* **31**, 1192–1201 (1988).
10. P. L'Ecuyer, "Random Numbers for Simulation," *Commun. ACM* **33**, 85–97 (1990).
11. G. Marsaglia, "A Current View of Random Number Generators," in *Computing Science and Statistics: Proceedings of the XVIth Symposium on the Interface*, 1985, pp. 3–10.
12. P. Coddington, "Random Number Generators for Parallel Computers," April 28, 1997, http://www.npac.syr.edu/users/paulc/papers/NHSEreview1.1/PRNGreview.ps.
13. M. Mascagni, S. A. Cuccaro, D. V. Pryor, and M. L. Robinson, "Recent Developments in Parallel Pseudorandom Number Generation," in *Proceedings of the Sixth SIAM Conference on Parallel Processing for Scientific Computing*, Vol. II, D. E. Keyes, M. R. Leuze, L. R. Petzold, and D. A. Reed, eds., SIAM, Philadelphia, 1993, pp. 524–529.
14. O. E. Percus and M. H. Kalos, "Random Number Generators for MIMD Parallel Processors," *J. Par. Distr. Comp.* **6**, 477–497 (1989).
15. R. P. Brent, "Uniform Random Number Generators for Supercomputers," in *Proceedings Fifth Australian Supercomputer Conference*, 5ASC Organizing Committee, 1992, pp. 95–104.
16. M. Mascagni, "Parallel Linear Congruential Generators with Prime Moduli," *Parallel Computing* (in press).
17. M. Mascagni, S. A. Cuccaro, and D. V. Pryor, "Techniques for Testing the Quality of Parallel Pseudorandom Number Generators," in *Proceedings of the Seventh SIAM Conference on Parallel Processing for Scientific Computing*, SIAM, Philadelphia, Pennsylvania, 1995, pp. 279–284.
18. M. Kalos and P. Whitlock, *Monte Carlo Methods*, Vol. I: *Basics*, Wiley-Interscience, New York, 1986.
19. D. M. Ceperley, "Path Integrals in the Theory of Condensed Helium," *Rev. Mod. Phys.* **67**(2), 279–355 (1995).
20. D. H. Lehmer, "Mathematical Methods in Large-Scale Computing Units," in *Proceedings of the 2nd Symposium on Large-Scale Digital Calculating Machinery*, Harvard Univ. Press, Cambridge, MA, 1949, pp. 141–146.
21. T. G. Lewis and W. H. Payne, "Generalized Feedback Shift Register Pseudorandom Number Algorithms," *J. ACM* **20**, 456–468 (1973).
22. R. C. Tausworthe, "Random Numbers Generated by Linear Recurrence Modulo Two," *Math. Comp.* **19**, 201–209 (1965).
23. S. W. Golomb, *Shift Register Sequences*, rev. ed., Aegean Park Press, Laguna Hills, CA, 1982.
24. J. L. Massey, "Shift-Register Synthesis and bch Decoding," *IEEE Trans. Inform. Theory* **15**, 122–127 (1969).
25. H. Niederreiter, *Random Number Generation and Quasi-Monte Carlo Methods*, SIAM, Philadelphia, 1992.
26. R. P. Brent, "On the Periods of Generalized Fibonacci Recurrences," *Math. Comp.* **63**,

389–401 (1994).

27. G. Marsaglia and L.-H. Tsay, "Matrices and the Structure of Random Number Sequences," *Linear Alg. Applic.* **67**, 147–156 (1985).
28. M. Mascagni, S. A. Cuccaro, D. V. Pryor, and M. L. Robinson, "A Fast, High-Quality, and Reproducible Lagged-Fibonacci Pseudorandom Number Generator," *J. Comp. Phys.* **15**, 211–219 (1995).
29. M. Mascagni, "A Parallel Non-Linear Fibonacci Pseudorandom Number Generator," Abstract, 45th SIAM Annual Meeting, 1997.
30. J. Eichenauer and J. Lehn, "A Nonlinear Congruential Pseudorandom Number Generator," *Stat. Hefte* **37**, 315–326 (1986).
31. H. Niederreiter, "Statistical Independence of Nonlinear Congruential Pseudorandom Numbers," *Montash. Math.* **106**, 149–159 (1988).
32. H. Niederreiter, "On a New Class of Pseudorandom Numbers for Simulation Methods," *J. Comput. Appl. Math.* **56**, 159–167 (1994).
33. G. Marsaglia, "Random Numbers Fall Mainly in the Planes," *Proc. Natl. Acad. Sci. USA* **62**, 25–28 (1968).
34. G. Marsaglia, "The Structure of Linear Congruential Sequences," in *Applications of Number Theory to Numerical Analysis*, S. K. Zaremba, ed., Academic Press, New York, 1972, pp. 249–285.
35. P. L'Ecuyer, "Efficient and Portable Combined Random Number Generators," *Commun. ACM* **31**, 742–774 (1988).
36. G. Marsaglia, Diehard, ftp://stat.fsu.edu/pub/diehard.
37. P. Coddington, "Tests of Random Number Generators Using Ising Model Simulations," *Int. J. Mod. Phys. C* **7**(3), 295–303 (1996).
38. E. Hlwaka, "Funktionen von Beschränkter Variation in der Theorie der Gleichverteiling," *Ann. Mat. Pura Appl.* **54**, 325–333 (1961).
39. K. F. Roth, "On Irregularities of Distribution," *Mathematika* **1**, 73–79 (1954).
40. J. H. Halton, "On the Efficiency of Certain Quasi-Random Sequences of Points in Evaluating Multi-dimensional Integrals," *Num. Math.* **2**, 84–90 (1960).
41. H. Faure, "Using Permutations to Reduce Discrepancy," *J. Comp. Appl. Math.* **31**, 97–103 (1990).
42. B. L. Fox, P. Bratley, and H. Niederreiter, "Implementation and Tests of Low-Discrepancy Point Sets," *ACM Trans. Model. Comp. Simul.* **2**, 195–213 (1992).
43. W. J. Morokoff and R. E. Caflisch, "Quasi-Monte Carlo Integration," *J. Comp. Phys.* **122**, 218–230 (1995).

BETWEEN CLASSICAL AND QUANTUM MONTE CARLO METHODS: "VARIATIONAL" QMC

DARIO BRESSANINI*

Istituto di Scienze Matematiche Fisiche e Chimiche, Università di Milano, sede di Como, I-22100 Como, Italy

PETER J. REYNOLDS*

Physical Sciences Division, Office of Naval Research, Arlington, VA 22217

CONTENTS

I. INTRODUCTION

The variational Monte Carlo method is reviewed here. It is in essence a

* Also Department of Physics, Georgetown University, Washington, DC (USA).

Advances in Chemical Physics, Volume 105, Monte Carlo Methods in Chemical Physics, edited by David M. Ferguson, J. Ilja Siepmann, and Donald G. Truhlar. Series Editors I. Prigogine and Stuart A. Rice.
ISBN 0-471-19630-4

classical statistical mechanics approach, yet allows the calculation of quantum expectation values. We give an introductory exposition of the theoretical basis of the approach, including sampling methods and acceleration techniques; its connection with trial wavefunctions; and how in practice it is used to obtain high quality quantum expectation values through correlated wavefunctions, correlated sampling, and optimization. A thorough discussion is given of the different methods available for wavefunction optimization. Finally, a small sample of recent works is reviewed, giving results and indicating new techniques employing variational Monte Carlo.

Variational Monte Carlo (or VMC, as it is now commonly called) is a method that allows one to calculate quantum expectation values given a trial wavefunction [1,2]. The actual Monte Carlo methodology used for this is almost identical to the usual classical Monte Carlo methods, particularly those of statistical mechanics. Nevertheless, quantum behavior can be studied with this technique. The key idea, as in classical statistical mechanics, is the ability to write the desired property $\langle O \rangle$ of a system as an average over an ensemble

$$\langle O \rangle = \frac{\int P(\mathbf{R}) O(\mathbf{R}) \, d\mathbf{R}}{\int P(\mathbf{R}) \, d\mathbf{R}} \tag{1.1}$$

for some specific probability distribution $P(\mathbf{R})$. In classical equilibrium statistical mechanics this would be the Boltzmann distribution. If $\langle O \rangle$ is to be a quantum expectation value, $P(\mathbf{R})$ must be the square of the wavefunction $\Psi(\mathbf{R})$. True quantum Monte Carlo methods (see, e.g., the following chapter) allow one to actually sample $\Psi(\mathbf{R})$. Nevertheless, classical Monte Carlo is sufficient (although approximate) through the artifact of sampling from a *trial* wavefunction. How to obtain such a wavefunction is *not* directly addressed by VMC. However, optimization procedures, which will be discussed below, and possibly feedback algorithms, enable one to modify an existing wavefunction choice once made.

A great advantage of Monte Carlo for obtaining quantum expectation values is that wavefunctions of great functional complexity are amenable to this treatment, since analytic integration is not being done. This greater complexity, including, for example, explicit two-body and higher-order correlation terms, in turn allows for a far more compact description of a many-body system than possible with most non–Monte Carlo methods, with the benefit of high absolute accuracy being possible. The primary disadvantage of using a Monte Carlo approach is that the calculated quantities contain a statistical uncertainty, which needs to be made small. This can always be done in VMC, but at the cost of CPU (central processing unit) time, since statistical uncertainty decreases as $N^{-1/2}$ with increasing number of samples N.

A quantity often sought with these methods is the expectation value of the Hamiltonian, that is, the total energy. As with all total energy methods, whether Monte Carlo or not, one needs to consider scaling. That is, as the systems being treated increase in size, how does the computational cost rise? Large-power polynomial scaling offers a severe roadblock to the treatment of many physically interesting systems. With such scaling, even significantly faster computers would leave large classes of interesting problems untouchable. This is the motivation behind the so-called order-N methods in, for instance, density functional theory, where in that case N is the number of electrons in the system. While density functional theory is useful in many contexts, often an exact treatment of electron correlation, or at least a systematically improvable treatment, is necessary or desirable. Quantum chemical approaches of the latter variety are unfortunately among the class of methods that scale with large powers of system size. This is another advantage of Monte Carlo methods, which scale reasonably well, generally between N^2 and N^3; moreover, algorithms with lower powers are possible to implement (e.g., using fast multipole methods to evaluate the Coulomb potential, and the use of localized orbitals together with sparse matrix techniques for the wavefunction computation).

The term "variational" Monte Carlo derives from the use of this type of Monte Carlo in conjunction with the variational principle; this provides a bounded estimate of the total energy together with a means of improving the wavefunction and energy estimate. Despite the inherent statistical uncertainty, a number of very good algorithms have been created that allow one to optimize trial wavefunctions in this way [3–5], and we discuss this at some length below. The best of these approaches go beyond simply minimizing the energy, and exploit the minimization of the energy variance as well, this latter quantity vanishing for energy eigenfunctions.

Before getting into details, let us begin with a word about notation. The position vector $\mathbf{R}$ which we use, lives in the $3M$-dimensional coordinate space of the M (quantum) particles comprising the system. This vector is, for example, the argument of the trial wavefunction $\Psi_T(\mathbf{R})$; however, sometimes we will omit the explicit dependence on $\mathbf{R}$ to avoid cluttering the equations, and simply write Ψ_T. Similarly, if the trial wavefunction depends on some parameter α (this may be, e.g., the exponent of a Gaussian or Slater orbital), we may write $\Psi_T(\mathbf{R}; \alpha)$, or simply $\Psi_T(\alpha)$, again omitting the explicit $\mathbf{R}$ dependence.

The essence of VMC is the creation and subsequent sampling of a distribution $P(\mathbf{R})$ proportional to $|\Psi_T(\mathbf{R})|^2$. Once such a distribution is established, expectation values of various quantities may be sampled.

Expectation values of nondifferential operators may be obtained simply as

$$\langle \hat{O} \rangle = \frac{\int |\Psi_T(\mathbf{R})|^2 \hat{O}(\mathbf{R})\, d\mathbf{R}}{\int |\Psi_T(\mathbf{R})|^2\, d\mathbf{R}} \cong \frac{1}{N} \sum_{i=1}^{N} \hat{O}(\mathbf{R}_i) \tag{1.2}$$

Differential operators are only slightly more difficult to sample, since we can write

$$\langle \hat{O} \rangle = \frac{\int |\Psi_T(\mathbf{R})|^2 \dfrac{\hat{O}\Psi_T(\mathbf{R})}{\Psi_T(\mathbf{R})}\, d\mathbf{R}}{\int |\Psi_T(\mathbf{R})|^2\, d\mathbf{R}} \cong \frac{1}{N} \sum_{i=1}^{N} \frac{\hat{O}\Psi_T(\mathbf{R}_i)}{\Psi_T(\mathbf{R}_i)} \tag{1.3}$$

II. MONTE CARLO SAMPLING OF A TRIAL WAVEFUNCTION

A. Metropolis Sampling

The key problem is how to create and sample the distribution $\Psi_T^2(\mathbf{R})$ (from now on, for simplicity, we consider only real trial wavefunctions). This is readily done in a number of ways, possibly familiar from statistical mechanics. Probably the most common method is simple Metropolis sampling [6]. Specifically, this involves generating a Markov chain of steps by "box sampling" $\mathbf{R}' = \mathbf{R} + \bar{\varsigma}\Delta$, with Δ the box size, and $\bar{\varsigma}$ a $3M$-dimensional vector of uniformly distributed random numbers $\varsigma \in [-1, +1]$. This is followed by the classic Metropolis accept/reject step, in which $(\Psi_T(\mathbf{R}')/\Psi_T(\mathbf{R}))^2$ is compared to a uniformly distributed random number between zero and unity. The new coordinate $\mathbf{R}'$ is accepted only if this ratio of trial wavefunctions squared exceeds the random number. Otherwise the new coordinate remains at $\mathbf{R}$. This completes one step of the Markov chain (or random walk). Under very general conditions, such a Markov chain results in an asymptotic equilibrium distribution proportional to $\Psi_T^2(\mathbf{R})$. Once established, the properties of interest can be "measured" at each point $\mathbf{R}$ in the Markov chain (which we refer to as a *configuration*) using Eqs. (1.2) and (1.3), and averaged to obtain the desired estimate. The more configurations that are generated, the more accurate the estimate one gets. As is normally done in standard applications of the Metropolis method, proper care must be taken when estimating the statistical error, since the configurations generated by a Markov chain are not statistically independent; they are *serially correlated* [7]. The device of dividing the simulation into *blocks* of sufficient length, and computing the statistical error only over the block averages, is usually sufficient to eliminate this problem.

B. Langevin Simulation

The sampling efficiency of the simple Metropolis algorithm can be improved when one switches to the Langevin simulation scheme [8]. The Langevin approach may be thought of as providing a kind of importance sampling which is missing from the standard Metropolis approach. One may begin by writing a Fokker–Planck equation whose steady-state solution is $\Psi_T^2(\mathbf{R})$. Explicitly, this is

$$\frac{\partial f(\mathbf{R}, t)}{\partial t} = D\nabla^2 f(\mathbf{R}, t) - D\nabla \cdot (f(\mathbf{R}, t)\mathbf{F}(\mathbf{R})) \tag{2.1}$$

where

$$\mathbf{F}(\mathbf{R}) = \nabla \ln \Psi_T^2(\mathbf{R}) \tag{2.2}$$

is an explicit function of Ψ_T generally known as either the *quantum velocity* or the *quantum force*. By direct substitution it is easy to check that $\Psi_T^2(\mathbf{R})$ is the exact steady-state solution.

The (time) discretized evolution of the Fokker–Planck equation may be written in terms of $\mathbf{R}$, and this gives the following Langevin-type equation

$$\mathbf{R}' = \mathbf{R} + D\tau\mathbf{F}(\mathbf{R}) + \sqrt{2D\tau}\,\chi \tag{2.3}$$

where τ is the step size of the time integration and χ is a Gaussian random variable with zero mean and unit width. Numerically one can use the Langevin equation to generate the path of a configuration, or random *walker* (more generally, an ensemble of such walkers), through position space. As with Metropolis, this path is also a Markov chain. One can see that the function $\mathbf{F}(\mathbf{R})$ acts as a drift, pushing the walkers toward regions of configuration space where the trial wavefunction is large. This increases the efficiency of the simulation, in contrast to the standard Metropolis move where the walker has the same probability of moving in every direction.

There is, however, a minor point that needs to be addressed: the time discretization of the Langevin equation, exact only for $\tau \to 0$, has introduced a time step *bias* absent in Metropolis sampling. This can be eliminated by performing different simulations at different time steps and extrapolating to $\tau \to 0$. However, a more effective procedure can be obtained by adding a Metropolis-like acceptance/rejection step after the Langevin move. The net result is a generalization of the standard Metropolis algorithm in which a Langevin equation, containing drift and diffusion (i.e., a quantum force term depending on the positions of all the electrons, plus white noise), is employed for the transition matrix carrying us from $\mathbf{R}$ to $\mathbf{R}'$. This is a specific generalization of Metropolis. We discuss the generic generalization next.

C. Generalized Metropolis

In the Metropolis algorithm, a single move of a walker starting at $\mathbf{R}$ can be split into two steps as follows: first a possible final point $\mathbf{R}'$ is selected; then an acceptance/rejection step is executed. If the first step is taken with a transition probability $T(\mathbf{R} \to \mathbf{R}')$, and if we denote for the acceptance/rejection step the probability $A(\mathbf{R} \to \mathbf{R}')$ that the attempted move from $\mathbf{R}$ to $\mathbf{R}'$ is accepted, then the total probability that a walker moves from $\mathbf{R}$ to $\mathbf{R}'$ is $T(\mathbf{R} \to \mathbf{R}')A(\mathbf{R} \to \mathbf{R}')$. Since we seek the distribution $P(\mathbf{R})$ using such a Markov process, we note that at equilibrium (and in an infinite ensemble), the fraction of walkers going from $\mathbf{R}$ to $\mathbf{R}'$, $P(\mathbf{R})T(\mathbf{R} \to \mathbf{R}')A(\mathbf{R} \to \mathbf{R}')$, must be equal to the fraction of walkers going from $\mathbf{R}'$ to $\mathbf{R}$, namely $P(\mathbf{R}')T(\mathbf{R}' \to \mathbf{R})A(\mathbf{R}' \to \mathbf{R})$. This condition, called *detailed balance*, is a *sufficient* condition to reach the desired steady state, and provides a constraint on the possible forms for T and A. For a given $P(\mathbf{R})$

$$P(\mathbf{R})T(\mathbf{R} \to \mathbf{R}')A(\mathbf{R} \to \mathbf{R}') = P(\mathbf{R}')T(\mathbf{R}' \to \mathbf{R})A(\mathbf{R}' \to \mathbf{R}) \tag{2.4}$$

thus the acceptance probability must satisfy

$$\frac{A(\mathbf{R} \to \mathbf{R}')}{A(\mathbf{R}' \to \mathbf{R})} = \frac{P(\mathbf{R}')}{P(\mathbf{R})} \frac{T(\mathbf{R}' \to \mathbf{R})}{T(\mathbf{R} \to \mathbf{R}')} \tag{2.5}$$

Since for our situation $P(\mathbf{R}) = \Psi_{\mathrm{T}}^2(\mathbf{R})$, a Metropolis solution for A is

$$A(\mathbf{R} \to \mathbf{R}') = \min\left(1, \; \left(\frac{\Psi_{\mathrm{T}}(\mathbf{R}')}{\Psi_{\mathrm{T}}(\mathbf{R})}\right)^2 \frac{T(\mathbf{R}' \to \mathbf{R})}{T(\mathbf{R} \to \mathbf{R}')}\right) \tag{2.6}$$

The original Metropolis scheme moves walkers in a rectangular box centered at the initial position; in this case the ratio of the T values is simply equal to unity, and the standard Metropolis algorithm is recovered. This is readily seen to be less than optimal if the distribution to be sampled is very different from uniform, such as, rapidly varying in regions of space. It makes sense to use a transition probability for which the motion towards a region of increasing $\Psi_{\mathrm{T}}^2(\mathbf{R})$ is enhanced. Toward this goal there are many possible choices for T; the Langevin choice presented above is a particular, very efficient, choice of the transition probability:

$$T(\mathbf{R} \to \mathbf{R}') = (4\pi D\tau)^{-3N/2} \; e^{-(\mathbf{R}' - \mathbf{R} - D\tau\mathbf{F}(\mathbf{R}))^2/4D\tau} \tag{2.7}$$

This transition probability is a displaced Gaussian, and may be sampled using the Langevin equation [Eq. (2.3)] by moving the walker first with a drift and then with a random Gaussian step representing diffusion.

III. TRIAL WAVEFUNCTIONS

The exact wavefunction is a solution to the Schrödinger equation. For any except the simplest systems the form of the wavefunction is unknown. However, it can be approximated in a number of ways. Generally this can be done systematically through series expansions of some sort, such as basis set expansions or perturbation theory. The convergence of these series depends on the types of terms included. Most variational electronic structure methods rely on a double basis-set expansion for the wavefunction: one in single electron orbitals and the other in M-electron Slater determinants. This is in no way the most general form of expansion possible. At a minimum, it omits explicit two-body (and many-body) terms. This omission results in generally slow convergence of the resultant series. An important characteristic of Monte Carlo methods is their ability to use arbitrary wavefunction forms, including ones having explicit interelectronic distance and other many-body dependencies. This enables greater flexibility and hence more compact representation than is possible with forms constructed solely with one-electron functions.

The one-electron form, however, provides a useful starting point for constructing the more general forms we desire. The one-electron form comes from the widely used methods of traditional ab initio electronic structure theory, based on molecular-orbital (MO) expansions and the Hartree–Fock approximation. As a first approximation, the M-electron wavefunction is represented by a single Slater determinant of spin orbitals. This independent-particle approximation completely ignores the many-body nature of the wavefunction, incorporating quantum exchange, but no correlation; within this approach correlation is later built in through a series expansion of Slater determinants.

The MOs are themselves expressed as linear combinations of atomic orbitals (AOs); the latter are usually a basis set of known functions. With a given basis set, the problem of variationally optimizing the energy transforms into that of finding the coefficients of the orbitals. Expressed in matrix form in an AO basis, and in the independent particle approximation of Hartree–Fock theory, this leads to the well-known self-consistent field (SCF) equations.

There are two broad categories of methods that go beyond Hartree–Fock in constructing wavefunctions: configuration interaction (CI), and many-body perturbation theory. In CI one begins by noting that the exact

M-electron wavefunction can be expanded as a linear combination of an infinite set of Slater determinants that span the Hilbert space of electrons. These can be any complete set of M-electron antisymmetric functions. One such choice is obtained from the Hartree–Fock method by substituting all excited states for each MO in the determinant. This, of course, requires an infinite number of determinants, derived from an infinite AO basis set, possibly including continuum functions. As in Hartree–Fock, there are no many-body terms explicitly included in CI expansions either. This failure results in an extremely slow convergence of CI expansions [9]. Nevertheless, CI is widely used, and has sparked numerous related schemes that may be used, in principle, to construct trial wavefunctions.

What is the physical nature of the many-body correlations needed to accurately describe the many-body system? Insight into this question might provide us with a more compact representation of the wavefunction. There are essentially two kinds of correlation: *dynamical* and *nondynamical.* An example of the former is angular correlation. Consider He (helium), where the Hartree–Fock determinant places both electrons uniformly in spherical symmetry around the nucleus: the two electrons are thus uncorrelated. One could add a small contribution of a determinant of S symmetry, built using $2p$ orbitals, to increase the wavefunction when the electrons are on opposite sides of the nucleus and decreases it when they are on the same side. Likewise, radial correlation can be achieved by adding a $2s$ term. Both of these dynamical correlation terms describe (in part) the instantaneous positions taken by the two electrons. On the other hand, nondynamical correlation results from geometry changes and near degeneracies. An example is encountered in the dissociation of a molecule. It also occurs when, for instance, a Hartree–Fock excited state is close enough in energy to mix with the ground state. These nondynamical correlations result in the well-known deficiency of the Hartree–Fock method that dissociation is not into two neutral fragments, but rather into ionic configurations. Thus, for a proper description of reaction pathways, a multideterminant wavefunction is required: one containing a determinant or a linear combination of determinants corresponding to all fragment states.

Hartree–Fock and post-Hartree–Fock wavefunctions, which do not *explicitly* contain many-body correlation terms lead to molecular integrals that are substantially more convenient for numerical integration. For this reason, the vast majority of (non-Monte Carlo) work is done with such independent-particle-type functions. However, given the flexibility of Monte Carlo integration, it is very worthwhile in VMC to incorporate many-body correlation *explicitly*, as well as incorporating other properties a wavefunction ideally should possess. For example, we know that because the true wavefunction is a solution of the Schrödinger equation, the local energy must be a constant for an eigenstate. (Thus, for approximate wavefunctions, the variance of the local energy becomes an important measure of wavefunction quality.) Because the local energy should be a constant everywhere in space, each singularity of the Coulomb potential must be canceled by a corresponding term in the local kinetic energy. This condition results in a *cusp*: a discontinuity in the first derivative of Ψ_T, where two charged particles meet [10]. Satisfying this leads, in large measure, to more rapidly convergent expansions. With a sufficiently flexible trial wavefunction, one can include appropriate parameters, which can then be determined by the cusp condition. For the electron–nucleus cusp, this condition is

$$\left.\frac{1}{\Psi}\frac{\partial\Psi}{\partial r}\right|_{r=0} = -Z \tag{3.1}$$

where r is any single electron–nucleus coordinate. If we solve for Ψ we find that, locally, it must be exponential in r. The extension to the many-electron case is straightforward. As any single electron (with all others fixed) approaches the nucleus, the exact wavefunction behaves asymptotically as in the one-electron case, for each electron individually. An extension of this argument to the electron–electron cusp is also readily done. In this case, as electron i approaches electron j, one has a two-body problem essentially equivalent to the hydrogenic atom. Therefore, in analogy to the preceding electron–nucleus case, one obtains the cusp conditions

$$\begin{aligned} \left.\frac{1}{\Psi}\frac{\partial\Psi}{\partial r_{ij}}\right|_{r_{ij}=0} &= \frac{1}{2} \qquad \text{unlike spin} \\ \left.\frac{1}{\Psi}\frac{\partial\Psi}{\partial r_{ij}}\right|_{r_{ij}=0} &= \frac{1}{4} \qquad \text{like spin} \end{aligned} \tag{3.2}$$

For like spins, the vanishing of the Slater determinant at $r_{ij} = 0$ contributes partially to satisfying the cusp condition. (Another factor of 2 results from the indistinguishability of the electrons.) From these equations we see the need for *explicit* two-body terms in the wavefunction, for with a sufficiently flexible form of Ψ, we can then satisfy the cusp conditions, thereby matching the Coulomb singularity for any particle pair with terms from the kinetic energy. Note also that while Slater-type (exponential) orbitals (STOs) have the proper hydrogenic cusp behavior, Gaussian-type orbitals (GTOs) do not. Thus, basis sets consisting of GTO's, although computationally expedient for non-Monte Carlo integral evaluation, cannot directly satisfy the electron-nucleus cusp condition, and are therefore less desirable as Monte Carlo trial wavefunctions.

Three-particle coalescence conditions also have been studied. These singularities are not a result of the divergence of the potential, but are entirely due to the kinetic energy (i.e., to the form of the wavefunction). To provide a feel for the nature of these terms, we note that Fock [11] showed by an examination of the helium atom in hyperspherical coordinates that terms of the form $(r_1^2 + r_2^2)\ln(r_1^2 + r_2^2)$ are important when r_1 and $r_2 \to 0$ simultaneously. Additional higher-order terms, describing correlation effects and higher n-body coalescences, also have been suggested.

Since *explicit* many-body terms are critical for a compact description of the wavefunction, let us review some early work along these lines. Hylleraas and Pekeris had great success for He with wavefunctions of the form

$$\Psi_{\text{Hylleraas}} = \sum_{k=1}^{k} d_k\, r_{12}^{a_k} s^{b_k} t^{e_k}\, e^{-s/2} \tag{3.3}$$

where $s = r_1 + r_2$, and $t = r_1 - r_2$. Here r_1 and r_2 are the scalar distances of the electrons from the nucleus. The electron–nucleus and the electron–electron cusp conditions can be satisfied by choosing the proper values for the coefficients. Because *all* the interparticle distances (for this simple two-electron case) are represented, very accurate descriptions of the He wavefunction may be obtained with relatively few terms. Moreover, this form may be readily generalized to larger systems, as has been done by Umrigar [4]. Despite a great improvement over single-particle expansions, a 1078-term Hylleraas function, which yields an energy of -2.903724375 hartrees, is surprisingly not all that much superior to a nine-term function which already yields -2.903464 hartrees [12,13]. On the other hand, by adding terms with powers of $\ln s$ and negative powers of s, one can obtain an energy of -2.9037243770326 hartrees with only 246 terms. The functional form clearly is very important. Recently, terms containing $\cosh t$ and $\sinh t$ have been added as well [14] to model "in–out" correlation (the tendency for one electron to move away from the nucleus as the other approaches the nucleus).

We can distinguish between two broad classes of explicitly correlated wavefunctions: polynomials in r_{ij} and other inter-body distances, and exponential or Jastrow forms [15,16]

$$\Psi_{\text{corr}} = e^{-U} \tag{3.4}$$

In this latter form, U contains all the r_{ij} and many-body dependencies, and the full wavefunction is given by

$$\Psi = \Psi_{\text{corr}}\Psi_D \tag{3.5}$$

where the second factor is the determinant(s) discussed earlier. The Jastrow forms contain one or more parameters that can be used to represent the cusps. As an example, consider a commonly used form, the Pade–Jastrow function:

$$-U = \sum_{i=1}^{N-1} \sum_{j>i}^{N} \frac{a_1 r_{ij} + a_2 r_{ij}^2 + \cdots}{1 + b_1 r_{ij} + b_2 r_{ij}^2 + \cdots} \tag{3.6}$$

The general behavior of e^{-U} is that it begins at unity (for $r_{ij} = 0$) and asymptotically approaches a constant value for large r_{ij}. One can verify that the electron–electron cusp condition simply requires a_1 to be $\frac{1}{2}$ for unlike spins and $\frac{1}{4}$ for like spins. The linear Pade–Jastrow form has only one free parameter, namely, b_1, with which to optimize the wavefunction. Creating a correlated trial wavefunction as above, by combining Ψ_{corr} with an SCF determinant, causes a global expansion of the electron density [17]. If we assume that the SCF density is relatively accurate to begin with, then we need to rescale this combined trial wavefunction to readjust the density. This can be accomplished simply by multiplying by an additional term, an *electron–nucleus* Jastrow function. If this Jastrow function is written with

$$-U = \sum_{i=1}^{N} \sum_{a=1}^{N_{\text{nuc}}} \frac{\lambda_1 r_{ia} + \lambda_2 r_{ia}^2 + \cdots}{1 + v_1 r_{ia} + v_2 r_{ia}^2 + \cdots} \tag{3.7}$$

then, in analogy to the electron–electron function, λ_1 is determined by the cusp condition. More general forms of U have been explored in the literature, including ones with electron–electron–nucleus terms, and powers of s and t [4,18]. These lead to greatly improved functional forms as judged by their rates of convergence and VMC variances.

A different approach towards building explicitly correlated wavefunctions is to abandon the distinction between the correlation part and the determinantal part. In such an approach one might try to approximate the exact wavefunction using a linear combination of many-body terms [19]. A powerful approach is to use explicitly correlated Gaussians [20,21]. Such a functional form, although very accurate for few electron systems, is difficult to use with more electrons (because of the rapidly expanding number of terms needed), and does not exploit the more general (integral) capabilities of VMC. A form amenable to VMC computation is a linear expansion in correlated exponentials [22], which shows very good convergence properties.

For such correlated wavefunctions, one can optimize the molecular-orbital coefficients, the atomic orbital exponents, the Jastrow parameters, and any other nonlinear parameters. Clearly, practical limitations will be reached for very large systems; but such optimization is generally practical for moderate-sized systems, and has been done for several. The next section discusses the means by which such optimizations may be performed by Monte Carlo.

IV. OPTIMIZATION OF A TRIAL WAVEFUNCTION USING VMC

In the previous sections we have seen how VMC can be used to estimate expectation values of an operator given a trial wavefunction $\Psi_T(\mathbf{R})$. Despite the "logic" used to select trial wavefunction forms, as described in the previous section, for a realistic system it is extremely difficult to know a priori the proper analytic form. Thus it is a challenge to generate a good trial wavefunction "out of thin air." Of course, one can choose a trial form that depends on a number of parameters; then, within this "family" one would like to be able to choose the "best" wavefunction. Moreover, if possible, one would like to be able to arbitrarily improve the "goodness" of the description, in order to approach the true wavefunction. It is clear that we need first to clarify what we mean by a "good" $\Psi_T(\mathbf{R})$, otherwise the problem is ill-posed.

If for the moment we restrict our attention to trial wavefunctions that approximate the lowest state of a given symmetry, a possible (and surely the most-used in practice) criterion of goodness is provided by the beloved variational principle

$$E[\Psi_T] = \langle H \rangle = \frac{\int \Psi_T^*(\mathbf{R}) H \Psi_T(\mathbf{R}) \, d\mathbf{R}}{\int \Psi_T^*(\mathbf{R}) \Psi_T(\mathbf{R}) \, d\mathbf{R}} \geq E_0 \tag{4.1}$$

which provides an *upper* bound to the exact ground-state energy E_0. According to this principle, $\Psi_{T1}(\mathbf{R})$ is better than $\Psi_{T2}(\mathbf{R})$ if $E[\Psi_{T1}] < E[\Psi_{T2}]$. This is a reasonable criterion since, in this sense, the exact wavefunction $\phi_0(\mathbf{R})$ is "better" than any other function. However, before discussing the practical application of Eq. (4.1), we point out that the variational principle is not the only available criterion of goodness for trial wavefunctions. We will see more about this later on; for now we simply point out that other criteria may well have advantages. A second remark worth making here is that the condition $\langle H \rangle \to E_0$ is not sufficient to guarantee that $\Psi_T \to \phi_0$ (i.e., for all points). This implies, for example, that although the energy is improving, expectation values of some other operators might appear to converge to an incorrect value, or may not converge at all (although this latter is a rather pathological case, and not very common for reasonably well-behaved trial wavefunctions).

Let us chose a family of trial wavefunctions $\Psi_T(\mathbf{R}; \mathbf{c})$, where $\mathbf{c}$ is a vector of parameters $\mathbf{c} \equiv \{c_1, c_2, \ldots, c_n\}$ on which the wavefunction depends parametrically. The best function in the family is selected by solving the problem

$$\min_{\mathbf{c}} \langle H(\mathbf{c}) \rangle \tag{4.2}$$

In most standard ab initio approaches, the parameters to minimize are the linear coefficients of the expansion of the wavefunction in some basis set. To make the problem tractable, one is usually forced to choose a basis set for which the integrals of Eq. (4.1) are analytically computable. However, as we have seen, it is practical to use very accurate explicitly correlated wavefunctions with VMC.

A typical VMC computation to estimate the energy or other expectation values for a given $\Psi_T(\mathbf{R})$ might involve the calculation of the wavefunction value, gradient, and Laplacian at several millions points distributed in configuration space. Computationally this is the most expensive part. So a desirable feature of $\Psi_T(\mathbf{R})$, from the point of view of Monte Carlo, is its compactness. It would be highly impractical to use a trial wavefunction represented, for example, as a CI expansion of thousands (or more) of Slater determinants.

For this reason, optimization of the wavefunction is a crucial point in VMC (and likewise in QMC as well; see the next chapter). To do this optimization well, and to allow a compact representation of the wavefunction, it is absolutely necessary to optimize the nonlinear parameters, most notably the orbital exponents and the parameters in the correlation factor(s), in addition to the linear coefficients.

A. ENERGY OPTIMIZATION

The naive application of the variational principle to the optimization problem is limited by the statistical uncertainty inherent in every Monte Carlo calculation. The magnitude of this statistical error has a great impact on the convergence of the optimization, and on the ability to find the optimal parameter set as well.

Consider the following algorithm to optimize a one parameter function $\Psi_T(\mathbf{R}, \alpha)$:

```
0)  Choose the initial parameter α^old
1)  Do a VMC run to estimate E^old=⟨H(α^old)⟩
2)  Repeat:
3)    Somehow select a new parameter α^new
        (maybe α^new=α^old+ζ)
4)    Do a VMC run to estimate E^new=⟨H(α^new)⟩
5)    If (E^new<E^old) α^old=α^new; E^old=E^new;
6)  Until the energy no longer diminishes.
```

Because the energy E^{new} is only an *estimate* of the true expectation value $\langle H(\alpha^{\mathrm{new}})\rangle$, step 5 is a weak point of this algorithm. Using a 95% confidence interval we can say only that

$$\begin{aligned} E^{\mathrm{old}} - 2\sigma^{\mathrm{old}} \le \langle H(\alpha^{\mathrm{old}})\rangle \le E^{\mathrm{old}} + 2\sigma^{\mathrm{old}} \\ E^{\mathrm{new}} - 2\sigma^{\mathrm{new}} \le \langle H(\alpha^{\mathrm{new}})\rangle \le E^{\mathrm{new}} + 2\sigma^{\mathrm{new}} \end{aligned} \tag{4.3}$$

Thus it is clear that sometimes we cannot decide which of E^{old} and E^{new} is the lower, and thus which parameter is the better.

B. Correlated Sampling

One way to treat this problem begins by noting that during the optimization we are not really interested in the energies associated with the different wavefunctions, but only in the energy differences. A key observation is that one can use the *same* random walk to estimate the energies of different wavefunctions, and that such energy estimates will be statistically correlated. Thus, their difference can be estimated with much greater precision than can be the energies themselves. (Loosely speaking, a part of the fluctuations that result from the particular walk chosen will cancel when computing the difference.)

Let us then proceed to estimate the energies of a whole set of wavefunctions with a single VMC run. For simplicity, let us consider a family of K functions that depend on a single parameter:

$$\{\Psi(\mathbf{R}, \alpha_k)\}_{k=1}^{K} \equiv \{\Psi(\mathbf{R}, \alpha_1), \Psi(\mathbf{R}, \alpha_2), \ldots, \Psi(\mathbf{R}, \alpha_k)\} \tag{4.4}$$

One can estimate the energy of *all* the $\Psi_k = \Psi(\mathbf{R}, \alpha_\kappa)$ by VMC simulation, simply by sampling from the function $\Psi_0 = \Psi(\mathbf{R}, \alpha_0)$ and noting that

$$\begin{aligned} \langle H(\alpha_k)\rangle &= \frac{\int \Psi_k H \Psi_k}{\int \Psi_k \Psi_k} = \frac{\int \Psi_0^2 \left(\frac{\Psi_k}{\Psi_0}\right)^2 \frac{H\Psi_k}{\Psi_k}\, d\mathbf{R}}{\int \Psi_0^2 \left(\frac{\Psi_k}{\Psi_0}\right)^2 d\mathbf{R}} \\ &\cong \frac{\sum_i \omega_k^2(\mathbf{R}_i) E_{\mathrm{L},k}(\mathbf{R}_i)}{\sum_i \omega_k^2(\mathbf{R}_i)} = E_k \end{aligned} \tag{4.5}$$

where the points $\mathbf{R}_i$ are distributed according to Ψ_0^2, $E_{\mathrm{L},k}$ is the local energy of the function Ψ_k, and the reweighting factor ω_k^2 is defined as the square of the ratio of the two wavefunctions. After only a single run we can choose the best parameter. This method is effective only if the functions Ψ_k are not too different from Ψ_0; that is, if the reweighting factors are not too small or too large, but rather are of the order of magnitude of unity. Otherwise a few points will dominate the sum, reducing the statistical quality of the sample and resulting in a larger error.

This method, sometimes called "differential Monte Carlo" [23], has been used in the past to compute energy differences coming from small perturbations of the Hamiltonian, or from small changes of the geometry of a molecule (i.e., for computing potential-energy surfaces). The main disadvantage of this method is that it becomes more and more cumbersome to apply as the number of parameters increases. One can see this by noting that we are, in essence, evaluating the energy on a grid in *parameter space*. Using a function that depends on say three parameters, and choosing 10 possible values for each parameter, we must evaluate $10^3 = 1000$ energy points.

C. Optimization Using Derivatives

Although the direct application of correlated sampling can be inefficient in a high-dimensional parameter space, it can be fruitfully combined with a standard minimization technique such as the steepest-descent method [24]. Consider again the trial wavefunction $\Psi_{\mathrm{T}}(\mathbf{R}; \mathbf{c})$ containing a parametric dependence on $\mathbf{c}$. Again we construct K "perturbed" functions, each one now differing from the reference function by the value of a single parameter,

$$\Psi_k = \Psi_{\mathrm{T}}(\mathbf{R}, \mathbf{c}_k), \qquad k = 1, 2, \ldots, K \tag{4.6}$$

where

$$\mathbf{c}_k = \{c_1, c_2, \ldots, c_k + \delta c_k, \ldots, c_K\} \tag{4.7}$$

Once again defining the weights

$$\omega_k^2 = \left(\frac{\Psi_k}{\Psi_0}\right)^2 \tag{4.8}$$

the variational energies of the various functions Ψ_k can be evaluated as in Eq. (4.5) using only the unperturbed wavefunction Ψ_0. If the variations are small, we can numerically estimate the partial derivatives of the energy with respect to the parameters using a finite-difference formula

$$\frac{\partial E}{\partial c_k} \cong \frac{\langle H(\mathbf{c}_k)\rangle - \langle H(\mathbf{c}_0)\rangle}{\delta c_k} = \frac{E_k - E_0}{\delta c_k} \tag{4.9}$$

Equipped with such a method to estimate the gradient of the energy with respect to the parameters, we can describe a simple steepest descent algorithm that tries to optimize the parameters in a single VMC run.

```
0)  Choose the initial parameters c and the magnitude
      of the perturbations δc_k
1)  Repeat:
2)    Estimate E_0=⟨H(c)⟩ and E_k=⟨H(c_k)⟩ (averaging over
      a block)
3)    Estimate the gradient ∇_c={∂E/∂c_1, ∂E/∂c_2, ...,
      ∂E/∂c_k}
4)    Update the parameter vector c←c−s·∇_c where s is
      a step size vector;
5)  Until the energy no longer diminishes.
```

It is also possible to estimate the derivative of the wavefunction with respect to a parameter *without* resorting to the finite-difference approximation. Consider again our one-parameter wavefunction $\Psi(\alpha) \equiv \Psi_T(\mathbf{R};\alpha)$. We would like to express the expectation value of the derivative of the energy in a form amenable to computation with VMC. By differentiating the expression for the energy, we get

$$\frac{\partial\langle H(\alpha)\rangle}{\partial\alpha} = \frac{2\int \frac{\partial\Psi(\alpha)}{\partial\alpha} H\Psi(\alpha)\, d\mathbf{R}}{\int \Psi^2(\alpha)\, d\mathbf{R}} - \frac{2\int \Psi(\alpha)H\Psi(\alpha)\, d\mathbf{R}}{\int \Psi^2(\alpha)\, d\mathbf{R}} \frac{\int \frac{\partial\Psi(\alpha)}{\partial\alpha}\Psi(\alpha)\, d\mathbf{R}}{\int \Psi^2(\alpha)\, d\mathbf{R}} \tag{4.10}$$

and using the trick of multiplying and dividing inside the integral by our probability distribution $\Psi^2(\alpha)$, we obtain

$$\frac{\partial\langle H(\alpha)\rangle}{\partial\alpha} = \left\langle \frac{\partial}{\partial\alpha} \ln|\Psi(\alpha)|^2 E_L(\alpha)\right\rangle - \left\langle \frac{\partial}{\partial\alpha} \ln|\Psi(\alpha)|^2\right\rangle \langle E_L(\alpha)\rangle \tag{4.11}$$

This means that we can estimate the derivative of the energy with respect to α exactly, without using finite-difference methods. Still, the computational power needed to optimize more than a few parameters remains large, even including the savings from correlated sampling, because the fluctuations on this quantity can be large.

D. Optimization of the Variance of the Local Energy

The statistical error in any VMC estimate of the energy obtained using N sample points is related to the fluctuation of the local-energy function. Specifically

$$E = \langle E_{\mathrm{L}} \rangle \cong \frac{1}{N} \sum_{i=1}^{N} E_{\mathrm{L}}(\mathbf{R}_i) \pm \frac{\sigma(E_{\mathrm{L}})}{\sqrt{N}} \tag{4.12}$$

where

$$\sigma^2(E_{\mathrm{L}}) = \langle E_{\mathrm{L}}^2 \rangle - \langle E_{\mathrm{L}} \rangle^2 \tag{4.13}$$

Suppose we knew the exact wavefunction: in that case the local energy would be a constant over all configuration space, and the error in its estimate would be rigorously zero. Similarly, if we could obtain very good wavefunctions the fluctuations of the local energy would be very small. (This is an amusing property of VMC: the more accurate the trial wavefunction, the "easier" the simulations. However, if accuracy of the wavefunction entails evaluation of massive expressions, such an approach makes life "harder" and is counterproductive.) The error thus provides a measure of the quality of the trial wavefunction. This leads naturally to the idea of optimizing the wavefunction to *minimize the error* of the estimate.

We have previously shown that $\langle E_{\mathrm{L}} \rangle = \langle H \rangle$; it is not difficult to show also, using the hermiticity of the Hamiltonian operator, that $\langle E_{\mathrm{L}}^2 \rangle = \langle H^2 \rangle$, so that

$$\langle H^2 \rangle = \langle E_{\mathrm{L}}^2 \rangle \cong \frac{1}{N} \sum_{i=1}^{N} E_{\mathrm{L}}^2(\mathbf{R}_i) \tag{4.14}$$

We can now rewrite Eq. (4.13) as

$$\sigma^2(H) = \langle H^2 \rangle - \langle H \rangle^2 \tag{4.15}$$

and, noting that $\sigma^2(H)$ is positive, we can formulate the following variational principle:

For a stationary state, the quantity $\sigma^2(H)$ is at a local minimum, and is equal to zero.

Optimizing a wavefunction using this principle means solving the problem

$$\min_{\mathbf{c}} \sigma^2(H(\mathbf{c})) \tag{4.16}$$

This version of the variational principle, although very old, is quite possibly not known to readers more familiar with standard ab initio methods; so before applying it to Monte Carlo optimization, let us discuss its properties. The reason it is little known and rarely applied in standard computational methods, is that Eq. (4.15) involves the expectation value of the square of the Hamiltonian, a very difficult quantity to compute analytically. A second problem is that for some (albeit, very poor) trial wavefunctions this quantity might diverge. A third problem is that sometimes, again for very poor wavefunctions, a local minimum can not be found [for example, the reader might easily check that using a single Gaussian function to describe the ground state of the hydrogen atom, a minimum of $\sigma^2(H)$ does not exist]. The main point—the difficulty in its computation—is not an issue in VMC, whereas the other two problems are not encountered in practical calculations. Moreover, there are advantages in using the "sigma variational principle" instead of the more common "energy variational principle." The first advantage is that we know exactly the minimum possible value, namely, zero; and this value is obtained only for the exact wavefunction. The minimum value of the energy, on the other hand, is not known. A second quality is that this variational principal can also be used for excited states, and not just for the first state of a given symmetry: $\sigma^2(H)$ is a minimum for *any* state. This is in contrast with the energy variational principle, where for an arbitrary stationary state the energy is not necessarily required to be a local minimum.

As a byproduct of the computation of $\sigma^2(H)$, after a VMC simulation we have both an *upper* bound to the energy (the usual $\langle H \rangle \geq E_0$), and a *lower* bound. The latter is a quantity that common techniques do not give. Under reasonable assumptions [25,26]

$$\langle H \rangle \geq E_0 \geq \langle H \rangle - \sqrt{\langle H^2 \rangle - \langle H \rangle^2} = \langle H \rangle - \sigma(H) \tag{4.17}$$

However, because the upper bound depends quadratically on the error of the trial wavefunction, while the lower bound depends only linearly on it, the lower bound is usually a significantly worse approximation to the exact energy than is the upper bound.

One might thus consider optimizing the wavefunction to *maximize the lower bound*. Because of the presence of $\langle H \rangle$ in Eq. (4.17), however, this usually leads to the same problems we have seen when trying to directly optimize the energy. If one instead decides to minimize the distance between the upper and the lower bounds, the sigma variational principle is recovered. [In practice, one optimizes $\sigma^2(H)$, not $\sigma(H)$, since they have the same minima, and one avoids computing a square root.]

Instead of minimizing $\sigma^2(H)$, sometimes it is preferable to optimize a related quantity, namely the second moment of the local energy with respect to a fixed (or reference) energy $E_{\mathbf{R}}$

$$\mu^2_{E_{\mathbf{R}}}(H) = \frac{\int \Psi(\mathbf{R})(E_{\mathrm{L}}(\mathbf{R}) - E_{\mathbf{R}})^2 \Psi(\mathbf{R})\, d\mathbf{R}}{\int \Psi(\mathbf{R})\Psi(\mathbf{R})\, d\mathbf{R}} \tag{4.18}$$

The constant $E_{\mathbf{R}}$ should be close to the energy of the state being sought, although the optimization does not depend strongly on its value. Minimizing this quantity [which many authors call $\sigma^2(H)$ without making the distinction] is *almost* equivalent to minimizing the variance. A little algebra shows that

$$\mu^2_{E_{\mathbf{R}}}(H) = \sigma^2(H) + (\langle H \rangle - E_{\mathbf{R}})^2 \tag{4.19}$$

The second term on the right-hand side can be made small. In fact, having the second term present allows one to balance minimization of the variance with minimization of the energy, the value of the reference energy dictating the nature of the balance. This last form of minimization procedure is preferable when looking for an excited state, since by a careful choice of the reference energy one can prevent falling to the ground state.

If it were only for the above properties, optimization using the variance would not be as widespread as it is today in VMC calculations. The property that improves the optimization by orders of magnitude is what we might call *local boundedness*. In practice, all estimates in VMC are done using a finite set of configuration points (the *walkers* $\mathbf{R}$); however, for a finite number of such points, the estimate of the *energy* can be unbounded from below *with respect to variations of the parameters*, whereas the *variance* is always bounded. One can demonstrate this with a toy problem in which there are only two (fixed) walkers being used for the optimization. It is relatively easy to find such a pair of walkers for which the local-energy surface in parameter space has singularities (i.e., values for the parameters at which the local energy goes to minus infinity); the variance, of course, must always be positive.

E. Optimization with a Fixed Ensemble

Since $\sigma^2(H)$ is a bounded quantity, even when estimated with a finite number of points, it becomes possible to use a *fixed* (relatively), small number of walkers to estimate the necessary integrals during the optimization [27–29]. Suppose then that we have an ensemble of N walkers distributed according to $\Psi(\mathbf{c}_0)$, where $\mathbf{c}_0$ is some initial set of parameters. The optimization algorithm for a fixed ensemble will use these N points to estimate the integrals even when, later during the minimization, the *parameter vector* is modified. This is achieved by the same reweighting method we previously discussed for correlated sampling; in fact, we are using correlated sampling to compute expectation values for different values of $\mathbf{c}$, only now with a fixed ensemble. Because the ensemble is fixed, the optimization process has become completely deterministic (apart from the input of random configuration points). One can thus choose whatever optimization procedure one is fond of (steepest descent, Powell's method, simulated annealing, etc.), and not have to worry about the uncertainties for now. We repeat here the relevant reweighting formulas

$$\begin{aligned}\langle H(\mathbf{c}^{\text{new}})\rangle &\cong \frac{\sum_{i=1}^{N} \omega_{\text{new}}^2 E_{\text{L}}(\mathbf{R}_i)}{\sum_{i=1}^{N} \omega_{\text{new}}^2} \\ \langle H^2(\mathbf{c}^{\text{new}})\rangle &\cong \frac{\sum_{i=1}^{N} \omega_{\text{new}}^2 E_{\text{L}}^2(\mathbf{R}_i)}{\sum_{i=1}^{N} \omega_{\text{new}}^2}\end{aligned} \tag{4.20}$$

where the ω terms are the ratios of new to original trial wavefunctions, and the E_{L} terms are evaluated with respect to the original wavefunction (parameters).

After the optimization procedure has found the best set of parameters (for the given set of walkers), it is a good idea to run a new VMC simulation, to produce an ensemble of walkers distributed according to the *new* wavefunction; then one should reoptimize again, until self-consistency is attained—specifically, until $\sigma^2(H)$, or $\mu_{E_{\text{R}}}^2(\mathbf{R})$, does not change significantly anymore. Another practical point: it is sometimes found that a more numerically stable optimization can be achieved by initially setting all the weights to unity [see Eq. (4.5)]. This is clearly an approximation, with the added *dis*advantage that it can prevent movements of the nodes during the optimization. (This can be seen by noting that, on approaching a node, the local energy diverges; but, in the weighted optimization, this is counterbalanced by a vanishing weight.) Nevertheless, the approximation is only to the *direction* of motion in parameter space, whereas the approach can frequently have the advantage of faster convergence, particularly when one is starting reasonably far from the optimum set of parameters.

With ensembles consisting of a few thousand points, this algorithm can reasonably be expected to optimize a hundred or so nonlinear parameters. Usually only two or three reoptimization iterations are necessary to reach convergence. Attempting to directly optimize the *energy* using this algorithm, however, might require one to two orders of magnitude more walkers, and a correspondingly increased computational cost.

Because ensuring that cusp conditions are met helps to minimize fluctuations, at least in the energy, optimization will be aided by satisfying these conditions. However, note that optimization also can be used to *obtain* good cusps; this can be done by including a *penalty* factor into the minimization procedure to force the various parameters of the trial wavefunction to build the exact or nearly exact cusps.

It is also worth mentioning that other functionals can (and have been) used for the minimization process. One can consider modifications of the variance functional [30], or even using a completely different approach to solving the problem [31]. Consider, for example, the following minimization;

$$\min_{\mathbf{c}}(\varepsilon(\mathbf{c})) = \min_{\mathbf{c}}\left(\int (\Psi_{\mathrm{T}}(\mathbf{R}) - \phi_0(\mathbf{R}))^2 \, d\mathbf{R}\right) \tag{4.21}$$

i.e., minimizing the least square error between the trial and the exact wavefunction. Although we do not know the exact wavefunction, one *can* sample it using quantum Monte Carlo techniques (such as those discussed in the next chapter).

V. ACCELERATION TECHNIQUES

Having obtained the best possible trial wavefunction that we are capable of, we are perhaps now ready to compute chemical properties. A good trial wavefunction will provide us with not only good expectation values but, as discussed earlier, also a low degree of statistical error. However, the efficiency of the VMC method depends not only on the quality of the trial wavefunction used but also on the *procedure* for generating the distribution to be sampled (i.e., for generating the trial wavefunction squared). We have seen that there are a number of possible algorithms, ranging from simple Metropolis, to Langevin, to generalized Metropolis schemes, which can all be used to obtain the same distribution. The latter improve efficiency over the simple Metropolis scheme. The reason, in part, is that in all these approaches, the Markov chain (or random-walk) nature of the generation procedure leads to correlations between sample points. Such correlations, depending on their magnitude, can greatly affect the rate of convergence.

Although we have identified the (mathematical) source of inefficiency to be the correlations present in the Markov chain, there is a deeper underlying physical cause. This is a fundamental problem of many simulations, namely, that of multiple scales, which, specifically, as is follows. Although

VMC scales well with M, it scales much more poorly with atomic number Z. A common estimate is that computational time T rises as $Z^{5.5}$ [32]. On reflection it is clear that the problem is the differing timescales (as well as distance and energy scales) for core and valence electrons. As Z increases, this range of timescales increases as well. In fact, $Z \to \infty$ is in many ways analogous to a critical point [33,34]. As in critical slowing down, an unending hierarchy of time scales ensues as the critical point is approached. This is the problem that must be addressed. In critical phenomena the problem has been effectively treated through a class of acceleration methods, particularly so-called cluster acceleration methods. These take advantage of the self-similarity that occurs in the vicinity of a critical point. In electronic structure problems there do exist analogous critical points [34,35]. However, typical systems are not near the regime of those critical points. Instead, a common way to address the large-Z problem has been through the use of effective-core potentials or pseudo-potentials that eliminate the large Z at the outset. This is the standard approach in quantum chemistry and solid-state physics. It is also becoming widely (and effectively) used in quantum Monte Carlo simulations [7].

However, instead of (or in addition to) effective-core potentials, we can cope with the differing timescales at least in part by purely Monte Carlo means. The Markov chain correlations can be vastly different depending on the explicit sampling algorithm employed. Several methods have been explored to change the sampling (or, again, the details of the Markov chain) in conjunction with modifications of the Metropolis scheme. These have met with differing degrees of success.

For example, one can render the attempted moves position-dependent. This, for example, might allow one to move valence electrons further than core electrons, corresponding to their great available "phase space." However, naively doing so does not recognize that exchange symmetry does not allow us to know which electrons are core or which are valence; any a priori assumptions are likely to lead to a situation in which core electrons end up in the valence region after some time, and vice versa. There are a number of correct ways to implement this idea, however. One allows step sizes to change dynamically, depending on relative positions, but entails the need for a modified coordinate system to maintain detailed balance [36]. Another implementation of this idea does it straightforwardly, but explicitly prevents exchange by carrying out the VMC integration only over a subset of the full space. Because the subset is permutationally equivalent to the full space, the integrals are equal (up to a constant factor). Another very different approach, borrowed from high-energy theory, has been to modify the VMC dynamics at the level of the Fokker–Planck equation. This allows more rapid approach to time-independent behavior while keeping the steady-state unchanged [37]. A very nice variant of this has recently been proposed by Mella et al. [38]. In this algorithm the time step size of the Langevin transition matrix is allowed to be an explicit function of position, shrinking as one approaches either nodes of the wavefunction or two-body cusps. Since the functional dependence of the time step on position is put explicitly into a Gaussian transition matrix (rather than a standard Metropolis "box"), the forward and reverse transition matrices overlap, enabling one to preserve detailed balance. By having the time step shrink to a minimum at the electron–nucleus cusp, one effectively treats the core region differently from the valence region. Moreover, the time step behavior obtained near the nodes is *also* advantageous. Similar improvements to the Metropolis algorithm in the vicinity of cusps and nodes have been proposed previously by Umrigar and colleagues [39]. Yet other schemes exist for "accelerating" convergence. For example, one can radically change the algorithm by mixing a molecular dynamics approach with VMC [40].

To get a sense for this type of approach, we will discuss one of these in slightly greater detail. This is the method of dividing the space. As a result of the antisymmetry of the electronic wavefunction there are multiple regions of ($3M$-dimensional) space that are equivalent. Specifically, up to a sign the value of the wavefunction is the same when any two coordinates representing like-spin electrons are interchanged. This results in $N_{\text{up}}!\,N_{\text{down}}!$ equivalent volumes or domains. We are *not* talking about *nodal volumes* here. The volumes we are distinguishing are the following: given a point $\mathbf{R}$ in configuration space there are another $N_{\text{up}}!\,N_{\text{down}}! - 1$ points generated by permutations of the indices. We can think of these points as belonging to different regions, or subspaces, of the full space. If we can explicitly construct such subspaces, then the integration over the entire $3M$-dimensional space is redundant, since for any operator $\hat{O}$ that is totally symmetric with respect to the exchange of two identical particles

$$\begin{aligned}\langle \hat{O} \rangle &= \int_{\text{all space}} \Psi_{\text{T}}^{*}(\mathbf{R})\hat{O}\Psi_{\text{T}}(\mathbf{R})\, d\mathbf{R} \Big/ \int_{\text{all space}} \Psi_{\text{T}}^{*}(\mathbf{R})\Psi_{\text{L}}(\mathbf{R})\, \text{d}\mathbf{R} \\ &= \int_{\text{any subspace}} \Psi_{\text{T}}^{*}(\mathbf{R})\hat{O}\Psi_{\text{T}}(\mathbf{R})\, d\mathbf{R} \Big/ \int_{\text{any subspace}} \Psi_{\text{T}}^{*}(\mathbf{R})\Psi_{\text{T}}(\mathbf{R})\, d\mathbf{R} \qquad (5.1)\end{aligned}$$

meaning we only need to integrate over a single subspace. Note that such subspaces are *not* uniquely defined.

Now we will see how this simple fact can be used to our advantage, to help alleviate the multiple time scales problem. Let us concentrate on an atomic system; molecules can be treated by a simple extension of this approach [41]. Subdividing the space allows us to treat the electrons as "distinguishable" within the simulation. Unlike the standard algorithm in which electrons exchange; that is, they can always cross the subspaces boundaries, in this approach—by integrating only over a subspace—we in effect enforce the boundaries, and constrain particles to stay within subspaces. Any starting configuration (e.g., random walker) is a single point in the $3M$-dimensional space, and thus resides in a single subspace. Subsequent moves need only enforce the boundaries by rejecting any attempts to cross them. Thus, by constructing subspaces such that electrons in the outer region of 3-space, away from the nucleus, stay far from the nucleus, and likewise electrons close in, and near the nucleus, stay close-in, we can assign different time steps to these different electrons. This allows them to explore their respective regions of configuration space with the most appropriate step sizes.

As the number of electrons increase, so does the number of equivalent subspaces and our freedom in choosing them. However, we can also combine some of the electrons (say, into shells) and further increase efficiency [41]. All the electrons within the same shell can explore their entire "shell" space. This is more efficient because we avoid unnecessary rejections resulting from crossings of electrons having the same timescale.

VI. RECENT RESULTS AND NEW DIRECTIONS

It is difficult, in such a short chapter, to review all the published results and all the theoretical developments in the VMC field. Although the method has only recently become anything resembling widespread, it already does have a substantial literature, with more and more papers published every year. Another difficulty in reviewing "results" is that in many instances VMC is used only as a first step, that of generating a highly optimized trial wavefunction, to be used in a "diffusion" or "Green's function" quantum Monte Carlo simulation. This is, in fact, the major use of VMC, and the variational expectation values so obtained are not the point, and sometimes not even reported. To the extent that there are VMC "results" in their own right, these have been recently reported in QMC reviews [7,42,43]. Here, in the spirit of Monte Carlo, we only *sample* the recent literature. We hope this will convey to the reader a feeling of how VMC is being used in actual works, and of the new directions.

Any method, particularly one just coming into the mainstream, needs benchmarks to be established: a series of calculations on "standard" systems, to be a reference when confronting untried systems. One paper that might serve this purpose is the work of Lüchow and Anderson [44], in which they calculate the ground-state energies of atoms from Li to F, and the ground state energies and dissociation energies of the first-row hydrides, using both VMC and QMC. They use simple standard trial wavefunctions (in the VMC sense) composed of a Slater determinant of near Hartree–Fock quality multiplied by a correlation factor of the kind used by Schmidt and Moskowitz [18], a generalized Jastrow factor. The correlation energies recovered with VMC range from 65% to 89% for atoms, while for the hydrides they range from 51% to 78%. Subsequent diffusion Monte Carlo simulations with these functions recovered up to 99% of the correlation energy. Along the same lines, Filippi and Umrigar [45] studied the first-row homonuclear diatomics in a systematic way, using correlated wavefunctions composed of a determinantal part multiplied by a generalized Jastrow factor. They used both single- and multiconfiguration functions, recovering up to 94% of the correlation energy by VMC and up to 99% when the trial wavefunctions were used in a subsequent quantum Monte Carlo simulation.

As mentioned earlier, the treatment of heavy atoms can be approximated using a pseudo-potential approach, both in standard ab initio methods and in VMC as well. With this approach Flad and Dolg [46] studied the Hg atom and the Hg_2 molecule. They used a 20-valence-electron pseudo-potential for the atom, while for the molecule they used a 2-valence-electron pseudo-potential together with a core polarization potential, recovering 84% of the correlation energy. Also using a pseudo-potential approach, Mitas and co-workers studied heavy atoms [47,48] with very good results, obtaining an electron affinity for the Sc atom in agreement with the experimental value, and values of the ionization potential and excitation energies for the Fe atom with an average error, with respect to the experimental value, of less than 0.15 eV. For still larger systems Grossman et al. studied the structure and stability of silicon [49] and carbon [50] clusters, ranging from a few to 20 atoms. Compared with standard quantum chemical approaches like HF, local density approximation to density functional theory (LDA), and coupled-cluster expansions (CCSD[T]), Monte Carlo proved superior and efficient in predicting the most stable isomer. In the case of Si_n for $n < 8$, the binding energies agree within 4% with experiment.

The VMC and QMC methods can be successfully applied to vibrational problems as well. For example, Blume et al. [51] calculated the vibrational frequency shift of HF molecules embedded in helium clusters as a function of cluster size, for up to 198 helium atoms. They obtained good agreement with experimental results. Quantum clusters have also been studied by Rick et al. [52] and Barnett and Whaley [53], with good agreement between the calculated energies and the exact values.

On the theoretical/algorithmic development side, Bueckert et al. [54] showed how to estimate the relativistic energy of atoms and molecules in VMC, without having to resort to perturbation theoretic approximations.

Alexander et al. [55] used VMC to compute cross sections for the elastic and inelastic scattering of fast electrons and X rays by H_2. Novel trial wavefunction optimization schemes has been proposed by Huang and Cao [56] in which they sample $\Psi_T^2(E_L - E_R)^2$, and by Tanaka [57], who introduces a fictitious Lagrangian, resembling that in the Car–Parrinello method, to be used in the simultaneous optimization of the wavefunction parameters and in the geometry optimization. This brief list is, as mentioned, only a sample.

VII. CONCLUSIONS

We hope to have convinced the reader that the VMC approach to obtaining quantum expectation values of interest in both chemical and physical problems is a very powerful one. We believe that it, in combination with fully quantum Monte Carlo procedures, will be the preferred choice in the near future, for many of the calculations performed these days by more traditional nonstochastic means. VMC is the first, and a necessary step, toward a complete quantum simulation of a system. It has the very desirable feature (often a rare one) that it can be learned, implemented, and tested in a short period of time. (It is now part of the folklore in the quantum Monte Carlo community, that the original code, written by J. B. Anderson and used for his first QMC paper, was only 77 lines of FORTRAN code.) We hope our readers will be inspired to write their own (toy or otherwise) VMC code; possibly thereby contributing to and enlarging the growing Monte Carlo community.

REFERENCES

1. W. L. McMillan, *Phys. Rev.* **138**, A442 (1965).
2. D. Ceperley, G. V. Chester, and M. H. Kalos, *Phys. Rev. B* **16**, 3081 (1977).
3. R. L. Coldwell, *Int. J. Quantum Chem. Symp.* **11**, 215 (1977).
4. C. J. Umrigar, K. G. Wilson, and J. W. Wilkins, *Phys. Rev. Lett.* **60**, 1719 (1988).
5. S. Huang, Z. Sun and W. A. Lester, Jr., *J. Chem. Phys.* **92**, 597 (1990).
6. N. Metropolis, A. W. Rosenbluth, M. N. Rosenbluth, A. H. Teller, and E. Teller, *J. Chem. Phys.* **21**, 1087 (1953).
7. B. L. Hammond, W. A. Lester, Jr., and P. J. Reynolds, *Monte Carlo Methods in Ab Initio Quantum Chemistry*, World Scientific, Singapore, 1994.
8. P. J. Reynolds, D. M. Ceperley, B. J. Alder, and W. A. Lester, Jr., *J. Chem. Phys.* **77**, 5593 (1982).
9. J. D. Morgan III, *Numerical Determination of the Electronic Structure of Atoms, Diatomic and Polyatomic Molecules*. M. Defranceschi and J. Delhalle, eds., NATO ASI Series C: Mathematical and Physical Sciences, Vol. 271, Kluwer, Dordrecht 1989.
10. T. Kato, *Commun. Pure Appl. Math.* **10**, 151 (1957).

11. V. Fock, *Izv. Akad. Nauk SSSR, Ser. Fiz.* **18**, 161 (1954).
12. E. A. Hylleraas, *Z. Physik* **54**, 347 (1929).
13. C. L. Pekeris, *Phys. Rev.* **112**, 1649 (1958).
14. J. D. Baker, J. D. Morgan, D. E. Freund, and R. N. Hill, *Phys. Rev. A* **43**, 1247 (1990).
15. R. Jastrow, *Phys. Rev.* **98**, 1479 (1955).
16. Z. Sun, P. J. Reynolds, R. K. Owen, and W. A. Lester, Jr., *Theor. Chim. Acta* **75**, 353 (1989).
17. R. N. Barnett, P. J. Reynolds, and W. A. Lester, Jr., *J. Chem. Phys.* **82**, 2700 (1985).
18. K. E. Schmidt and J. Moskowitz, *J. Chem. Phys.* **93**, 4178 (1990).
19. J. Rychlewski, *Int. J. Quantum Chem.* **49**, 477 (1994).
20. S. A. Alexander, H. J. Monkhorst, R. Roeland, and K. Szalewicz, *J. Chem. Phys.* **93**, 4230 (1990).
21. W. Cencek and J. Rychlewski, *J. Chem. Phys.* **98**, 1252 (1993).
22. D. Bressanini, M. Mella, and G. Morosi, *Chem. Phys. Lett.* **240**, 566 (1995).
23. B. H. Wells, *Chem. Phys. Lett.* **115**, 89 (1985).
24. S. Huang, Z. Sun, and W. A. Lester, Jr., *J. Chem. Phys.* **92**, 597 (1990).
25. D. H. Weinstein, *Proc. Natl. Acad. Sci. USA* **20**, 529 (1934).
26. L. R. Pratt, *Phys. Rev. A* **40**, 6077 (1989).
27. R. L. Coldwell and R. E. Lowther, *Int. J. Quantum Chem. Symp.* **12**, 329 (1978).
28. H. Conroy, *J. Chem. Phys.* **47**, 5307 (1967).
29. A. A. Frost, *J. Chem. Phys.* **10**, 242 (1942).
30. S. A. Alexander, R. L. Coldwell, H. J. Monkhorst, and J. D. Morgan III, *J. Chem. Phys.* **95**, 6622 (1991).
31. R. Bianchi, D. Bressanini, P. Cremaschi, M. Mella, and G. Morosi, *Int. J. Quantum Chem.* **57**, 321 (1996).
32. D. M. Ceperley, *J. Stat. Phys.* **43**, 815 (1986).
33. P. J. Reynolds, in *Computational Physics and Cellular Automata*, A. Pires, D. P. Landau, and H. J. Herrmann, eds., World Scientific, Singapore, 1990, p. 144.
34. P. Serra and S. Kais, *Phys Rev. Lett.* **77**, 466 (1996); *Phys. Rev. A* **55**, 238 (1997).
35. P. Serra and S. Kais, *Chem. Phys. Lett.* **260**, 302 (1996); *J. Phys A: Math. Gen.* **30**, 1483 (1997).
36. C. J. Umrigar, *Phys. Rev. Lett.* **71**, 408 (1993).
37. P. J. Reynolds, *Int. J. Quantum Chem. Symp.* **24**, 679 (1990).
38. M. Mella, A. Lüchow, and J. B. Anderson, *Chem. Phys. Lett.* **265**, 467 (1996).
39. C. J. Umrigar, M. P. Nightingale, and K. J. Runge, *J. Chem. Phys.* **99**, 2865 (1993).
40. D. Bressanini, M. Mella, and G. Morosi, manuscript in preparation.
41. D. Bressanini and P. J. Reynolds, manuscript in preparation.
42. D. M. Ceperley and L. Mitas, *Advances in Chemical Physics XCIII*, I. Prigogine and S. A. Rice, eds., Wiley, New York, 1996.
43. W. A. Lester and B. L. Hammond, *Annu. Rev. Phys. Chem.* **41**, 283 (1990).
44. A. Lüchow and J. B. Anderson, *J. Chem. Phys.* **105**, 7573 (1996).
45. C. Filippi and C. J. Umrigar, *J. Chem. Phys.* **105**, 213 (1996).
46. H.-J. Flad and M. Dolg, *J. Phys. Chem.* **100**, 6152 (1996).

47. L. Mitas, *Phys. Rev. A* **49**, 4411 (1994).
48. L. Mitas, *Comp. Phys. Commun.* **96**, 107 (1996).
49. J. C. Grossman and L. Mitas, *Phys. Rev. Lett.* **74**, 1323 (1995).
50. J. C. Grossman, L. Mitas, and K. Raghavachari, *Phys. Rev. Lett.* **75**, 3870 (1995).
51. D. Blume, M. Lewerenz, F. Huisken, and M. Kaloudis, *J. Chem. Phys.* **105**, 8666 (1996).
52. S. W. Rick, D. L. Lynch, and J. D. Doll, *J. Chem. Phys.* **95**, 2506 (1991).
53. R. N. Barnett and K. B. Whaley, *J. Chem. Phys.* **96**, 2953 (1992).
54. H. Bueckert, S, Rothstein, and J. Vrbik, *Chem. Phys. Lett.* **190**, 413 (1992).
55. S. A. Alexander, R. L. Coldwell, R. E. Hoffmeyer, and A. J. Thakkar, *Int. J. Quantum Chem. Symp.* **29**, 627 (1995).
56. H. Huang and Z. Cao, *J. Chem. Phys.* **104**, 200 (1996).
57. S. Tanaka, *J. Chem. Phys.* **100**, 7416 (1994).

MONTE CARLO EIGENVALUE METHODS IN QUANTUM MECHANICS AND STATISTICAL MECHANICS

M. P. NIGHTINGALE

Department of Physics
University of Rhode Island, Kingston, RI 02881

C. J. UMRIGAR

Cornell Theory Center and Laboratory of Atomic and Solid State Physics
Cornell University, Ithaca, NY 14853

CONTENTS

Advances in Chemical Physics, Volume 105, Monte Carlo Methods in Chemical Physics, edited by David M. Ferguson, J. Ilja Siepmann, and Donald G. Truhlar. Series Editors I. Prigogine and Stuart A. Rice.
ISBN 0-471-19630-4

I. INTRODUCTION

Many important problems in computational physics and chemistry can be reduced to the computation of dominant eigenvalues of matrices of high or infinite order. We shall focus on just a few of the numerous examples of such matrices, namely, quantum mechanical Hamiltonians, Markov matrices, and transfer matrices. Quantum Hamiltonians, unlike the other two, probably can do without introduction. Markov matrices are used both in equilibrium and nonequilibrium statistical mechanics to describe dynamical phenomena. Transfer matrices were introduced by Kramers and Wannier in 1941 to study the two-dimensional Ising model [1], and ever since, important work on lattice models in classical statistical mechanics has been done with transfer matrices, producing both exact and numerical results [2].

The basic Monte Carlo methods reviewed in this chapter have been used in many different contexts and under many different names for many decades, but we emphasize the solution of eigenvalue problems by means of Monte Carlo methods and present the methods from a unified point of view. A vital ingredient in the methods discussed here is the use of optimized trial functions. Section IV deals with this topic briefly, but in general we suppose that optimized trial functions are given. We refer the reader to Ref. 3 for more details on their construction.

The analogy of the time-evolution operator in quantum mechanics on the one hand, and the transfer matrix and the Markov matrix in statistical mechanics on the other, allows the two fields to share numerous techniques. Specifically, a transfer matrix G of a statistical mechanical lattice system in d dimensions often can be interpreted as the evolution operator in discrete, imaginary time t of a quantum mechanical analog in $d-1$ dimensions. That is, $G \approx \exp(-t\mathscr{H})$, where $\mathscr{H}$ is the Hamiltonian of a system in $d-1$ dimensions, the quantum mechanical analog of the statistical mechanical system. From this point of view, the computation of the partition function and of the ground-state energy are essentially the same problems: finding the largest eigenvalue of G and of $\exp(-t\mathscr{H})$, respectively. As far as the Markov matrix is concerned, this simply is the time-evolution operator of a system evolving according to stochastic dynamics. The largest eigenvalue of such matrices equals unity, as follows from conservation of probability, and for systems in thermal equilibrium, the corresponding eigenstate is also known, namely the Boltzmann distribution. Clearly, the dominant eigenstate in this case poses no problem. For nonequilibrium systems, the stationary state is unknown and one might use the methods described in this chapter in dealing with them. Another problem is the computation of the relaxation time of a system with stochastic dynamics. This problem is equivalent to the computational of the second largest eigenvalue of the Markov matrix, and again the current methods apply.

The emphasis of this chapter is on methods rather than applications, but the reader should have a general idea of the kind of problems for which these methods can be employed. Therefore, we start off by giving some specific examples of the physical systems one can deal with.

A. Quantum Systems

In the case of a quantum mechanical system, the problem in general is to compute expectation values, in particular the energy, of bosonic or fermionic ground or excited eigenstates. For systems with n electrons, the spatial coordinates are denoted by a $3n$-dimensional vector $\mathbf{R}$. In terms of the vectors $\mathbf{r}_i$ specifying the coordinates of electron number i, this reads $\mathbf{R} = (\mathbf{r}_1, \ldots, \mathbf{r}_n)$. The dimensionless Hamiltonian is of the form

$$\langle \mathbf{R} | \mathscr{H} | \mathbf{R}' \rangle = \left[-\frac{1}{2\mu} \vec{\nabla}^2 + \mathscr{V}(\mathbf{R}) \right] \delta(\mathbf{R} - \mathbf{R}') \tag{1.1}$$

For atoms or molecules atomic units are used $\mu = 1$, and $\mathscr{V}$ is the usual Coulomb potential acting between the electrons and between the electrons and nuclei:

$$\mathscr{V}(\mathbf{R}) = \sum_{i<j} \frac{1}{r_{ij}} - \sum_{\alpha,\, i} \frac{Z_\alpha}{r_{\alpha i}} \tag{1.2}$$

where for arbitrary subscripts a and b we define $r_{ab} = |\mathbf{r}_a - \mathbf{r}_b|$; indices i and j label the electrons; we assume that the nuclei are of infinite mass and that nucleus α has charge Z_α and is located at position $\mathbf{r}_\alpha$.

In the case of quantum mechanical van der Waals clusters [4,5], μ is the reduced mass—$\mu = 2^{1/3} m\varepsilon\sigma/\hbar^2$ in terms of the mass m, Planck's constant $\hbar$

and the conventional Lennard–Jones parameters ε and σ—and the potential is given by

$$\mathscr{V}(\mathbf{R}) = \sum_{i<j} \frac{1}{r_{ij}^{12}} - \frac{2}{r_{ij}^{6}} \tag{1.3}$$

The quantum nature of the system increases with $1/\mu^2$, which is proportional to the conventional de Boer parameter.

The ground state wavefunction of a bosonic system is positive everywhere, which is very convenient in a Monte Carlo context and allows one to obtain results with an accuracy that is limited only by practical considerations. For fermionic systems, the ground-state wave function has nodes, and this places more fundamental limits on the accuracy one can obtain with reasonable effort. In the methods discussed in this chapter, this bound on the accuracy takes the form of the so-called *fixed-node approximation.* Here one assumes that the nodal surface is given, and computes the ground-state wavefunction subject to this constraint.

The time-evolution operator $\exp(-\tau\mathscr{H})$ in the position representation is the Green function

$$G(\mathbf{R}', \mathbf{R}, \tau) = \langle \mathbf{R}' | e^{-\tau\mathscr{H}} | \mathbf{R} \rangle \tag{1.4}$$

For both bosonic systems and fermionic systems in the fixed-node approximation, G has only nonnegative elements. This is essential for the Monte Carlo methods discussed here. A problem specific to quantum mechanical systems is that G is known only asymptotically for short times, so that the finite-time Green function has to be constructed by the application of the generalized Trotter formula [6,7], $G(\tau) = \lim_{m\to\infty} G(\tau/m)^m$, where the position variables of G have been suppressed.

B. Transfer Matrices

Our next example is the transfer matrix of statistical mechanics. The largest eigenvalue yields the free energy, from which all thermodynamic properties follow. As a typical transfer matrix, one can think of the one-site, Kramers–Wannier transfer matrix for a two-dimensional model of Ising spins, $s_i = \pm 1$. Such a matrix takes a particularly simple form for a square lattice wrapped on a cylinder with helical boundary conditions with pitch one. This produces a mismatch of one lattice site for a path on the lattice around the cylinder. This geometry has the advantage that a two-dimensional lattice can be built one site at a time and that the process of adding each single site is identical each time. Suppose we choose a lattice of M sites, wrapped on a cylinder with a circumference of L lattice spaces. Imagine that we are adding sites so that the lattice grows toward the left. We can then define a conditional partition function $Z_M(\mathbf{S})$, which is a sum over those states (also referred to as configurations) for which the leftmost edge of the lattice is in a given state $\mathbf{S}$. The physical interpretation of $Z_M(\mathbf{S})$ is the relative probability of finding the leftmost edge in a state $\mathbf{S}$ with which the rest of the lattice to its right is in thermal equilibrium.

If one has helical boundary conditions and spins that interact only with their nearest neighbors, one can repeatedly add just a single site and the bonds connecting it to its neighbors above and to the right. Analogously, the transfer matrix G can be used to compute recursively the conditional partition function of a lattice with one additional site

$$Z_{M+1}(\mathbf{S}') = \sum_{\mathbf{S}} G(\mathbf{S}' \mid \mathbf{S}) Z_M(\mathbf{S}) \tag{1.5}$$

with

$$G(\mathbf{S}' \mid \mathbf{S}) = \exp[K(s'_1 s_1 + s'_1 s_L)] \prod_{i=2}^{L} \delta_{s'_i, s_{i-1}} \tag{1.6}$$

with $\mathbf{S} = (s_1, s_2, \ldots, s_L)$ and $\mathbf{S}' = (s'_1, s'_2, \ldots, s'_L)$, and the s_i, $s'_i = \pm 1$ are Ising spins. With this definition of the transfer matrix, the matrix multiplication in Eq. (1.5) accomplishes the following: (1) a new site, labeled 1, is appended to the lattice at the left edge; (2) the Boltzmann weight is updated to that of the lattice with increased size; (3) the old site L is thermalized; and finally (4) old sites $1, \ldots, L-1$ are pushed down on the stack and are renamed to $2, \ldots, L$. Sites have to remain in the stack until all interacting sites have been added, which determines the minimal depth of the stack. It is clear from Figure 1 that the transfer matrix is nonsymmetric and indeed symmetry is not required for the methods discussed in this chapter. It is of some interest that transfer matrices usually have the property that a right eigenvector can be transformed into a left eigenvector by a simple permutation and reweighting transformation. The details are not important here and let it suffice to mention that this follows from an obvious symmetry of the diagram shown in Figure 1: (1) rotate over π about an axis perpendicular to the paper, which permutes the states; and (2) move the vertical bond back to its original position, which amounts to reweighting by the Boltzmann weight of a single bond.

Equation (1.5) implies that for large M and generic boundary conditions at the right-hand edge of the lattice, the partition function approaches a

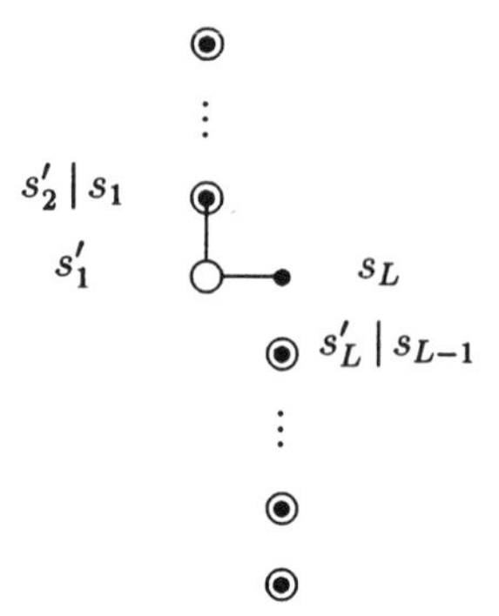

Figure 1. Graphical representation of the transfer matrix. The primed variables are associated with the circles and combine into the left index of the matrix; the dots go with the right index and the unprimed variables. Coincidence of a circle and a dot produces a δ function. An edge indicates a contribution to the Boltzmann weight. Repeated application of this matrix constructs a lattice with nearest-neighbor bonds and helical boundary conditions.

dominant right eigenvector ψ_0 of the transfer matrix G

$$Z_M(\mathbf{S}) \propto \lambda_0^M \psi_0(\mathbf{S}) \tag{1.7}$$

where λ_0 is the dominant eigenvalue. Consequently, for $M \to \infty$ the free energy per site is given by

$$f = -kT \ln \lambda_0 \tag{1.8}$$

The problem relevant to this paper is the computation of the eigenvalue λ_0 by Monte Carlo [8].

C. Markov Matrices

Discrete-time Markov processes are a third type of problem we shall discuss. One of the challenges in this case is to compute the correlation time of such a process in the vicinity of a critical point, where the correlation time goes to infinity, a phenomenon called "critical slowing down." Computationally, the problem amounts to the evaluation of the second largest eigenvalue of the Markov matrix, or more precisely its difference from unity. The latter goes to zero as the correlation time approaches infinity.

The Markov matrix defines the stochastic evolution of the system in discrete time. That is, suppose that at time t the probability of finding the system in state $\mathbf{S}$ is given by $\rho_t(\mathbf{S})$. If the probability of making a transition from state $\mathbf{S}$ to state $\mathbf{S}'$ is $\hat{P}(\mathbf{S}' \mid \mathbf{S})$ (sorry about the hat, we shall take it off soon!), then

$$\rho_{t+1}(\mathbf{S}') = \sum_{\mathbf{S}} \hat{P}(\mathbf{S}' \mid \mathbf{S})\rho_t(\mathbf{S}) \tag{1.9}$$

In the case of interest here, the Markov matrix $\hat{P}$ is constructed so that its stationary state is the Boltzmann distribution $\psi_{\mathrm{B}}^2 = \exp(-\beta\mathscr{H})$. Sufficient conditions are that (1) each state can be reached from every state in a finite number of transitions and that (2) $\hat{P}$ satisfies detailed balance

$$\hat{P}(\mathbf{S}' \mid \mathbf{S})\psi_{\mathrm{B}}(\mathbf{S})^2 = \hat{P}(\mathbf{S} \mid \mathbf{S}')\psi_{\mathrm{B}}(\mathbf{S}')^2 \tag{1.10}$$

It immediately follows from detailed balance that the matrix P defined by

$$P(\mathbf{S}' \mid \mathbf{S}) = \frac{1}{\psi_{\mathrm{B}}(\mathbf{S}')} \hat{P}(\mathbf{S}' \mid \mathbf{S})\psi_{\mathrm{B}}(\mathbf{S}) \tag{1.11}$$

is symmetric and equivalent by similarity transformation to the original Markov matrix. Because of this symmetry, expressions tend to take a simpler form when P is used, as we shall do, but one should keep in mind that P itself is not a Markov matrix, since the sum on its left index does not yield unity identically.

Again, to provide a specific example, we mention that the methods discussed below have been applied [9,10] to an Ising model on a square lattice with the heat bath or Yang [11] transition probabilities and random site selection. In that case, time evolution takes place by single spin–flip transitions that occur between a given state $\mathbf{S}$ to one of the states $\mathbf{S}'$ that differ only at one randomly selected site. For any such pair of states, the transition probability is given by

$$\hat{P}(\mathbf{S}' \mid \mathbf{S}) = \begin{cases} \dfrac{1}{2N} \dfrac{e^{-(1/2)\beta\,\Delta\mathscr{H}}}{\cosh(1/2)\beta\,\Delta\mathscr{H}} & \text{for} \quad \mathbf{S}' \neq \mathbf{S} \\ 1 - \sum_{\mathbf{S}'' \neq \mathbf{S}} \hat{P}(\mathbf{S}'' \mid \mathbf{S}) & \text{for} \quad \mathbf{S}' = \mathbf{S} \end{cases} \tag{1.12}$$

for a system of N sites with Hamiltonian

$$-\beta\mathscr{H} = K \sum_{(i,\,j)} s_i s_j + K' \sum_{(i,\,j)'} s_i s_j \tag{1.13}$$

where (i, j) denotes nearest-neighbor pairs, and $(i, j)'$ denotes next-nearest-neighbor pairs and $\Delta\mathscr{H} \equiv \mathscr{H}(\mathbf{S}') - \mathscr{H}(\mathbf{S})$.

We note that whereas the transfer matrix is used to deal with systems that are infinite in one direction, the systems used in the dynamical computations are of finite spatial dimensions only.

II. THE POWER METHOD

Before discussing technical details of Monte Carlo methods to compute eigenvalues and expectation values, we introduce the mathematical ideas and the types of expressions for which statistical estimates are sought. We formulate the problem in terms of an operator G of which one wants to compute the dominant eigenstate and eigenvalue, $|\psi_0\rangle$ and λ_0. Mathematically, but not necessarily in a Monte Carlo setting, dominant may mean dominant relative to eigenstates of a given symmetry only.

The methods to be discussed are variations of the power method, which relies on the fact that for a generic initial state $|u^{(0)}\rangle$ of the appropriate symmetry, the states $|u^{(t)}\rangle$ defined by

$$|u^{(t+1)}\rangle = \frac{1}{c_{t+1}} G\,|u^{(t)}\rangle \tag{2.1}$$

converge to the dominant eigenstate $|\psi_0\rangle$ of G, if the constants c_t are chosen so that $|u^{(t)}\rangle$ assumes a standard form, in which case the constants c_t converge to the dominant eigenvalue. This follows immediately by expanding the initial state $|u^{(0)}\rangle$ in eigenstates of G. One possible standard form is that, in some convenient representation, the component of $|u^{(t)}\rangle$ largest in magnitude equals unity.

For quantum mechanical systems, G usually is the imaginary-time evolution operator $\exp(-\tau\mathscr{H})$. As mentioned above, a technical problem in that case is that an explicit expression is known only asymptotically for short times τ. In practice, this asymptotic expression is used for a small but finite τ and this leads to systematic, time-step errors. We shall deal with this problem below at length, but ignore it for the time being.

The exponential operator $\exp(-\tau\mathscr{H})$ is one of various alternatives that can be employed to compute the ground-state properties of the Hamiltonian. If the latter is bounded from above, one may be able to use $\mathbb{1} - \tau\mathscr{H}$, where τ should be small enough that $\lambda_0 \equiv 1 - \tau E_0$ is the dominant eigenvalue of $\mathbb{1} - \tau\mathscr{H}$. In this case, there is no time-step error and the same holds for yet another method of inverting the spectrum of the Hamiltonian: the Green function Monte Carlo method. There one uses $(\mathscr{H} - E)^{-1}$, where E is a constant chosen so that the ground state becomes the dominant eigenstate of this operator. In a Monte Carlo context, matrix elements of the respective operators are proportional to transition probabilities and therefore have to be non negative, which, if one uses either of the last two methods, may impose further restrictions on the values of τ and E.

For the statistical mechanical applications, the operators G are indeed evolution operators by construction. The transfer matrix evolves the physical system in a spatial rather than time direction, but this spatial direction corresponds to time from the point of view of a Monte Carlo time series. With this in mind, we shall refer to the operator G as the evolution operator, or the *Monte Carlo* evolution operator, if it is necessary to distinguish it from the usual time-evolution operator $\exp(-\tau\mathscr{H})$.

Suppose that X is an operator of which one wants to compute an expectation value. Particularly simple to deal with are the cases in which the operators X and G are the same or commute. We introduce the following notation. Suppose that $|u_\alpha\rangle$ and $|u_\beta\rangle$ are two states, then $X_{\alpha\beta}^{(p',p)}$ denotes the matrix element

$$X_{\alpha\beta}^{(p',p)} = \frac{\langle u_\alpha | G^{p'} X G^{p} | u_\beta \rangle}{\langle u_\alpha | G^{p'+p} | u_\beta \rangle} \tag{2.2}$$

This definition is chosen to simplify the discussion, and generalization to physically relevant expressions, such as Eq. (3.30) is straightforward.

Various Monte Carlo methods are designed to estimate particular instances of $X_{\alpha\beta}^{(p',p)}$, and often the ultimate goal is to compute the expectation value in the dominant eigenstate

$$X_0 = \frac{\langle \psi_0 | X | \psi_0 \rangle}{\langle \psi_0 | \psi_0 \rangle} \tag{2.3}$$

which reduces to an expression for the dominant eigenvalue of interest if one chooses for X, in the applications discussed in the introduction, the Hamiltonian, transfer or Markov matrix.

The simplest method is the variational Monte Carlo method, discussed in the next section. Here an approximate expectation value is computed by employing an approximate eigenvector of G. Typically, this is an optimized trial state, say $|u_T\rangle$, in which case variational Monte Carlo yields $X_{TT}^{(0,0)}$, which is simply the expectation value of X in the trial state. Clearly, variational Monte Carlo estimates of X_0 have both systematic and statistical errors.

The variational error can be removed asymptotically by projecting out the dominant eigenstate, that is, by reducing the spectral weight of subdominant eigenstates by means of the power method. The simplest case is obtained if one applies the power method only to the right on the state $|u_\beta\rangle$ but not to the left on $\langle u_\alpha|$ in Eq. (2.2). Mathematically, this is the essence of diffusion and transfer matrix Monte Carlo, and in this way one obtains the desired result X_0 if the operator X commutes with the Monte Carlo evolution operator G. In our notation, this means that X_0 is given by the statistical estimate of $X_{\mathrm{TT}}^{(0,\,\infty)}$. In principle, this yields an unbiased estimate of X_0, but in practice one has to choose p finite but large enough that the estimated systematic error is less than the statistical error. In some practical situations it can in fact be difficult to ascertain that this indeed is the case. If one is interested in the limiting behavior for infinite p or p', the state $|u_\alpha\rangle$ or $|u_\beta\rangle$ need not be available in closed form. This freedom translates into the flexibility in algorithm design exploited in diffusion and transfer matrix Monte Carlo.

If G and X do not commute, the mixed estimator $X_{\mathrm{TT}}^{(0,\,\infty)}$ is not the desired result, but the residual systematic error can be reduced by combining the variational and mixed estimates by means of the expression

$$X_0 = 2X_{\mathrm{TT}}^{(0,\,\infty)} - X_{\mathrm{TT}}^{(0,\,0)} + \mathcal{O}[(|\psi_0\rangle - |u_{\mathrm{T}}\rangle)^2] \tag{2.4}$$

To remove the variational bias systematically, if G and X do not commute, the power method must be used to both the left and the right in Eq. (2.2). Thus one obtains from $X_{\mathrm{TT}}^{(\infty,\,\infty)}$ an exact estimate of X_0 subject only to statistical errors. Of course, one has to pay the price of the implied double limit in terms of loss of statistical accuracy. In the context of the Monte Carlo algorithms discussed below, such as diffusion and transfer matrix Monte Carlo, this double projection technique to estimate $X_{\mathrm{TT}}^{(\infty,\,\infty)}$ is called *forward walking* or *future walking*.

We end this section on the power method with a brief discussion of the computational complexity of using the Monte Carlo method for eigenvalue problems. In Monte Carlo computations one can distinguish operations of three levels of computational complexity, depending on whether the operations have to do with single particles or lattice sites, the whole system, or state space summation or integration. The first typically involves a fixed number of elementary arithmetic operations, whereas this number clearly is at least proportional to the system size in the second case. Exhaustive state-space summation grows exponentially in the total system size, and for these problems Monte Carlo is often the only viable option.

Next, the convergence of the power method itself comes into play. The number of iterations required to reach a certain given accuracy is proportional to $\log|\lambda_0/\lambda_1|$, where the λ_0 and λ_1 are the eigenvalues of largest and second largest magnitude. If one is dealing with a single-site transfer matrix of a critical system, that spectral gap is proportional to L^{-d} for a system in d dimensions with a cross section of linear dimension L. In this case, a single matrix multiplication is of the complexity of a one-particle problem. In contrast, both for the Markov matrix defined above, and the quantum mechanical evolution operator, the matrix multiplication itself is of system-size complexity. Moreover, both of these operators have their own specific problems. The quantum evolution operator of $G(\tau)$ has a gap on the order of τ, which means that τ should be chosen large for rapid convergence, but one does not obtain the correct results, because of the time-step error, unless τ is small. Finally, the spectrum of the Markov matrix displays critical slowing down. This means that the gap of the single spin-flip matrix is on the order of L^{-d-z}, where z is typically a little bigger than two [9]. These convergence properties are well understood in terms of the mathematics of the power method. Not well understood, however, are problems that are specific to the Monte Carlo implementation of this method, which in some form or another introduces multiplicative fluctuating weights that are correlated with the quantities of interest [12,13].

III. SINGLE-THREAD MONTE CARLO

In the previous section we have presented the mathematical expressions that can be evaluated with the Monte Carlo algorithms to be discussed next. The first algorithm is designed to compute an approximate statistical estimate of the matrix element X_0 by means of the variational estimate $X_{\mathrm{TT}}^{(0,0)}$. We write[1] $\langle \mathbf{S} | u_{\mathrm{T}} \rangle = \langle u_{\mathrm{T}} | \mathbf{S} \rangle \equiv u_{\mathrm{T}}(\mathbf{S})$ and for nonvanishing $u_{\mathrm{T}}(\mathbf{S})$ define the *configurational eigenvalue* $X_{\mathrm{T}}(\mathbf{S})$ by

$$X_{\mathrm{T}}(\mathbf{S}) u_{\mathrm{T}}(\mathbf{S}) \equiv \langle \mathbf{S} | X | u_{\mathrm{T}} \rangle = \sum_{\mathbf{S}'} \langle \mathbf{S} | X | \mathbf{S}' \rangle \langle \mathbf{S}' | u_{\mathrm{T}} \rangle \tag{3.1}$$

This yields

$$X_{\mathrm{TT}}^{(0,0)} = \frac{\sum_{\mathbf{S}} u_{\mathrm{T}}(\mathbf{S})^2 X_{\mathrm{T}}(\mathbf{S})}{\sum_{\mathbf{S}} u_{\mathrm{T}}(\mathbf{S})^2} \tag{3.2}$$

which shows that X_{TT} can be evaluated as a time average over a Monte Carlo time series of states $\mathbf{S}_1, \mathbf{S}_2, \ldots$ sampled from the probability distribution $u_{\mathrm{T}}(\mathbf{S})^2$, that is, a process in which prob($\mathbf{S}$), the probability of finding a state $\mathbf{S}$ at any time is given by

$$\mathrm{prob}(\mathbf{S}) \propto u_{\mathrm{T}}(\mathbf{S})^2. \tag{3.3}$$

[1] We assume throughout that the states we are dealing with are represented by real numbers and that X is represented by a real, symmetric matrix. In many cases, generalization to complex numbers is trivial, but for some physical problems, while formal generalization may still be possible, the resulting Monte Carlo algorithms may be too noisy to be practical.

For such a process, the ensemble average in Eq. (3.2) can be written in the form of a time average

$$X_{\mathrm{TT}}^{(0,0)} = \lim_{L \to \infty} \frac{1}{L} \sum_{t=1}^{L} X_{\mathrm{T}}(\mathbf{S}_t) \tag{3.4}$$

For this to be of practical use, it has to be assumed that the configurational eigenvalue $X_{\mathrm{T}}(\mathbf{S})$ can be computed efficiently, which is the case if the sum over states $\mathbf{S}'$ in $\langle \mathbf{S} | X | u_{\mathrm{T}} \rangle = \Sigma_{\mathbf{S}'} \langle \mathbf{S} | X | \mathbf{S}' \rangle \langle \mathbf{S}' | u_{\mathrm{T}} \rangle$ can be performed explicitly. For discrete states this means that X should be represented by a sparse matrix; if the states $\mathbf{S}$ form a continuum, $X_{\mathrm{T}}(\mathbf{S})$ can be computed directly if X is diagonal or *near-diagonal*, that is, involves no or only low-order derivatives in the representation used. The more complicated case of an operator X with arbitrarily nonvanishing off-diagonal elements will be discussed at the end of this section.

An important special case—relevant, for example, to electronic structure calculations—is to choose for the operator X the Hamiltonian $\mathscr{H}$ and for $\mathbf{S}$ the $3N$-dimensional real-space configuration of the system. Then, the quantity X_{T} is called the *local energy*, denoted by E_{L}. Clearly, in the ideal case that $| u_{\mathrm{T}} \rangle$ is an exact eigenvalue of the evolution operator G, and if X commutes with G then the configurational eigenvalue $X_{\mathrm{T}}(\mathbf{S})$ is a constant independent of $\mathbf{S}$ and equals the true eigenvalue of X. In this case the variance of the Monte Carlo estimator in Eq. (3.4) goes to zero, which is an important zero-variance principle satisfied by variational Monte Carlo. The practical implication is that the efficiency of the Monte Carlo computation of the energy can be improved arbitrarily by improving the quality of the trial function. Of course, usually the time required for the computation of $u_{\mathrm{T}}(\mathbf{S})$ increases as the approximation becomes more sophisticated. For the energy the optimal choice minimizes the product of variance and time; no such optimum exists for an operator that does not commute with G or if one makes the fixed node approximation, described in Section VI.A.2, since in these cases the results have a systematic error that depends on the quality of the trial wavefunction.

A. Metropolis Method

A Monte Carlo process sampling the probability distribution $u_{\mathrm{T}}(\mathbf{S})^2$ is usually generated by means of the generalized Metropolis algorithm, as follows. Suppose a configuration $\mathbf{S}$ is given at time t of the Monte Carlo process. A new configuration $\mathbf{S}'$ at time $t + 1$ is generated by means of a stochastic process that consists of two steps: (1) an intermediate configuration $\mathbf{S}''$ is proposed with probability $\Pi(\mathbf{S}'' | \mathbf{S})$; (2) (a) $\mathbf{S}' = \mathbf{S}''$ with probability $p \equiv A(\mathbf{S}'' | \mathbf{S})$, that is, the proposed configuration is accepted; (b)

$\mathbf{S}' = \mathbf{S}$ with probability $q \equiv 1 - A(\mathbf{S}'' \mid \mathbf{S})$, that is, the proposed configuration is rejected and the old configuration $\mathbf{S}$ is promoted to time $t + 1$. More explicitly, the Monte Carlo sample is generated by means of a Markov matrix P with elements $P(\mathbf{S}' \mid \mathbf{S})$ of the form

$$P(\mathbf{S}' \mid \mathbf{S}) = \begin{cases} A(\mathbf{S}' \mid \mathbf{S})\Pi(\mathbf{S}' \mid \mathbf{S}) & \text{for} \quad \mathbf{S}' \neq \mathbf{S} \\ 1 - \sum_{\mathbf{S}'' \neq \mathbf{S}} A(\mathbf{S}'' \mid \mathbf{S})\Pi(\mathbf{S}'' \mid \mathbf{S}) & \text{for} \quad \mathbf{S}' = \mathbf{S} \end{cases} \tag{3.5}$$

if the states are discrete and

$$P(\mathbf{S}' \mid \mathbf{S}) = A(\mathbf{S}' \mid \mathbf{S})\Pi(\mathbf{S}' \mid \mathbf{S}) + \left[1 - \int d\mathbf{S}'' \, A(\mathbf{S}'' \mid \mathbf{S})\Pi(\mathbf{S}'' \mid \mathbf{S})\right] \delta(\mathbf{S}' - \mathbf{S}) \tag{3.6}$$

if the states are continuous. Correspondingly, in Eq. (3.5) Π and P are probabilities whereas in Eq. (3.6) they are probability densities.

The Markov matrix P is designed to satisfy detailed balance

$$P(\mathbf{S}' \mid \mathbf{S})u_{\mathrm{T}}(\mathbf{S})^2 = P(\mathbf{S} \mid \mathbf{S}')u_{\mathrm{T}}(\mathbf{S}')^2 \tag{3.7}$$

so that, if the process has a unique stationary distribution, this will be $u_{\mathrm{T}}(\mathbf{S})^2$, as desired. In principle, one has great freedom in the choice of the proposal matrix Π, but it is necessary to satisfy the requirement that transitions can be made between (almost) any pair of states with nonvanishing probability (density) in a finite number of steps.

Once a proposal matrix Π is selected, an acceptance matrix is defined so that detailed balance, Eq. (3.7), is satisfied:

$$A(\mathbf{S}' \mid \mathbf{S}) = \min\left[1, \frac{\Pi(\mathbf{S} \mid \mathbf{S}')u_{\mathrm{T}}(\mathbf{S}')^2}{\Pi(\mathbf{S}' \mid \mathbf{S})u_{\mathrm{T}}(\mathbf{S})^2}\right] \tag{3.8}$$

For a given choice of Π, infinitely many choices can be made for A that satisfy detailed balance but the choice given above is the one with the largest acceptance. We note that if the preceding algorithm is used, then $X_{\mathrm{T}}(\mathbf{S}_t)$ in the sum in Eq. (3.4), can be replaced by the expectation value conditional on $\mathbf{S}_t''$ having been proposed:

$$X_{\mathrm{TT}}^{(0,\,0)} = \lim_{L \to \infty} \frac{1}{L} \sum_{t=1}^{L} [p_t X_{\mathrm{T}}(\mathbf{S}_t'') + q_t X_{\mathrm{T}}(\mathbf{S}_t)] \tag{3.9}$$

where $p_t = 1 - q_t$ is the probability of accepting $\mathbf{S}_t''$. This has the advantage of reducing the statistical error somewhat, since now $X_{\mathrm{T}}(\mathbf{S}_t'')$ contributes to the average even for rejected moves, and this will increase efficiency if $X_T(\mathbf{S}_t'')$ is readily available.

If the proposal matrix Π is symmetric, as is the case if one samples from a distribution uniform over a cube centered at $\mathbf{S}$, such as in the original Metropolis method [14], the factors of Π in the numerator and denominator of Eq. (3.8) cancel.

Finally we note that it is not necessary to sample the distribution u_T^2 to compute X_{TT}; any distribution that has sufficient overlap with u_T^2 will do. To make this more explicit, let us introduce the average of some stochastic variable X with respect to an arbitrary distribution ρ:

$$\langle X \rangle_\rho \equiv \frac{\sum_{\mathbf{S}} X(\mathbf{S})\rho(\mathbf{S})}{\sum_{\mathbf{S}} \rho(\mathbf{S})} \tag{3.10}$$

The following relation shows that the desired results can be obtained by reweighting; in other words, any distribution ρ will suffice as long as the ratio $u_T(\mathbf{S})^2/\rho(\mathbf{S})$ does not fluctuate too wildly:

$$X_{TT}^{(0,\,0)} = \langle X \rangle_{u_T{}^2} = \frac{\langle X u_T^2/\rho \rangle_\rho}{\langle u_T^2/\rho \rangle_\rho} \tag{3.11}$$

This is particularly useful for calculation of the difference of expectation values with respect to two closely related distributions. An example of this [15,16] is the calculation of the energy of a molecule as a function of the internuclear distance.

B. Projector Monte Carlo and Importance Sampling

The generalized Metropolis method is a very powerful way to sample an arbitrary *given* distribution, and it allows one to construct infinitely many Markov processes with the desired distribution as the stationary state. None of these, however, may be appropriate to design a Monte Carlo version of the power method to solve eigenvalue problems. In this case, the evolution operator G is *given*, possibly in approximate form, and its dominant eigenstate may *not* be known. To construct an appropriate Monte Carlo process, the first problem is that G itself is not a Markov matrix, specifically, it may violate one or both of the properties $G(\mathbf{S}' \mid \mathbf{S}) \geq 0$ and $\Sigma_{\mathbf{S}'} G(\mathbf{S}' \mid \mathbf{S}) = 1$. This problem can be solved if we can find a factorization of the evolution matrix G into a Markov matrix P and a weight matrix g with nonnegative elements such that

$$G(\mathbf{S}' \mid \mathbf{S}) = g(\mathbf{S}' \mid \mathbf{S}) P(\mathbf{S}' \mid \mathbf{S}) \tag{3.12}$$

The weights g must be finite, and this almost always precludes use of the Metropolis method for continuous systems, as can be understood as follows. Since there is a finite probability that a proposed state will be rejected, the Markov matrix $P(\mathbf{S}' \mid \mathbf{S})$ will contain terms involving $\delta(\mathbf{S} - \mathbf{S}')$, but generically, G will not have the same structure and will not allow the definition of finite weights g according to Eq. (3.12). However, the Metropolis algorithm can be incorporated as a component of an algorithm if an approximate stationary state is known and if further approximations are made, as in the diffusion Monte Carlo algorithm discussed in Section VI.B.

As a comment on the side we note that violation of the condition that the weight g be positive results in the notorious *sign problem*, which is in most cases unavoidable in the treatment of fermionic or frustrated systems. Many ingenious attempts [17] have been made to solve this problem, but this is still a topic of active research. However, as mentioned, we restrict ourselves in this chapter to the case of evolution operators G with nonnegative matrix elements only.

We resume our discussion of the factorization given in Eq. (3.12). Suppose for the sake of argument that the left eigenstate $\hat{\psi}_0$ of G is known and that its elements are positive:

$$\sum_{\mathbf{S}'} \hat{\psi}_0(\mathbf{S}')G(\mathbf{S}' \mid \mathbf{S}) = \lambda_0 \hat{\psi}_0(\mathbf{S}) \tag{3.13}$$

If, in addition, the matrix elements of G are nonnegative, the following matrix $\hat{P}$ is a Markov matrix:

$$\hat{P}(\mathbf{S}' \mid \mathbf{S}) = \frac{1}{\lambda_0} \hat{\psi}_0(\mathbf{S}')G(\mathbf{S}' \mid \mathbf{S}) \frac{1}{\hat{\psi}_0(\mathbf{S})} \tag{3.14}$$

Unless one is dealing with a Markov matrix from the outset, the left eigenvector of G is seldom known, but it is convenient, in any event, to perform a so-called *importance sampling* transformation on G. For this purpose we introduce a guiding function u_{g} and define

$$\hat{G}(\mathbf{S}' \mid \mathbf{S}) = u_g(\mathbf{S}')G(\mathbf{S}' \mid \mathbf{S}) \frac{1}{u_g(\mathbf{S})} \tag{3.15}$$

We shall return to the issue of the guiding function, but for the time being the reader can think of it either as an arbitrary, positive function, or as an approximation to the dominant eigenstate of G. From a mathematical point of view, anything that can be computed with the original Monte Carlo evolution operator G can also be computed with $\hat{G}$, since the two represent the same abstract operator in a different basis. The representations differ only by normalization constants. All we have to do is to write all expressions derived above in terms of this new basis.

We continue our discussion in terms of the transform $\hat{G}$ and replace Eq. (3.12) by the factorization

$$\hat{G}(\mathbf{S}' \mid \mathbf{S}) = \hat{g}(\mathbf{S}' \mid \mathbf{S})\hat{P}(\mathbf{S}' \mid \mathbf{S}) \tag{3.16}$$

and we assume that $\hat{P}$ has the explicitly known distribution u_g^2 as its stationary state. The guiding function u_g appears in those expressions, and it should be kept in mind that they can be reformulated by means of the reweighting procedure given in Eq. (3.11) to apply to processes with different explicitly known stationary states. On the other hand, one might be interested in the infinite projection limit $p \to \infty$. In that case, one might use a Monte Carlo process for which the stationary distribution is not known explicitly. Then, the expressions below should be rewritten so that the unknown distribution does not appear in expressions for the time averages. The function u_g will still appear, but only as a transformation known in closed form and no longer as the stationary state of P. Clearly, a process for which the distribution is not known in closed form cannot be used to compute the matrix elements $X_{\alpha\beta}^{(p',\,p)}$ for finite p and p' and given states $|\,u_\alpha\rangle$ and $|\,u_\beta\rangle$.

One possible choice for $\hat{P}$ that avoids the Metropolis method and produces finite weights is the following *generalized heat-bath* transition matrix:

$$\hat{P}(\mathbf{S}' \mid \mathbf{S}) = \frac{\hat{G}(\mathbf{S}' \mid \mathbf{S})}{\sum_{\mathbf{S}_1} \hat{G}(\mathbf{S}_1 \mid \mathbf{S})} \tag{3.17}$$

If $G(\mathbf{S}' \mid \mathbf{S})$ is symmetric, this transition matrix has a known stationary distribution, *viz.*, $G_g(\mathbf{S})u_g^2(\mathbf{S})$, where $G_g(\mathbf{S}) = \langle \mathbf{S} \mid G \mid u_g \rangle / u_g(\mathbf{S})$, the configurational eigenvalue of G in state $\mathbf{S}$. $\hat{P}$ must be chosen such that the corresponding transitions can be sampled directly. This is usually not feasible unless $\hat{P}$ is sparse or near-diagonal, or can be transformed into a form involving non-interacting degrees of freedom. We note that if $\hat{P}$ is defined by Eq. (3.17), the weight matrix $\hat{g}$ depends only on $\mathbf{S}$.

C. Matrix Elements

We now address the issue of computing the matrix elements $X_{\alpha\beta}^{(p',\,p)}$, assuming that the stationary state u_g^2 is known explicitly and that the weight matrix $\hat{g}$ has finite elements. We shall discuss the following increasingly complex possibilities: (1) $[X, G] = 0$ and X is near-diagonal in the $\mathbf{S}$

representation, (2) X is diagonal in the $\mathbf{S}$ representation, or (3) $X(\mathbf{S}|\mathbf{S}')$ is truly off-diagonal. The fourth case, $[X, G] = 0$ and X is not near-diagonal, is omitted since it can easily be constructed from the three cases discussed explicitly. When discussing case (3) we introduce the concept of *side walks* and explain how these can be used to compute matrix elements of a more general nature than discussed up to that point. After deriving the expressions, we shall discuss the practical problems they give rise to, and ways to reduce the variance of the statistical estimators. Since this yields expressions in a more symmetric form, we introduce the transform

$$\hat{X}(\mathbf{S}'|\mathbf{S}) = u_g(\mathbf{S}')X(\mathbf{S}'|\mathbf{S})\frac{1}{u_g(\mathbf{S})} \tag{3.18}$$

1. $[X, G] = 0$ *and* X *Near-Diagonal*

In this case, $X_{\alpha\beta}^{(0,\, p'+p)} = X_{\alpha\beta}^{(p',\, p)}$, and it suffices to consider the computation of $X_{\alpha\beta}^{(0,\, p)}$. By repeated insertion in Eq. (2.2) of the resolution of the identity in the $\mathbf{S}$ basis, one obtains the expression

$$X_{\alpha\beta}^{(0,\, p)} = \frac{\sum_{\mathbf{S}_p, \ldots, \mathbf{S}_0} u_\alpha(\mathbf{S}_p)X_\alpha(\mathbf{S}_p)[\prod_{i=0}^{p-1} G(\mathbf{S}_{i+1}|\mathbf{S}_i)]u_\beta(\mathbf{S}_0)}{\sum_{\mathbf{S}_p, \ldots, \mathbf{S}_0} u_\alpha(\mathbf{S}_p)[\prod_{i=0}^{p-1} G(\mathbf{S}_{i+1}|\mathbf{S}_i)]u_\beta(\mathbf{S}_0)} \tag{3.19}$$

In the steady state, a series of subsequent states $\mathbf{S}_t, \mathbf{S}_{t+1}, \ldots, \mathbf{S}_{t+p}$ occurs with probability

$$\mathrm{prob}(\mathbf{S}_t, \mathbf{S}_{t+1}, \ldots, \mathbf{S}_{t+p}) \propto \left[\prod_{i=0}^{p-1} \hat{P}(\mathbf{S}_{t+i+1}|\mathbf{S}_{t+i})\right] u_g(\mathbf{S}_t)^2 \tag{3.20}$$

To relate products of the matrix P to those of G, it is convenient to introduce the following definitions

$$\hat{W}_t(p, q) = \prod_{i=q}^{p-1} \hat{g}(\mathbf{S}_{t+i+1}|\mathbf{S}_{t+i}) \tag{3.21}$$

Also, we define

$$\hat{u}_\omega(\mathbf{S}) = \frac{u_\omega(\mathbf{S})}{u_g(\mathbf{S})} \tag{3.22}$$

where ω can be any of a number of subscripts.

With these definitions, combining Eqs. (3.12), (3.19), and (3.20), one finds

$$X_{\alpha\beta}^{(0,\,p)} = \lim_{L\to\infty} \frac{\sum_{t=1}^{L} \hat{u}_\alpha(\mathbf{S}_{t+p}) X_\alpha(\mathbf{S}_{t+p}) \hat{W}_t(p,\,0) \hat{u}_\beta(\mathbf{S}_t)}{\sum_{t=1}^{L} \hat{u}_\alpha(\mathbf{S}_{t+p}) \hat{W}_t(p,\,0) \hat{u}_\beta(\mathbf{S}_t)} \tag{3.23}$$

2. *Diagonal X*

The preceding discussion can be generalized straightforwardly to the case in which X is diagonal in the $\mathbf{S}$ representation. Again by repeated insertion of the resolution of the identity in the $\mathbf{S}$-basis in the Eq. (2.2) for $X_{\alpha\beta}^{(p',\,p)}$, one obtains the identity

$$X_{\alpha\beta}^{(p',p)} = \frac{\sum_{\mathbf{S}_{p'+p},\ldots,\mathbf{S}_0} u_\alpha(\mathbf{S}_{p'+p}) \left[\prod_{i=p}^{p'+p-1} G(\mathbf{S}_{i+1}|\mathbf{S}_i)\right] X(\mathbf{S}_p|\mathbf{S}_p) \left[\prod_{i=0}^{p-1} G(\mathbf{S}_{i+1}|\mathbf{S}_i)\right] u_\beta(\mathbf{S}_0)}{\sum_{\mathbf{S}_{p'+p},\ldots,\mathbf{S}_0} u_\alpha(\mathbf{S}_{p'+p}) \left[\prod_{i=0}^{p'+p-1} G(\mathbf{S}_{i+1}|\mathbf{S}_i)\right] u_\beta(\mathbf{S}_0)} \tag{3.24}$$

Again by virtue of Eq. (3.20), we find

$$X_{\alpha\beta}^{(p',\,p)} = \lim_{L\to\infty} \frac{\sum_{t=1}^{L} \hat{u}_\alpha(\mathbf{S}_{t+p'+p}) \hat{W}_t(p'+p,\,p) X(\mathbf{S}_{t+p}\,|\,\mathbf{S}_{t+p}) \hat{W}_t(p,\,0) \hat{u}_\beta(\mathbf{S}_t)}{\sum_{t=1}^{L} \hat{u}_\alpha(\mathbf{S}_{t+p'+p}) \hat{W}_t(p'+p,\,0) \hat{u}_\beta(\mathbf{S}_t)} \tag{3.25}$$

3. *Nondiagonal X*

If the matrix elements of G vanish only when also those of X do, the method described above can be generalized immediately to the final case in which X is nondiagonal. Then, the analog of Eq. (3.25) is

$$X_{\alpha\beta}^{(p',p)} = \lim_{L\to\infty} \frac{\sum_{t=1}^{L} \hat{u}_\alpha(\mathbf{S}_{t+p'+p}) \hat{W}_t(p'+p,\,p+1) x(\mathbf{S}_{t+p+1}|\mathbf{S}_{t+p}) \hat{W}_t(p,\,0) \hat{u}_\beta(\mathbf{S}_t)}{\sum_{t=1}^{L} \hat{u}_\alpha(\mathbf{S}_{t+p'+p}) \hat{W}_t(p'+p,\,0) \hat{u}_\beta(\mathbf{S}_t)} \tag{3.26}$$

where the x matrix elements are defined by

$$x(\mathbf{S}'\,|\,\mathbf{S}) = \frac{X(\mathbf{S}'\,|\,\mathbf{S})}{P(\mathbf{S}'\,|\,\mathbf{S})} = \frac{\hat{X}(\mathbf{S}'\,|\,\mathbf{S})}{\hat{P}(\mathbf{S}'\,|\,\mathbf{S})} \tag{3.27}$$

Clearly, the preceding definition of $x(\mathbf{S}'\,|\,\mathbf{S})$ fails when $\hat{P}(\mathbf{S}'\,|\,\mathbf{S})$ vanishes but $\hat{X}(\mathbf{S}'\,|\,\mathbf{S})$ does not. If that can happen, a more complicated scheme can be employed in which one introduces *side walks*. This is done by interrupting the continuing stochastic process at time $t + p$ by introducing a finite series of auxiliary states $\mathbf{S}'_{t+p+1}, \ldots, \mathbf{S}'_{t+p'+p}$. The latter are generated by a separate stochastic process so that in equilibrium, the sequence of subsequent states $\mathbf{S}_t, \mathbf{S}_{t+1}, \ldots, \mathbf{S}_{t+p}, \mathbf{S}'_{t+p+1}, \ldots, \mathbf{S}'_{t+p'+p}$ occurs with probability

$$\text{prob}[(\mathbf{S}_t, \mathbf{S}_{t+1}, \ldots, \mathbf{S}_{t+p}, \mathbf{S}'_{t+p+1}, \ldots, \mathbf{S}'_{t+p'+p})] \propto \tag{3.28}$$

$$\left[\prod_{i=p+1}^{p'+p-1} \hat{P}(\mathbf{S}'_{t+i+1} \,|\, \mathbf{S}_{t+i})\right] \hat{P}_X(\mathbf{S}'_{t+p+1} \,|\, \mathbf{S}_{t+p}) \left[\prod_{i=0}^{p-1} \hat{P}(\mathbf{S}'_{t+i+1} \,|\, \mathbf{S}_{t+i})\right] u_g(\mathbf{S}_t)^2 \tag{3.29}$$

where $\hat{P}_X$ is a Markov matrix chosen to replace $\hat{P}$ in Eq. (3.27) so as to yield finite weights x. In this scheme, one generates a continuing thread identical to the usual Monte Carlo process in which each state $\mathbf{S}_t$ is sampled from the stationary state of $\hat{P}$, at least if one ignores the initial equilibration. Each state $\mathbf{S}_t$ of this backbone forms the beginning of a side walk, the first step of which is sampled from $\hat{P}_X$, while $\hat{P}$ again generates subsequent ones. Clearly, with respect to the side walk, the first step disrupts the stationary state, so that the p' states $\mathbf{S}'_{t'}$, which form the side walk, do not sample the stationary state of the original stochastic process generated by $\hat{P}$, unless $\hat{P}_X$ coincidentally has the same stationary state as $\hat{P}$.

A problem with the matrix elements we dealt with up to now is that in the limit p' or $p \to \infty$ all of them reduce to matrix elements involving the dominant eigenstate, although symmetries might be used to yield other eigenstates besides the absolute dominant one. However, if symmetries fail, one has to employ the equivalent of an orthogonalization scheme, such as, discussed in the next section, or one is forced to resort to evolution operators that contain, in exact or in approximate form, the corresponding projections. An example of this are matrix elements computed in the context of the fixed-node approximation [18], discussed in Section VI.A.2. Within the framework of this approximation, one considers quantities of the form

$$X_{\alpha\beta}^{(p',\,p)} = \frac{\langle u_\alpha \,|\, G_1^{p_1} X G_2^{p_2} \,|\, u_\beta \rangle}{\sqrt{\langle u_\alpha \,|\, G_1^{2p_1} \,|\, u_\alpha \rangle \langle u_\beta \,|\, G_2^{2p_2} \,|\, u_\beta \rangle}} \tag{3.30}$$

where the G_i are evolution operators combined with appropriate projectors, which in the fixed node approximation are defined by the nodes of the states $u_\alpha(\mathbf{S})$ and $u_\beta(\mathbf{S})$. We shall describe how the above expression, Eq. (3.30), can be evaluated, but rather than writing out all the expressions explicitly, we present just the essence of the Monte Carlo method.

To deal with these expressions, one generates a backbone time series of states sampled from any distribution, say, $u_g(\mathbf{S})^2$, that has considerable overlap with the states $|u_\alpha(\mathbf{S})|$ and $|u_\beta(\mathbf{S})|$. Let us distinguish those backbone states by a superscript 0. Consider any such state $S^{(0)}$ at some given time. It forms the starting point of two side walks. We denote the states of these side walks side by $\mathbf{S}_{t_i}^{(i)}$ where $i = 1, 2$ identifies the side walk and t_i labels the side steps. The side walks are generated from factorizations of the usual form, defined in Eq. (3.16), say $\hat{G}_i = \hat{g}_i \hat{P}_i$. A walk

$$\mathscr{S} = [\mathbf{S}^{(0)}, (\mathbf{S}_1^{(1)}, \mathbf{S}_2^{(1)}, \ldots), (\mathbf{S}_1^{(2)}, \mathbf{S}_2^{(2)}, \ldots)] \tag{3.31}$$

occurs with probability

$$\text{prob}(\mathscr{S}) = u_g(\mathbf{S}(0))^2 \hat{P}_1(\mathbf{S}^{(0)} | \mathbf{S}_1^{(1)}) \ldots \hat{P}_2(\mathbf{S}^{(0)} | \mathbf{S}_1^{(2)}) \ldots . \tag{3.32}$$

We leave it to the reader to show that this probability suffices for the computation of all expressions appearing in numerator and denominator of Eq. (3.30), in the case that X is diagonal, and to generate the appropriate generalizations to other cases.

In the expressions derived above, the power method projections precipitate products of reweighting factors $\hat{g}$, and, as the projection times p and p' increase, the variance of the Monte Carlo estimators grows at least exponentially in the square root of the projection time. Clearly, the presence of the fluctuating weights $\hat{g}$ is due to the fact that the evolution operator $\hat{G}$ is not Markovian in the sense that it fails to conserve probability. The importance sampling transformation Eq. (3.15) was introduced to mitigate this problem. In Section V, an algorithm involving branching walks will be introduced, which is a different strategy devised to deal with this problem. In diffusion and transfer matrix Monte Carlo, both strategies, importance sampling and branching, are usually employed simultaneously.

D. Excited States

Given a set of basis states, excited eigenstates can be computed variationally by solving a linear variational problem, and the Metropolis method can be used to evaluate the required matrix elements. The methods involving the power method, as described above, can then be used to remove the variational bias systematically [13,19,20].

In this context matrix elements appear in the solution of the following variational problem. As was mentioned several times before, the price paid for reducing the variational bias is increased statistical noise, a problem that appears in this context with a vengeance. Again, the way to keep this problem under control is the use of optimized trial vectors.

The variational problem to be solved is the following one. Given n basis states $|u_i\rangle$, find the $n \times n$ matrix of coefficients $d_i^{(j)}$ such that

$$|\tilde{\psi}_j\rangle = \sum_{i=1}^{n} d_i^{(j)} |u_i\rangle \tag{3.33}$$

are the best variational approximations for the n lowest eigenstates $|\psi_i\rangle$ of some Hamiltonian $\mathscr{H}$. In this problem we shall use the language of the quantum mechanical systems, where one has to distinguish the Hamiltonian from the evolution operator $\exp(-\tau\mathscr{H})$. In the statistical mechanical applications, one has only the equivalent of the latter. In the expressions to be derived below the substitution $\mathscr{H}G^p \to G^{p+1}$ will produce the expressions required for the statistical mechanical applications, at least if we assume that the nonsymmetric matrices that appear in that context have been symmetrized.[2]

One seeks a solution to the linear variational problem in Eq. (3.34) in the sense that for all i the Rayleigh quotient $\langle\tilde{\psi}_i|\mathscr{H}|\tilde{\psi}_i\rangle/\langle\tilde{\psi}_i|\tilde{\psi}_i\rangle$ is stationary with respect to variation of the coefficients d. The solution is that the matrix of coefficients d has to satisfy the following generalized eigenvalue equation:

$$\sum_{i=1}^{n} H_{ki} d_i^{(j)} = \tilde{E}_j \sum_{i=1}^{n} N_{ki} d_i^{(j)} \tag{3.34}$$

where

$$H_{ki} = \langle u_k|\mathscr{H}|u_i\rangle \tag{3.35}$$

and

$$N_{ki} = \langle u_k|u_i\rangle \tag{3.36}$$

Before discussing Monte Carlo issues, we note a number of important properties of this scheme. Firstly, the basis states $|u_i\rangle$ in general are not orthonormal, and this is reflected by the fact that the matrix elements of N have to be computed. Secondly, it is clear that any nonsingular linear combination of the basis vectors will produce precisely the same results, obtained from the correspondingly transformed version of Eq. (3.34). The

[2] This is not possible in general for transfer matrices of systems with helical boundary conditions, but the connection between left and right eigenvectors of the transfer matrix (see Section I.B) can be used to generalize the approach discussed here.

final comment is that the variational eigenvalues bound the exact eigenvalues: $\tilde{E}_i \geq E_i$. One recovers exact eigenvalues E_i and the corresponding eigenstates, if the $|u_i\rangle$ span the same space as the exact eigenstates.

The required matrix elements can be computed using the variational Monte Carlo method discussed in the previous section. Furthermore, the power method can be used to reduce the variational bias. Formally, one simply defines new basis states

$$|u_i^{(p)}\rangle = G^p |u_i\rangle \tag{3.37}$$

and substitutes these new basis states for the original ones. The corresponding matrices

$$H_{ki}^{(p)} = \langle u_k^{(p)} | \mathscr{H} | u_i^{(p)} \rangle \tag{3.38}$$

and

$$N_{ki}^{(p)} = \langle u_k^{(p)} | u_i^{(p)} \rangle \tag{3.39}$$

can again be computed by applying the methods introduced in Section III for the computation of general matrix elements by a Monte Carlo implementation of the power method.

As an explicit example illustrating the nature of the Monte Carlo time averages one has to evaluate in this approach, we write down the expression for $N_{ij}^{(p)}$ as used for the computation of eigenvalues of the Markov matrix relevant to the problem of critical slowing down:

$$N_{ij}^{(p)} \approx \sum_t \frac{u_i(\mathbf{S}_t)}{\psi_{\mathrm{B}}(\mathbf{S}_t)} \frac{u_j(\mathbf{S}_{t+p})}{\psi_{\mathrm{B}}(\mathbf{S}_{t+p})} \tag{3.40}$$

where the $\mathbf{S}_t$ are configurations forming a time series that, as we recall, is designed to sample the distribution of a system in thermodynamic equilibrium, namely, the Boltzmann distribution ψ_{B}^2. The expression given in Eq. (3.41) yields the u/ψ_{B} autocorrelation function at lag p. The expression for $H_{ij}^{(p)}$ is similar, and represents a cross-correlation function involving the configurational eigenvalues of the Markov matrix in the various basis states. Compared to the expressions derived in Section III, Eq. (3.40) takes a particularly simple form in which products of fluctuating weights are absent, because in this particular problem one is dealing with a probability conserving evolution operator from the outset.

Equation (3.40) shows why this method is sometimes called "correlation function Monte Carlo," but it also illustrates a new feature, namely, that it is efficient to compute all required matrix elements simultaneously. This can be done by generating a Monte Carlo process with a known distribution that has sufficient overlap with all $|u_i(\mathbf{S})| \equiv |\langle \mathbf{S}| u_i \rangle|$. This can be arranged, for example, by sampling a guiding function $u_g(\mathbf{S})$ defined by

$$u_g(\mathbf{S}) = \sqrt{\sum_{i=1}^{n} a_i \, u_i(\mathbf{S})^2} \tag{3.41}$$

where the coefficients a_i approximately normalize the basis states $|u_i\rangle$ which may require a preliminary Monte Carlo run. See Ceperley and Bernu [13] for an alternative choice for a guiding function. In the computations [10] to obtain the spectrum of the Markov matrix in critical dynamics, as illustrated by Eq. (3.40), the Boltzmann distribution, is used as a guiding function. It apparently has enough overlap with the decaying modes that no special purpose distribution has to be generated.

E. How to Avoid Reweighting

Before discussing the branching algorithms designed to deal more efficiently with the reweighting factors appearing in the expressions discussed above, we briefly mention an alternative that has surfaced occasionally without being studied extensively, to our knowledge. The idea will be illustrated in the case of the computation of the matrix element $X_{\alpha\beta}^{(0,\,p)}$, and we take Eq. (3.19) as our starting point. In statistical mechanical language, we introduce a reduced Hamiltonian

$$\mathscr{H} = \ln u_g(\mathbf{S}_p) + \sum_{i=0}^{p-1} \ln G(\mathbf{S}_{i+1} | \mathbf{S}_i) + \ln u_g(\mathbf{S}_0) \tag{3.42}$$

and the corresponding Boltzmann distribution $\exp -\mathscr{H}(\mathbf{S}_p, \ldots, \mathbf{S}_0)$. One can now use the standard Metropolis algorithm to sample this distribution for this system consisting of $p+1$ layers bounded by the layers 0 and p. For the evaluation of Eq. (3.19) by Monte Carlo, this expression then straightforwardly becomes a ratio of correlation functions involving quantities defined at the boundaries. To see this, all one has to do is divide the numerator and denominator by the partition function

$$Z = \sum_{\mathbf{S}_p, \ldots, \mathbf{S}_0} e^{-\mathscr{H}(\mathbf{S}_p, \ldots, \mathbf{S}_0)} \tag{3.43}$$

Note that in general, boundary terms involving some appropriately defined u_g should be introduced to ensure the nonnegativity of the distribution. For the simultaneous computation of matrix elements for several values of the indices α and β, a guiding function u_g should be chosen that has considerable overlap with the corresponding $|u_\alpha|$ and $|u_\beta|$.

The Metropolis algorithm can of course be used to sample any probability distribution, and the introduction of the previous Hamiltonian illustrates just one particular point of view. If one applies the above idea to the case of the imaginary-time quantum mechanical evolution operator, one obtains a modified version of the standard path-integral Monte Carlo method, in which case the layers are usually called time slices. Clearly, this method has the advantage of suppressing the fluctuating weights in estimators. However, the disadvantage is that sampling the full, layered system yields a longer correlation time than sampling the single-layer distribution u_g^2. This is a consequence of the fact that the microscopic degrees of freedom are more strongly correlated in a layered system than in a single layer. Our limited experience suggests that for small systems reweighting is more efficient, whereas the Metropolis approach tends to become more efficient as the system grows in size [61].

IV. TRIAL FUNCTION OPTIMIZATION

In the previous section it was shown that eigenvalue estimates can be obtained as the eigenvalues of the matrix $N^{(p)^{-1}}H^{(p)}$. The variational error in these estimates decreases as p increases. In general, these expressions involve weight products of increasing length, and consequently the errors grow exponentially, but even in the simple case of a probability-conserving evolution operator, errors grow exponentially. This is a consequence of the fact that the autocorrelation functions in $N^{(p)}$, and the cross-correlation functions in $H^{(p)}$, in the limit $p \to \infty$ reduce to quantities that contain an exponentially vanishing amount of information about the subdominant or excited-state eigenvalues, since the spectral weight of all but the dominant eigenstate is reduced to zero by the power method.

The practical implication is that this information has to be retrieved with sufficient accuracy for small values of p, before the signal disappears in the statistical noise. The projection time p can be kept small by using optimized basis states constructed to reduce the overlap of the linear space spanned by the basis states $|u_i\rangle$ with the space spanned by the eigenstates beyond the first n of interest. We shall describe, mostly qualitatively, how this can be done by a generalization of a method used for optimization of individual basis states [3,21–23], namely, minimization of variance of the configurational eigenvalue, the local energy in quantum Hamiltonian problems.

Suppose that $u_T(\mathrm{S}, v)$ is the value of the trial function u_T for configuration S and some choice of the parameters v to be optimized. As in Eq. (3.1), the *configurational eigenvalue* $\lambda(S, v)$ of configuration $\mathbf{S}$ is defined by

$$u'_{\mathrm{T}}(\mathbf{S}, v) \equiv \lambda(\mathbf{S}, v) u_{\mathrm{T}}(\mathbf{S}, v) \tag{4.1}$$

where a prime is used to denote, for arbitrary $|f\rangle$, the components of $G|f\rangle$, or $\mathscr{H}|f\rangle$ as is more convenient for quantum mechanical applications. The optimal values of the variational parameters are obtained by minimization of the variance of $\lambda(\mathbf{S}, v)$, estimated as an average over a small Monte Carlo sample. In the ideal case, that is, if an exact eigenstate can be reproduced by some choice of the parameters of u_{T}, the minimum of the variance yields the exact eigenstate not only if it were to be computed exactly, but even if it is approximated by summation over a Monte Carlo sample. A similar zero-variance principle holds for the method of simultaneous optimization of several trial states to be discussed next. This is in sharp contrast with the more traditional Rayleigh–Ritz extremization of the Rayleigh quotient, which frequently can produce arbitrarily poor results if minimized over a small sample of configurations.

For conceptual simplicity, we first generalize the preceding method to the more general ideal case that reproduces the exact values of the desired n eigenstates of the evolution operator G. As a byproduct, our discussion will produce an alternative to the derivation of Eq. (3.34). To compute n eigenvalues, we have to optimize the n basis states $|u_i\rangle$, where we have dropped the index "T," and again we assume that we have a sample of M configurations $\mathbf{S}_\alpha$, $\alpha = 1, \ldots, M$ sampled from u_{g}^2. The case we consider is ideal in the sense that we assume that these basis states $|u_i\rangle$ span an n-dimensional invariant subspace of G. In that case, by definition there exists a matrix Λ of order n such that

$$\tilde{\psi}'_i(\mathbf{S}_\alpha) = \sum_{j=1}^{n} \Lambda_{ij} u_j(\mathbf{S}_\alpha) \tag{4.2}$$

Again, the prime on the left-hand side of this equation indicates multiplication by G or by $\mathscr{H} = -\tau^{-1} \ln G$. If the number of configurations is large enough, Λ is for all practical purposes determined uniquely by the set of equations (4.2) and one finds

$$\Lambda = N^{-1} H \tag{4.3}$$

where

$$N_{ij} = \frac{1}{M} \sum_{\alpha=1}^{M} \frac{u_i(S_\alpha) u_j(S_\alpha)}{u_{\mathrm{g}}(S_\alpha)^2} \qquad H_{ij} = \frac{1}{M} \sum_{\alpha=1}^{M} \frac{u_i(S_\alpha) u'_j(S_\alpha)}{u_{\mathrm{g}}(S_\alpha)^2} \tag{4.4}$$

In the limit $M \to \infty$ this indeed reproduces the matrices N and H in Eq. (3.34). In the nonideal case, the space spanned by the n basis states $|u_i\rangle$ is not an invariant subspace of the matrix G. In that case, even though Eq. (4.2) generically has no true solution, Eqs. (4.3) and (4.4) still constitute a solution in the least-squares sense, as the reader is invited to show for himself by solving the normal equations.

Next, let us consider the construction of a generalized optimization criterion. As mentioned before, if a set of basis states span an invariant subspace, so does any nonsingular linear combination. In principle, the optimization criterion should have the same invariance. The *matrix* Λ lacks this property, but its *spectrum* is invariant. Another consideration is that, while the local eigenvalue is defined by a single configuration $\mathbf{S}$, it takes at least n configurations to determine the "local" matrix Λ. This suggests that one subdivide the sample into subsamples of at least n configurations each and minimize the variance of the *local spectrum* over these *subsamples*. Again in principle, this has the advantage that the optimization can exploit the fact that linear combinations of the basis states have more variational freedom to represent the eigenstates than does each variational basis function separately. In practice, however, this advantage seems to be negated by the difficulty of finding good optimal parameters. This is a consequence of the fact that invariance under linear transformation usually can be mimicked by variation of the parameters of the basis states. In other words, a linear combination of basis states can be represented accurately, at least relative to the noise in the local spectrum, by a single basis state with appropriately chosen parameters. Consequently, intrinsic flaws of the trial states exceed what can be gained in reducing the variance of the local spectrum by exploiting the linear variational freedom, most of which is already used anyway in the linear variational problem that was discussed at the beginning of this section. This means that one has to contend with a near-singular nonlinear optimization problem. In practice, to avoid the concomitant slow convergence, it seems to be more efficient to break the "gauge symmetry" and select a preferred basis, which most naturally is done by requiring that each basis state *itself* is a good approximate eigenstate.

The preceding considerations, of course, leave us with two criteria: minimization of the variance of the local spectrum as a whole and minimization of the variance of the configurational eigenvalues separately. To be of practical use, both criteria have to be combined, since if one were to proceed just by minimization of the variance of the configurational eigenvalues separately, one would simply keep reproducing the same eigenstate. In a non–Monte Carlo context this can be solved simply by some orthogonalization scheme, but as far as Monte Carlo is concerned, that is undesirable since it fails to yield a zero-variance optimization principle.

V. BRANCHING MONTE CARLO

In Section III we discussed a method to compute Monte Carlo averages by exploiting the power method to reduce the spectral weight of undesirable, subdominant eigenstates. We saw that this leads to products of weights of subsequent configurations sampled by a Monte Carlo time series. To suppress completely the systematic errors due to finite projection times, namely, the variational bias, one has to take averages of infinite products of weights. This limit would produce an "exact" method with infinite variance, which is of no practical use.

We have also discussed how optimized trial states can be used to reduce the variance of this method. The variance reduction may come about in two ways. Firstly, by starting with optimized trial states of higher quality, the variational bias is smaller to begin with so that fewer power method projections are required. In practical terms, this leads to a reduction of the number of factors in the fluctuating products. Secondly, a good estimate of the dominant eigenstate can be used to reduce the amount by which the evolution operator, divided by an appropriate constant, violates conservation of probability, which reduces the variance of the individual fluctuating weight factors. All these considerations also apply to the branching Monte Carlo algorithm discussed in this section, which can be modified accordingly and in complete analogy with our previous discussion.

Before discussing the details of the branching algorithm, we mention that the algorithm presented here [8] contains the mathematical essence of both the diffusion and transfer matrix Monte Carlo algorithms. A related algorithm—Green function Monte Carlo—adds yet another level of complexity due to the fact that the evolution operator is known only as an infinite series. This series is stochastically summed at each step of the power method iterations. In practice, this implies that even the time step becomes stochastic and intermediate Monte Carlo configurations are generated that do not contribute to expectation values. Neither Green function Monte Carlo, nor its generalization designed to compute quantities at non-zero temperature [24], will be discussed in this chapter, and we refer the interested reader to the literature for further details [25–28].

Let us consider in detail the mechanism that produces large variance. This will allow us to explain what branching accomplishes if one has to compute products of many (ideally infinitely many) fluctuating weights. The time average over these products will typically be dominated by only very few large terms; the small terms are equally expensive to compute, but play no significant role in the average. This problem can be solved by performing many simultaneous Monte Carlo walks. One evolves a collection of walkers from one generation to the next and the key idea is to eliminate the lightweight walkers which produce relatively small contributions to the time average. To keep the number of walkers reasonably constant, heavy-weight walkers are duplicated and the clones are subsequently evolved (almost) independently.

An algorithm designed according to this concept does not cut off the products over weights and therefore seems to correspond to infinite projection time. It would therefore seem that the time average over a stationary branching process corresponds to an average over the exact dominant eigenstate of the Monte Carlo evolution operator, but, as we shall see, this is rigorously the case only in the limit of an infinite number of walkers [29,30]; for any finite number of walkers, the stationary distribution has a bias inversely proportional to the number of walkers, the *population-control bias*. If the fluctuations in the weights are small and correlations (discussed later) decay rapidly, this bias tends to be small. In many applications this appears to be the case and the corresponding bias is statistically insignificant. However, if these methods are applied to statistical mechanical systems at the critical point, significant bias can be introduced. We shall discuss a simple method of nearly vanishing computational cost to detect this bias and correct for it in all but the worst-case scenarios.

To discuss the branching Monte Carlo version of the power method, we continue to use the notation introduced above and again consider the Monte Carlo evolution operator $G(\mathbf{S}' \,|\, \mathbf{S})$. As above, the states $\mathbf{S}$ and $\mathbf{S}'$ will be treated here as discrete, but in practice the distinction between continuous and discrete states is a minor technicality, and generalization to the continuous case follows immediately by replacing sums by integrals and by replacing Kronecker δ's by Dirac δ functions.

To implement the power method iterations in Eq. (2.1) by a branching Monte Carlo process, $|u^{(t)}\rangle$ is represented by a collection of N_t walkers, where a walker by definition is a state–weight pair $(\mathbf{S}_\alpha, w_\alpha)$, $\alpha = 1, \ldots, N_t$. As usual, the state variable $\mathbf{S}_\alpha$ represents a possible configuration of the system evolving according to G, and w_α represents the statistical weight of walker α. These weights appear in averages and the efficiency of the branching Monte Carlo algorithm is realized by maintaining the weights in some range $w_\mathrm{l} < w_\alpha < w_\mathrm{u}$, where w_l and w_u are bounds introduced so as to keep all weights w_α of the same order of magnitude.

The first idea is to interpret a collection of walkers that make up generation t as a representation of the (sparse) vector $|\underline{\mathrm{u}}^{(t)}\rangle$ with components

$$\langle \mathbf{S} \,|\, \underline{\mathrm{u}}^{(t)} \rangle \equiv \underline{\mathrm{u}}^{(t)}(\mathbf{S}) = \sum_{\alpha=1}^{N_t} w_\alpha \, \delta_{\mathbf{S}, \mathbf{S}_\alpha} \tag{5.1}$$

where δ is the usual Kronecker δ function. The underbar is used to indicate that the $\underline{\mathrm{u}}^{(t)}(\mathbf{S})$ represents a stochastic vector $|\underline{\mathrm{u}}^{(t)}\rangle$. Of course, the same is true formally for the single-thread Monte Carlo. The new feature is that one can think of the collective of walkers as a reasonably accurate representation of the stationary state at each time step, rather than in the long run.

The second idea is to define a stochastic process in which the walkers evolve with transition probabilities such that the expectation value of $c_{t+1} | \underline{u}^{(t+1)} \rangle$, as represented by the walkers of generation $t + 1$, equals $G | \underline{u}^{(t)} \rangle$ for any given collection of walkers representing $| \underline{u}^{(t)} \rangle$. It is tempting to conclude that, owing to this construction, the basic recursion relation of the power method, Eq. (2.1), is satisfied in an average sense, but this conclusion is not quite correct. The reason is that in practice, the constant c_t are defined on the fly. Consequently, c_{t+1} and $| \underline{u}^{(t+1)} \rangle$ are correlated random variables and therefore there is no guarantee that the stationary state expectation value of $| \underline{u}^{(t)} \rangle$ is *exactly* an eigenstate of G, except in the limit of nonfluctuating normalization constants c_t, which, as we shall see, is tantamount to an infinite number of walkers. More explicitly, the problem is that if one takes the time average of Eq. (2.1) and if the fluctuations of the c_{t+1} are correlated with $| \underline{u}^{(t)} \rangle$ or $| \underline{u}^{(t+1)} \rangle$, one does not produce the same state on the left- and right-hand sides of the time-averaged equation and therefore the time-averaged state need not satisfy the eigenvalue equation. The resulting bias has been discussed in the various contexts [12,30,31].

One way to define a stochastic process is to rewrite the power method iteration Eq. (2.1) as

$$u^{(t+1)}(\mathbf{S}') = \frac{1}{c_{t+1}} \sum_{\mathbf{S}} P(\mathbf{S}' \,|\, \mathbf{S}) g(\mathbf{S}) u^{(t)}(\mathbf{S}) \tag{5.2}$$

where

$$g(\mathbf{S}) = \sum_{\mathbf{S}'} G(\mathbf{S}' \,|\, \mathbf{S}) \quad \text{and} \quad P(\mathbf{S}' \,|\, \mathbf{S}) = \frac{G(\mathbf{S}' \,|\, \mathbf{S})}{g(\mathbf{S})} \tag{5.3}$$

This is in fact what how transfer matrix Monte Carlo is defined. Referring the reader back to the discussion of Eq. (3.12), we note that in diffusion Monte Carlo the weight g is defined so that it is not just a function of the initial state S, but also depends on the final S′. The algorithm given below can trivially be generalized to accommodate this by making the substitution $g(\mathbf{S}) \rightarrow g(\mathbf{S}' \,|\, \mathbf{S})$.

Equation (5.2) describes a process represented by a Monte Carlo run which, after a few initial equilibration steps, consists of a time series of M_0 updates of all walkers at times labeled by $t = \ldots, 0, 1, \ldots, M_0$. The update at time t consists of two steps designed to perform stochastically the matrix multiplications in Eq. (5.2). Following Nightingale and Blöte [30], the process is defined by the following steps. Let us consider one of these updates, the one that transforms the generation of walkers at time t into the generation at time $t + 1$. We denote variables pertaining to times t and $t + 1$ respectively by unprimed and primed symbols.

1. Update the old walker $(\mathbf{S}_i, w_i)$ to yield a temporary walker $(\mathbf{S}'_i, w'_i)$ according to the transition probability $P(\mathbf{S}'_i \mid \mathbf{S}_i)$, where $w'_i = g(\mathbf{S}_i)w_i/c'$, for $i = 1, \ldots, N_t$. Step 2, defined below, can change the number of walkers. To maintain their number close to a target number, say, N_0, choose $c' = \hat{\lambda}_0(N_t/N_0)^{1/s}$, where $\hat{\lambda}_0$ is a running estimate of the eigenvalue λ_0 to be calculated, where $s \geq 1$.
2. From the temporary walkers construct the new generation of walkers as follows
   a. Split each walker $(\mathbf{S}', w')$ for which $w' > b_{\mathrm{u}}$ into two walkers $(\mathbf{S}, \frac{1}{2}w')$. The choice $b_{\mathrm{u}} = 2$ is a reasonable one.
   b. Join pairs $(\mathbf{S}'_i, w'_i)$ and $(\mathbf{S}'_j, w'_j)$ with $w'_i < b_1$ and $w'_j < b_1$ to produce a single walker $(\mathbf{S}'_k, w'_i + w'_j)$, where $\mathbf{S}'_k = \mathbf{S}'_i$ or $\mathbf{S}'_k = S'_j$ with relative probabilities w'_i and w'_j. Here $b_1 = \frac{1}{2}$ is reasonable.
   c. Walkers for which $b_1 < w'_i < b_{\mathrm{u}}$, or left single in step 2b, become members of the new generation of walkers.

Note that, if the weights $g(\mathbf{S})$ fluctuate on average more than by a factor of 2, multiple split and join operations may be needed.

It may help to explicate why wildly fluctuating weights adversely impact the efficiency of the algorithm. In that case, some walkers will have multiple descendents in the next generations, whereas others will have none. This leads to an inefficient algorithm since any generation will have several walkers that are either identical or are closely related, which will produce strongly correlated contributions to the statistical time averages. In its final analysis, this is the same old problem that we encountered in a single-thread algorithm, where averages would be dominated by few terms with relatively large, explicitly given statistical weights. Branching mitigates this problem since walkers that are descendants of a given walker eventually decorrelate, but, as discussed in Sections III and VI, the best cure is importance sampling and in practice both strategies are used simultaneously.

The algorithm described above was constructed so that for any given realization of $|\underline{\mathrm{u}}^{(t)}\rangle$, the expectation value of $c_{t+1}|\underline{\mathrm{u}}^{(t+1)}\rangle$, in accordance with Eq. (2.1), satisfies

$$\mathrm{E}[c_{t+1}|\underline{\mathrm{u}}^{(t+1)}\rangle] = G|\underline{\mathrm{u}}^{(t)}\rangle \tag{5.4}$$

where $\mathrm{E}(\cdot)$ denotes the conditional average over the transitions defined by the preceding stochastic process. More generally, by p-fold iteration one finds [32]

$$\mathrm{E}\left[\left(\prod_{b=1}^{p} c_{t+b}\right)|\underline{u}^{(t+p)}\rangle\right] = G^p\,|\underline{u}^{(t)}\rangle \tag{5.5}$$

The stationary state average of $|\underline{u}^{(t)}\rangle$ is close to the dominant eigenvector of G, but, as mentioned above, it has a systematic bias, proportional to $1/N_0$, when the number N_t of walkers is finite. If, as is the case in some applications, this bias exceeds the statistical errors, one can again rely on the power method to reduce this bias by increasing p. If that is done, one is back to the old problem of having to average products of fluctuating weights, and, as usual, the variance of the corresponding estimators increases as their bias decreases. Fortunately, in practice the population control bias of the stationary state is quite small, if at all detectable, but even in those cases, expectation values involving several values of p should be computed to verify the absence of population control bias. The reader is referred to Refs. 2, 12, 31–33 for a more detailed discussion of this problem. Suffice it to mention here, firstly, that s, as defined in the first step of the branching algorithm given above, is the expected number of time steps it takes to restore the number of walkers to its target value N_0 and, secondly, that strong population control ($s = 1$) tends to introduce a stronger bias than weaker control ($s > 1$).

With Eq. (5.5) one constructs an estimator [32] of the dominant eigenvector $|u^{(\infty)}\rangle$ of the evolution operator G:

$$|\check{u}^{(p)}\rangle = \frac{1}{M_0}\sum_{t=1}^{M_0}\left(\prod_{b=0}^{p-1} c_{t-b}\right)|\underline{u}^{(t)}\rangle \tag{5.6}$$

For $p = 0$, in which case the product over b reduces to unity, this yields the stationary state of the branching Monte Carlo, which frequently is treated as the dominant eigenstate of G.

Clearly, this branching Monte Carlo algorithm can be used to compute the right-projected mixed estimators that were denoted by $X_{\mathrm{TT}}^{(0,\,\infty)}$ in Section III. For this purpose one sets $p' = 0$ in Eq. (2.2) and makes the substitution $\langle u_\alpha| = \langle u_{\mathrm{T}}|$ and $G^p|u_\beta\rangle| = |\check{u}^{(p)}\rangle$. Expressions for several values of p can be computed simultaneously and virtually for the price of one. We explicitly mention the important special case obtained by choosing for the general operator X the evolution operator G itself. This yields the following estimator for the dominant eigenvalue λ_0 of G:

$$\lambda_0 \approx \frac{\sum_{t=1}^{M_0}\left(\prod_{b=0}^{p} c_{t-b}\right)W^{(t)}}{\sum_{t=1}^{M_0}\left(\prod_{b=0}^{p-1} c_{t-b}\right)W^{(t-1)}} \tag{5.7}$$

where

$$W^{(t)} = \sum_{i=1}^{N_t} w_i^{(t)} u_{\rm T}(\mathbf{S}_i) \tag{5.8}$$

In diffusion Monte Carlo this estimator can be used to construct the *growth estimate* of the ground state energy. That is, since in that special case $G \approx \exp(-\tau\mathscr{H})$, eigenvalues of the evolution operator and the Hamiltonian are related by

$$E_0 = -\frac{1}{\tau} \ln \lambda_0 \tag{5.9}$$

Besides expressions such as Eq. (5.7) one can construct expressions with reduced variance. These involve the configurational eigenvalue of G or $\mathscr{H}$ in the same way this was done in our discussion of these single-thread algorithm.

Again, in practical applications it is important to combine the raw branching algorithm defined above with importance sampling. Mathematically, this works in precisely the same way as in Section III in that one reformulates the same algorithm in terms of the similarity transform $\hat{G}$ with $u_{\rm g} = u_{\rm T}$ chosen to be an accurate, approximate dominant eigenstate [see Eq. (3.15)]. In the single-thread algorithm, the result is that the fluctuations of the weights g and their products are reduced. In the context of the branching algorithm, this yields reduced fluctuations in the weight of walkers individually and in the size of the walker population. One result is that the population control bias is reduced. If we ignore this bias, a more fundamental difference is that the steady state of the branching algorithm is modified. That is, in the raw algorithm the walkers sample the dominant eigenstate of G, namely, $\psi_0(\mathbf{S})$, but, if the trial state $|u_{\rm T}\rangle$ is used for importance sampling, the distribution is $u_{\rm T}(\mathbf{S})\psi_0(\mathbf{S})$, which, of course, is simply the dominant eigenstate of $\hat{G}$.

So far, we have only discussed how mixed expectation values can be computed with the branching Monte Carlo algorithm and, as was mentioned before, this yields the desired result only if one deals with operators that commute with the evolution operator G. This algorithm can, however, also be used to perform power method projections to the left. In fact, most of the concepts discussed in Sections III.C.I, III.C.2, and III.C.3 can be implemented straightforwardly. To illustrate this point we shall show how one can compute the left- and right-projected expectation value of a diagonal operator X. Since the branching algorithm is designed to explicitly perform the multiplication by G including all weights, all that is required is the following generalization [34], called *forward* or *future walking*.

Rather than defining a walker to be the pair formed by a state and a weight, for forward walking we define the walker to be of the form [**S**, w,

$X(\mathbf{S}_{-1}), \ldots, X(\mathbf{S}_{-p'})]$, where $\mathbf{S}_{-1}, \mathbf{S}_{-2}, \ldots$ are previous states of the walker. In other words, each walker is equipped with a finite stack of depth p' of previous values of the diagonal operator X. In going from one generation of walkers to the next, the state and weight of a walker are updated just as before to $\mathbf{S}'$ and w'. The only new feature is that the value $X(\mathbf{S})$ is pushed on the stack: $[\mathbf{S}, w, X(\mathbf{S}_{-1}), \ldots, X(\mathbf{S}_{-p'})] \to [\mathbf{S}', w', X(\mathbf{S}), X(\mathbf{S}_{-1}), \ldots, X(\mathbf{S}_{-p'+1})]$. In this way, the p' times left-projected expectation value of X is obtained simply by replacing $X(\mathbf{S})$ by $X(\mathbf{S}_{-p'})$. Note that one saves the history of X rather than a history of configurations only for the purpose of efficient use of computer memory.

VI. DIFFUSION MONTE CARLO

A. Simple Diffusion Monte Carlo Algorithm

The diffusion Monte Carlo method [36–40] discussed in this section is an example of the general class of algorithms discussed in this chapter, all of which rely on stochastic implementation of the power method to increase the relative spectral weight of the eigenstates of interest. In the quantum mechanical context of the current section, the latter is usually the ground state. In this general sense, there is nothing new in this section. However, a number of features enter that lead to technical challenges that cannot be dealt with in a general setting. There is the problem that the projector, the evolution operator $G = \exp(-\tau\mathscr{H})$, is only known in the limit of small imaginary time τ, and in addition, for applications to electronic structure problems, there are specific practical difficulties associated with nodes and cusps in the wavefunctions. Bosonic ground states are simpler in that they lack nodes, and we shall deal with systems only in passing. As in the rest of this chapter, we deal only with the basic concepts. In particular, we focus on considerations relevant to the design of an efficient diffusion Monte Carlo algorithm. The reader is referred to Refs. 41–43 for applications and other issues not covered here.

For quantum mechanical problems, the power method iteration Eq. (2.1) takes the form

$$|\psi(t+\tau)\rangle = e^{-(\mathscr{H}-E_{\mathrm{T}})\tau}|\psi(t)\rangle \tag{6.1}$$

Here E_{T} is a shift in energy such that $E_0 - E_{\mathrm{T}} \approx 0$, where E_0 is the ground-state energy. In the real-space representation we have

$$\begin{aligned}\psi(\mathbf{R}', t+\tau) &= \int d\mathbf{R}\, \langle \mathbf{R}' | e^{-(\mathscr{H}-E_{\mathrm{T}})\tau} | \mathbf{R}\rangle\langle \mathbf{R} | \psi(t)\rangle \\ &= \int d\mathbf{R}\, G(\mathbf{R}', \mathbf{R}, \tau)\psi(\mathbf{R}, t),\end{aligned} \tag{6.2}$$

In practical Monte Carlo settings, the shift E_T is computed on the fly and consequently is a slowly varying, nearly constant function of time, but for the moment we take it to be constant. The wavefunction $\psi(\mathbf{R}, t)$ is the solution to the Schrödinger equation in imaginary time

$$-\frac{1}{2\mu}\nabla^2\psi(\mathbf{R}, t) + [\mathscr{V}(\mathbf{R}) - E_T]\psi(\mathbf{R}, t) = -\frac{\partial\psi(\mathbf{R}, t)}{\partial t} \tag{6.3}$$

To make contact with Eq. (3.12), we should factor the Green function into a probability conserving part and a weight. For small times, this can be accomplished by the following approximation. If, on the left-hand side we just had the first term, Eq. (6.3) would reduce to a diffusion equation, whence the method gets its name. For diffusion, the Green function is probability conserving and is given by

$$P(\mathbf{R}', \mathbf{R}, \tau) = \frac{e^{-\mu(\mathbf{R}'-\mathbf{R})^2/2\tau}}{(2\pi\tau/\mu)^{3n/2}} \tag{6.4}$$

If, on the other hand, we just had the second term, then Eq. (6.3) would reduce to a rate equation for which the Green function is $\delta(\mathbf{R}' - \mathbf{R})$ with prefactor

$$g(\mathbf{R}', \mathbf{R}, \tau) = e^{\tau\{E_T - [\mathscr{V}(\mathbf{R}) + \mathscr{V}(\mathbf{R}')]/2\}} \tag{6.5}$$

In combining these ingredients, we have to contend with the following general relation for noncommuting operators $\mathscr{H}_i$:

$$e^{(\mathscr{H}_1 + \mathscr{H}_2)\tau} = e^{(1/2)\mathscr{H}_1\tau}\, e^{\mathscr{H}_2\tau}\, e^{(1/2)\mathscr{H}_1\tau} + \mathscr{O}(\tau^3) = e^{\mathscr{H}_1\tau}\, e^{\mathscr{H}_2\tau} + \mathscr{O}(\tau^2) \tag{6.6}$$

As long as one is dealing with a δ function, the weight in Eq. (6.5) is evaluated always at $\mathbf{R}' = \mathbf{R}$, and therefore the expression on the right can be written also in nonsymmetric form. However, Eq. (6.6) suggests that the exponent in Eq. (6.5) should be used in symmetric fashion as written. This is indeed the form we shall employ for the time being, but we note that the final version of the diffusion algorithm employs a nonsymmetric split of the exponent, since proposed moves are not always accepted. Since there are other sources of $\mathscr{O}(\tau^2)$ contributions anyway this does not degrade the performance of the algorithm. Combination of the preceding ingredients yields the following approximate, short-time Green function for Eq. (6.3)

$$
\begin{aligned}
G(\mathbf{R}', \mathbf{R}, \tau) &= \frac{e^{-\mu(\mathbf{R}'-\mathbf{R})^2/2\tau}}{(2\pi\tau/\mu)^{3n/2}} \, e^{\tau\{E_{\mathrm{T}}-[\mathscr{V}(\mathbf{R}')+\mathscr{V}(\mathbf{R})]/2\}} \\
&\quad + \mathcal{O}(\tau^3) \\
&= \frac{e^{-\mu(\mathbf{R}'-\mathbf{R})^2/2\tau}}{(2\pi\tau/\mu)^{3n/2}} \, e^{\tau[E_{\mathrm{T}}-\mathscr{V}(\mathbf{R})]} + \mathcal{O}(\tau^2)
\end{aligned} \tag{6.7}
$$

For this problem, the power method can be implemented by Monte Carlo by means of both the single-thread scheme discussed in Section III and the branching algorithm of Section V. We shall use the latter option in our discussion. In this case, one performs the following steps. Walkers of the zeroth generation are sampled from $\psi(\mathbf{R}, 0) = \psi_{\mathrm{T}}(\mathbf{R})$ using the Metropolis algorithm. The walkers are propagated forward an imaginary time τ by sampling a new position $\mathbf{R}'$ from a multivariate Gaussian centered at the old position $\mathbf{R}$ and multiplying the weights of the walkers by the second factor in Eq. (6·7). Then the split/join step of the branching algorithm is performed to obtain the next generation of walkers, with weights in the targeted range.

1. *Diffusion Monte Carlo with Importance Sampling*

For many problems of interest, the potential energy $\mathscr{V}(\mathbf{R})$ exhibits large variations over coordinate space and in fact may diverge at particle coincidence points. As we have discussed in the general case, the fluctuations of weights g, produce noisy statistical estimates. As described in Sections III and V, this problem can be greatly mitigated by applying the similarity (or importance sampling) transformation [36,27] to the evolution operator. Employing the general mathematical identity $S\ \exp(-\tau\mathscr{H})S^{-1} = \exp(-\tau S\mathscr{H}S^{-1})$, this transformation can be applied conveniently to the Hamiltonian. That is, given a trial function $\psi_{\mathrm{T}}(\mathbf{R})$, one can introduce a distribution $f(\mathbf{R}, t) = \psi_{\mathrm{T}}(\mathbf{R})\psi(\mathbf{R}, t)$. If $\psi(\mathbf{R}, t)$ satisfies the Schrödinger equation [Eq. (6.3)], it is a simple calculus exercise to show that f is a solution of the equation [39,40]

$$
\begin{aligned}
&\psi_{\mathrm{T}}(\mathbf{R})(\mathscr{H} - E_{\mathrm{T}})\psi_{\mathrm{T}}(\mathbf{R})^{-1} f(\mathbf{R}, t) \\
&\qquad = -\frac{1}{2\mu}\vec{\nabla}^2 f(\mathbf{R}, t) + \vec{\nabla}\cdot[\mathbf{V}(\mathbf{R}) f(\mathbf{R}, t)] - S(\mathbf{R}) f(\mathbf{R}, t) \\
&\qquad = -\frac{\partial f(\mathbf{R}, t)}{\partial t}
\end{aligned} \tag{6.8}
$$

Here the *velocity* $\mathbf{V}$ is a *function* (not an operator), often referred to in the literature as the quantum force, and is given by

$$\mathbf{V}(\mathbf{R}) = (\mathbf{v}_1, \ldots, \mathbf{v}_n) = \frac{1}{\mu} \frac{\vec{\nabla}\psi_{\mathrm{T}}(\mathbf{R})}{\psi_{\mathrm{T}}(\mathbf{R})} \tag{6.9}$$

The coefficient of the source term, which is responsible for branching in the diffusion Monte Carlo context, is

$$S(\mathbf{R}) = E_{\mathrm{T}} - E_{\mathrm{L}}(\mathbf{R}) \tag{6.10}$$

which is defined in terms of the *local energy*

$$E_{\mathrm{L}}(\mathbf{R}) = \frac{\mathscr{H}\psi_{\mathrm{T}}(\mathbf{R})}{\psi_{\mathrm{T}}(\mathbf{R})} = -\frac{1}{2\mu} \frac{\vec{\nabla}^2\psi_{\mathrm{T}}(\mathbf{R})}{\psi_{\mathrm{T}}(\mathbf{R})} + \mathscr{V}(\mathbf{R}) \tag{6.11}$$

the equivalent of the configurational eigenvalue introduced in Eq. (3.1).

Compared to the original Schrödinger equation, to which, of course, it reduces for the case $\psi_{\mathrm{T}} \equiv 1$, the second term in Eq. (6.8) is new, and corresponds to drift. Again, one can explicitly write down the Green function of the equation with just a single term on the left-hand side. The drift Green function G_{D} of the equation obtained by suppressing all except this term is

$$G_{\mathrm{D}}(\mathbf{R}', \mathbf{R}, \tau) = \delta[\mathbf{R}' - \tilde{\mathbf{R}}(\tau)] \tag{6.12}$$

where $\tilde{\mathbf{R}}(\tau)$ satisfies the differential equation

$$\frac{d\tilde{\mathbf{R}}}{dt} = \mathbf{V}(\tilde{\mathbf{R}}) \tag{6.13}$$

subject to the boundary condition $\tilde{\mathbf{R}}(0) = \mathbf{R}$. Again, at short times the noncommutativity of the various operators on the left-hand side of the equation can be ignored and thus one obtains the following short-time Green function.

$$\begin{aligned} G(\mathbf{R}', \mathbf{R}, \tau) &= \int d\mathbf{R}''\, \delta[\mathbf{R}'' - \tilde{\mathbf{R}}(\tau)] \frac{e^{-\mu(\mathbf{R}'-\mathbf{R}'')^2/2\tau}}{(2\pi\tau/\mu)^{3n/2}} \\ &\quad \times e^{\tau\{E_{\mathrm{T}} - [E_{\mathrm{L}}(\mathbf{R}') + E_{\mathrm{L}}(\mathbf{R})]/2\}} + \mathcal{O}(\tau^2) \\ &= \frac{e^{-\mu[\mathbf{R}' - \tilde{\mathbf{R}}(\tau)]^2/2\tau}}{(2\pi\tau/\mu)^{3n/2}} e^{\tau\{E_{\mathrm{T}} - [E_{\mathrm{L}}(\mathbf{R}') + E_{\mathrm{L}}(\mathbf{R})]/2\}} + \mathcal{O}(\tau^2) \end{aligned} \tag{6.14}$$

Equation (6.14) again can be viewed in our general framework by defining the probability conserving generalization of Eq. (6.4)

$$P(\mathbf{R}', \mathbf{R}, \tau) = \frac{e^{-\mu[\mathbf{R}' - \tilde{\mathbf{R}}(\tau)]^2/2\tau}}{(2\pi\tau/\mu)^{3n/2}} \tag{6.15}$$

and the remainder of the Green function is $\delta[\mathbf{R}' - \tilde{\mathbf{R}}(\tau)]$ with prefactor

$$g(\mathbf{R}', \mathbf{R}, \tau) = e^{\tau\{E_T - [E_L(\mathbf{R}') + E_L(\mathbf{R})]/2\}} \tag{6.16}$$

which is the analog of Eq. (6.5).

When employed for the branching Monte Carlo algorithm, the factorization given in Eqs. (6.15) and (6.16) differs from the original factorization Eqs. (6.4) and (6.5) in two respects: (1) the walkers do not only diffuse but also drift toward the important regions, *i.e.*, in the direction in which $|\psi_T(\mathbf{R})|$ is increasing; and (2) the branching term is better behaved since it depends on the local energy rather than the potential. In particular, if the trial function $\psi_T(\mathbf{R})$ obeys cusp conditions [44], then the local energy at particle coincidences is finite even though the potential may be infinite. If one were so fortunate that $\psi_T(\mathbf{R})$ is the exact ground state, the branching factor would reduce to a constant, which can be chosen to be unity by choosing E_T to be the exact energy of that state.

The expressions, as written, explicitly contain the expression $\tilde{\mathbf{R}}(\tau)$, which has to be obtained by integration of the velocity $\mathbf{V}(\mathbf{R})$. Since we are using a short-time expansion anyway, this exact expression may be replaced by the approximation

$$\tilde{\mathbf{R}}(\tau) = \mathbf{R} + \mathbf{V}(\mathbf{R})\tau + \mathcal{O}(\tau^2) \tag{6.17}$$

In the improvements discussed below, designed to reduce time-step error, this expression is improved on so that regions where $\mathbf{V}$ diverges do not make large contributions to the overall time-step error.

2. *Fixed-Node Approximation*

The absolute ground state of a Hamiltonian with particle exchange symmetry is bosonic. Consequently, unless one starts with a fermionic, that is, antisymmetric wavefunction, and implements the power method in a way that respects this antisymmetry, the power method will reduce the spectral weight of the fermionic eigenstate to zero relative to the weight of the bosonic state. The branching algorithm described above assumes that all weights are positive and therefore is incompatible with the requirement of preserving antisymmetry. The algorithm needs modification, if we are interested in the fermionic ground state.

If the nodes of the fermionic ground state were known, they could be imposed as boundary conditions [37] and the problem could be dealt with by solving the Schrödinger equation within a single, connected region bounded by the nodal surface, a region we shall refer to as a *nodal pocket*. Since all the nodal pockets of the ground state of a fermionic system are equivalent, this would yield the exact solution of the problem everywhere. Unfortunately, the nodes of the wave function of an n-particle system form a $(3n - 1)$-dimensional surface, which should not be confused with the nodes of single-particle orbitals. Of this full surface, in general, only the $(3n - 3)$-dimensional subset, corresponding to the coincidence of two like-spin particles, is known. Hence, we are forced to employ an approximate nodal surface as a boundary condition to be satisfied by the solution of the Schrödinger equation. This is called the *fixed-node* approximation. Usually, one chooses for this purpose the nodal surface of an optimized trial wavefunction [37], and such nodes can at times yield a very accurate results if sufficient effort is invested in optimizing the trial wavefunction.

Since the imposition of the boundary condition constrains the solution, it is clear that the fixed-node energy is an upper bound on the true fermionic ground state energy. In diffusion Monte Carlo applications, the fixed-node energy typically has an error that is 5–10 times smaller than the error of the variational energy corresponding to the same trial wavefunction, although this can vary greatly depending on the system and the trial wavefunction employed.

For the Monte Carlo implementation of this approach one has to use an approximate Green function, which, as we have seen, may be obtained by iteration of a short-time approximation. To guide the choice of an approximant accurate over a wide time range, it is useful to consider some general properties of the fixed-node approximation. Mathematically, the latter amounts to solution of the Schrödinger equation in a potential that equals the original physical potential inside the nodal pocket of choice, and is infinite outside. The corresponding eigenstates of the Hamiltonian are continuous and vanish outside the nodal pocket. Note that the solution can be analytically continued outside the initial nodal pocket only if the nodal surface is exact; otherwise there is a derivative discontinuity at the nodal surface. These properties are shared by the Green function consistent with the boundary conditions. This can be seen by writing down the spectral decomposition for the evolution operator in the position representation

$$G(\mathbf{R}', \mathbf{R}, \tau) = \langle \mathbf{R}' | \, e^{-\tau \mathscr{H}} \, | \mathbf{R} \rangle = \sum_i \psi_i(\mathbf{R}') \, e^{-\tau E_i} \psi_i(\mathbf{R}) \tag{6.18}$$

where the ψ_i factors are the eigenstates satisfying the required boundary conditions. For notational convenience only, we have assumed that the spectrum has a discrete part only and that the wavefunctions can be chosen to be real. The Green function vanishes outside the nodal pocket and generically vanishes linearly at the nodal surface, as do the wavefunctions.

The Green function of interest in practical applications is the one corresponding to importance sampling, the similarity transform of Eq. (6.18)

$$\hat{G}(\mathbf{R}', \mathbf{R}, \tau) = \psi_{\mathrm{T}}(\mathbf{R}')\langle\mathbf{R}'|\, e^{-\tau\mathscr{H}}\,|\mathbf{R}\rangle \frac{1}{\psi_{\mathrm{T}}(\mathbf{R})} \tag{6.19}$$

This Green function vanishes at the nodes quadratically in its first index, which, in the Monte Carlo context, is the one that determines to which state a transition is made from a given initial state.

The approximate Green functions of Eqs. (6.7) and (6.14) have tails that extend beyond the nodal surface and, consequently, walkers sampled from these Green functions have a finite probability of attempting to cross the node. Since expectation values ought to be calculated in the $\tau \to 0$ limit the relevant quantity to consider is what fraction of walkers attempt to cross the node *per unit time* in the $\tau \to 0$ limit. If the Green function of Eq. (6.7) is employed, this is a finite number, whereas, if the importance-sampled Green function of Eq. (6.14) is employed, no walkers cross the surface since the velocity $\mathbf{V}$ is directed away from the nodes and diverges at the nodes. In practice, of course, the calculations are performed for finite τ, but the preceding observation leads to the conclusion that in the former case it is necessary to reduce to zero the weight of a walker that has crossed a node, that is, to kill the walker, while in the latter case one can either kill the walkers or reject the proposed move, since in the $\tau \to 0$ limit they yield the same result.

We now argue that, for finite τ, rejecting moves is the choice with the smaller time-step error when the importance-sampled Green function is employed. Sufficiently close to a node, the component of the velocity perpendicular to the node dominates all other terms in the Green function and it is illuminating to consider a free particle in one dimension subject to the boundary condition that ψ have a node at $x = 0$. The exact Green function for this problem is

$$G(x', x, \tau) = \frac{1}{\sqrt{2\pi\tau/\mu}} \left[e^{-\mu(x'-x)^2/2\tau} - e^{-\mu(x'+x)^2/2\tau}\right] \tag{6.20}$$

while the corresponding importance sampled Green function is

$$\hat{G}(x', x, \tau) = \frac{x'}{x\sqrt{2\pi\tau/\mu}} [e^{-\mu(x'-x)^2/2\tau} - e^{-\mu(x'+x)^2/2\tau}] \tag{6.21}$$

We note that the integral of the former, over the region $x > 0$, is less than one and decreases with time. In terms of the usual language used for diffusion problems, this is because of absorption at the $x = 0$ boundary. In our case, this provides the mathematical justification for killing the walkers that cross. On the other hand, the integral of the Green function of Eq. (6.21) equals one. Consequently, for finite τ it seems likely that algorithms that reject moves across the node, such as the one discussed in Section VI.B yield a better approximation than algorithms that kill the walkers that cross the node.

As mentioned, it can be shown that all the nodal pockets of the ground state of a fermionic system are equivalent and trial wavefunctions are constructed to have the same property. Consequently, the Monte Carlo averages will not depend on the initial distribution of the walkers over the nodal pockets. The situation is more complicated for excited states, since different nodal pockets of excited-state wavefunctions are not necessarily equivalent, for neither bosons nor fermions. Any initial state with walkers distributed randomly over nodal pockets pockets, will evolve to a steady-state distribution with walkers only in the pocket with the lowest average local energy, at least if we ignore multiple-node crossings and assume a sufficiently large number of walkers, so that fluctuations in the average local energy can be ignored.

3. *Problems with Simple Diffusion Monte Carlo*

The diffusion Monte Carlo algorithm corresponding to Eq. (6.14) is in fact not viable for a wavefunction with nodes for the following two reasons:

1. In the vicinity of the nodes the local energy of the trial function ψ_T diverges inversely proportional to the distance to the nodal surface. For nonzero τ, there is a nonzero density of walkers at the nodes. Since the nodal surface for a system with n electrons is $(3n - 1)$-dimensional, the variance of the local energy diverges for any finite τ. In fact, the expectation value of the local energy also diverges, but only logarithmically.
2. The velocity of the electrons at the nodes diverges inversely as the distance to the nodal surface. The walkers that are close to a node at one time step, drift at the next time step to a distance inversely proportional to the distance from the node. This results in a charge distribution with a component that falls off as the inverse square of

distance from the atom or molecule, whereas in reality the decay is exponential.

These two problems are often remedied by introducing cutoffs in the values of the local energy and the velocity [45,46], chosen such that they have no effect in the $\tau \to 0$ limit, so that the results extrapolated to $\tau = 0$ are correct. In the next section better remedies are presented.

B. Improved Diffusion Monte Carlo Algorithm

1. *The Limit of Perfect Importance Sampling*

In the limit of perfect importance sampling, that is if $\psi_T(\mathbf{R}) = \psi_0(\mathbf{R})$, the energy shift E_T can be chosen such that the branching term in Eq. (6.8) vanishes identically for all $\mathbf{R}$. In this case, even though the energy can be obtained with zero variance, the steady-state distribution of the diffusion Monte Carlo algorithm discussed above is only approximately the desired distribution ψ_T^2, because of the time-step error in the Green function. However, since one has an explicit expression, ψ_T^2, for the distribution to be sampled, it is possible to use the Metropolis algorithm, described in Section III.A, to sample the desired distribution exactly. Although the ideal $\psi_T(\mathbf{R}) = \psi_0(\mathbf{R})$ is never achieved in practice, this observation leads one to devise an improved algorithm that can be used when moderately good trial wavefunctions are known.

If for the moment we ignore the branching term in Eq. (6.8), then we have the equation

$$-\frac{1}{2\mu}\vec{\nabla}^2 f + \vec{\nabla} \cdot (\mathbf{V}f) = -\frac{\partial f}{\partial t} \tag{6.22}$$

This equation has a known steady-state solution $f = \psi_T^2$ for any ψ_T, which in the limit of perfect importance sampling is the desired distribution. However, the approximate drift–diffusion Green function used in the Monte Carlo algorithm defined by Eq. (6.14) without the branching factor, is not the exact Green function of Eq. (6.22). Therefore, for any finite time step τ, we do not obtain ψ_T^2 as a steady state, even in the ideal case. Following Reynolds et al. [39], we can change the algorithm in such a way that it *does* sample ψ_T^2 in the ideal case, which also reduces the time-step error in non-ideal, practical situations. This is accomplished by using a generalized [47–49] Metropolis algorithm [14]. The approximate drift–diffusion Green function is used to propose moves, which are then accepted with probability

$$p = \min\left(\frac{\psi_T(\mathbf{R}')^2 \tilde{G}(\mathbf{R}, \mathbf{R}', \tau)}{\psi_T(\mathbf{R})^2 \tilde{G}(\mathbf{R}', \mathbf{R}, \tau)}, 1\right) \equiv 1 - q \tag{6.23}$$

in accordance with the detailed balance condition.

As was shown above, the true fixed-node Green function vanishes outside the nodal pocket of the trial wavefunction. However, since we are using an approximate Green function, moves across the nodes will be proposed for any finite τ. To satisfy the boundary conditions of the fixed-node approximation, these proposed moves are always rejected.

If we stopped here, we would have an exact and efficient variational Monte Carlo algorithm to sample from ψ_T^2. Now, we reintroduce the branching term to convert the steady-state distribution from ψ_T^2 to $\psi_T \psi_0$. This is accomplished by reweighting the walkers with the branching factor [see Eq. (6.14)]

$$\Delta w = \begin{cases} \exp\{\frac{1}{2}[S(\mathbf{R}') + S(\mathbf{R})]\tau_{\text{eff}}\} & \text{for an accepted move} \\ \exp[S(\mathbf{R})\tau_{\text{eff}}] & \text{for a rejected move} \end{cases} \tag{6.24}$$

where S is defined in Eq. (6.10). An effective time step τ_{eff}, which will be defined presently, is required because the Metropolis algorithm introduces a finite probability of not moving forward and rejecting the proposed configuration. Before defining τ_{eff}, we note that an alternative to expression (6.24) is obtained by replacing the two reweighting factors by a single expression:

$$\Delta w = \exp\left[\left\{\frac{p}{2}\left(S(\mathbf{R}') + S(\mathbf{R})\right) + qS(\mathbf{R})\right\}\tau_{\text{eff}}\right] \quad \text{for all moves} \tag{6.25}$$

This expression, written down with Eq. (3.9) in mind, yields somewhat smaller fluctuations and time-step error than expression (6.24).

Following Reynolds et al. [39], an effective time step τ_{eff} is introduced in Eq. (6.25) to account for the changed rate of diffusion. We set

$$\tau_{\text{eff}} = \tau \frac{\langle p \, \Delta R^2 \rangle}{\langle \Delta R^2 \rangle} \tag{6.26}$$

where the angular brackets denote the average over all attempted moves, and ΔR are the displacements resulting from diffusion. This equals

$\langle \Delta R^2 \rangle_{\text{accepted}} / \langle \Delta R^2 \rangle$ but again has some what smaller fluctuations.

An estimate of τ_{eff} is readily obtained iteratively from sets of equilibrium runs. During the initial run, τ_{eff} is set equal to τ. For the next runs, the value of τ_{eff} is obtained from the values of τ_{eff} computed with Eq. (6.26) during the previous equilibration run. In practice, this procedure converges in two iterations, which typically consume less than 2% of the total computation time. Since the statistical errors in τ_{eff} affect the results obtained, the number of Monte Carlo steps performed during the equilibration phase needs to be sufficiently large that this is not a major component of the overall statistical error.

The value of τ_{eff} is a measure of the rate at which the Monte Carlo process generates uncorrelated configurations, and thus a measure of the efficiency of the computation. Since the acceptance probability decreases when τ increases, τ_{eff} has a maximum as a function of τ. However, since the time-step error increases with τ, it is advisable to use values of τ that are smaller than this "optimum."

Algorithms that do not exactly simulate the equilibrium distribution of the drift–diffusion equation if the branching term is suppressed, in other words, algorithms that do not use the Metropolis accept/reject mechanism, can for sufficiently large τ, have time-step errors that make the energy estimates higher than the variational energy. On the other hand, if the drift–diffusion terms are treated exactly by including an accept/reject step, the energy, evaluated for any τ, must lie below the variational energy, since the branching term enhances the weights of the low-energy walkers relative to that of the high-energy walkers.

2. *Persistent Configurations*

As mentioned above, the accept/reject step has the desirable feature of yielding the exact electron distribution in the limit that the trial function is the exact ground state. However, in practice the trial function is less than perfect and as a consequence the accept/reject procedure can lead to the occurrence of persistent configurations, as we now discuss.

For a given configuration $\mathbf{R}$, consider the quantity $Q = \langle q \Delta w \rangle$, where q and Δw are the rejection probability and the branching factor given by Eqs. (6.23) and (6.25). The average in the definition of Q is over all possible moves for the configuration $\mathbf{R}$ under consideration. If the local energy at $\mathbf{R}$ is relatively low and if τ_{eff} is sufficiently large, Q may be in excess of one. In that case, the weight of the walker at $\mathbf{R}$, or more precisely, the total weight of all walkers in that configuration, will increase with time, except for fluctuations, until the time-dependent trial energy E_{T} adjusts downward to stabilize the total population. This population contains on average a certain number of copies of the persistent configuration. Since persistent configurations must necessarily have an energy that is lower than the true fixed-node energy, this results in a negatively biased energy estimate. The persistent configuration may disappear because of fluctuations, but the more likely occurrence is that it is replaced by another configuration that is even more strongly persistent, that is, one that has an even larger value of $Q = \langle q\Delta w \rangle$. This process produces a cascade of configurations of ever-decreasing energies. Both sorts of occurrences are demonstrated in Figure 2. Persistent configurations are most likely to occur near nodes, or near nuclei if ψ_T does not obey the cusp conditions. Improvements to the approximate Green function in these regions, as discussed in the next section, help to reduce greatly the probability of encountering persistent configurations to the point that they are never encountered even in long Monte Carlo runs.

Despite the fact that the modifications described in the next section eliminate persistent configurations for the systems we have studied, it is clearly desirable to have an algorithm that cannot display this pathology even in principle. One possible method is to replace τ_{eff} in Eq. (6.24) by τ for an accepted move and by zero for a rejected move. This ensures that Δw

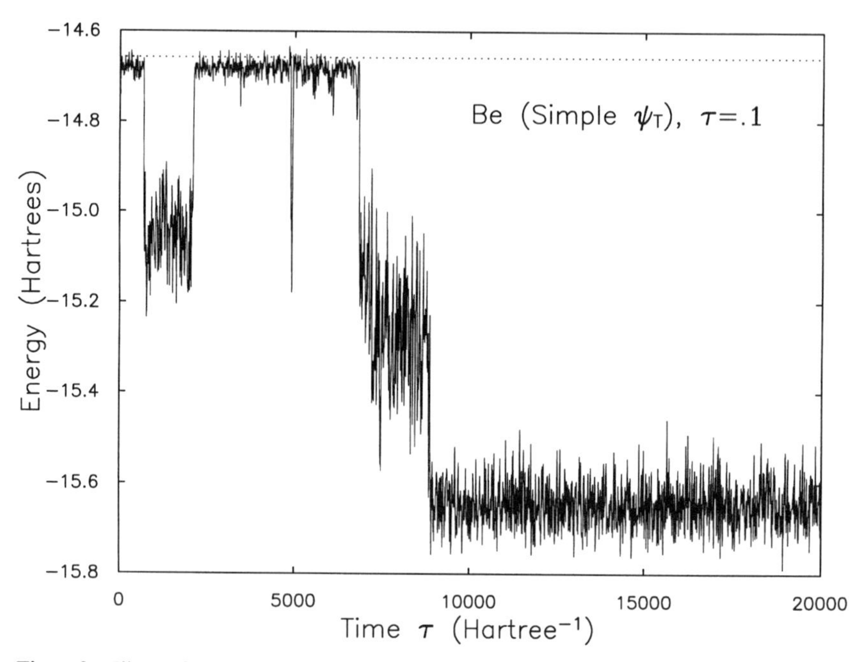

Figure 2. Illustration of the persistent configuration catastrophe. The dotted horizontal line is the true fixed-node energy for a simple Be wavefunction extrapolated to $\tau = 0$.

never exceeds unity for rejected moves, hence eliminating the possibility of persistent configurations. Further, this has the advantage that it is not necessary to determine τ_{eff}. Other ways to eliminate the possibility of encountering persistent configurations are discussed in Ref. 50.

3. *Singularities*

The number of iterations of Eq. (6.2) required for the power method to converge to the ground state grows inversely with the time step τ. Thus, the statement made above, specifically that the Green function of Eq. (6.7) is in error only to $\mathcal{O}(\tau^2)$, would seem to imply that the errors in the electron distribution and the averages calculated from the short-time Green function are of $\mathcal{O}(\tau)$. However, the presence of nonanalyticities and divergences in the local energy and the velocity may invalidate this argument; the short-time Green function may lack uniform convergence in τ over $3n$-dimensional configuration space. Consequently, an approximation that is designed to be better behaved in the vicinity of singularities and therefore behaves more uniformly over space may outperform an approximation that is correct to a higher order in τ for generic points in configuration space, but ignores these singularities.

Next we discuss some important singularities that one may encounter and their implications for the diffusion Monte Carlo algorithm. The behavior of the local energy E_L and velocity $\mathbf{V}$ near nodes of ψ_T are described in Table I. Although the true wavefunction ψ_0 has a constant local energy E_0 at all points in configuration space, the local energy of ψ_T diverges at most points of the nodal surface of ψ_T for almost all the ψ_T that are used in practice, and it diverges at particle coincidences (either electron–nucleus or electron–electron) for wave functions that fail to obey cusp conditions [44].

TABLE I
Behavior of Local Energy E_L and Velocity v as a Function of the Distance $R_\perp$ of an Electron to the Nearest Singularity[a]

Region	Local Energy	Velocity
Nodes	$E_L \sim \pm\dfrac{1}{R_\perp}$ for ψ_T	$v \sim \dfrac{1}{R_\perp}$
	$E_L = E_0$ for ψ_0	
Electron–nucleus/electron	$E_L \sim \dfrac{1}{x}$ for some ψ_T	v has a discontinuity for both ψ_T and ψ_0
	$E_L = E_0$ for ψ_0	

[a] The behavior of various quantities is shown for an electron approaching a node or another particle, either a nucleus or an electron. The singularity in the local energy at particle overlap is only present for a ψ_T that fails to satisfy the cusp conditions.

The velocity $\mathbf{V}$ diverges at the nodes and for the Coulomb potential has a discontinuity at particle coincidences both for approximate and for the true wavefunction. For the nodeless wavefunction of Lennard-Jones particles in their ground state [5], $\mathbf{V}$ diverges as r^{-6} in the interparticle distance r. Since the only other problem for these bosonic systems is the divergence of E_L at particle coincidences, we continue our discussion for electronic systems and refer the interested reader to the literature [5] for details.

The divergences in the local energies cause large fluctuations in the population size: negative divergences lead to large local population densities, and positive divergences lead to small ones. The divergence of $\mathbf{V}$ at the nodes typically leads to a proposed next state of a walker in a very unlikely region of configuration space and is therefore likely to be rejected. The three-dimensional velocity of an electron that is close to a nucleus is directed toward the nucleus. Hence the true Green function, for sufficiently long time, exhibits a peak at the nucleus, but the approximate Green function of Eq. (6.14) causes electrons to overshoot the nucleus. This is illustrated in Figure 3, where we show the qualitative behavior of the true Green function and of the approximate Green function for an electron that starts at $x = -0.2$ in the presence of a nearby nucleus at $x = 0$. At short time t_1, the approximate Green function of Eq. (6.14) agrees very well with the true Green function. At a longer time t_2, the true Green function begins to develop a peak at the nucleus that is absent in the approximate Green function, whereas at a yet longer time t_3, the true Green function is peaked at the nucleus while the approximate Green function has overshot the nucleus.

The combined effect of these nonanalyticities is a large time-step error, which can be of either sign in the energy, and large statistical uncertainty in the computed expectation values. We now give a brief description of how these nonanalyticities are treated. The divergence of the local energy at particle coincidences is cured simply by employing wavefunctions that obey cusp conditions [44]. The other nonanalyticities are addressed by employing a modification of the Green function of Eq. (6.14) such that it incorporates the divergence of E_L and $\mathbf{V}$ at nodes and the discontinuity in $\mathbf{V}$ at particle coincidences but smoothly reduces to Eq. (6.14) in the short-time limit or in the limit that the nearest nonanalyticity is far away. The details can be found in Ref. 50. The modified algorithm has a time-step error that is two to three orders of magnitude smaller [50] than the simple algorithm corresponding to Eq. (6.14) with cutoffs imposed on E_L and $\mathbf{V}$.

We have used the application to all-electron electronic structure calculations to illustrate the sort of problems that can lead to large time-step errors and their solution. Other systems may exhibit only a subset of these problems or a modified version of them. For example, in calculations of

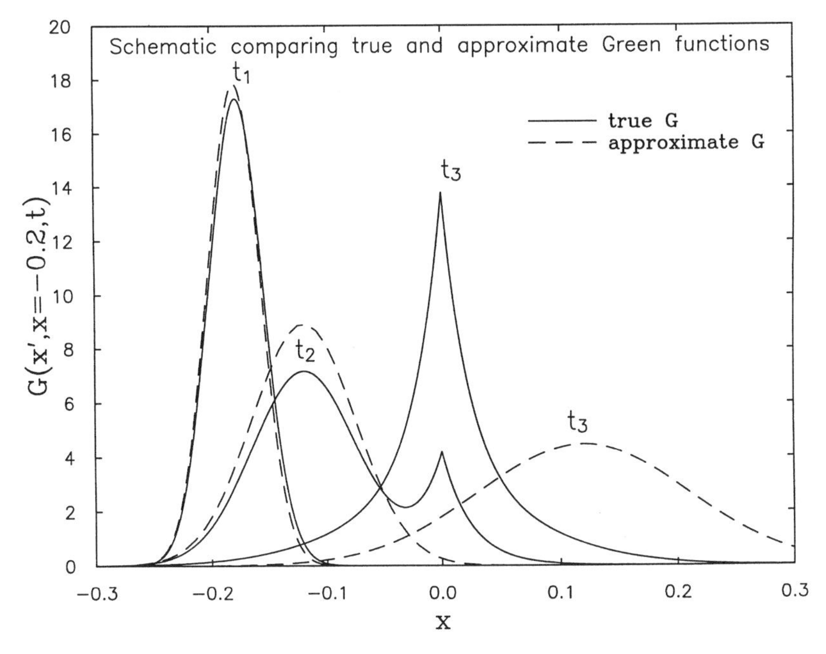

Figure 3. Schematic comparing the qualitative behaviors of the true G with the approximate G of Eq. (6.14) of an electron that is located at $x = -0.2$ at time $t = 0$ and evolves in the presence of a nearby nucleus located at $x = 0$. The Green functions are plotted for three times: $t_1 < t_2 < t_3$.

bosonic clusters [5] there are no nodes to contend with, whereas in electronic structure calculations employing pseudopotentials [51–54], or pseudo-Hamiltonians [55], the potential need not diverge at electron–nucleus coincidences. We note, in passing, that the use of the latter methods has greatly extended the practical applicability of quantum Monte Carlo methods to relatively large systems of practical interest [53,56], but at the price of introducing an additional approximation [57].

VII. CLOSING COMMENTS

The material presented above was selected to describe from a unified point of view Monte Carlo algorithms as employed in seemingly unrelated areas in quantum and statistical mechanics. Details of applications were given only to explain general ideas or important technical problems, such as encountered in diffusion Monte Carlo. We ignored a whole body of literature, but we wish to just mention a few topics. Domain Green function Monte Carlo [25–28] is one that comes very close to topics that were covered. In this method the Green function is sampled exactly by iterating on an approximate Green function. Infinite iteration, which the Monte Carlo method does in principle, produces the exact Green function. Consequently, this method lacks a time-step error, and in this sense has the advantage of being exact. In practice, there are other reasons, besides the time-step error that force the algorithm to move slowly through state space, and currently the available algorithms seem to be less efficient than diffusion Monte Carlo, even when one accounts for the effort required to perform the extrapolation to vanishing time step.

Another area that we just touched on in passing is path integral Monte Carlo [58]. Here we remind the reader that path integral Monte Carlo is a particularly appealing alternative for the evaluation of matrix elements such as $X_{\alpha\beta}^{(p',p)}$ in Eq. (2.2). The advantage of this method is that no products of weights appear, but the disadvantage is that it seems to be more difficult to move rapidly through state space. This is a consequence of the fact that branching algorithms propagate just a single time slice through state space, whereas path integral methods deal with a whole stack of slices, which for sampling purposes tends to produce a more rigid object. Finally, it should be mentioned that we ignored the vast literature on quantum lattice systems [59].

In splitting the evolution operator into weights and probabilities [see Eq. (3.16)] we assumed that the weights were nonnegative. To satisfy this requirement, the fixed-node approximation was employed for fermionic systems. An approximation in the same vein is the fixed-phase approximation [60], which allows one to deal with systems in which the wavefunction is necessarily complex-valued. The basic idea here is analogous to that underlying the fixed-node approximation. In the latter, a trial function is used to approximate the nodes while diffusion Monte Carlo recovers the magnitude of the wavefunction. In the fixed phase approximation, the trial function is responsible for the phase and Monte Carlo produces the magnitude of the wavefunction.

ACKNOWLEDGMENTS

This work was supported by the (U.S.) National Science Foundation through Grants DMR-9725080 and CHE-9625498 and by the Office of Naval Research. The authors thank Richard Scalettar and Andrei Astrakharchik for their helpful comments.

REFERENCES

1. H. A. Kramers and G. H. Wannier, *Phys. Rev.* **60**, 252 (1941).
2. M. P. Nightingale, in *Finite-Size Scaling and Simulation of Statistical Mechanical Systems*, V. Privman, ed., World Scientific, Singapore 1990, pp. 287–351.

3. M. P. Nightingale and C. J. Umrigar, in *Recent Advances in Quantum Monte Carlo Methods*, W. A. Lester, Jr., ed., World Scientific, Singapore, 1997, p. 201. (See also http://xxx.lanl.gov/abs/chem-ph/9608001).
4. A. Mushinski and M. P. Nightingale, *J. Chem. Phys.* **101**, 8831 (1994).
5. M. Meierovich, A. Mushinski and M. P. Nightingale, *J. Chem. Phys.* **105**, 6498 (1996).
6. M. Suzuki, *Commun. Math. Phys.* **51**, 183 (1976).
7. M. Suzuki, *Prog. Theor. Phys.* **58**, 1377 (1977).
8. M. P. Nightingale and H. W. J. Blöte, *Phys. Rev. B* **54**, 1001 (1996).
9. M. P. Nightingale and H. W. J. Blöte, *Phys. Rev. Lett.* **76**, 4548 (1996).
10. M. P. Nightingale and H. W. J. Blöte, *Phys. Rev. Lett.* **80**, 1007 (1998). (See also http://xxx.lanl.gov/abs/cond-mat/9708063.)
11. C. P. Yang, *Proc. Symp. Appl. Math.* **15**, 351 (1963).
12. J. H. Hetherington, *Phys. Rev. A* **30**, 2713 (1984).
13. D. M. Ceperley and B. Bernu, *J. Chem. Phys.* **89**, 6316 (1988).
14. N. Metropolis, A. W. Rosenbluth, M. N. Rosenbluth, A. M. Teller, and E. Teller, *J. Chem. Phys.* **21**, 1087 (1953).
15. P. J. Reynolds, R. N. Barnett, B. L. Hammond, and W. A. Lester, *Stat. Phys.* **43**, 1017 (1986).
16. C. J. Umrigar, *Int. J. Quantum. Chem. Symp.* **23**, 217 (1989).
17. D. Ceperley and B. J. Alder, *J. Chem. Phys.* **81**, 5833 (1984); D. Arnow, M. H. Kalos, M. A. Lee, and K. E. Schmidt, *J. Chem. Phys.* **77**, 5562 (1982); J. Carlson and M. H. Kalos, *Phys. Rev. C* **32**, 1735 (1985); S. Zhang and M. H. Kalos, *Phys. Rev. Lett.* **67**, 3074 (1991); J. B. Anderson, in *Understanding Chemical Reactivity*, S. R. Langhoff, ed., Kluwer, 1994; R. Bianchi, D. Bressanini, P. Cremaschi, and G. Morosi, *Chem. Phys. Lett.* **184**, 343 (1991); V. Elser, *Phys. Rev. A* **34**, 2293 (1986); P. L. Silvestrelli, S. Baroni, and R. Car, *Phys. Rev. Lett.* **71**, 1148 (1993).
18. R. N. Barnett, P. J. Reynolds, and W. A. Lester, *J. Chem. Phys.* **96**, 2141 (1992).
19. B. Bernu, D. M. Ceperley, and W. A. Lester, Jr., *J. Chem. Phys.* **93**, 552 (1990).
20. W. R. Brown, W. A. Glauser, and W. A. Lester, Jr., *J. Chem. Phys.* **103**, 9721 (1995).
21. R. L. Coldwell, *Int. J. Quantum Chem. Symp.* **11**, 215 (1977).
22. C. J. Umrigar, K. G. Wilson, and J. W. Wilkins, *Phys. Rev. Lett.* **60**, 1719 (1988); in *Computer Simulation Studies in Condensed Matter Physics*, D. P. Landau, K. K. Mon, and H.-B. Schüttler, eds., *Springer Proceedings in Physics*, Vol. 33, Springer-Verlag, Berlin, 1988, p. 185.
23. M. P. Nightingale, "Computer Simulation Studies in Condensed Matter Physics IX," in *Springer Proc. Phys. 82*, D. P. Landau, K. K. Mon, and H. B. Schüttler, eds., Springer, Berlin, 1997.
24. P. A. Whitlock and M. H. Kalos, *J. Comp. Phys.* **30**, 361 (1979).
25. D. M. Ceperley and M. H. Kalos, in *Monte Carlo Methods in Statistical Physics*, K. Binder, ed., Topics Current Phys., Vol. 7, Springer, Berlin, Heidelberg 1979, Chapter 4.
26. K. E. Schmidt and J. W. Moskowitz, *J. Stat. Phys.* **43**, 1027 (1986); J. W. Moskowitz and K. E. Schmidt, *J. Chem. Phys.* **85**, 2868 (1985).
27. M. Kalos, D. Leveque, and L. Verlet, *Phys. Rev. A* **9**, 2178 (1974).
28. D. M. Ceperley, *J. Comp. Phys.* **51**, 404 (1983).

29. J. H. Hetherington, *Phys. Rev. A* **30**, 2713 (1984).
30. M. P. Nightingale and H. W. J. Blöte, *Phys. Rev. B* **33**, 659 (1986).
31. C. J. Umrigar, M. P. Nightingale, and K. J. Runge, *J. Chem. Phys.* **99**, 2865 (1993).
32. M. P. Nightingale and H. W. J. Blöte, *Phys. Rev. Lett.* **60**, 1662 (1988).
33. K. J. Runge, *Phys. Rev. B* **45**, 12292 (1992).
34. M. H. Kalos, *J. Comput. Phys.* **1**, 257 (1966); the original idea of "forward walking" predates this paper (M. H. Kalos, private communication). For further references, see Ref. 11 in Ref. 35.
35. K. J. Runge, *Phys. Rev. B* **45**, 7229 (1992).
36. R. Grimm and R. G. Storer, *J. Comp. Phys.* **4**, 230 (1969); **7**, 134 (1971); **9**, 538 (1972).
37. J. B. Anderson, *J. Chem. Phys.* **63**, 1499 (1975); *J. Chem. Phys.* **65**, 4121 (1976).
38. D. M. Ceperley and B. J. Alder, *Phys. Rev. Lett.* **45**, 566 (1980).
39. P. J. Reynolds, D. M. Ceperley, B. J. Alder, and W. A. Lester, *J. Chem. Phys.* **77**, 5593 (1982).
40. J. W. Moskowitz, K. E. Schmidt, M. A. Lee, and M. H. Kalos, *J. Chem. Phys.* **77**, 349 (1982).
41. D. M. Ceperley and L. Mitas, in *New Methods in Computational Quantum Mechanics, Advances in Chemical Physics*, XCIII, I. Prigogine and S. A. Rice, eds., Wiley, New York, 1996.
42. J. B. Anderson, *Int. Rev. Phys. Chem.* **14**, 85 (1995).
43. B. L. Hammond, W. A. Lester, and P. J. Reynolds, *Monte Carlo Methods in Ab Initio Quantum Chemistry*, World Scientific, 1994.
44. T. Kato, *Commun. Pure Appl. Math.* **10**, 151 (1957).
45. M. F. DePasquale, S. M. Rothstein, and J. Vrbik, *J. Chem. Phys.* **89**, 3629 (1988).
46. D. R. Garmer and J. B. Anderson, *J. Chem. Phys.* **89**, 3050 (1988).
47. W. K. Hastings, *Biometrika* **57**, 97 (1970).
48. D. Ceperley, G. V. Chester, and M. H. Kalos, *Phys. Rev. B* **16**, 3081 (1977).
49. M. H. Kalos and P. A. Whitlock, *Monte Carlo Methods*, Vol. 1, Wiley, New York, 1986.
50. C. J. Umrigar, M. P. Nightingale, and K. J. Runge, *J. Chem. Phys.* **99**, 2865 (1993).
51. M. M. Hurley and P. A. Christiansen, *J. Chem. Phys.* **86**, 1069 (1987); P. A. Christiansen, *J. Chem. Phys.* **88**, 4867 (1988); P. A. Christiansen, *J. Chem. Phys.* **95**, 361 (1991).
52. B. L. Hammond, P. J. Reynolds, and W. A. Lester, *J. Chem. Phys.* **87**, 1130 (1987).
53. S. Fahy, X. W. Wang, and Steven G. Louie, *Phys. Rev. Lett.* **61**, 1631 (1988); *Phys. Rev. B* **42**, 3503 (1990).
54. H.-J. Flad, A. Savin, and H. Preuss, *J. Chem. Phys.* **97**, 459 (1992).
55. G. B. Bachelet, D. M. Ceperley, and M. G. B. Chiocchetti, *Phys. Rev. Lett.* **62**, 2088 (1989); A. Bosin, V. Fiorentini, A. Lastri, and G. B. Bachelet, in *Materials Theory and Modeling Symposium*, J. Broughton, P. Bristowe, and J. Newsam, eds., Materials Research Society, Pittsburgh, 1993.
56. L. Mitas, *Phys. Rev. A* 49, 4411 (1994); L. Mitas, in *Electronic Properties of Solids Using Cluster Methods*, T. A. Kaplan and S. D. Mahanti, eds., Plenum Press, New York, 1994; J. C. Grossman and L. Mitas, *Phys. Rev. Lett.* **74**, 1323 (1995); J. C. Grossman, L. Mitas, and K. Raghavachari, *Phys. Rev. Lett.* **75**, 3870 (1995); J. C. Grossman and L. Mitas, *Phys. Rev. B* **52**, 16735 (1995).

57. P. A. Christiansen and L. A. Lajohn, *Chem. Phys. Lett.* **146**, 162 (1988); M. Menchi, A. Bosin, F. Meloni, and G. B. Bachelet, in *Materials Theory and Modeling Symposium*, J. Broughton, P. Bristowe, and J. Newsam, eds., Materials Research Society, Pittsburgh, 1993.

58. D. M. Ceperley, *Rev. Mod. Phys.* **67**, 279 (1995); D. L. Freeman and J. D. Doll, *Adv. Chem. Phys. B* **70**, 139 (1988); J. D. Doll, D. L. Freeman, and T. L. Beck, *Adv. Chem. Phys.*, **78**, 61 (1990); B. J. Berne and D. Thirumalai, *Annu. Rev. Phys. Chem.* **37**, 401 (1986).

59. J. Hirsch, *Phys. Rev. B* **28**, 4059 (1983); G. Sugiyama and S. E. Koonin, *Ann. Phys.* **168**, 1 (1986); Masuo Suzuki, *J. Stat. Phys.* **43**, 883 (1986); S. Sorella, S. Baroni, R. Car, and M. Parrinello, *Europhys. Lett.* **8**, 663 (1989); R. Blankenbecler and R. L. Sugar, *Phys. Rev. D* **27**, 1304 (1983); S. Sorella, E. Tosatti, S. Baroni, R. Car, and M. Parrinello, *Int. J. Mod. Phys. B* **1**, 993 (1989); S. R. White, D. J. Scalapino, R. L. Sugar, E. Y. Loh, Jr., J. E. Gubernatis, and R. T. Scalettar, *Phys. Rev. B* **40**, 506 (1989); N. Trivedi and D. M. Ceperley, *Phys. Rev. B* **41**, 4552 (1990); D. F. B. ten Haaf, J. M. van Bemmel, J. M. J. van Leeuwen, W. van Saarloos, and D. M. Ceperley, *Phys. Rev. B* **51**, 13039 (1995); A. Muramatsu, R. Preuss, W. von der Linden, P. Dietrich, F. F. Assaad, and W. Hanke, in *Computer Simulation Studies in Condensed Matter Physics*, D. P. Landau, K. K. Mon, and H.-B. Schüttler, *Springer Proceedings in Physics*, Springer-Verlag, Berlin, 1994; H. Betsuyaku, in *Quantum Monte Carlo Methods in Equilibrium and Nonequilibrium Systems*, M. Suzuki, ed., *Proceedings of the Ninth Taniguchi International Symposium*, Springer-Verlag, Berlin, 1987 pp. 50–61; M. Takahashi, *Phys. Rev. Lett.* **62**, 2313 (1989); H. De Raedt and A. Lagendijk, *Phys. Rep.* **127**, 233 (1985); W. von der Linden, *Phys. Rep.* **55**, 220, 53 (1992); S. Zhang, J. Carlson, and J. E. Gubernatis, *Phys. Rev. Lett.* **74**, 3652 (1995). See also Refs. 30 and 35.

60. G. Ortiz, D. M. Ceperley, and R. M. Martin, *Phys. Rev. Lett.* **71**, 2777 (1993); G. Ortiz and D. M. Ceperley, *Phys. Rev. Lett.* **75**, 4642 (1995).

61. M. P. Nightingale, Y. Ozelci, and Y. Ye, unpublished.

ADAPTIVE PATH-INTEGRAL MONTE CARLO METHODS FOR ACCURATE COMPUTATION OF MOLECULAR THERMODYNAMIC PROPERTIES

ROBERT Q. TOPPER

Department of Chemistry, School of Engineering, The Cooper Union for the Advancement of Science and Art, New York, 10003

CONTENTS

I. MOTIVATION

One of the challenges of modeling combustion processes in the vapor phase is that a number of thermodynamic and kinetic parameters must be determined [1,2]. The rates of the various chemical processes that constitute the mechanism, as well as the thermodynamic properties of the species

Advances in Chemical Physics, Volume 105, Monte Carlo Methods in Chemical Physics, edited by David M. Ferguson, J. Ilja Siepmann, and Donald G. Truhlar. Series Editors I. Prigogine and Stuart A. Rice.
ISBN 0-471-19630-4

involved, must be specified or estimated over a wide range of temperatures. However, experimental data may or may not be available for all the species and processes of interest. Similar considerations can sometimes apply to the modeling of chemical manufacturing processes for design purposes [3]. Our interest lies in developing methods that can be used to predict thermodynamic properties and reaction rates as accurately as possible over a wide range of temperatures and pressures. This is the focus of this chapter.

For certain compounds in the vapor phase, we may have extensive knowledge of all thermodynamic properties as a function of temperature and pressure. As an example, chemical engineers may refer to "steam tables" for thermodynamic information about vapors at high temperatures [4–7]. However, for other, less well characterized compounds we have less experimental information and so must generally turn to theory to provide the needed data for steam tables. It is well established that for small molecules at moderate temperatures, a successful theoretical treatment must take into account (either exactly or approximately) the quantum mechanical nature of the vibrational–rotational state distribution as a function of temperature [8–12]. Thus, classical Monte Carlo calculations of the type described elsewhere in this volume will not suffice to completely solve this problem. However, the converged calculation of the vibration–rotation energy level distribution for excited rotational states ($J > 0$) is currently a state-of-the-computational-art proposition even for small molecules [13–15]. In particular, fully converged vibration–rotation eigenvalue calculations have not (to our knowledge) yet been used to calculate converged quantum free energies for molecules with more than three atoms [15]. Thus, approximate methods for thermodynamic property prediction are generally used.

There are two basic approaches for thermodynamic property prediction most commonly employed in the literature, which we coin as "spectroscopic" and "first-principles" perturbative approaches. In the spectroscopic perturbative (SP) approach one develops a model based on rigorous perturbation theory and then parameterizes the model with spectroscopic data. For example, one might treat the molecule as a rigid rotator–harmonic oscillator, and use spectroscopically measured rotation constants and harmonic oscillator frequencies to obtain the moments of inertia and vibrational frequencies needed to predict the molecular partition function [4–7]. Improvements on this basic model include the use of perturbative approaches to include anharmonicity effects, Coriolis coupling constants, and Darling–Dennison resonances, again using spectroscopic data as the input. The SP approach has proven to be quite successful for prediction of the properties of many small, rigid polyatomic molecules where the accurate determination of all the required spectroscopic constants is a practical proposition. The SP approach is, for example, the one taken in the well-known *JANAF Thermochemical Tables* [6]. However, this method cannot be used in cases where spectroscopic data are not available.

At this point the first-principles perturbative (FP) approach becomes valuable. The same kinds of perturbative models are used to describe the vibrational–rotational motions as in the SP approach. However, data from electronic structure theory computations or potential energy functions are used to parameterize the formulas instead of spectroscopically obtained data. The FP approach has for example, been pursued by Martin et al. [16–18] and by Isaacson, Truhlar, and co-workers [19–25]. This avenue is especially valuable when spectroscopic data are not available for a molecule of interest. Codes are available that can carry out vibrational perturbation theory computations, using a grid of ab initio data as input; SURVIBTM [24,26] and SPECTRO [24,27] are notable examples. Alternatively, most available electronic structure theory packages provide options to compute molecular thermodynamic properties, treating the molecular rotations as those of a classical rigid body and the vibrations as those of a separable normal-mode quantum harmonic oscillator. This ansatz is well known and is thoroughly described in several useful texts and references [4-7].

However useful these approaches have proved to be, we are convinced that there is still a need for alternatives. In particular, when a molecular system is "floppy" or undergoes significant conformational changes at the temperature of interest (the $HCN \rightleftharpoons HNC$ interconversion might be an example of this [28]), the use of perturbation theory may not be appropriate or sufficiently accurate. Moreover, the energy-level expressions generated by perturbative methods do not always converge thermodynamic properties at high temperatures. Also, as stated previously, energy-level computations are not always a practical proposition. Finally, there is some need to evaluate in detail the appropriateness of the various dynamical assumptions made in perturbation theory. For example, one may question whether the separation of rotational motions from vibrational motions is sufficiently accurate at moderate to high temperatures. At low temperatures, the methods used to treat (or neglect) anharmonicity and mode-mode coupling effects may be at issue [24]. The resolution of these technical questions is important for the further development of codes to predict absolute rate constants from first principles [29–32].

We therefore ask, given currently available computational resources, whether we can carry out direct, accurate calculations of molecular thermodynamic properties from first principles of quantum mechanics without resorting to perturbative methods. Can these calculations be carried out with accuracy that is competitive with the most accurate perturbative treatments at low temperatures while extending this accuracy into temperatures greater than 1000 K? We have found that this can be achieved by using the Fourier path-integral Monte Carlo techniques developed by Feynman [33–35] and by Doll, Freeman, Coalson, and co-workers [36–44]. The methods we have used for this purpose [25,45,46] are the main subject of this paper. Of course, path-integral calculations have also been presented for computing thermodynamic and dynamical properties of clusters [44,47–51], surface adsorbates [52], liquids [53], solids [54], biological molecules [55], and activated complexes [56–60] at low and moderate temperatures. However, our interest lies in developing accurate methods for calculating thermodynamic properties (specifically, absolute free energies) within the combustion regime [25,45,46,61]. We will not attempt to review the vast array of applications in which path-integral techniques have been found useful in other contexts; we instead refer the reader to several recent reviews [41,42,44,60,62–65].

In Section II we develop the theoretical background of the path-integral method. Again, experts in this area will find nothing new; we have developed our discussion in order to instruct those who are new to the field. Our experience is that few nonspecialists are intimately familiar with the path-integral formulation of quantum mechanics developed by Feynman. We begin with a brief discussion of the path-integral version of quantum dynamics. We then establish the path-integral representation of statistical mechanics, and develop and discuss the discretized [62] and Fourier [42,44] path representations. The reader interested in the centroid path-integral representation should consult a recent article in this journal by Voth [60]. Section III describes the path-integral Monte Carlo methods we have developed for computing molecular free energies, and reviews some applications. In Section IV we summarize the contents of this manuscript and look forward toward possible routes for future improvements of both the Monte Carlo methods and the path-integral formalism.

II. THEORETICAL BACKGROUND

A. Path Integrals in Quantum Dynamics

Some discussion of exactly what a path integral is may be helpful to those unfamiliar with the path-integral representation of quantum mechanics [34,66]. In one sentence, a path integral is a summation of a path-dependent quantity over all possible paths that could possibly connect two fixed endpoints, with each path's contribution to the sum weighted according to its relative importance. In the classical limit, only one path (the classical path) contributes to the path integral. The path-integral representation of quantum dynamics, originally conceived of and developed by Feynman

[34], actually has its formal roots in classical mechanics. Accordingly, a brief review of relevant concepts in classical mechanics follows.

1. *Classical Dynamics and Action Integrals*

Action integrals, which are at the heart of the Feynman path-integral formalism, are employed as a formal tool in classical mechanics for generating Lagrange's equations of motion. A brief review of this postulate and its application is appropriate here and allows us to introduce some notation. Of course, the reader may consult any of a number of excellent texts for more detailed treatments of this subject [67–69].

Let the vector of coordinates describing a system's position at a particular time t be labeled $\mathbf{x}(t)$. Consider an arbitrary path that connects a fixed initial place and time, $\mathbf{x}(0)$, to a particular final place and time, $\mathbf{x}'(\tau)$. We label such a path $\bar{\mathbf{x}}(t)$ (see Fig. 1). In principle, there are an infinite number of paths connecting the two endpoints $\mathbf{x}(0)$ and $\mathbf{x}'(\tau)$. When the total energy is conserved, only one of these paths corresponds to the classical–mechanical path, $\bar{\mathbf{x}}_{\mathrm{CM}}(t)$ [69]. Which one?

Hamilton's principle [68] establishes a procedure for generating the classical path. One establishes the Lagrangian function $\mathscr{L}(\mathbf{x}, \dot{\mathbf{x}})$:

$$\mathscr{L}(\mathbf{x}, \dot{\mathbf{x}}) = T - V \tag{2.1}$$

where T is the kinetic energy, V is the potential energy, and $\dot{\mathbf{x}}$ is the rate of change of $\mathbf{x}$ with respect to t. Note that if we integrate $\mathscr{L}$ along an arbitrary

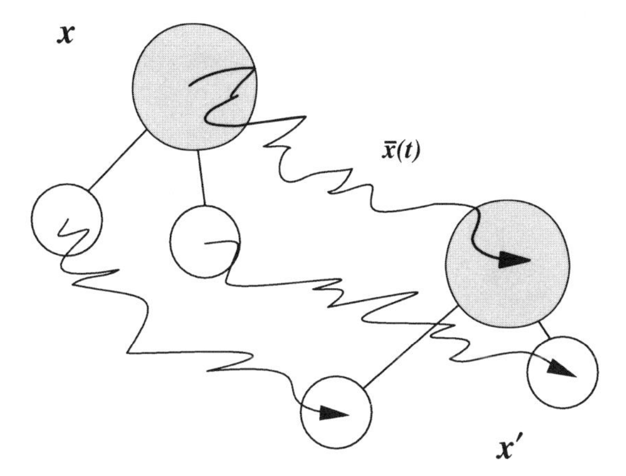

Figure 1. Illustration of a molecule moving along a path $\bar{\mathbf{x}}(t)$ which connects an initial molecular geometry $\mathbf{x}(0)$ to a final geometry $\mathbf{x}'(\tau)$. An actual path may be considerably more oscillatory than the one depicted here.

path $\bar{\mathbf{x}}(t)$ connecting the endpoints, the result will be path-dependent. Accordingly, we define the integral of $\mathscr{L}$ along a particular path to be the path's Lagrangian action:

$$S_{\mathscr{L}}(\mathbf{x} \to \mathbf{x}') = \int_0^{\tau} \mathscr{L}(\bar{\mathbf{x}}, \dot{\bar{\mathbf{x}}})\, dt \tag{2.2}$$

where the arrow in the symbol $S_{\mathscr{L}}(\mathbf{x} \to \mathbf{x}')$ is meant to denote the path $\bar{\mathbf{x}}(t)$ connecting the two endpoints. Hamilton's principle postulates that $\bar{\mathbf{x}}_{\mathrm{CM}}(t)$ is a unique path that leaves $S_{\mathscr{L}}$ stationary if the path is subjected to an infinitesimal nonclassical perturbation. Application of the calculus of variations to $S_{\mathscr{L}}$ yields the well-known Lagrangian equations of classical motion:

$$\frac{d}{dt}\left(\frac{\partial L}{\partial \dot{x}_j}\right) = \left(\frac{\partial L}{\partial x_j}\right) \tag{2.3}$$

where x_j is the jth component of $\mathbf{x}$ and $\dot{x}_j = dx_j/dt$. For an N-particle system, $\mathbf{x}$ has $3N$ components.

2. *Path-Integral Quantum Dynamics*

Path-integral statistical mechanics is obtained in strict analogy to path-integral quantum dynamics [71]. The path-integral version of quantum dynamics focuses on the fact that a quantum mechanical system is allowed to violate Hamilton's principle. For example, tunneling (a nonclassical motion) is possible within quantum mechanics. Seizing on a comment by Dirac [72] on the relationship between $S_{\mathscr{L}}$ and the quantum propagator [66], Feynman obtained a formula for the probability that a quantum system initially localized at $\mathbf{x}$ will be found later at $\mathbf{x}'$ at some later time τ. This probability must include contributions from all possible paths that could lead between the two points within the specified time interval τ, and the probability of nonclassical paths must vanish in the limit $\hbar \to 0$. Moreover, since not all paths will contribute equally, each must be appropriately weighted. On the basis of these considerations, Feynman took the probability to be given by $P = |\mathscr{K}(\mathbf{x}, \mathbf{x}')|^2$, where $\mathscr{K}(\mathbf{x}, \mathbf{x}')$ is a "kernel" of the probability defined by

$$\mathscr{K}(\mathbf{x}, \mathbf{x}') = \int_x^{x'} \mathscr{D}\bar{\mathbf{x}} \exp\left\{\frac{iS_{\mathscr{L}}(\mathbf{x} \to \mathbf{x}')}{\hbar}\right\} \tag{2.4}$$

The symbol $\int_x^{x'} \mathscr{D}\bar{\mathbf{x}}$ does not denote an ordinary (Riemann) integral, but instead indicates a path integral over all paths connecting $\mathbf{x}$ to $\mathbf{x}'$ in the time interval τ. To evaluate the path integral, imagine systematically generating all possible paths that could possibly connect the fixed space–time endpoints $\mathbf{x}(0)$ and $\mathbf{x}'(\tau)$. A distinct value of the integrand $\exp\{iS_{\mathscr{L}}/\hbar\}$ would be generated by each path. These values would then be weighted according to their relative importance and then summed up to yield the path integral. In other words, we can write $\mathscr{K}(\mathbf{x}, \mathbf{x}')$ as

$$\mathscr{K}(\mathbf{x}, \mathbf{x}') = \lim_{N_p \to \infty} \sum_{m=1}^{N_p} \frac{1}{W_m} \exp\left\{\frac{iS_{\mathscr{L}}(\mathbf{x} \xrightarrow{m} \mathbf{x}')}{\hbar}\right\} \tag{2.5}$$

The symbol $\xrightarrow{m}$ indicates the mth path in an arbitrary sequence of (distinct) paths connecting the two endpoints, W_m is the "weight" assigned to the contribution of the mth path to the path integral, and N_p is the number of paths. The most effective way to generate paths is still somewhat of an open question, but there are some formal tools that have proven useful. In particular, the "discretized path-integral" (DPI) representation described by Feynman and Hibbs [34] provides a relatively straightforward formal framework for path-generation.

Without loss of generality, we may consider the motion of a single particle along one direction $\mathbf{x}$ and eventually generalize what we learn to many dimensions. From elementary principles of quantum mechanics, we know that $\mathscr{K}(\mathbf{x}, \mathbf{x}')$ has the composition property [66]. That is, $\mathscr{K}(\mathbf{x}, \mathbf{x}')$ can be expressed as

$$\mathscr{K}(x, x') = \int_{x^2} dx^2\, \mathscr{K}(x, x^2)\mathscr{K}(x^2, x') \tag{2.6}$$

where x^2 is a space–time point at a time halfway between the times of the two endpoints (x, x'). The integral in the above expression is a Riemann integral that we evaluate over all possible intermediate space–time points. If we next insert a large number $P - 1$ of intermediate points at equal time intervals $\Delta t = \tau/P$, we have

$$\mathscr{K}(x, x') \equiv \mathscr{K}(x^1, x^{P+1}) = \int dx^2 \cdots \int dx^P \prod_{n=1}^{P} \mathscr{K}(x^n, x^{n+1}) \tag{2.7}$$

where for convenience we have defined $x \equiv x^1$ and $x' \equiv x^{P+1}$.

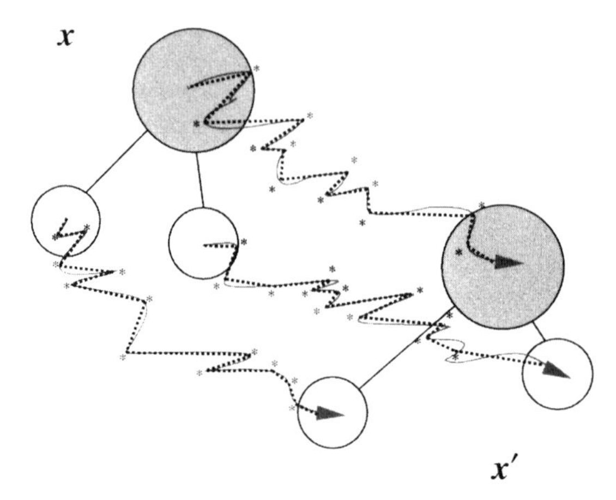

Figure 2. Illustration of a discretized representation of a path linking **x**(0) to **x**′(τ) (see Fig. 1). Straight lines (dashed) connecting 12 intermediate points (denoted *) are used to represent the path. Note that the smooth components of the path are not represented as well as are the jagged components.

Next, we note that if P is large, the distance between space–time points is very small and it is reasonable to make a straight-line approximation to the path connecting any two intermediate points (see Fig. 2). This approximation is convergent if we weight each straight line so that it contributes the proper amount to the kernel; thus we obtain

$$\mathscr{K}(x^1, x^{P+1}) = \lim_{P \to \infty} \int dx^2 \cdots \int dx^P \prod_{n=1}^{P} \frac{1}{W_n^{\mathrm{DPI}}} R(x^n, x^{n+1}) \tag{2.8}$$

where $R(x^n, x^{n+1}) = \exp\{iS_{\mathscr{L}}(x^n \xrightarrow{\mathrm{DPI}} x^{+1})/\hbar\}$ is the exponentiated action evaluated over a straight line path connecting two successive points, and W_n^{DPI} is an appropriate weight factor (yet to be determined) for this "discretized" representation of the path integral.

We note that the Lagrangian action for a particle of mass μ moving along a path connecting two neighboring points is simply

$$S_{\mathscr{L}}(x^n \to x^{n+1}) = \int_{t^n}^{t^{n+1}} \left[\frac{\mu \dot{x}^2}{2} - V(x)\right] ds \tag{2.9}$$

When the path is a straight line between two very closely spaced points, the action may be expressed as

$$S_{\mathscr{L}}\left(x^n \xrightarrow{\text{DPI}} x^{n+1}\right) = \Delta t\left\{\frac{\mu}{2}\left[\frac{x^{n+1}-x^n}{\Delta t}\right]^2 - \left[\frac{V(x^n)+V(x^{n+1})}{2}\right]\right\}$$
$$= \frac{\mu[x^{n+1}-x^n]^2}{2\,\Delta t} - [V(x^n)+V(x^{n+1})]\,\frac{\Delta t}{2} \qquad (2.10)$$

where, as noted in a very useful article by Coalson [40], a "trapezoid rule" has been used to integrate $\mathscr{L}$. Thus we have

$$R(x^n, x^{n+1}) = \exp\left\{\frac{i\mu}{2\hbar\,\Delta t}(x^{n+1}-x^n)^2 - \frac{i\,\Delta t}{2\hbar}[V(x^n)+V(x^{n+1})]\right\} \qquad (2.11)$$

To specify the weight factor, we realize that W_n^{DPI} must be independent of the specific potential V (since V is held constant over each interval). Moreover, the kernels at successive space–time points must be orthonormal to one another (otherwise the probabilities of localization at each space–time point are not mutually exclusive). These facts taken together lead to $W_n^{\text{DPI}} = \sqrt{\mu/2\pi i\hbar\,\Delta t}$ (which is independent of P because the time interval is the same between all successive space–time points) [66]. After a little algebra, we eventually obtain

$$\mathscr{K}(x, x') = \lim_{P\to\infty}\left(\frac{\mu P}{2\pi i\hbar\tau}\right)^{P/2}\int dx^2 \cdots \int dx^P$$
$$\times \exp\left[\frac{i\mu P}{2\hbar\tau}\sum_{n=1}^{P}(x^{n+1}-x^n)^2 - \frac{i\tau}{\hbar P}\sum_{n=1}^{P}V(x^n)\right] \qquad (2.12)$$

Thus, for a system of N particles moving in $3N$ degrees of freedom, if μ_j is the mass associated with the jth degree of freedom, we have

$$\mathscr{K}(\mathbf{x}, \mathbf{x}') = \lim_{P\to\infty}\left[\prod_{j=1}^{3N}\left(\frac{\mu_j P}{2\pi i\hbar\tau}\right)^{P/2}\right]\int d\mathbf{x}^2 \cdots \int d\mathbf{x}^P$$
$$\times \exp\left[\frac{iP}{2\hbar\tau}\sum_{n=1}^{P}\sum_{j=1}^{3N}\mu_j(x_j^{n+1}-x_j^n)^2 - \frac{i\tau}{\hbar P}\sum_{n=1}^{P}V(\mathbf{x}^n)\right] \qquad (2.13)$$

assuming that all particles' paths are discretized using the name number of intermediate points. By generating all possible x_j^n, we can generate all possible paths connecting the endpoints.

The formulas just developed are clearly relevant to quantum dynamics, but their relevance to the Monte Carlo computation of molecular thermodynamic properties has not yet been developed. It turns out that we can develop a theory of quantum statistical mechanics [33] that is completely analogous to the Feynman path-integral version of quantum dynamics.

B. Quantum Statistical Mechanics and Density Matrices

Here we review well-known principles of quantum statistical mechanics as necessary to develop a path-integral representation of the partition function. The equations of quantum statistical mechanics are, like so many equations, easy to write down and difficult to implement (at least, for interesting systems). Our purpose here is not to solve these equations but rather to write them down as integrals over configuration space. These integrals can be seen to have a form that is isomorphic to the discretized path-integral representation of the kernel developed in the previous section.

The simplest representations of statistical–mechanical quantities can be obtained by working in the familiar canonical ensemble, that is, systems with well-specified values of the number of atoms, the volume, and the temperature. We will assume that the system is composed of N_m noninteracting molecules, composed of distinguishable particles. This requires the use of appropriately chosen statistical correction factors, such as molecular symmetry numbers [71] and factors of $N_m!$ [8,10,11].

Systems of distinguishable particles in the canonical ensemble will admit a set of discrete quantum states, specified by a set of normalized spatial wavefunctions $\psi_{\mathbf{n}}(\mathbf{x})$. The wavefunctions satisfy the time-independent Schrödinger equation; $\hat{H}\psi_{\mathbf{n}} = E_{\mathbf{n}}\psi_{\mathbf{n}}$, with $E_{\mathbf{n}}$ an energy eigenvalue. Now $\psi_{\mathbf{n}}(\mathbf{x})$ may also happen to be an eigenstate of an operator $\hat{Y}$, whose expectation value corresponds to some other mechanical property. Then the thermodynamic average of $\hat{Y}$ is given by

$$\langle Y \rangle = \frac{1}{Q} \sum_{\mathbf{n}} Y_{\mathbf{n}} \, e^{-\beta E_{\mathbf{n}}} \tag{2.14}$$

where $Y_{\mathbf{n}}$ is the eigenvalue of $\hat{Y}$, $\beta = 1/k_B T$ with k_{B} the Boltzmann constant and T the thermodynamic temperature, and

$$Q = \sum_{\mathbf{n}} e^{-\beta E_{\mathbf{n}}} \tag{2.15}$$

is the canonical partition function, a sum over Boltzmann factors. The Helmholtz free energy A is then given by

$$A = -k_B T \ln Q \tag{2.16}$$

Assuming a perfect gas of N_m identical molecules, one can easily show [8,10,11] that the Gibbs free energy within the Born–Oppenheimer approximation is

$$G = -N_m k_B T \ln\left(\frac{q_{vr} q_t}{N_m}\right) \tag{2.17}$$

where q_{vr} is the vibrational–rotational partition function for a single molecule and q_t is the translational partition function, given by

$$q_t = \frac{\lambda^3 N_m k_B T}{P} \tag{2.18}$$

$$\lambda = \sqrt{\frac{2\pi M k_B T}{h^2}} \tag{2.19}$$

where q_t is the mass of a single molecule, and P is the pressure. Simple manipulations eventually lead to formulas for the heat capacity, entropy, and enthalpy of the vapor; thus, knowledge of q_{vr} leads to a complete determination of all the thermodynamic properties of the system. However, the preceding relations are deceptively complex. In order to directly compute q_{vr}, we would need to know the complete spectrum of eigenvalues and eigenfunctions for $\hat{H}$, a daunting task even for individual small molecules in the gas phase. For this reason, most practical work in molecular thermodynamics currently involves approximating q_{vr} according to a harmonic oscillator–rigid rotator model that is then improved by using perturbation theory with spectroscopic or ab initio data as input [6,18]. However, our goal is to directly compute q_{vr} from first principles.

We return to Q in our discussion, as all the resulting formulae can be directly applied to q_{vr} as well. As it happens, Q can be easily shown to be the trace of a Hermitian matrix [11]. Since an exponential function can be written as a power series, we have

$$e^{-\beta\hat{H}}\psi_{\mathbf{n}} = e^{-\beta E_{\mathbf{n}}}\psi_{\mathbf{n}} \tag{2.20}$$

The $\psi_{\mathbf{n}}$ are an orthonormal set, so multiplying both sides by $\psi_{\mathbf{n}}^*$ and integrating over all space yields

$$\int_{\mathbf{x}} d\mathbf{x}\, \psi_{\mathbf{n}}^* \, e^{-\beta \hat{H}} \psi_{\mathbf{n}} = e^{-\beta E_{\mathbf{n}}} \tag{2.21}$$

Summing both sides over all quantum states, we obtain

$$Q = \sum_{\mathbf{n}} \int_{\mathbf{x}} d\mathbf{x}\, \psi_{\mathbf{n}}^* \, e^{-\beta \hat{H}} \psi_{\mathbf{n}} \tag{2.22}$$

Thus Q is the trace, or sum over diagonal elements, of the matrix $\int_{\mathbf{x}} d\mathbf{x} \psi_{\mathbf{m}}^*$ $\exp[-\beta \hat{H}]\psi_{\mathbf{n}}$, which we recognize as the density matrix. It is conventional to denote the trace using the notation $Q = \mathrm{Tr}(\hat{\rho})$, with $\hat{\rho} \equiv \exp(-\beta \hat{H})$ the "density operator."

Introduction of Dirac notation [66,71] at this point helps us transform the trace of the density operator into an integral over configuration space, which ultimately gives rise to the path-integral representation. We let $|\mathbf{n}\rangle$ represent a state such that the system is "found" to have a particular set of quantum numbers $\mathbf{n}$ (it is an eigenstate of the measurement of $\mathbf{n}$); similarly, we let $|\mathbf{x}\rangle$ represent a state in which the system is surely "found" at a particular position $\mathbf{x}$. According to this picture, the wavefunction $\psi_{\mathbf{n}}(\mathbf{x})$ is a projection of the quantum number amplitude upon the position amplitude, specifically, $\psi_{\mathbf{n}}(\mathbf{x}) = \langle \mathbf{n} | \mathbf{x} \rangle$ and $\psi_{\mathbf{n}}^*(\mathbf{x}) = \langle \mathbf{x} | \mathbf{n} \rangle$. In this notation the partition function becomes

$$Q = \sum_{\mathbf{n}} \int_{x} d\mathbf{x}\, \langle \mathbf{x} | \mathbf{n} \rangle \hat{\rho} \langle \mathbf{n} | \mathbf{x} \rangle \tag{2.23}$$

Since the $|\mathbf{n}\rangle$ are an orthonormal set of discrete functions, they satisfy the completeness relation $\sum_{\mathbf{n}} |\mathbf{n}\rangle\langle\mathbf{n}| = 1$ and we have the well-known relation [33]

$$Q = \int_{\mathbf{x}} d\mathbf{x}\, \langle \mathbf{x} | \hat{\rho} | \mathbf{x} \rangle \tag{2.24}$$

where $\rho(\mathbf{x}, \mathbf{x}'; \beta) \equiv \langle \mathbf{x} | \hat{\rho} | \mathbf{x}' \rangle$ is the configuration–space representation of the density matrix. Thus, if we can integrate the density matrix over all nuclear coordinates, which is a "trace" (a sum over $\mathbf{x} = \mathbf{x}'$ elements) of $\hat{\rho}$ in configuration space, we can obtain the partition function. As we shall see, the density matrix is isomorphic to $K(\mathbf{x}, \mathbf{x}')$ and thus can be represented as a path integral.

We complete our discussion of density matrices by mentioning that the density matrix in any representation satisfies the well-known Bloch equation [11,33,66]:

$$\frac{\partial \rho}{\partial \beta} = -\hat{H}\rho \tag{2.25}$$

The significance of this is that the solutions to the Bloch equation are known for certain simple systems. For example, for a "free particle" ($V = 0$) in a one-dimensional space we find from the Bloch equation that the density matrix is [66]

$$\rho_{\text{fp}}(x, x'; \beta) = \left(\frac{\mu}{2\pi\hbar^2\beta}\right)^{1/2} \exp\left[-\frac{\mu}{2\hbar^2\beta}(x - x')^2\right] \tag{2.26}$$

which is strikingly reminiscent of the integrand in the DPI representation of the kernel in quantum dynamics. The density matrix for the quantum harmonic oscillator, with Hamiltonian $\hat{H} = (-\hbar^2/2\mu)\nabla^2 + (\mu\omega^2/2)x^2$, can also be written down in closed form [33,45].

C. Path-Integral Statistical Mechanics

In order to develop the path-integral representation of the density matrix, we use a well-established formal approach [70,73] that, incidentally, yields insight and useful algorithmic approaches to computing Q. Again, we choose to work in a one-dimensional space, which we can do without sacrificing generality. We momentarily relabel x as x^1 so that $Q = \int dx^1 \langle x^1 | \exp[-\beta\hat{H}] | x^1 \rangle$. Regardless of how we label them, any set of continuous states $| x^n \rangle$ obeys a completeness relation:

$$\int dx^n | x^n \rangle \langle x^n | = 1 \tag{2.27}$$

The density operator happens to have the useful property

$$\hat{\rho} = [e^{-\beta\hat{H}/P}]^P \tag{2.28}$$

where P is an integer [74]. For example, if $P = 2$, then

$$Q = \int dx^1 \langle x^1 | e^{-\beta\hat{H}/2} e^{-\beta\hat{H}/2} | x^1 \rangle \tag{2.29}$$

We may insert a completeness relation and obtain

$$Q = \int dx^1 \int dx^2 \, \langle x^1 | \, e^{-\beta\hat{H}/2} \, | x^2 \rangle \langle x^2 | \, e^{-\beta\hat{H}/2} \, | x^1 \rangle \tag{2.30}$$

For arbitrary P, when P completeness relations are used, we see that

$$Q = \int dx^1 \int dx^2 \cdots \int dx^P \prod_{n=1}^{P} \langle x^n | \, e^{-\beta\hat{H}/P} \, | x^{n+1} \rangle \tag{2.31}$$

where we have defined the index n to be periodic in P, that is, $x^{P+1} = x^1$.

We now write an explicit formula for the integrand. There is no unique prescription for how to do this, and significant effort has been expended on developing alternative approaches. However, we follow a relatively simple and well-traveled route that shows the clear connection to the quantum dynamical path integral. If P is sufficiently large, then the matrix elements in the above expression can be approximated by [73–76]

$$\langle x^n | \, e^{-\beta\hat{H}/P} \, | x^{n+1} \rangle \approx e^{-\beta V(x^n)/2P} \langle x^n | \, e^{-\beta\hat{T}/P} \, | x^{n+1} \rangle \, e^{-\beta V(x^{n+1})/2P} \tag{2.32}$$

Again, this expression is not unique, and alternatives continue to be explored [39,65,77–80]. An expansion of the $|x^n\rangle$ as linear combinations of momentum states, which are themselves also eigenkets of the kinetic-energy operator, leads to $Q = \lim_{P\to\infty} Q_P^{\text{DPI}}$ with [62]

$$Q_P^{\text{DPI}} = \left(\frac{\mu P}{2\pi\beta\hbar^2}\right)^{P/2} \int dx^1 \cdots \int dx^P \times \exp\left\{-\beta\left[\frac{\mu P}{2\hbar^2\beta^2} \sum_{n=1}^{P} (x^{n+1} - x^n)^2 + \frac{1}{P} \sum_{n=1}^{P} V(x^n)\right]\right\} \tag{2.33}$$

Thus, for N particles we would have

$$Q_P^{\text{DPI}} = \left[\prod_{j=1}^{3N} \left(\frac{\mu_j P}{2\pi\beta\hbar^2}\right)^{P/2}\right] \int dx^1 \cdots \int dx^P \times \exp\left\{-\beta\left[\frac{P}{2\hbar^2\beta^2} \sum_{n=1}^{P} \sum_{j=1}^{3N} \mu_j (x_j^{n+1} - x_j^n)^2 + \frac{1}{P} \sum_{n=1}^{P} V(x^n)\right]\right\} \tag{2.34}$$

Referring back to Section II.A we find that this expression is isomorphic to a DPI representation of $\mathscr{K}(\mathbf{x}, \mathbf{x})$, specifically, a kernel for paths that begin and end at the same point x over the time interval τ. Incidentally, such paths are a subset of all possible periodic orbits that could pass through a given point in configuration space (even in classical mechanics, multiple periodic orbits can cross a given configuration–space point) [81]. One is reminded of Gutzwiller's semiclassical formula for the density of states, which involves a sum over all quantizable periodic orbits [82].

By comparing analogous terms in $\mathscr{K}(\mathbf{x}, \mathbf{x})$ and Q, we see that we can think of the partition function as a path integral over periodic orbits that recur in a "complex time interval" equal to $\tau^* \equiv \beta\hbar/i = -i\beta\hbar$. There is no claim here that the closed paths used to generate Q correspond to actual quantum dynamics, but simply that there is an isomorphism. We therefore can refer to the equation above as the *discretized path-integral* (DPI) representation of the partition function. Using Feynman's notation, we have in the infinite-P limit

$$Q = \int d\mathbf{x} \int_{\mathbf{x}}^{\mathbf{x}} \mathscr{D}\bar{\mathbf{x}} \exp\left\{\frac{-iS_E}{\hbar}\right\} \tag{2.35}$$

with S_E the action of the classical energy functional $E = T(\dot{\mathbf{x}}) + V(\mathbf{x})$ along a closed "thermodynamic path" parameterized by the "thermodynamic time" s

$$S_E = \int_0^{\beta\hbar} ds\; E[\bar{\mathbf{x}}(s), \dot{\bar{\mathbf{x}}}(s)] \tag{2.36}$$

where s has been rescaled so as to vary from 0 to $i\tau = \beta\hbar$. A closed thermodynamic path is illustrated for a one-dimensional system in Fig. 3. The physical meaning of the thermodynamic path $\bar{\mathbf{x}}(s)$ may still be less than clear, since the "time" evolution along a path does not correspond to actual dynamics. However, it can actually serve as a useful starting point for improved understanding of the configuration–space representation of Q. For example, consider Eq. (2.35) as the temperature increases (i.e., as we decrease β). Since $\beta\hbar$ tends to vanish in the high-temperature limit, the paths have no time in which to fluctuate away from x and return. The paths thus coalesce onto the configuration space points when the system is hot. The contribution of configuration–space points to a quantum partition function must generally be "fuzzy," and in a semiclassical limit, the paths will be somewhat (but not completely) localized. Thus "warm" systems

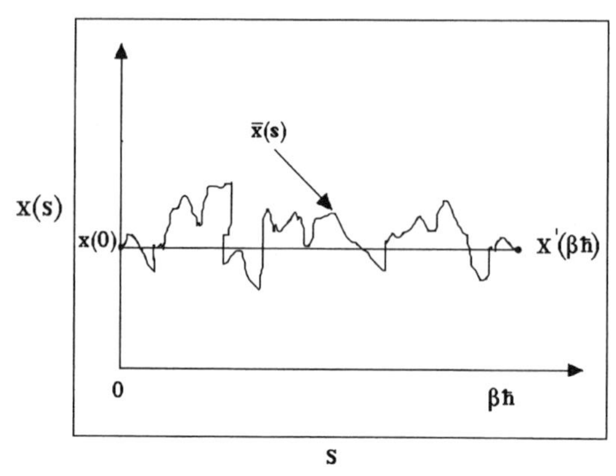

Figure 3. Illustration of a one-dimensional "thermodynamic path" $\bar{x}(s)$ that returns to its initial position after a "thermodynamic time" interval $= \beta\hbar$. This kind of thermodynamic path, which appears in the trace of the density matrix, is analogous to a classical periodic orbit.

should be well described by paths that are less diffuse, whereas the properties of "cold" systems will generally be dominated by integration paths that are long (and potentially, more complex).

The DPI representation of the partition function also exhibits an interesting and useful classical isomorphism, pointed out by Chandler and Wolynes [83]. Consider for a moment the classical–statistical mechanics of a linear, periodic chain of A atoms with identical masses m, using periodic boundary conditions to link the last atom with the first. The chain is assumed to be linked by nearest-neighbor harmonic forces characterized by a single force constant k. Furthermore, we allow the beads on each necklace to interact with an external potential field, U. The classical partition function for such a system would be

$$Q_{\mathrm{CM}} = \left(\frac{2\pi m}{\beta h^2}\right)^{A/2} \int dx_1 \cdots \int dx_{3N} \times \exp\left\{-\beta\left[\frac{k}{2}\sum_{m=1}^{A}(x_{m+1} - x_m)^2 + \sum_{m=1}^{A} U(x_m)\right]\right\} \tag{2.37}$$

This expression is clearly isomorphic to Q_P^{DPI}; $\mu P/\hbar^2\beta^2$ plays the same role (and has the same dimensions) as the force constant k, and $V(x^j)/P$ plays the same role as $U(x_m)$. In many dimensions, the quantum partition function becomes isomorphic to a solution of "necklaces," the beads of which experience an external potential created by the positions of all the other beads (effectively, a bead–bead "interaction" potential). This observation has led to the use of molecular dynamics techniques for computing quantum thermodynamic averages of many-body fluids, by integrating the equations of motion for the classical system that corresponds to the quantum system [63]. However, as pointed out by Hall and Berne [84], the increasing stiffness of the force constants with increasing P can lead to sampling problems if P is large because of the nonergodic nature of the polymer dynamics, which mostly lies on KAM surfaces [81]. Berne and Thirumalai [62] discussed the issues surrounding the lack of ergodicity in "path-integral molecular dynamics" work and suggested the use of periodic resampling of the "momentum variables" of the chain from a Maxwell distribution to ensure ergodic sampling of the "phase space." However, our focus will be on the use of Monte Carlo methods, and within a different representation of the paths.

D. Fourier Representation of the Partition Function

The DPI representation of the path integral that was developed in the preceding section is not unique. Another path-integral representation is often used that has come to be known as the Fourier representation [33,34,36–42,44,85]. Like the DPI representation, the Fourier representation transforms the path integral into an infinite-dimensional Riemann integral. In this formalism, we consider the paths to be periodic "signals" that can be represented as a Fourier series. Consider the density matrix $\rho(\mathbf{x}, \mathbf{x}'; \beta)$. Since the partition function is the trace of the density matrix, we have

$$\rho(\mathbf{x}, \mathbf{x}'; \beta) = \int_{\mathbf{x}}^{\mathbf{x}'} \mathscr{D}\bar{\mathbf{x}} \exp\left\{\frac{-iS_E}{\hbar}\right\} \tag{2.38}$$

Consider the jth component of the thermodynamic path $\bar{\mathbf{x}}(s)$, $\bar{x}_j$. We expand each component of $\bar{\mathbf{x}}(s)$ in a Fourier series, using a straight line with slope $= (x'_j - x_j)/\beta\hbar$, and intercept x_j as a "free particle" ($V = 0$) reference path for the expansion (see Fig. 4A):

$$\bar{x}_j = x_j + \frac{(x'_j - x_j)s}{\beta\hbar} + \sum_{l=1}^{\infty} a_{j,l} \sin\left(\frac{l\pi s}{\beta\hbar}\right) \tag{2.39}$$

An illustration of a path constructed by truncating this series to a finite number of terms is given in Figure 4B. The use of other reference paths besides a straight-line path (e.g., the path of a harmonic oscillator) has been explored elsewhere in the literature [85]. Alternatively, one might also consider using other signal representations to describe the paths. For example,

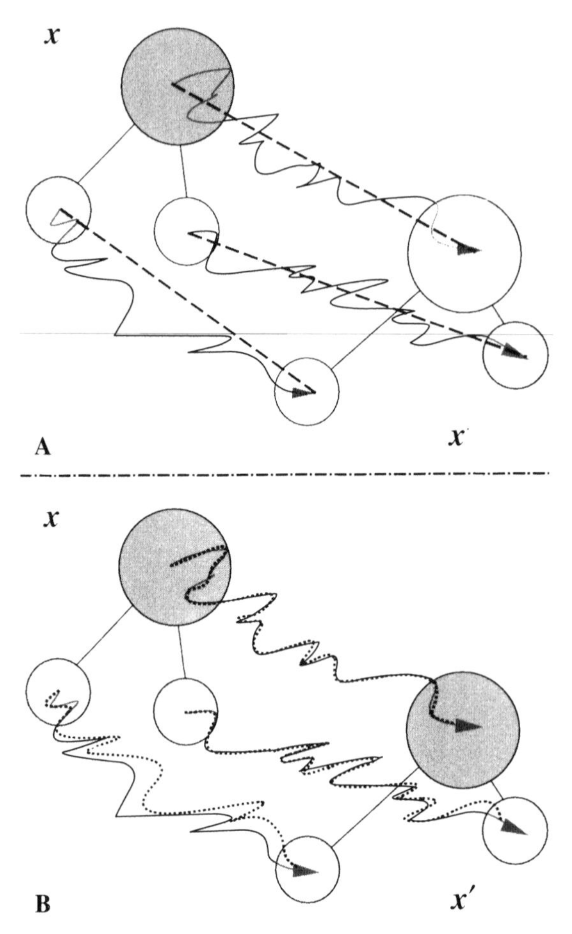

Figure 4. Schematics illustrating the Fourier representation of a path linking $\mathbf{x}(0)$ to $\mathbf{x}'(\tau)$ (see Figs. 1 and 2). (*A*) Straight lines (dashed) connecting $\mathbf{x}(0)$ and $\mathbf{x}'(\tau)$ form the reference path for the Fourier expansion. The actual path may or may not actually oscillate about a straight-line reference path. (*B*) Fourier representation of the path (dashed lines). When one or more components of the actual path do not oscillate about the straight-line reference path, more Fourier coefficients may be necessary to represent the path accurately.

we are currently exploring the possibility of using a wavelet representation [86,87].

Using the Fourier representation of the paths, we differentiate $\bar{x}_j$ and substitute the resulting "velocities" into E. We find that the action of the kinetic energy's contribution to the integrand can be calculated exactly, and we may readily transform the density matrix into

$$\rho(\mathbf{x}, \mathbf{x}'; \beta) = J(\beta) \exp\left\{-\frac{\sum_{j=1}^{3N} \mu_j(x_j - x_j')^2}{2\beta\hbar^2}\right\}\left(\prod_{j=1}^{3N}\prod_{l=1}^{\infty}\int_{-\infty}^{\infty} da_{j,l}\right)$$
$$\times \exp\left\{-\sum_{j=1}^{3N}\sum_{l=1}^{\infty}\frac{a_{j,l}^2}{2\sigma_{j,l}^2} - \frac{S_V}{\hbar}\right\} \tag{2.40}$$

Here we have defined $S_V = \int_0^{\beta\hbar} ds\ V[\bar{\mathbf{x}}(s)]$ as the action of the potential energy. The action is computed by integrating the potential energy's variations along the path generated by a particular set of values of configurations $\mathbf{x}$ and Fourier coefficients $\mathbf{a}$, all of which now play the role of independent variables. The prefactor $J(\beta)$ is the Jacobian of the transformation, given by [41,85]

$$J(\beta) = \prod_{j=1}^{3N}\left[\left(\frac{\mu_j}{2\pi\beta\hbar^2}\right)^{1/2}\prod_{l=1}^{\infty}\left(\frac{1}{\sigma_{j,l}\sqrt{2\pi}}\right)\right] \tag{2.41}$$

and $\sigma_{j,\ell}$ is an abbreviation for the following combination of parameters;

$$\sigma_{j,l} = \sqrt{\frac{2\beta\hbar^2}{\mu_j \pi^2 l^2}} \tag{2.42}$$

To obtain the partition function, we simply take the trace of the density matrix by setting $\mathbf{x} = \mathbf{x}'$ and integrating over all $\mathbf{x}$. Realizing that for practical work we will generally truncate all of the Fourier series at some maximum number of Fourier coefficients (K), we obtain the result $Q = \lim_{K\to\infty} Q^K$ such that

$$Q^K = J^K(\beta)\left(\prod_{j=1}^{3N}\int dx_j\right)\left(\prod_{j=1}^{3N}\prod_{l=1}^{K}\int_{-\infty}^{\infty} da_{j,l}\right)$$
$$\times \exp\left\{-\sum_{j=1}^{3N}\sum_{l=1}^{K}\frac{a_{j,l}^2}{2\sigma_{j,l}^2} - \frac{S_V}{\hbar}\right\} \tag{2.43}$$

where $J^K(\beta)$ is the Jacobian with all product series truncated to finite K. The paths are given by the truncated expression

$$\tilde{x}_j = x_j + \sum_{l=1}^{K} a_{j,l} \sin\left(\frac{l\hbar s}{\beta\hbar}\right) \tag{2.44}$$

where we use the symbol $\tilde{x}$ to indicate truncation of the Fourier expansion of an exact path $\bar{\mathbf{x}}$ to finite K. Thus, at a particular sampled configuration $\mathbf{x}$, one would generate all possible Fourier coefficients to generate all possible periodic paths emanating from $\mathbf{x}$ and returning in the "thermodynamic period" $= \beta\hbar$. (*Note*: The thermodynamic period is often referred to as a "complex time interval." We avoid this terminology, to emphasize that the paths considered are not dynamical paths from the propagator, but statistical–mechanical paths used to compute thermodynamic properties). The variations of the potential energy along each generated thermodynamic path would then be recorded and integrated to obtain a value of the action for that particular set of Fourier coefficients. Clearly, a well-designed Monte Carlo method can provide a useful route for generating all possible Fourier coefficients.

E. Partial Averaging of the Fourier Representation

The Fourier representation of the partition function has one limitation; it converges fairly slowly with respect to K when β is large. To improve the situation, a "partial averaging" strategy was developed by Doll, Coalson, and Freeman [38,39,41,42]. Like the basic Fourier path-integral representation, the partial-averaged (PA) Fourier path-integral representation converges to the exact result, but more rapidly as a function of K. This improvement is at the expense of a moderate increase in numerical effort; as we will see, one must calculate both V and some components of $\nabla^2 V$ along each path to compute the necessary action integrals. Partial averaging is very well described in an article by Doll et al. [42] our discussion and notation closely (but not exactly) follows theirs. Also, we present an explicit formula here for the partition function using a practical approximation to the PA representation introduced by Coalson et al. [39].

The physical reason for the success of partial averaging is that the principal contribution of the high-order Fourier coefficients, which correspond to rapidly oscillating path components, is to produce a broadened envelope of oscillation about the "core" paths generated by the low-order coefficients [39]. The effect of this envelope can be approximated by first averaging $\exp[-S_V/\hbar]$ over the high-order Fourier coefficients and then using the Gibbs inequality to establish a new expression for the density matrix. The new expression is smoothly convergent and (as a bonus) also forms a strict lower bound to $\rho(\mathbf{x}, \mathbf{x}; \beta)$, and thus the partition function.

We follow Doll et al. [42] and begin by writing a Fourier path-integral (FPI) formula for the ratio of the density matrix of the full system to that of a free-particle system (zero potential energy). The Jacobian and other prefactors cancel in the ratio, and we obtain

$$\frac{\rho(\mathbf{x}, \mathbf{x}'; \beta)}{\rho_{\mathrm{fp}}(\mathbf{x}, \mathbf{x}'; \beta)} = \frac{(\prod_{j=1}^{3N} \prod_{l=1}^{\infty} \int_{-\infty}^{\infty} da_{j,l}) \exp\{-\sum_{j=1}^{3N} \sum_{l=1}^{\infty} a_{j,l}^2/2\sigma_{j,l}^2 - S_V/\hbar\}}{(\prod_{j=1}^{3N} \prod_{l=1}^{\infty} \int_{-\infty}^{\infty} da_{j,l}) \exp\{-\sum_{j=1}^{3N} \sum_{l=1}^{\infty} a_{j,l}^2/2\sigma_{j,l}^2\}} \tag{2.45}$$

For convenience, we now change the thermodynamic time variable s to a new, dimensionless variable $u \equiv s/\beta\hbar$. This yields

$$\frac{\rho(\mathbf{x}, \mathbf{x}'; \beta)}{\rho_{\mathrm{fp}}(\mathbf{x}, \mathbf{x}'; \beta)} = \frac{(\prod_{j=1}^{3N} \prod_{l=1}^{\infty} \int_{-\infty}^{\infty} da_{j,l}) \exp\{-\sum^{3N}{}_{j=1} \sum_{l=1}^{\infty} a_{j,l}^2/2\sigma_{j,l}^2 - \beta S_{V[\bar{\mathbf{x}}(u)]}\}}{(\prod_{j=1}^{3N} \prod_{l=1}^{\infty} \int_{-\infty}^{\infty} da_{j,l}) \exp(-\sum_{j=1}^{3N} \sum_{l=1}^{\infty} a_{j,l}^2/2\sigma_{j,l}^2\}} \tag{2.46}$$

with $S_{V[\bar{\mathbf{x}}(u)]} = \int_0^1 du\, V[\bar{\mathbf{x}}(u)]$ and

$$\bar{x}_j = x_j + (x_j' - x_j)u + \sum_{l=1}^{\infty} a_{j,l} \sin(l\pi u) \tag{2.47}$$

It is worth noting at this point that expressions (2.46) and (2.47), taken together, inform us that the paths $\bar{\mathbf{x}}$ are linear combinations of Gaussian-distributed random variables $\mathbf{a}$, as pointed out by Coalson et al. [39].

Next, we divide the set of all possible Fourier coefficients $\mathbf{a}$ into two subsets; a "lower" set with $l < K$ and an "upper" set with $l > K$. Introducing a new notation, we refer to the lower set as $\lfloor \mathbf{a} \rfloor \equiv (a_{1,1} \cdots a_{3N,K})$ and to the upper set as $\lceil \mathbf{a} \rceil \equiv (a_{1,K+1} \cdots a_{3N,\infty})$. Thus we can write the paths as

$$\bar{x}_j = \tilde{x}_j + \sum_{l=K+1}^{\infty} a_{j,l} \sin(l\pi u) \tag{2.48}$$

where $\tilde{\mathbf{x}}$ is a path from $\mathbf{x}$ to $\mathbf{x}'$ including only the lower set of Fourier components. We then define the partial average of a function $g(\mathbf{x}, \mathbf{x}', \mathbf{a})$ to be its average over the upper set of Fourier coefficients;

$$\langle g \rangle_{\lceil \mathbf{a} \rceil} = \frac{\int d\lceil \mathbf{a} \rceil g(\mathbf{x}, \mathbf{x}', \mathbf{a}) \exp[-\sum_{j=1}^{3N} \sum_{l=K+1}^{\infty} a_{j,l}^2/2\sigma_{j,l}^2]}{\int d\lceil \mathbf{a} \rceil \exp[-\sum_{j=1}^{3N} \sum_{l=K+1}^{\infty} a_{j,l}^2/2\sigma_{j,l}^2]}. \tag{2.49}$$

It is important to note that $\langle g \rangle_{\lceil \mathbf{a} \rceil}$ is a function of the lower set of Fourier coefficients and the Cartesian coordinates, $\langle g \rangle_{\lceil \mathbf{a} \rceil} = \langle g \rangle \lceil \mathbf{a} \rceil\ (\mathbf{x}, \mathbf{x}', \lfloor \mathbf{a} \rfloor)$. If we let $\exp\{-\beta S_V\}$ play the role of $g(\mathbf{x}, \mathbf{x}', \mathbf{a})$, we find that we can write the density matrix ratio as

$$\frac{\rho(\mathbf{x}, \mathbf{x}'; \beta)}{\rho_{\text{fp}}(\mathbf{x}, \mathbf{x}'; \beta)} = \frac{\int_{-\infty}^{\infty} d\lfloor \mathbf{a} \rfloor \langle \exp\{-\beta S_V\} \rangle_{\lceil \mathbf{a} \rceil} \exp\{-\sum_{j=1}^{3N} \sum_{l=1}^{K} a_{j,l}^2/2\sigma_{j,l}^2\}}{\int_{-\infty}^{\infty} d\lfloor \mathbf{a} \rfloor \exp\{-\sum_{j=1}^{3N} \sum_{l=1}^{K} a_{j,l}^2/2\sigma_{j,l}^2\}} \tag{2.50}$$

The partial average of the action can be effectively evaluated by employing the Gibbs inequality [33], which tells us that the exponentiated action integral, averaged over any probability distribution function, obeys the relation

$$\left\langle \exp\left\{-\beta \int_0^1 du\ V[\bar{\mathbf{x}}(u)]\right\}\right\rangle \geq \exp\left\{-\beta \int_0^1 du \langle V[\bar{\mathbf{x}}(u)]\rangle\right\} \tag{2.51}$$

a remarkable result that is of great utility here. Substituting this result yields $\rho(\mathbf{x}, \mathbf{x}'; \beta) \geq \rho_{\text{PA}}^K(\mathbf{x}, \mathbf{x}'; \beta)$, where $\rho_{\text{PA}}^K(\mathbf{x}, \mathbf{x}'; \beta)$ is given by

$$\frac{\rho_{\text{PA}}^K(\mathbf{x}, \mathbf{x}'; \beta)}{\rho_{\text{fp}}(\mathbf{x}, \mathbf{x}'; \beta)} = \frac{\begin{array}{r}\int_{-\infty}^{\infty} d\lfloor \mathbf{a} \rfloor \exp\{-\beta \int_0^1 du \langle V[\bar{\mathbf{x}}(u)]\rangle_{\lceil \mathbf{a} \rceil}\} \\ \exp\{-\sum_{j=1}^{3N} \sum_{l=1}^{K} a_{j,l}^2/2\sigma_{j,l}^2\}\end{array}}{\int_{-\infty}^{\infty} d\lfloor \mathbf{a} \rfloor \exp\{-\sum_{j=1}^{3N} \sum_{l=1}^{K} a_{j,l}^2/2\sigma_{j,l}^2\}} \tag{2.52}$$

and $\langle V[\bar{\mathbf{x}}(u)]\rangle_{\lceil \mathbf{a} \rceil}$, which plays the role of an "effective" potential, is the partial average of the potential energy:

$$V_{\text{eff}} \equiv \langle V[\vec{\mathbf{x}}(u)]\rangle_{\lceil \mathbf{a} \rceil} = \frac{\int_{-\infty}^{\infty} d\lceil \mathbf{a} \rceil V[\bar{\mathbf{x}}(u)] \exp\{-\sum_{j=1}^{3N} \sum_{l=K+1}^{\infty} a_{j,l}^2/2\sigma_{j,l}^2\}}{\int_{-\infty}^{\infty} \mathrm{d}\lceil \mathbf{a} \rceil \exp\{-\sum_{j=1}^{3N} \sum_{l=K+1}^{\infty} a_{j,l}^2/2\sigma_{j,l}^2\}} \tag{2.53}$$

Thus we have established $\rho_{\text{PA}}^K(\mathbf{x}, \mathbf{x}'; \beta)$ as a lower bound to the true density matrix. Clearly $\rho_{\text{PA}}^K(\mathbf{x}, \mathbf{x}'; \beta)$ is an average of the exponentiated effective potential over the lower set of Fourier coefficients.

In order to write an expression for V_{eff} in closed form, one uses the fact that a linear combination of the upper set of Gaussian-distributed random variables to form the jth component of $\tilde{\mathbf{x}}$ generates a single Gaussian-distributed random process with a single width parameter (ξ_j). Then V_{eff} may be shown to be a one-dimensional (1D) Gaussian transform of the potential energy for each Cartesian degree of freedom. The details of the derivation are given by Doll et al. [42]. Letting the set of transform variables be $\mathbf{y}$, the result is

$$V_{\text{eff}}[\tilde{\mathbf{x}}(u)] = \left[\prod_{j=1}^{3N} \frac{1}{\sqrt{2\pi\xi_j^2(u)}} \int_{-\infty}^{\infty} dy_j\right] \times V(\tilde{\mathbf{x}} + \mathbf{y}) \exp\left\{-\sum_{j=1}^{3N} \frac{y_j^2}{2\xi_j^2}\right\} \tag{2.54}$$

with

$$\xi_j^2(u) = \frac{\hbar^2\beta}{\mu_j} u(1-u) - \sum_{l=1}^{K} \sigma_l^2 \sin^2(l\pi u) \tag{2.55}$$

For most systems of interest the Gaussian transforms are not analytic, and so this expression for the effective potential is inconvenient. However, a Taylor expansion about the average value of $\mathbf{y}$ yields (to terms of order ξ_j^2 and lower)

$$V_{\text{eff}} = V(\tilde{\mathbf{x}}) + \frac{1}{2} \sum_{j=1}^{3N} \left(\frac{\partial^2 V}{\partial x_j^2}\right)_{\tilde{\mathbf{x}}} \xi_j^2 + \cdots \tag{2.56}$$

If the series expansion of V_{eff} is truncated to the number of terms shown above, we have what Doll et al. [42] coined the "primitive PA" (PPA) expression for the density matrix:

$$\frac{\rho_{\text{PPA}}^K(\mathbf{x}, \mathbf{x}'; \beta)}{\rho_{\text{fp}}(\mathbf{x}, \mathbf{x}'; \beta)} = \frac{\begin{array}{c}\int_{-\infty}^{\infty} d\lfloor\mathbf{a}\rfloor \exp\{-\beta \int_0^1 du\ V_{\text{prim}}(\tilde{\mathbf{x}})\} \\ \exp\{-\sum_{j=1}^{3N} \sum_{l=1}^{K} a_{j,k}^2/2\sigma_{j,l}^2\}\end{array}}{\int_{-\infty}^{\infty} d\lfloor\mathbf{a}\rfloor \exp\{-\sum_{j=1}^{3N} \sum_{l=1}^{K} a_{j,l}^2/2\sigma_{j,l}^2\}}. \tag{2.57}$$

with

$$V_{\text{prim}} = V(\tilde{\mathbf{x}}) + \frac{1}{2} \sum_{j=1}^{3N} \left(\frac{\partial^2 V}{\partial x_j^2}\right)_{\tilde{\mathbf{x}}} \xi_j^2 \tag{2.58}$$

To obtain the partition function, we first set $\mathbf{x} = \mathbf{x}'$ and then multiply both sides by $\rho_{\mathrm{fp}}(\mathbf{x}, \mathbf{x}; \beta)$. Note that

$$\rho_{\mathrm{fp}}(\mathbf{x}, \mathbf{x}; \beta) = J(\beta) \int_{-\infty}^{\infty} d\lfloor \mathbf{a} \rfloor \exp\left\{ -\sum_{j=1}^{3K} \sum_{l=1}^{K} \frac{a_{j,l}^2}{2\sigma_{j,k}^2} \right\}$$

$$\times \int_{-\infty}^{\infty} d\lceil \mathbf{a} \rceil \exp\left\{ -\sum_{j=1}^{3N} \sum_{l=K+1}^{\infty} \frac{a_{j,l}^2}{2\sigma_{j,l}^2} \right\} \tag{2.59}$$

and thus there is some cancellation of terms between numerator and denominator. The remaining terms are integrals of Gaussian functions of the upper set, which are analytic. We thus obtain

$$\rho_{\mathrm{PPA}}^{K}(\mathbf{x}, \mathbf{x}; \beta) = J(\beta) \left[\prod_{j=1}^{3N} \prod_{l=K+1}^{\infty} (\sigma_{j,l} \sqrt{2\pi}) \right] \int_{-\infty}^{\infty} d\lfloor \mathbf{a} \rfloor$$

$$\times \exp\left\{ -\beta \int_{0}^{1} du\, V_{\mathrm{prim}}(\tilde{\mathbf{x}}) - \sum_{j=1}^{3N} \sum_{l=1}^{K} \frac{a_{j,l}^2}{2\sigma_{j,l}^2} \right\} \tag{2.60}$$

Using Eq. (2.37) for $J(\beta)$ yields

$$\rho_{\mathrm{PPA}}^{K}(\mathbf{x}, \mathbf{x}; \beta) = J^{K}(\beta) \left[\prod_{j=1}^{3N} \prod_{l=K+1}^{\infty} (\sigma_{j,l} \sqrt{2\pi}) \right] \int_{-\infty}^{\infty} d\lfloor \mathbf{a} \rfloor$$

$$\times \exp\left\{ -\beta \int_{0}^{1} du\, V_{\mathrm{prim}}(\tilde{\mathbf{x}}) - \sum_{j=1}^{3N} \sum_{l=1}^{K} \frac{a_{j,l}^2}{2\sigma_{j,l}^2} \right\} \tag{2.61}$$

Integrating over all $\mathbf{x}$, we finally obtain a PPA formula for the partition function

$$Q_{\mathrm{PPA}}^{K} = J^{K}(\beta) \int d\mathbf{x} \int_{-\infty}^{\infty} d\lfloor \mathbf{a} \rfloor$$

$$\times \exp\left\{ -\beta\, S_{V_{\mathrm{prim}}(\tilde{x})} - \sum_{j=1}^{3N} \sum_{l=1}^{K} \frac{a_{j,l}^2}{2\sigma_{j,l}^2} \right\} \tag{2.62}$$

which is exactly the same as the FPI partition function, but replacing $V(\tilde{\mathbf{x}})$ with $V_{\mathrm{prim}}(\tilde{\mathbf{x}})$. Note that when the same value of K is chosen, the paths generated in the PPA approach are exactly the same as those generated in the basic FPI approach; the difference is that the effect of higher-order coefficients on the action is taken into account in the process of partial averaging. Also note that the convergence properties of the PPA formula are quite different from those of the basic FPI formula. It has been shown that the PPA representation of the density matrix yields expressions for the average energy, heat capacity, and free energy of atomization of atomic clusters that converge much more rapidly with respect to K than do the FPI formulas [41]. Partition functions of one-dimensional anharmonic oscillators also have been observed to converge more rapidly when partial averaging is used [38,78]. Similar improvements are to be expected for molecular partition functions.

We leave the topic of partial averaging by mentioning that the PPA approximation is not a unique recipe for computing the density matrix within the PA formalism. Coalson et al. have also presented a "cumulant approach" to partial averaging which appears to have significant value in formal work [39]. However, all practical implementations of partial averaging to date appear to have made use of the PPA approximation to the effective potential. We also note that partial averaging itself has been shown to be a powerful tool in formal work. For example, partial averaging has been used in the development of an alternative to the "small-argument propagator" used in the DPI representation [39].

III. MONTE CARLO APPROACHES TO THE MOLECULAR PARTITION FUNCTION

Because of the isomorphism between the DPI representation of the partition function and the classical configuration integral of a solution of harmonic polymer loops, both Monte Carlo and molecular dynamics techniques have been used to evaluate average energies using the basic DPI representation [62,63]. However, using the Fourier representation has some advantages for path-integral computations of partition functions. This can be seen by examining the partition function of a particle confined to a large, finite domain D by an infinitely strong potential with zero potential energy within D. As conventional (but somewhat imprecisely), we refer to this as a "particle in a box" (pb). In the basic DPI representation we have

$$Q_{\text{pb}} = \lim_{P\to\infty} \left(\frac{\mu P}{2\pi\beta\hbar^2}\right)^{P/2} \int dx^1 \cdots \int dx^P \times \exp\left\{-\frac{\mu P}{2\hbar^2\beta} \sum_{n=1}^{P} (x^{n+1} - x^n)^2\right\} \tag{3.1}$$

A change of variables to "necklace normal modes" transforms the integrand into a function which is separable and integrable [63]. This is, of course, straightforward to achieve numerically for a fixed value of P. For $3N$ degrees of freedom and P "beads," a square eigenvector matrix of order $(3N + P)$ is generated which is a canonical transformation to the new normal coordinates z. The eigenvalues of this procedure are the squared necklace normal-mode frequencies. After diagonalization, we would have

$$Q_{\text{pb}} = \lim_{P \to \infty} \left(\frac{\mu P}{2\pi \beta \hbar^2} \right)^{P/2} \int dz^1 \cdots \int dz^P \exp\left\{ -\beta \sum_{n=1}^{P} (\bar{\omega}^n z^n)^2 \right\} \tag{3.2}$$

with $\bar{\omega}^n$ the frequency of the nth polymer normal mode. This change of variables can, of course be carried out for any system with nonzero potential energy as well. The potential is then evaluated in this new coordinate system by using the canonical transformation between the two coordinate systems generated by the normal-mode analysis at each sampled point to obtain the potential energy in the original Cartesian coordinate system.

The reason for using necklace normal modes is relatively simple to see. If one used a Metropolis walk to generate configurations in the old coordinates, it would be analogous to making single-particle moves in a classical Metropolis Monte Carlo simulation of a polymer. It is well known that single-particle moves in polymer systems tends to produce extremely low acceptance ratios, resulting in severe sampling problems. For this reason a number of "move strategies" are used in polymer studies. However, when polymer normal coordinates are used to make Metropolis moves, moves along the new coordinates z become "collective" rather than "local," which helps to partially overcome this problem. If P need only be moderately large to converge the system property of interest, this procedure is effective. For situations where P must be large, Sprik, Klein and Chandler have developed a "staging" Monte Carlo algorithm, wherein they build up each polymer configuration in stages [88]. They report that staging is quite effective at generating configurations for even large-P computations [89].

In the Fourier representation one need not carry out the normal-mode transformation above or resort to a staging algorithm. The partition function for the particle in a box in $3N$ dimensions is

$$Q_{\text{pb}} = J(\beta) \int_D d\mathbf{x} \int_{-\infty}^{\infty} d\mathbf{a} \exp\left\{ -\sum_{j=1}^{3N} \sum_{l=1}^{\infty} \left(\frac{a_{j,l}^2}{2\sigma_{j,l}^2} \right) \right\} \tag{3.3}$$

Clearly the integrand (which is a multidimensional Gaussian) is already separable and integrable. Moreover, the frequency spectrum of the Fourier components has been shown to be mappable to the polymer normal-mode spectrum within the DPI representation [40]. However, as previously noted, this form of the partition function also converges to the exact value quite slowly as a function of K when β is large. On the other hand, we have found that the variance of the partition function in the Fourier representation typically decreases rapidly as a function of K [25,45,46], whereas the variance of Q tends to increase with the dimensionality in discretized representations [63]. The variance controls the rate at which Monte Carlo (and MD) calculations converge; consequently, although many Fourier coefficients must be included to converge the partition function, as K increases increasingly fewer Monte Carlo samples are needed to achieve a specified precision. To a certain extent, these two properties counterbalance one another in terms of the computational effort necessary to achieve a converged calculation of Q. Moreover, if partial averaging is used the minimum value of K needed to achieve a desired level of convergence is greatly decreased. However, partial averaging also involves more work in evaluating the action at each sampled configuration. In future work we plan to extend our methods to implementation of the partial averaged Fourier representation of Q in order to compare the efficacy of the two approaches in this context. As we have shown in the previous section, such an extension is straightforward.

A. Uncorrelated Monte Carlo Sampling of Fourier Coefficients

In order to use Monte Carlo techniques to directly calculate partition functions of molecules, we cast the Fourier representation of Q into the form of an average over the Fourier coefficient variables. Our basic philosophy in previous work has been to completely avoid the use of the Metropolis algorithm [90]. There are two reasons for this approach. One reason is that in all Metropolis Monte Carlo work, one must equilibrate the system for a fairly large number of iterations before accumulating information for statistical analysis [12,90–93]. This is because the Metropolis algorithm is guaranteed to sample the target distribution only asymptotically. We reason that by using an importance sampling function that is integrable and invertible for the Fourier coefficient variables and by using adaptive importance sampling techniques (described later) for the nuclear variables, we can achieve fully converged calculations without wasting any precious configurations, or more to the point, expensive potential-energy calculations. Another reason for our approach is the knowledge that the Metropolis algorithm is a correlated random walk, with a "memory" associated with the generation of configurations [90,94–96]. This correlation between subsequent configurations results in an increase of the statistical error compared with uncorrelated sampling from the same distribution function; this increase can be substantial [45,97–99]. Thus, if we can use uncorrelated sampling from a useful probability distribution function, it will be more computationally effective than correlated sampling from the same distribution.

We let the coordinates $\mathbf{x}$ be a set of convenient Cartesian-type coordinates specifying the geometry of a single molecule (e.g., in previous work we have variously used mass-scaled Jacobi coordinates [25,100–105] and mass-scaled Cartesian coordinates [46]) with the origin at the molecular center of mass. The domain D thus confines the atoms to stay within some maximum distance from the center of mass. Truncating to finite K, we have

$$Q_{\mathrm{pb}}^{[K]} = \mathscr{V} \prod_{j=1}^{3N-3} \sqrt{\frac{\mu_j}{2\pi\beta\hbar^2}} \tag{3.4}$$

where $\mathscr{V}$ is the volume of D. If we divide this into the partition function of the molecular partition function, again truncated to finite K, the Jacobian prefactors cancel and we have

$$\frac{Q^{[K]}}{Q_{\mathrm{pb}}^{[K]}} \approx \frac{\int_D d\mathbf{x} \int_{-\infty}^{\infty} d\lfloor \mathbf{a} \rfloor \exp\{[-\sum_{j=1}^{3N-3} \sum_{l=1}^{K} a_{j,l}^2/2\sigma_{j,l}^2] - S_V(\mathbf{x}, \mathbf{a})/\hbar\}}{\int_D d\mathbf{x} \int_{-\infty}^{\infty} d\lfloor \mathbf{a} \rfloor \exp[-\sum_{j=1}^{3N-3} \sum_{l=1}^{K} a_{j,l}^2/2\sigma_{j,l}^2]} \tag{3.5}$$

In the preceding equation we have assumed that the domain D is large enough to enclose all of the most important $\mathbf{x}$ values. We now define a normalized Gaussian probability density function in Fourier space:

$$P(\mathbf{a}) = \frac{\exp\{-\sum_{j=1}^{3N-3} \sum_{l=1}^{K} a_{j,l}^2/2\sigma_{j,l}^2\}}{\int_{-\infty}^{\infty} d\lfloor \mathbf{a} \rfloor \exp\{-\sum_{j=1}^{3N-3} \sum_{l=1}^{K} a_{j,l}^2/2\sigma_{j,l}^2\}} \tag{3.6}$$

We can then evaluate the ratio of partition functions by a Monte Carlo calculation, drawing random samples uniformly from configuration space (this inefficient scheme will be improved on later) and nonuniformly in the Fourier coefficient space according to $P(\mathbf{a})$. We then simply multiply the average over many Monte Carlo samples by $Q_{\mathrm{pb}}^{[K]}$, which is an analytic function, to obtain $Q^{[K]}$. This result is then divided by σ, the symmetry number of the molecule, to finally obtain an approximation to the properly symmetrized molecular vibration-rotation partition function q. We have

$$q_{\mathrm{FPI}} = \frac{Q_{\mathrm{pb}}^{[K]}}{\sigma} \left\langle \exp\left[\frac{-S_V(\mathbf{x}, \mathbf{a})}{\hbar} \right] \right\rangle \tag{3.7}$$

where $\langle \ \rangle$ indicates an average of $\exp[-S_V(\mathbf{x}, \mathbf{a})/\hbar]$ over the nuclear and Fourier coefficient variables, drawing nuclear configurations uniformly within D and drawing Fourier coefficients from $P(\mathbf{a})$.

It also should be noted that the strategy outlined above is easily adapted should one wish to implement the PPA approach. One simply replaces the potential energy in the action by the primitive effective potential, obtaining

$$q_{\mathrm{PPA}} = \frac{Q_{\mathrm{pb}}^{[K]}}{\sigma} \left\langle \exp\left[\frac{-S_{V_{\mathrm{prim}}}(\mathbf{x}, \mathbf{a})}{\hbar} \right] \right\rangle \tag{3.8}$$

with V_{prim} replacing V in the action. Computing V_{prim} along each sampled path is, of course, more numerically demanding than computing V, but the number of Fourier coefficients needed for convergence is generally greatly reduced from those needed by the FPI prescription. All of the sampling techniques we have developed for use in the unaveraged Fourier representation may be used "out of the box" in the partial averaging representation.

Samples can be drawn from $P(\mathbf{a})$ in an uncorrelated way by a number of algorithms. For example, the Box–Mueller algorithm is a simple, effective and well-documented method for sampling numbers from a Gaussian distribution [90,106]. The real challenge remaining for practical work is devising a way to properly sample the configuration space variables while keeping the variance-reducing properties of uncorrelated Monte Carlo sampling. We will discuss a strategy we have developed for this purpose in the next subsection.

Note that in order for $S_V(\mathbf{x}, \mathbf{a})$ to be computed at a given configuration, it must be integrated along the closed path specified by an origin ($\mathbf{x}$) and a set of Fourier coefficients ($\mathbf{a}$). It has been shown that this integration must be carried out quite accurately for high-precision computations, especially when β is large. Trapezoidal rules and Gauss–Legendre quadrature schemes have previously been employed for this purpose. Doll, et al. reported a PPA-Monte Carlo study of a quartic oscillator's average potential and kinetic energies that more than 100 quadrature points were required for each integration, and that the best results were obtained using a trapezoidal rule [42]. In our studies of molecular partition functions we found that accurate results could be obtained using Gauss–Legendre integration and a similar number of quadrature points [25,46]. If n_{quad} quadrature points are required, the potential energy must be computed n_{quad} times for each path in order to obtain a single value of $S_V(\mathbf{x}, \mathbf{a})$. This should be contrasted to a classical computation of the configuration integral, in which one computes the potential-energy only once at each point in configuration space. Thus, efficient implementation of the potential energy surface is (as always) essential for practical work.

B. Adaptively Optimized Stratified Sampling of Configurations

We can markedly improve the properties of the algorithm described above by doing a better job of sampling the configuration variables **x**. Our goal is to do this in such a way as to retain the virtues of uncorrelated Monte Carlo sampling while overcoming the crudeness of uniform random sampling. Furthermore, the sampling strategy should be well suited to any particular molecular system of interest; a strategy that is effective for a relatively rigid triatomic molecule (such as H_2O) is not guaranteed to work well for a molecule that can isomerize at moderate temperatures (HCN/HNC).

With these design parameters in mind, we have developed a strategy we call "adaptively optimized stratified sampling" (AOSS) [45]. Stratified sampling is a time-honored technique for variance reduction that has enjoyed widespread use over the decades (especially by statisticians and scattering theorists), and is complementary to importance sampling techniques such as the Metropolis algorithm [90,104,107–110]. It has not however enjoyed widespread use in molecular Monte Carlo applications (with some exception [25,45,46,104]); accordingly, we provide a description of this useful method.

An excellent description of stratified sampling in the context of a statistical application is given by Bhattacharya and Johnson [109]. The goal in their example is to estimate the average cost of housing in a city, taking into account the fact that there is a cost associated with finding out the cost of each individual housing situation and that the total budget is limited. Rather than taking a uniform random sample of houses over the entire city, a stratification of the city into neighborhoods of roughly similar socioeconomic situations is made, and all of them are uniformly sampled. The individual average incomes of the neighborhoods (strata) are then used to estimate the average income of the city. For a fixed number of samples, the variance is substantially lower using stratified sampling than uniform random sampling, because the neighborhood boundaries were established so that the income has a low variance within each neighborhood. Moreover, the information gleaned about the variance within each neighborhood can be used to allocate future samples in a more optimal way, using more samples to sample neighborhoods with high variance than those with low variance.

The application of these principles to the integration of functions is well developed, although not widely used in the chemical physics community. Here we briefly describe the application of stratified sampling to the evaluation of an integral. We first develop some basic notation for Monte Carlo integration. Our goal is to evaluate the integral ϕ:

$$\phi = \int_D d\mathbf{z}\ f(\mathbf{z}) \tag{3.9}$$

where ϕ is an integral over a domain D with finite volume $\mathscr{V}$ given by

$$\mathscr{V} = \int_D d\mathbf{z} \tag{3.10}$$

The second central moment of $f(\mathbf{z})$ is relevant to the rate of convergence of a Monte Carlo computation of ϕ, and is given by

$$\mu = \mathscr{V} \int_D d\mathbf{z}\ [f(\mathbf{z}) - \Phi]^2 \tag{3.11}$$

where $\Phi = \phi/\mathscr{V}$ is the average value of $f(\mathbf{z})$ over the domain. If $(\varsigma_1, \varsigma_2, \ldots, \varsigma_n)$ is a set of n mutually independent randomly generated configurations drawn uniformly on D, an unbiased estimate of ϕ is obtained from the well-known formula [90,110]

$$\phi_n = \mathscr{V} \langle f \rangle_n \tag{3.12}$$

with $\langle f \rangle_n$ the average value of $f(\mathbf{z})$;

$$\langle f \rangle_n = \frac{1}{n} \sum_{i=1}^{n} f(\varsigma_i) \tag{3.13}$$

Naturally, $\langle f \rangle_n$ is, in turn, an approximation to Φ. This approximation generally improves with increasing n; the law of large numbers [90,107–109] assures us that this procedure asymptotically converges $\langle f \rangle_n$ [and all other central moments of $f(\mathbf{z})$]; thus we are assured that

$$\lim_{n \to \infty} \phi_n = \phi \tag{3.14}$$

Moreover, the central-limit theorem guarantees us that in the limit of "sufficiently large n" the standard error of this procedure is given by $w_n = \sqrt{m_n/n}$, with

$$m_n = \frac{\mathscr{V}^2}{n-1} \sum_{i=1}^{n} [f(\varsigma_i) - \langle f \rangle_n]^2 \tag{3.15}$$

Note that "sufficiently large" refers to the asymptotic limit in which $m_n \to \mu$. Also note that the foregoing expressions for the standard error are not valid for Markov processes used to estimate integrals (such as the Metropolis algorithm) because the ς_i would not be mutually independent random samples. The standard error of a Metropolis Monte Carlo calculation can be estimated through estimations of the covariance of the integrand [45,97–99] or through "blocking" the data into sets that are significantly longer than the correlation length (this latter strategy is often employed) [111]. The term m_n^2 is often referred to as the "variance" of ϕ_n.

The standard error obtained from the procedure described above can be diminished, for the same n, by using stratified sampling. The domain is first divided up into N_s subdomains, or "strata," according to some prescription; for example, the division could be into a cubic grid. Each subdomain is sampled uniformly with n_k samples (which will generally not be equal to one another). The integral is then given by a sum of stratum estimates [104,108]:

$$\phi_n = \sum_{k=1}^{N_s} \mathscr{V}_k \langle f \rangle_{n_k} \tag{3.16}$$

where $\langle f \rangle_{n_k}$ is the average of $f(\mathbf{z})$ within the kth stratum and $\mathscr{V}_k$ is the stratum volume. Similarly, the standard error is a sum of stratum standard errors, which after some simplification yields [104,108]

$$w_n^2 = \sum_{k=1}^{N_s} \frac{\mathscr{V}_k^2}{n_k - 1} [\langle f^2 \rangle_{n_k} - \langle f \rangle_{n_k}^2] \tag{3.17}$$

We see that, in contrast to ordinary uniform random sampling, we now have a number of parameters to choose and optimize. First and foremost is the number and placement of the stratum boundaries; second is the choice of how many samples to take within each fixed stratum. Although there is no general solution to the first problem, the second problem (choosing the set of n_k that minimizes w_n^2) has been solved [90,104,107–110]. One simply uses the well-known method of Lagrange multipliers in the asymptotic limit (large n_k). The constraints that characterize the optimization are (1) $\sum_{k=1}^{N_s} n_k = n$ and (2) $\sum_{k=1}^{N_s} \mathscr{V}_k = \mathscr{V}$. One finds that n_k*, the set of optimum n_k, are those that obey

$$n_k^* = n\sqrt{\frac{\mu_k}{\Gamma}} \tag{3.18}$$

where μ_k is the second moment of $f(\mathbf{z})$ within the kth stratum and

$$\Gamma = \left(\sum_{k=1}^{N_s} \sqrt{\mu_k}\right)^2 \tag{3.19}$$

We see that n_k^* corresponds to placing the largest fraction of samples where the fluctuations of $f(\mathbf{z})$ about its stratum average are largest. In Monte Carlo work we estimate the μ_k by computing m_k :

$$m_k = \frac{\mathscr{V}_k^2}{n_k - 1} \sum_{i_k=1}^{n_k} [f(\varsigma_{i_k}) - \langle f \rangle_{n_k}]^2 \approx \mu_k \tag{3.20}$$

Of course, accurately computing the m_k would necessitate an accurate Monte Carlo calculation within each stratum. However, if one can estimate the m_k from small preliminary samples within all of the strata well enough to determine how many additional samples (if any) should be generated and in which strata, it is possible to use the m_k in practical work.

The preceding discussion assumed that we knew an optimal way to stratify the domain. However, for many-dimensional functions such determinations can be problematic. Our adaptively optimized version of stratified sampling uses information gained from Monte Carlo sampling about the variance of different subdomains within a molecule's configuration space to divide it up it in an optimal way and then sample it according to the formula derived above for n_k^*. We do not know a priori where the stratum boundaries should be initially placed, what shape they should have, or how many boundaries should be placed. Our situation is like that of a statistician from New York sent to make an housing cost survey of the city of Denver but given no information about the neighborhoods making up the city, and provided with only a limited budget. The statistician might choose a rectangular grid of Denver as a starting point (with streets as stratum boundaries), gather information within each grid, and then use the information to set up a new, better grid. This new grid would then be sampled as best as could be done with the remaining budget. In a perfect world one would naturally like to use the data generated from the first "probe" sample as part of the final set of statistics, since each sample is expensive. The initial probe sample must therefore be carried out in such a way as to yield the information needed about the variance in each grid without skewing the final statistics.

The example discussed in the preceding paragraph conveys the essence of our adaptively optimized stratified sampling strategy. Our goal is to compute the partition function of the molecule as accurately as possible given our fixed computational "budget." As noted above, our budget is generally determined by the number of times we must compute the potential energy. This number of times is, in turn, determined by n, the total number of configurations generated. We will achieve this by designing an algorithm that adaptively samples a stratification of the molecular configuration space, changing its tactics as necessary to approach an optimum sampling strategy. Other stratified sampling algorithms with a similar (but not identical) philosophy have been developed [110,112].

First we must establish the shape of D. Since the potential energy is only a function of the distance of the two particles from the center of mass, a spherical domain centered about the center of mass seems a reasonable choice. Of course, for larger molecules of increasing complexity this particular boundary shape will likely not be useful, and a different stratification geometry would need to be established. We have discussed elsewhere the logistics of uniformly sampling the interior of spherical domains with random numbers [25,45]. Notably, although there are direct methods for sampling the interior of spheres in two and three dimensions, we are restricted to the use of the rejection method for sampling the interior of a many-dimensional sphere (a "hypersphere") with fixed radius R. In the rejection method, a random sample is drawn from a uniform distribution within a hypercube just enclosing the hypersphere. The squared distance of the sample from the center of the sphere is computed. The sample is then rejected if its squared distance is greater than R^2.

The scaling of the rejection method with respect to the number of molecular degrees of freedom is relevant. This can be assessed by a simple argument, which follows. We let $N_{\text{vr}} \equiv 3N - 3$ be the number of vibrational–rotational degrees of freedom of the molecule. The volume of an N_{vr}-dimensional hypersphere with radius R is [11,113,114]

$$\mathscr{V}_{\text{sphere}}^{[N_{\text{vr}}]} = \frac{\pi^{N_{\text{vr}}/2} R^{N_{\text{vr}}}}{\Gamma(N_{\text{vr}}/2 + 1)} \tag{3.21}$$

where $\Gamma\{j\}$ is the gamma function. If we carry out a large number of Monte Carlo samples of the interior of the hypercube, the fraction of samples which happen to lie within the hypersphere will approach the volume ratio:

$$\frac{\mathscr{V}_{\text{sphere}}^{[N_{\text{vr}}]}}{\mathscr{V}_{\text{cube}}^{[N_{\text{vr}}]}} = \frac{\pi^{N_{\text{vr}}/2}}{2^{N_{\text{vr}}}\Gamma(N_{\text{vr}}/2 + 1)} \tag{3.22}$$

This ratio, and therefore the fraction of random numbers actually used to generate information in the Monte Carlo integration, decreases rapidly as a function of N_{vr}. Thus, when N_{vr} is large, one rejects many samples before finally accepting one for use in Monte Carlo integration. Generally, generating random numbers is much faster than computing the action of the potential energy of a molecule at a particular configuration (recall that the action will have to be computed n_{quad} times). Thus, if the integrand has a large degree of spherical symmetry, many contributions will be eliminated that would not contribute much to the final result. This can actually improve the chances for a successful integration. However, if N_{vr} is very large, the use of a spherical boundary might prove to be prohibitive if the rejection rate became too high. In this context it is important to use high-quality random-number generators, preferably with long periods, and to restart the random-number sequences with new seeds when the number of random numbers generated from a particular seed becomes a significant fraction of its period. Of course, such considerations apply to all Monte Carlo work.

Given our choice of a spherically shaped domain, we next must establish its size. This is one point where our molecular case varies from that of the statistician's problem, since the city limits of Denver are fixed and well known. The size of D will optimally be large enough to enclose all the configurations that provide a sizable contribution to the integrand, but not so large as to waste samples in regions of space so far away from the center of mass that they contribute little. In the spirit of trying to make as much use as possible of samples used in the initial stages of the calculation in the final statistics, we have in the past carried out this task in what we referred to as "stage 1" of the three-stage AOSS algorithm [45]. In this algorithm, T concentric "test spheres," each of which has a radius R_t [$R_t = (R_1, R_2 \ldots R_T)$ with $R_t > R_{t+1}$], were established. These radii were chosen so as to be equally spaced. The largest radius (R_1) was somewhat arbitrarily chosen so as to be "too large" on the basis of known distances of the nuclei of the molecule of interest from the center of mass in the equilibrium geometry. Then, starting with the largest radius R_1, we uniformly sampled the interior of each successively smaller sphere. The exponentiated action was generated at each sampled configuration for Monte Carlo averaging. A fixed number of samples p was used in the first test, and an approximate value of the molecular partition function, q_1, was generated from the Monte Carlo average. A fraction of those samples happened to lie within R_1; those samples were used, together with p additional samples, to generate a second value of the partition function q_2, and so forth. When the hyperradius became too small to enclose all of the important region, the q_t rapidly decreased. The algorithm responded by stepping back out to a value of $R_t \equiv R$ that was sufficiently large for the q_t to remain stationary within a specified tolerance. This method has the advantage that all samples generated with $R_t < R$ can be used in the final determination of the partition function. We call the hypersphere with hyperradius R the "great sphere."

The preceding algorithm is effective, but requires the user to specify a tolerance parameter that depends not only on the particular molecular system of interest but also on the values chosen for T, R_1, and R_T. In practice the tolerance parameter is obtained by trial and error. In addition, if one chooses a value for R_1 that is too small, the method will fail. Accordingly, we have recently pursued a second strategy that is simpler and equally effective in terms of numerical effort. We have observed that the optimum hyperradius obtained by the above procedure is generally numerically independent of the number of Fourier coefficients used in the calculation. Moreover, the action integrals can be evaluated rather crudely, with a small number of quadrature points, without affecting the optimized hyperradius. Finally, the optimum hyperradius is empirically observed to be a rather weak function of temperature. Thus, we can simply carry out a series of unconverged ($K = 1$) uniform-sampling partition function calculations at a high temperature, using each of the R_t, and then plot the results to visually determine the optimum hyperradius. This obviates the specification of a tolerance parameter and provides a visual check on whether an appropriate value of R_1 has been chosen. Representative calculations of this type for H_2S and D_2S at a particular temperature are shown in Figure 5. We see in both cases that as expected, q_{vr} rises to a characteristic plateau value.

Having determined R, we now proceed to stage 2 of the AOSS algorithm. In stage 2, we determine the best way to divide up the molecular configuration space into three concentric strata. A justification for using three strata is that we can identify three physically distinct regions of variation of the potential energy of a diatomic molecule as a function of distance from the center of mass; the short-range repulsive region, the intermediate-ranged covalent region, and the longer-ranged van der Waals region. One could certainly generalize the procedure to four or more strata, which would result in a more flexible algorithm; however, the bookkeeping required to adaptively construct the stratum boundaries becomes considerably more complex as the number of strata increases. Our choice of using three strata is a compromise between these considerations.

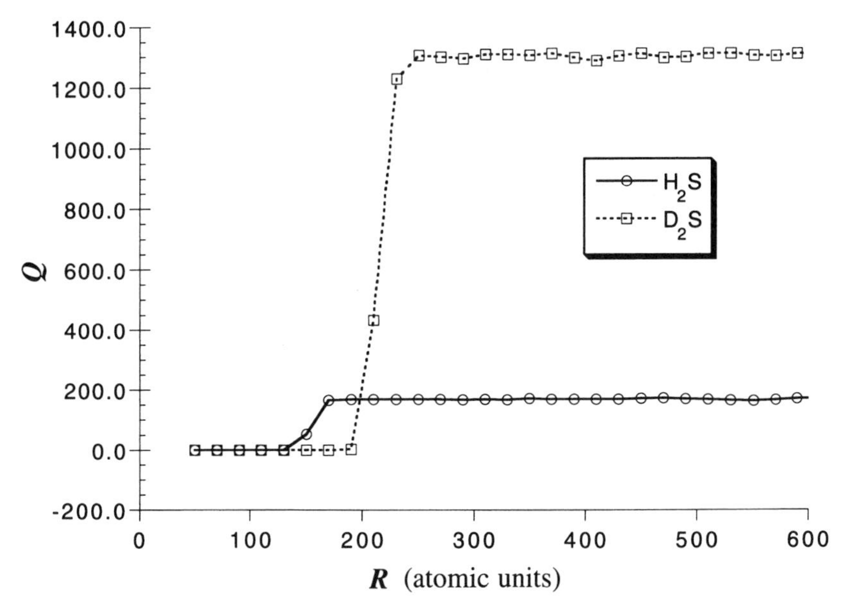

Figure 5. Plot of the partition function Q as a function of the hyperradius R of the great sphere enclosing the molecules H_2O and D_2O. A mass-weighted Jacobi coordinate system is used in each case. One Fourier coefficient is used per degree of freedom, and 10^5 Monte Carlo samples were used for each calculation. The Monte Carlo samples were drawn uniformly on the interior of the great sphere in coordinate space and from $P(\mathbf{a})$ in Fourier coefficient space.

To determine the boundaries of the three strata, the great sphere is divided up into B concentric hyperspheres with hyperradii $(r_1, r_2 \dots r_T)$ with $r_b < r_{b+1}$ and $r_b < R$. The hyperradii r_b are chosen such that they divide up the space within the great sphere into "bins" that all have equal hypervolumes; thus, each set of adjacent hyperspheres encloses a space with volume $\mathscr{V}_b^{[N_{\rm vr}]} = \mathscr{V}^{[N_{\rm vr}]}/B$. This ensures that a random sample of the interior of the great sphere will generate approximately the same number of "hits" within each bin, which, in turn, ensures that we will gain a useful amount of information about the function's variation within each bin. Note that the hyperradius of each sampled point uniquely specifies which bin it falls into.

Next, a "probe sample" of $n_{\rm pr}$ configurations is randomly generated on the interior of the great sphere and assigned to bins. This probe sample is a fraction of our total budget of samples, n. If the original version of the AOSS algorithm is used in stage 1, the m samples used to determine R can also be used as part of the probe sample to make stage 2 more affordable. Having probed the space, we then generate all possible combinations of the bins such that three concentric and neighboring strata are generated by each combination. We refer to a particular combination of bins as a "trial stratification." The number of possible trial stratifications is given by $(B-2)(B-1)/2$; if $B=100$, then 4851 stratifications are possible. This number is large but not prohibitively so. Each stratification is evaluated by computing m_k within each stratum; we choose the "best stratification" as the one wherein each stratum contributes approximately the same amount to the error as any other stratum. For a particular stratification, we evaluate

$$\mathrm{var}(m_k) = \frac{1}{2} \sum_{k=1}^{3} [m_k - \langle m_k \rangle]^2 \tag{3.23}$$

with

$$\langle m_k \rangle = \frac{1}{3} \sum_{k=1}^{3} m_k \tag{3.24}$$

and choose the stratification that yields the minimum value of $\mathrm{var}(m_k)$. We now have enough information to make an initial guess at the optimum number of samples to generate within each stratum. Letting the fraction of optimum samples be $\chi_k \equiv n_k^*/n$, we have an initial guess given by $\chi_k^{(1)} = \sqrt{m_k/\Gamma}$. Since the strata were chosen so that they contribute similarly to the error, all three $\chi_k^{(1)}$ will optimally be the same as one another (~ 0.33). Whether or not this occurs can be monitored as a measure of the success of stage 2 (and stage 1 as well; if the hyperradius of the great sphere is excessively large, then stage 2 optimization may fail).

Having fixed our stratification boundaries, we now proceed to stage 3 of the algorithm. In this final stage we use our estimate of the χ_k to allocate the remainder of our "budget" of n samples to the three strata in a near-optimum way. However, if the $\chi_k^{(1)}$ are not completely accurate, it might be a mistake to commit to them completely at this point in a calculation. Fortunately, there is little computational overhead involved in improving our estimates of the χ_k "on the fly" and adjusting the sampling frequencies accordingly. To achieve this, we divide the remaining $n_r = n - n_{pr}$ samples into r rounds of $n_s = n_r/r$ samples each. In the first round we use the $\chi_k^{(1)}$ from stage 2 to estimate what the relative number of samples should be within each stratum at the end of the round with the formula

$$n_k^{(1)} = \chi_k^{(1)}(n_{pr} + n_s) \tag{3.25}$$

Let the total number of samples accumulated within the kth stratum in stages 1 and 2 be ν_k. Then

$$N_k^{(1)} = n_k^{(1)} - \nu_k \tag{3.26}$$

is the number of samples generated during the first round (unless $N_k^{(1)} \leq 0$, in which case no samples are generated). After the strata have received $N_k^{(1)}$ additional samples, each stratum's m_k is recomputed using its accumulated samples. The new m_k are used to form $\chi_k^{(2)}$, a new estimate of χ_k but with improved statistics. Then $\chi_k^{(2)}$ is used to estimate $n_k^{(2)}$ with the formula

$$n_k^{(2)} = \chi_k^{(2)}(n_{\text{pr}} + 2n_s) \tag{3.27}$$

and we accumulate $N_k^{(2)} = n_k^{(2)} - N_k^{(1)}$ samples. This process is iterated until either a total of n samples are generated or a specified statistical error is achieved. This completes our description of the AOSS method.

C. Applications

Converged vibrational–rotational (vib–rot) partition functions have been presented for HCl, H_2O, D_2O, H_2S, and H_2Se using the FPI representation and the computational methods we have described here, that is, uncorrelated sampling from $P(\mathbf{a})$ in the Fourier coefficient space and AOSS in the configuration space [25,46]. We have referred to this combination of techniques as the "AOSS-U method" [45]. In the case of HCl [46], excellent agreement was found between partition functions obtained from AOSS-U calculations of the partition function and those obtained from summing Boltzmann factors obtained from a converged Rayleigh–Ritz variational basis-set computation of the vib–rot states [115]. In both calculations the potential-energy surface developed by Hulbert and Hirschfelder was employed [116,117]. The results are summarized in Table I. Calculations of the vib–rot free energy G_{vr} are also presented using the formula

$$G_{\text{vr}} = -RT \ln q_{\text{vr}} \tag{3.28}$$

In all cases the partition functions and free energies determined by the two methods are in close agreement with one another. We note that for diatomic molecules the two methods are complementary in terms of computational efficiency; variational calculations outperform AOSS-U at low temperatures whereas AOSS-U becomes more effective than the variational method at high temperatures. This is because the number of basis functions required to variationally converge the energies of the excited vib–rot states

TABLE I
Calculated Vibrational–Rotational Partition Functions and Free-Energy Contributions of HCl as a Function of Temperature[a]

T(K)	q(var)[b]	q(var)[c]	G(var) (kJ/mol)[e]	G_{vr}(AOSS-U) (kJ/mol)[e]
300	0.01651	0.0167 ± 0.0008[d]	10.24	10.21 ± 0.1
600	1.153	1.17 ± 0.02	−0.7102	-0.783 ± 0.09
900	5.713	5.76 ± 0.07	−13.04	-13.102 ± 0.09
1200	14.11	14.2 ± 0.2	−26.41	-26.47 ± 0.1
2000	53.55	53.6 ± 0.5	−66.19	-66.21 ± 0.2
4000	257.5	257 ± 2	−184.6	-184.55 ± 0.3
5000	421.0	420 ± 3	−251.2	-251.11 ± 0.3

[a] All calculations use the potential-energy surface due to Hulbert and Hirschfelder [116,117] as input. The equilibrium geometry of the molecule defines the zero of energy.

[b] Vibration–rotation partition function for HCl obtained via standard Rayleigh–Ritz variational (var) basis-set methods from Topper et al. [46].

[c] Vibration–rotation partition function for HCl obtained via Fourier path-integral AOSS-U Monte Carlo calculations from Topper et al. [46]. Error bars are given at 95% confidence level ($2w_n$). Unless otherwise noted, all calculations used $K = 128$ Fourier coefficients per degree of freedom and $n = 100\,000$ Monte Carlo samples.

[d] Same as footnote c, except $K = 1024$.

[e] Vibrational–rotational contribution to Gibbs free energy for HCl from variational and AOSS-U calculations.

becomes large, whereas the number of Fourier coefficients needed for convergence decreases as T increases (the paths become more free-particle-like). Conversely, a small basis set suffices to variationally converge enough states to converge the partition function at low temperature, but many Fourier coefficients must be used. This situation would presumably be greatly improved by using partial averaging, as many fewer Fourier coefficients would need to be included to converge the partition function.

A method for computing free-energy changes, presented by Reinhardt and co-workers [118,119], was recently adapted to quantum path integrals and applied to HCl at 300 K [61], using the same potential-energy surface as in our previous work. Their method uses a variational method (not the Rayleigh–Ritz variational method described above) to estimate lower and upper bounds for the free-energy change brought about by switching the Hamiltonian from one form to another. Hogenson and Reinhardt [61] calculated the change in quantum free energy brought about by switching the Hamiltonian from that of the HCl molecule to that of the same molecule confined to the interior of a sphere but with zero potential energy. Using the discretized path-integral representation, they found it necessary to use 512 discretization points per degree of freedom to achieve acceptable convergence levels. They employed the multigrid algorithm due to Jahnke and Sauer [120] and Loh [121] to overcome the Monte Carlo sampling problems commensurate with the use of this extremely large value of P. We thus refer to their combination of techniques as the DPI multigrid variational (MV) approach. In Table II we compare their results with our own. Note that the MV calculations are in very good agreement with our variational and AOSS-U calcuations, which bodes well for their approach as an alternative route to the free energy. The AOSS-U calculations establish a tighter confidence interval than do the MV calculations; however, it is really impossible to fairly compare the two algorithms to one another without testing them on the basis of the total computational effort required for a given level of convergence. We have no sense as to which method is more efficient in that regard. However, note that the AOSS-U method does not require an equilibration period; all Monte Carlo samples generated are thus used to form the final calculation of the partition function. We believe that this fact alone will tend to make the efficiency of the AOSS-U approach competitive with methods that rely on a Markov process to generate the configurations for Monte Carlo integration. In addition, the serial correlations of a Markov process tend to increase its variance compared to independent random sampling [45] (all things being equal, which they are admittedly not in this case). In the end, we are not completely "sold" that one method is really superior to the other for this particular application. The MV method has a great deal of potential and should be further explored in this context. A number of other approaches, such as the state integration method employed by Freeman and Doll [41], also hold promise in this regard. The state integration method, in conjunction with partial

TABLE II
Calculated Vibrational–Rotational Partition Functions and Free Energies for HCl at 300 K[a]

Calculation Type	q_{vr} [b]	G_{vr} (kJ/mol)[c]
Rayleigh–Ritz variational,[d]	0.01651	10.24
FPI AOSS-U[e]	0.0167 ± 0.0008	10.2 ± 0.1
DPI MV (lower bound)[f]	0.0156 ± 0.0016	10.4 ± 0.2
DPI MV (upper bound)[g]	0.0163 ± 0.0016	10.3 ± 0.2

[a] See footnote *a* of Table I.
[b] Vibration–rotation partition function for HCl.
[c] See footnote *e* of Table I.
[d] See footnote *b* of Table I.
[e] See footnotes *c* and *d* of Table I.
[f] Calculations using the discretized path-integral multigrid variational (DPI/MV) method from Hogensen and Reinhardt [61]. In order to facilitate comparison between methods, all error bars are given at the 95% confidence level ($2w_n$).

averaging techniques, has been used to compute quantum free energies of 20-atom argon clusters at 10 K [41].

The use of basis-set methods (such as the Rayleigh–Ritz variational method) for accurate computations of the vib–rot energies of triatomic molecules is much more demanding than for diatomic molecules, although converged calculations of highly excited vib–rot states of triatomics have been achieved [13–15]. However, fully converged computations of the energies of the manifold of vib–rot excited states for triatomic molecules are still the computational state of the art. For this reason, a number of approximate methods have been used for estimating vib–rot partition functions of polyatomic molecules [16–25]. The present method gives us a tool for assessing the reliability of the various approximate methods that have been used for this purpose.

In a previous study of several dihydrogen chalcogens, we compared AOSS-U computations of the free energy over a range of temperatures (600–4000 K) to various approximate treatments and to data from the *JANAF Thermochemical Tables* [6]. Our computations employed global potential-energy surfaces developed by Kauppi and Halonen [122]. Most of the approximate treatments involved the complete separation of vibrational motions from rotations and the treatment of the rotations as those of a classical rigid rotator. The vibrational motions were variously treated with a harmonic oscillator–normal-mode model (HO), with second-order vibrational perturbation theory, and with an approximate, "simple" perturbation theory treatment developed by Truhlar and Isaacson [23], all using the Kauppi–Halonen potential-energy surface as input. Comparison was also made to "first-principles perturbative" calculations by Martin, François, and Gijbels (MFG) [18]. They employed ab initio methods to generate a quartic vibrational force field. This was then used as input for a perturbation-theory calculation of the vibrational partition function. An approximate, nonseparable quantum mechanical formula was used to treat the rotational motions, including the effects of centrifugal distortion. We also compared our AOSS-U computations to "spectroscopic perturbative" calculations by Gurvich, Veyts, and Alcock (GVA) [7]. GVA employed higher-order vibrational perturbation theory, allowing coupling between vibrational and rotational degrees of freedom. To treat the rotational motions, GVA used a rigid asymmetric top expression with the Stripp–Kirkwood correction and included corrections for centrifugal distortion and Darling–Dennison resonance. The data in the *JANAF Thermochemical Tables* is obtained from a similar procedure. Some of these results are summarized in Table III; we refer the interested reader to the original paper for more detailed tables [25]. In order to compare our results with the standard treatments, we followed the *JANAF* format and tabulated gef(T),

TABLE III
Quantum Gibbs Free Energies of H_2O, D_2O, H_2S, and H_2Se as a Function of Temperature

	T (K)	gef(T) FPI (J mol^{-1} K^{-1})[a]	gef(T) MFG (J mol^{-1} K^{-1})[b]	gef(T) JANAF (J mol^{-1} K^{-1})[c]	gef(T) GVA (J mol^{-1} K^{-1})[d]	gef(T) Mills (J mol^{-1} K^{-1})[e]	gef(T) HO (J mol^{-1} K^{-1})[f]
H_2O	600	179.4 ± 0.8	179.1	179.0	179.0		178.7
	1000	196.9 ± 0.8	196.9	196.8	196.8		196.5
	1500	211.5 ± 0.5	212.0	212.0	212.0		211.4
	2400	231.2 ± 0.3	231.1	231.2	231.2		230.3
	4000	254.7 ± 0.3		254.9	254.9		253.2
D_2O	600	188.9 ± 0.4		188.7	188.7		188.5
	1000	207.0 ± 0.6		207.2	207.2		206.9
	1500	222.7 ± 0.3		223.2	223.3		222.7
	2400	243.7 ± 0.3		243.8	243.9		243.0
	4000	268.8 ± 0.2		268.6	268.7		267.3
H_2S	600	196.2 ± 0.3		196.1	196.2	196.1	195.9
	1000	215.2 ± 0.6		214.6	214.7	214.8	214.3
	1500	231.0 ± 0.4		230.7	230.9	230.9	230.3
	2400	251.2 ± 0.3		251.4	251.6		250.6
	4000	276.7 ± 0.2		276.3	276.7		274.9
H_2Se	600	209.8 ± 0.3				209.3	209.6
	1000	228.6 ± 0.4				228.3	228.4
	1500	245.6 ± 0.2				244.6	244.7
	2400	266.6 ± 0.3					265.6
	4000	292.4 ± 0.2					290.4

[a] AOSS-U FPI calculation of gef(T) from Topper et al. [25] using the Kauppi–Halonen [122] potential-energy surfaces as input.
[b] First-principles perturbative calculation of gef(T) from Martin et al. [18].
[c] Spectroscopic perturbative calculation of gef(T) from Chase et al. [6].
[d] Spectroscopic perturbative calculation of gef(T) from Gurvich et al. [7].
[e] Estimated values of gef(T) from Mills [4].
[f] Classical rigid rotator–quantum harmonic oscillator calculation of get(T) from Topper et al. [25] using the Kauppi–Halonen [122] potential-energy surfaces as input.

defined by

$$\mathrm{gef}(T) = \frac{-(G - E^G)}{T} \tag{3.29}$$

where E^G is the zero-point energy of the molecule. This establishes a different energy reference for each molecule, i.e., its enthalpy at 0 K. Since E^G cannot be measured experimentally, it is routinely calculated using vibrational perturbation theory, an approach we have also followed for the purposes of comparison (except in the case of the rigid rotator–harmonic oscillator model). It is important to emphasize that G is necessary for predicting equilibrium constants, and that the estimation of E^G is not necessary for practical applications of the AOSS-U method.

For a perfect gas, gef(T) is separable into translational and vib–rot contributions. Thus we obtain

$$\mathrm{gef}(T) = R \ln q_{\mathrm{trans}} + R \ln \tilde{q}_{\mathrm{vr}} \tag{3.30}$$

where $\tilde{q}_{\mathrm{trans}}$ is the translational partition function of the molecular center of mass and $\tilde{q}_{\mathrm{vr}}$ is the vib–rot partition function with the zero of energy at the molecular ground-state energy level E^G:

$$\tilde{q}_{\mathrm{vr}} = q_{\mathrm{vr}} \exp(\beta E^G) \tag{3.31}$$

Not surprisingly, in the case of the best-studied molecule (H_2O) we found that all methods were in substantial agreement with one another for temperatures ranging between 600 and 1500 K. Calculations at 2400 K began to show substantial deviation between the HO model and the more rigorous results. At 4000 K there was excellent agreement between the AOSS-U, GVA, and *JANAF* calculations (Martin et al. did not present calculations at this ultrahigh temperature). Similar conclusions applied to D_2O and H_2S. However, the thermodynamic and spectroscopic data available for comparison to our H_2Se computations are very limited; the data available are based on estimates of the molecular heat capacity and entropy [4]. Our free-energy values values for H_2Se seem to be (to the best of our knowledge) the most accurate currently available in the literature. A challenge for future work includes further decreasing the variance of the Monte Carlo calculations, so as to obtain even more precision in the calculations.

Encouraged by these results, we next present a demonstration of the calculation of the vib–rot contribution to the entropy for various chalcogen dihydrides using the AOSS-U method. We achieve this by generating Gibbs free-energy data over a range of temperatures on a fine grid and fitting the

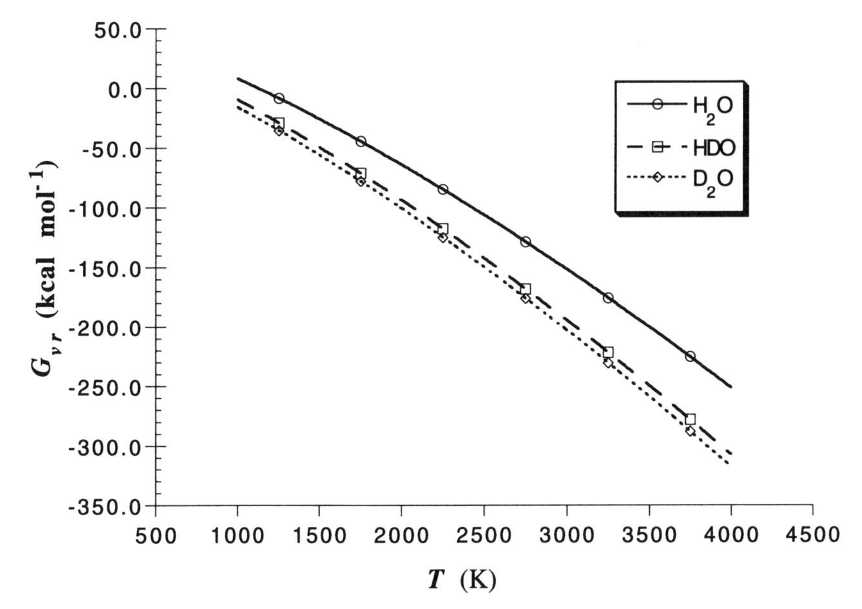

Figure 6. Computed vibration–rotation contributions to the Gibbs free energy and 4th-order polynomial curve fits for H_2O, HDO, and D_2O as a function of temperature. All computations were carried out using the AOSS-U Monte Carlo method in mass-weighted Jacobi coordinates. Three hundred Fourier coefficients were used per degree of freedom and 10^6 Monte Carlo samples were used for each calculation. Error bars at the 95% confidence level, as well as all free energy fluctuations, are smaller than the width of the lines showing the curve fits. An increment of 10 K was used over the temperature interval (1000–4000 K).

data to a polynomial function. A temperature grid of 10 K was used over the temperature range (1000 K $\leq T \leq$ 4000 K); 300 Fourier coefficients were used per degree of freedom, with 100 quadrature points used to evaluate the action at each configuration. We used 10^6 Monte Carlo samples to integrate the partition function at each temperature, achieving 95% confidence intervals of $.02q_{\mathrm{vr}}$ or less. This corresponds to a determination of the molecular free energy at each temperature with an uncertainty of approximately 0.2 kJ/mol. More precise calculations could, of course, be obtained by expending more computer time.

In order to obtain smooth derivatives from this relatively coarse data set, the vib–rot free energy was calculated at each temperature and the data set for each molecule was then fit to a polynomial of the form

$$G_{\mathrm{vr}} = a + bT + cT^2 + dT^3 + eT^4 + \cdots \tag{3.32}$$

truncated to include terms of fourth order or less in T. Note that truncating this series to include only first-order terms or less in T is equivalent to the assumption of constant molecular enthalpy (this can be seen by integration of the Gibbs–Helmholtz equation). Plots of the resulting free energy curves, along with the discrete data set used to fit them, are shown in Figures 6 and 7. The smooth variation of the free energy curves, as well as the accuracy with which the curve-fitting procedure represents the computed data, lead us to conclude that the molecular entropy S_{vr}, which is given by

$$S_{\mathrm{vr}} = -\left(\frac{\partial G_{\mathrm{vr}}}{\partial T}\right)_P \tag{3.33}$$

may be accurately computed from the derivative of the expansion of G_{vr}. Thus, we let

$$S_{\mathrm{vr}} = -b - 2cT - 3dT^2 - 4eT^3 + \cdots \tag{3.34}$$

In Figure 8 we display calculations of S_{vr} from the AOSS-U fitting procedure for H_2O and compare them to data obtained from the *JANAF*

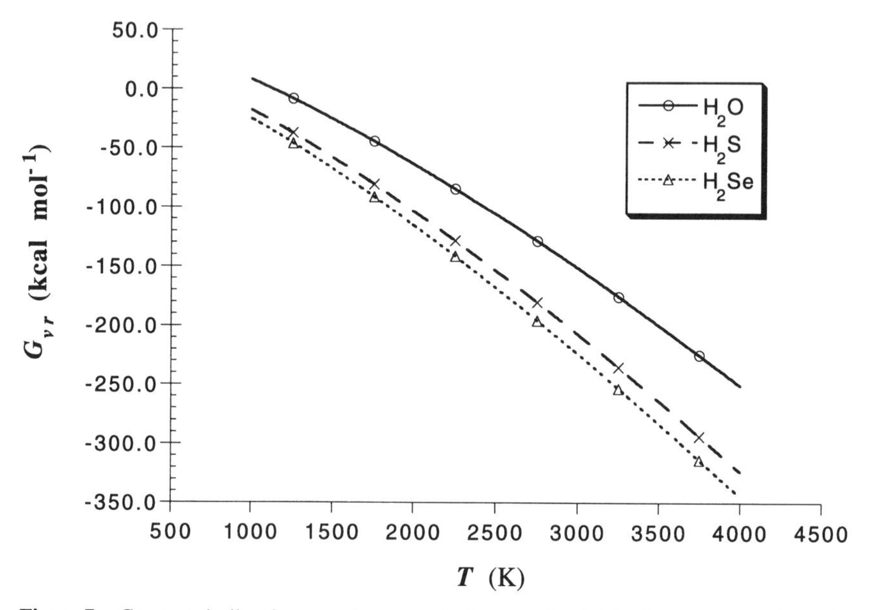

Figure 7. Computed vibration–rotation contributions to the Gibbs free energy and fourth-order polynomial curve fits for H_2O, H_2S, and H_2Se as a function of temperature. See Figure 6 for detailed description of the calculations.

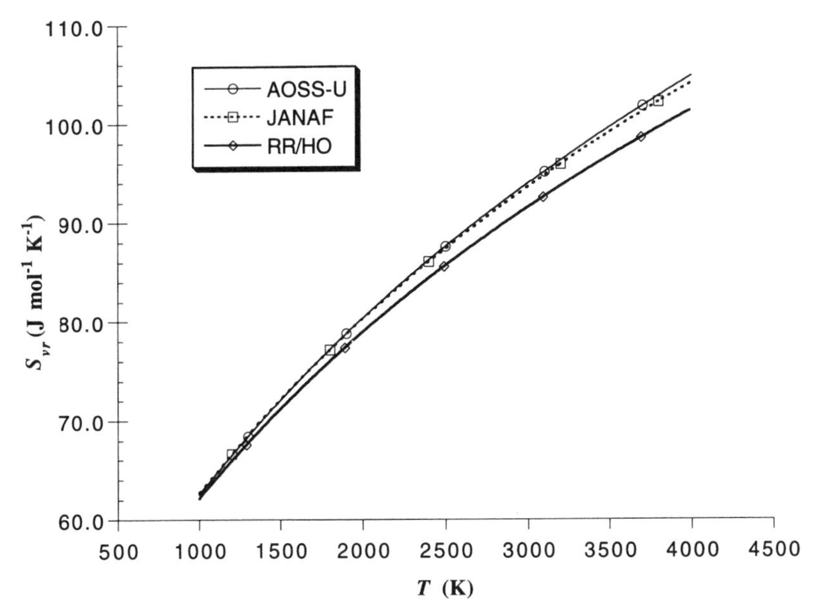

Figure 8. Vibration–rotation contributions to the entropy for H_2O as a function of temperature. The lower curve (◇, solid line) shows the entropy predicted by the classical rigid rotator–quantum harmonic oscillator (RR/HO) model, the intermediate curve (□, dashed line) shows data taken from the *JANAF Thermochemical Tables*, and the upper curve (○, solid line) is obtained from differentiating the fourth-order fit to the AOSS-U free energy calculations shown in Figure 6. Note that (as expected) all three curves coincide at low temperatures and that the *JANAF* and AOSS-U results are in close agreement throughout the entire temperature range.

Thermochemical Tables [6] and to values obtained from the rigid rotator–harmonic oscillator approximation. We note that the *JANAF* and AOSS-U results are in excellent agreement with one another throughout the entire temperature range. There is some slight deviation between the two at temperatures greater than 3100 K; however, this is in the same temperature regime that one might expect the dynamical assumptions underlying the quasi-empirical *JANAF* formulas to break down somewhat. More careful numerical analysis of the data is required to assess whether the deviation between the two curves is real or an artifact of the fitting procedure, but the close agreement with the *JANAF* data at low and intermediate temperatures indicates that this procedure holds promise. We also note that as expected, there are quantitative differences between the entropies obtained from the rigid rotator–harmonic oscillator approximation and the AOSS-U and *JANAF* entropies at temperatures above 2000 K. Figure 9 displays

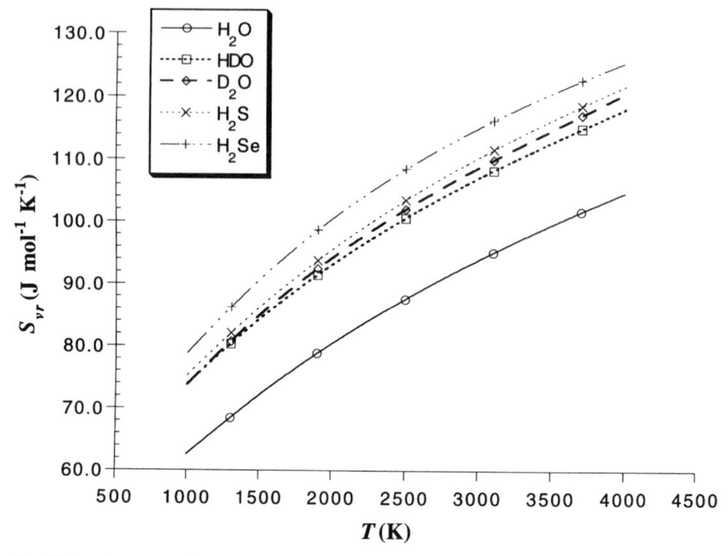

Figure 9. Vibration–rotation contributions to the entropy for H_2O, HDO, D_2O, H_2S, and H_2Se as a function of temperature. All curves are obtained by differentiating the fourth-order fits to the AOSS-U free-energy calculations shown in Figures 6 and 7.

AOSS-U entropy curves for H_2O, HDO, D_2O, H_2S, and H_2Se as a function of temperature.

IV. SUMMARY

We have reviewed the use of Fourier path-integral Monte Carlo methods to smoothly converge molecular free energies at high temperatures (i.e., in the combustion regime). The reader who is just starting to learn about path-integral Monte Carlo methods should be informed that our use of adaptively optimized stratified sampling Monte Carlo methods is a relatively unorthodox approach to the calculation of molecular thermodynamic properties. Generally speaking, variations on the Metropolis algorithm are typically used for the computation of classical free energies and free-energy changes and for the path-integral computation of thermodynamic energies, heat capacities, and free energy changes [42,61,63]. These approaches have proved extremely successful in their respective contexts. Our own approach to computing molecular free energies, the AOSS-U method, has certain strengths and weaknesses in comparison to the Metropolis algorithm for our particular application. One advantage is that all Monte Carlo samples generated in the calculation can be used in the final accumulation of statistics without the need to "throw out" unequilibrated samples, as is usual in Metropolis-type Monte Carlo algorithms. Another advantage is that the absence of serial correlations makes the error analysis rather simple. A third advantage is that the AOSS-U method is easily vectorizable and parallelizable, and thus can take full advantage of the properties of the new generation of random-number generators. The application of AOSS-U methods to the computation of other thermodynamic properties, such as the molecular heat capacity, remains to be explored.

Thus far, application of the AOSS-U method has been limited only to several well-characterized compounds (for benchmarking purposes and validification of the methods) and one poorly characterized group of compounds (the hydrogen selenides). This has less to do with the general applicability of the method than with the lack of highly accurate, closed-form molecular potential-energy functions. All path-integral methods require that the potential be computed many times at each sampled configuration. This feature has thus far necessitated the use of closed-form potential-energy functions, fit to ab initio and experimental data, as is the case in a great deal of Monte Carlo and molecular dynamics work. However, it would be most desirable to obtain information about the potential energy directly from accurate electronic structure computations, along the lines of the techniques introduced by Marx and Parrinello [123]. For example, the B3-LYP density functional method currently shows great promise in this context [124]. The direct use of electronic structure information would extend the range of possible systems considerably. In the meantime we are typically restricted to using potential-energy functions that are fit to electronic structure calculations and/or experimental data.

A number of interesting directions remain to be explored. We anticipate implementing partial averaging in a future version of our AOSS-U code, which should greatly decrease the dimensionality of the problem. We also hope to develop extensions of these methods for high-precision computation of molecular heat capacities, which are needed in order to truly compute "quantum steam tables." There are also other formal routes to explore. We have not yet explored the use of the centroid representation [60], which may prove to be a useful alternative in this context. Encouragingly, the centroid representation has recently found application in the computation of free energies of activation, which requires high precision. Also, as mentioned above, we are exploring the development of a new path representation using wavelet transforms [86,87]. The advent of "jump-walking" Metropolis Monte Carlo techniques [43,125–127], which provide a powerful route for overcoming the quasiergodicity problems inherent in the Metropolis algorithm [90,128], leads us to believe that a path-integral version of the optimal histogram method (described elsewhere in this volume) might be a feasible approach to quantum partition functions and higher-order cumulants such as the heat capacity. Variance reduction remains of great interest; accordingly, efforts are underway to include the use of importance sampling functions in the AOSS-U algorithm [129]. We are working on applications of the AOSS-U method to the prediction of equilibrium constants for atmospheric proton-transfer reactions and combustion reactions, and to free-energy calculations of conformationally flexible molecules and clusters such as HCN, H_2O_2, and $Cl^-(H_2O)$ [87]. Finally, we note that recent spectroscopic observations [130] in conjunction with theoretical calculations [131] have shown that water molecules exist within sunspot umbrae, which have temperatures close to 3200 K [132]. Our own calculations [25], which predict that the H_2O molecule is a thermodynamically stable species up to at least 4000 K, are in agreement with these subsequent results. We, therefore, believe that it is possible to employ the methods described in this paper to make fully converged quantum-mechanical predictions of equilibrium constants for chemical and isotope exchange reactions within a sunspot.

ACKNOWLEDGMENTS

I wish to thank my students Denise Bergin, for carrying out the calculations shown in Figure 5 and for a number of helpful discussions and suggestions; and Andrew Sweeney, for providing several useful references. Thanks to Prof. Donald Truhlar (University of Minnesota) for his encouragement and innovative ideas and contributions and to Prof. David Freeman (University of Rhode Island), who has shared his many original insights and experiences with the Fourier path-integral representation and Monte Carlo methods. I have benefited greatly from the contributions of my previous collaborators; Drs. Yi-Ping Liu, Manish Mehta, Steven Mielke, Gregory Tawa, and Qi Zhang. Prof. Suse Broyde (New York University) generously shared her computational resources and thus made the entropy calculations possible, using a vectorized adaptation of the AOSS-U algorithm by Steve Mielke and Yi-Ping Liu. Finally, I thank the Cooper Union for the Advancement of Science and Art, and especially the Department of Chemistry and the School of Engineering, for their support of this work.

REFERENCES

1. J. A. Miller, R. J. Kee, and C. K. Westbrook, *Annu. Rev. Phys. Chem.* **41**, 345 (1990).
2. V. S. Melissas and D. G. Truhlar, *J. Phys. Chem.* **98**, 875 (1994).
3. D. F. Rudd and C. C. Watson, *Strategy of Process Engineering*, Wiley, New York, 1968.
4. K. C. Mills, *Thermodynamic Data for Inorganic Sulphides, Selenides and Tellurides*, Butterworths, London, 1974.
5. K. Raznjevic, *Handbook of Thermodynamic Tables and Charts*, Hemisphere, Washington, DC, 1976.
6. M. W. Chase Jr., C. A. Davies, J. R. Downey, D. J. Frurip, R. A. McDonald and A. N. Syverud, *JANAF Thermochemical Tables*, 3rd ed., American Chemical Society and American Physical Society for the National Bureau of Standards, New York, 1985.

7. L. V. Gurvich, I. V. Veyts and C. B. Alcock, *Thermodynamic Properties of Individual Substances*, 4th ed., Hemisphere, New York, 1989–1991.
8. J. E. Mayer and M. G. Mayer, *Statistical Mechanics*, Wiley, New York, 1940.
9. K. S. Pitzer, *Quantum Chemistry*, Prentice-Hall, Englewood Cliffs, NJ, 1953.
10. J. E. Mayer, *Equilibrium Statistical Mechanics*, Pergamon, Oxford, 1968.
11. D. A. McQuarrie, *Statistical Thermodynamics*, University Science, Mill Valley, CA, 1973.
12. R. L. Rowley, *Statistical Mechanics for Thermophysical Property Calculations*, PTR Prentice-Hall, Englewood Cliffs, NJ, 1994.
13. N. C. Handy, *Int. Rev. Phys. Chem.* **8**, 275 (1989).
14. J. Tennyson, S. Miller and J. R. Henderson, in *Methods in Computational Chemistry*, Vol. 4, S. Wilson, ed., Plenum Press, New York, 1992.
15. H. Partridge and D. W. Schwenke, *J. Chem. Phys.* **106**, 4618 (1997).
16. J. M. L. Martin, J. P. Francois and R. Gijbels, *Z. Phys. D* **21**, 47 (1991).
17. J. M. L. Martin, J. P. Francois and R. Gijbels, *J. Chem. Phys.* **95**, 8374 (1991).
18. J. M. L. Martin, J. P. Francois and R. Gijbels, *J. Chem. Phys.* **96**, 7633 (1992).
19. A. D. Isaacson, D. G. Truhlar, K. Scanlon and J. Overend, *J. Chem. Phys.* **75**, 3017 (1981).
20. A. D. Isaacson and D. G. Truhlar, *J. Chem. Phys.* **75**, 4090 (1981).
21. A. D. Isaacson and X.-G. Zhang, *Theor. Chim. Acta* **74**, 493 (1988).
22. D. G. Truhlar, *J. Comp. Chem.* **12**, 266 (1991).
23. D. G. Truhlar and A. D. Isaacson, *J. Chem. Phys.* **94**, 357 (1991).
24. Q. Zhang, P. N. Day and D. G. Truhlar, *J. Chem. Phys.* **98**, 4948 (1993).
25. R. Q. Topper, Q. Zhang, Y.-P. Liu and D. G. Truhlar, *J. Chem. Phys.* **98**, 4991 (1993).
26. W. C. Ermler, H. C. Hsieh and L. B. Harding, *Comp. Phys. Commun.* **51**, 257 (1988).
27. A. Willetts, J. F. Gaw, W. H. Green and N. C. Handy, "SPECTRO, a Theoretical Spectroscopy Package," version 2.0, Cambridge, UK, 1990 (unpublished). See also J. F. Gaw, A. Willetts, W. H. Green and N. C. Handy, in *Advances in Molecular Vibrations and Collision Dynamics*, J. Bowman, ed. JAI, Greenwich, CT, 1990.
28. T. J. Lee, C. E. Dateo, B. Gazdy and J. M. Bowman, *J. Phys. Chem.* **97**, 8937 (1993).
29. M. R. Soto and M. Page, *J. Chem. Phys.* **97**, 7287 (1992).
30. Y.-P. Liu, G. C. Lynch, T. N. Truong, D.-H. Lu and D. G. Truhlar, *J. Am. Chem. Soc.* **115**, 2408 (1993).
31. D. G. Truhlar, B. C. Garrett and S. J. Klippenstein, *J. Phys. Chem.* **100**, 12771 (1996).
32. R. Steckler, Y.-Y. Chuang, E. L. Cotiño, P. L. Fast, J. C. Corchado, W.-P. Hu, Y.-P. Liu, G. C. Lynch, K. A. Nguyen, C. F. Jackels, M. Z. Gu, I. Rossi, S. Clayton, V. S. Melissas, B. C. Garrett, A. D. Isaacson and D. G. Truhlar, POLYRATE-7.2, University of Minnesota, Minneapolis, 1997.
33. R. P. Feynman, *Statistical Mechanics: A Series of Lectures*, Benjamin, Reading, MA, 1972.
34. R. P. Feynman and A. R. Hibbs, *Quantum Mechanics and Path Integrals*, McGraw-Hill, New York, 1965.
35. R. P. Feynman and H. Kleinert, *Phys. Rev. A* **34**, 5080 (1986).
36. J. D. Doll and D. L. Freeman, *J. Chem. Phys.* **80**, 2239, (1984).
37. D. L. Freeman and J. D Doll, *J. Chem. Phys.* **80**, 5709 (1984).
38. J. D. Doll, R. D. Coalson and D. L. Freeman, *Phys. Rev. Lett.* **55**, 1 (1985).

39. R. D. Coalson, D. L. Freeman and J. D. Doll, *J. Chem. Phys.* **85**, 4567 (1986).
40. R. D. Coalson, *J. Chem. Phys.* **85**, 926 (1986).
41. D. L. Freeman and J. D. Doll, *Adv. Chem. Phys.* **70B**, 139 (1988).
42. J. D. Doll, D. L. Freeman and T. L. Beck, *Adv. Chem. Phys.* **78**, 61 (1990).
43. D. D. Frantz, D. L. Freeman and J. D. Doll, *J. Chem. Phys.* **97**, 5713 (1992).
44. D. L. Freeman and J. D. Doll, *Annu. Rev. Phys. Chem.* **47**, 43 (1996).
45. R. Q. Topper and D. G. Truhlar, *J. Chem. Phys.* **97**, 3647 (1992).
46. R. Q. Topper, G. J. Tawa and D. G. Truhlar, *J. Chem. Phys.* **97**, 3668 (1992).
47. D. Scharf, M. L. Klein and G. J. Martyna, *J. Chem. Phys.* **97**, 3590 (1992).
48. C. Chakravarty, *J. Chem. Phys.* **102**, 956 (1995).
49. C. Chakravarty, *J. Chem. Phys.* **104**, 7223 (1996).
50. Y. Kwon, D. M. Ceperley and K. B. Whaley, *J. Chem. Phys.* **104**, 2341 (1996).
51. K. Kinugawa, P. B. Moore and M. L. Klein, *J. Chem. Phys.* **106**, 1154 (1997).
52. D. Kim, J. D. Doll and J. E. Gubernatis, *J. Chem. Phys.* **106**, 1641 (1997).
53. J. Lobaugh and G. A. Voth, *J. Chem. Phys.* **106**, 2400 (1997).
54. H. Gai, G. K. Schenter and B. C. Garrett, *J. Chem. Phys.* **104**, 680 (1996).
55. C. Zheng, J. A. McCammon and P. G. Wolynes, *Chem. Phys.* **158**, 261 (1991).
56. G. A. Voth, D. Chandler and W. H. Miller, *J. Chem. Phys.* **91**, 7749 (1989).
57. R. P. McRae, G. K. Schenter, B. C. Garrett, G. R. Haynes, G. A. Voth and G. C. Schatz, *J. Chem. Phys.* **97**, 7392 (1992).
58. E. Pollak, *J. Chem. Phys.* **103**, 973 (1995).
59. M. Messina, G. K. Schenter and B. C. Garrett, *J. Chem. Phys.* **103**, 3430 (1995).
60. G. A. Voth, *Adv. Chem. Phys.* **93**, 135 (1996).
61. G. J. Hogenson and W. P. Reinhardt, *J. Chem. Phys.* **102**, 4151 (1995).
62. B. J. Berne and D. Thirumalai, *Annu. Rev. Phys. Chem.* **37**, 401 (1986).
63. B. J. Berne, *J. Stat. Phys.* **43**, 911 (1986).
64. D. M. Ceperley and L. Mitas, *Adv. Chem. Phys.* **93**, 1 (1996).
65. C. H. Mak and R. Egger, *Adv. Chem. Phys.* **93**, 39 (1996).
66. J. J. Sakurai, *Modern Quantum Mechanics*, Benjamin/Cummings, Menlo Park, CA, 1985.
67. A. L. Fetter and J. D. Walecka, *Theoretical Mechanics of Particles and Continua*, McGraw-Hill, New York, 1980.
68. H. Goldstein, *Classical Mechanics*, 2nd ed., Addison-Wesley, Reading, MA, 1980.
69. P. W. Atkins and R. S. Friedman, *Molecular Quantum Mechanics*, 3rd ed., Oxford Univ. Press, Oxford, 1997.
70. V. A. Benderskii, D. E. Makarov and C. A. Wight, *Adv. Chem. Phys.* **88**, 1 (1994).
71. P. A. M. Dirac, *The Principles of Quantum Mechanics*, 4th ed., Clarendon Press, Oxford, 1958.
72. E. Pollak and P. Pechukas, *J. Am. Chem. Soc.* **100**, 2984 (1978).
73. J. A. Barker, *J. Chem. Phys.* **70**, 2914 (1979).
74. H. F. Trotter, *Proc. Am. Math. Soc.* **10**, 545 (1959).
75. M. Suzuki, *Commun. Math. Phys.* **51**, 183 (1976).
76. H. De Raedt and B. De Raedt, *Phys. Rev. A* **28**, 3575 (1983).

77. M. Takahashi and M. Imada, *J. Phys. Soc. Jpn.* **53**, 963 (1984).
78. H. Kono, A. Takasaka and S. H. Lin, *J. Chem. Phys.* **88**, 6390 (1988).
79. C. H. Mak and H. C. Andersen, *J. Chem. Phys.* **92**, 000 (1990).
80. J. Cao and B. J. Berne, *J. Chem. Phys.* **92**, 7531 (1990).
81. R. Q. Topper, *Rev. Comp. Chem.* **10**, 101 (1997).
82. M. C. Gutzwiller, *Chaos in Classical and Quantum Mechanics*, Springer-Verlag, New York, 1990.
83. D. Chandler and P. G. Wolynes, *J. Chem. Phys.* **74**, 4078 (1981).
84. R. W. Hall and B. J. Berne, *J. Chem. Phys.* **81**, 3641 (1984).
85. W. H. Miller, *J. Chem. Phys.* **63**, 1166 (1975).
86. G. Strang and T. Nguyen, *Wavelets and Filter Banks*, Wellesley-Cambridge Press, Wellesley, MA, 1996.
87. R. Q. Topper, work in progress.
88. M. Sprik, M. L. Klein, and D. Chandler, *Phys. Rev. B* **31**, 4234 (1985).
89. M. Sprik, M. L. Klein, and D. Chandler, *Phys. Rev. B.* **32**, 545 (1985).
90. M. H. Kalos and P. A. Whitlock, *Monte Carlo Methods*, Vol. I: *Basics*, Wiley, New York, 1986.
91. J. P. Valleau and G. M. Torrie, in *Statistical Mechanics*, Part A: *Equilibrium Techniques*, B. J. Berne, ed., Plenum Press, New York, 1977, pp. 169–191.
92. K. Binder, in *Monte Carlo Methods in Statistical Physics*; K. Binder, ed., Springer-Verlag, Berlin, 1979, pp. 1–36.
93. M. P. Allen and D. J. Tildesley, *Computer Simulation of Liquids*, Oxford Science Pub., London, 1987.
94. J. L. Doob, *Ann. Math. Stat.* **14**, 229 (1943).
95. W. J. Dixon, *Ann. Math. Stat.* **15**, 119 (1944).
96. U. Grenander, *Arkiv. Mat.* **1**, 195 (1950).
97. E. B. Smith and B. H. Wells, *Mol. Phys.* **53**, 701 (1984).
98. S. K. Schiferl and D. C. Wallace, *J. Chem. Phys.* **83**, 5203 (1985).
99. T. P. Straatsma, H. J. C. Berendsen, and A. J. Stam, *Mol. Phys.* **57**, 89 (1986).
100. J. O. Hirschfelder and J. S. Dahler, *Proc. Natl. Acad. Sci. USA* **42**, 363 (1956).
101. D. W. Jepsen and J. O. Hirschfelder, *J. Chem. Phys.* **30**, 1032 (1959).
102. J. O. Hirschfelder, *Int. J. Quantum Chem. Symp.* **3**, 17 (1969).
103. M. Karplus, R. N. Porter, and R. D. Sharma, *J. Chem. Phys.* **43**, 3259 (1965).
104. D. G. Truhlar and J. T. Muckerman, in *Atom-Molecule Collision Theory*; R. B. Bernstein, ed., Plenum, New York, 1979, pp. 505–566.
105. L. M. Raff and D. L. Thompson, in *Theory of Chemical Reaction Dynamics*; Vol. III, M. Baer, ed., CRC Press, Boca Raton, FL, 1985, pp. 1–121.
106. G. E. P. Box and M. E. Muller, *Ann. Math. Stat.* **29**, 610 (1958).
107. J. M. Hammersley and D. C. Handscomb, *Monte Carlo Methods*, Methuen, London, 1965.
108. M. H. Hansen, W. N. Harwitz and W. G. Madow, *Sample Survey Methods and Theory*, Vols. 1 and 2, Wiley, New York, 1953.
109. G. K. Bhattacharya and R. A. Johnson, *Statistical Concepts and Methods*, Wiley, New York, 1977.

110. W. H. Press, S. A. Teukolsky, W. T. Vetterling and B. P. Flannery, *Numerical Recipes: The Art of Scientific Computing*, 2nd ed., Cambridge Univ. Press, Cambridge, UK, 1992.
111. D. L. Freeman, private communication; G. J. Tawa, private communication.
112. W. H. Press and G. R. Farrar, *Comp. Phys.* **4**, 190 (1990).
113. J. Avery, *Hyperspherical Harmonics: Applications in Quantum Theory*, Kluwer, Dordrecht, 1989.
114. H. Flanders, *Differential Forms with Applications to the Physical Sciences*, Dover, New York, 1989.
115. L. Pauling and E. B. Wilson, Jr., *Introduction to Quantum Mechanics With Applications to Chemistry*, Dover, New York, 1963, pp. 188–190.
116. H. M. Hulbert and J. O. Hirschfelder, *J. Chem. Phys.* **9**, 61 (1941).
117. H. M. Hulbert and J. O. Hirschfelder, *J. Chem. Phys.* **35**, 1091(E) (1961).
118. W. P. Reinhardt and J. E. Hunter III, *J. Chem. Phys.* **97**, 1599 (1992).
119. J. E. Hunter III, W. P. Reinhardt and T. Davis, *J. Chem. Phys.* **99**, 6856 (1993).
120. W. Jahnke and T. Sauer, *Chem. Phys. Lett.* **201**, 499 (1993).
121. E. Loh, Jr., in *Computer Simulation Studies in Condensed Matter Physics*, Vol. 33, D. P. Landau, K. K. Mon, and H.-B. Schuettler, eds., Springer, Berlin, 1988, pp 19–30.
122. E. Kauppi and L. Halonen, *J. Phys. Chem.* **94**, 5779 (1990).
123. D. Marx and M. Parrinello, *J. Chem. Phys.* **104**, 4077 (1996).
124. A. P. Scott and L. Radon, *J. Phys. Chem.* **100**, 16502 (1996).
125. D. D. Frantz, D. L. Freeman and J. D. Doll, *J. Chem. Phys.* **93**, 2769 (1990).
126. D. D. Frantz, *J. Chem. Phys.* **102**, 3747 (1995).
127. A. Matro, D. L. Freeman and R. Q. Topper, *J. Chem. Phys.* **104**, 8690 (1996).
128. J. P. Valleau and S. G. Whittington, in *Statistical Mechanics*, Part A: *Equilibrium Techniques*; B. J. Berne, ed., Plenum Press, New York, 1977, pp. 137–168.
129. J. Srinivasan and D. G. Truhlar, work in progress.
130. L. Wallace, P. Bernath, W. Livingston, K. Hinkle, J. Busler, B. Guo, and K. Zhang, *Science* **268**, 1155 (1995).
131. O. L. Polyansky, N. F. Zobov, S. Viti, J. Tennyson, P. F. Bernath and L. Wallace, *Science* **277**, 346 (1997).
132. T. Oka, *Science* **277**, 328 (1997).

MONTE CARLO SAMPLING FOR CLASSICAL TRAJECTORY SIMULATIONS

GILLES H. PESLHERBE

Department of Chemistry and Biochemistry, University of Colorado at Boulder, Boulder, CO 80309-0215

HAOBIN WANG

Department of Chemistry, University of California, Berkeley, Berkeley, CA 94720

WILLIAM L. HASE

Department of Chemistry, Wayne State University, Detroit, MI 48202-3489

CONTENTS

Advances in Chemical Physics, Volume 105, Monte Carlo Methods in Chemical Physics, edited by David M. Ferguson, J. Ilja Siepmann, and Donald G. Truhlar. Series Editors I. Prigogine and Stuart A. Rice.
ISBN 0-471-19630-4

I. INTRODUCTION

Classical trajectory simulations [1–4] are widely used to study the unimolecular and intramolecular dynamics of molecules and clusters; reactive and inelastic collisions between atoms, molecules, and clusters; and collisions of these species with surfaces. In a classical trajectory study the motions of the individual atoms are simulated by solving the classical equations of motion, usually in the form of Hamilton's equations [5]:

$$\frac{\partial H}{\partial q_i} = \frac{-dp_i}{dt} \quad \text{and} \quad \frac{\partial H}{\partial p_i} = \frac{dq_i}{dt} \tag{1.1}$$

where H, the sum of the kinetic $T(\mathbf{p}, \mathbf{q})$ and potential energies $V(\mathbf{q})$, is the system's Hamiltonian:

$$H = T(\mathbf{p}, \mathbf{q}) + V(\mathbf{q}) \tag{1.2}$$

For the most general case T depends on both the momenta $\mathbf{p}$ and the coordinate $\mathbf{q}$.

There are several components to a classical trajectory simulation [1–4]. A potential-energy function $V(\mathbf{q})$ must be formulated. In the past $V(\mathbf{q})$ has been represented by an empirical function with adjustable parameters or an analytic fit to electronic structure theory calculations. In recent work [6] the potential energy and its derivatives $\partial V/\partial q_i$ have been obtained directly from an electronic structure theory, without an intermediate analytic fit. Hamilton's equations of motion [Eq. (1.1)] are solved numerically and numerous algorithms have been developed and tested for doing this is an efficient and accurate manner [1–4]. When the trajectory is completed, the final values for the momenta and coordinates are transformed into properties that may be compared with experiment, such as product vibrational, rotational, and relative translational energies.

The remaining aspect of a trajectory simulation is choosing the initial momenta and coordinates. These initial conditions are chosen so that the results from an ensemble of trajectories may be compared with experiment and/or theory, and used to make predictions about the chemical system's molecular dynamics. In this chapter Monte Carlo methods are described for sampling the appropriate distributions for initial values of the coordinates and momenta. Trajectories may be integrated in different coordinates and conjugate momenta, such as internal [7], Jacobi [8], and Cartesian. However, the Cartesian coordinate representation is most general for systems of any size and the Monte Carlo selection of Cartesian coordinates and momenta is described here for a variety of chemical processes. Many of the Monte Carlo procedures described are incorporated in the general chemical dynamics computer program VENUS [9].

II. UNIMOLECULAR DECOMPOSITION

In a unimolecular reaction a reactant A is excited above its unimolecular threshold E_0 so that it may dissociate to product(s):

$$A^* \rightarrow \text{product(s)} \tag{2.1}$$

where the (*) denotes vibrational–rotational excitation. A question of fundamental interest in unimolecular rate theory is whether the unimolecular dissociation is random during the complete unimolecular decomposition from $t = 0$ to ∞ [10–12]. This will be the case if A* is initially excited with a microcanonical ensemble, so there is an initial uniform sampling of its phase space, and if the intramolecular dynamics of A* is ergodic [13] within the timescale of the unimolecular reaction [14], so that the initial microcanonical ensemble is maintained.

For random unimolecular dissociation, the lifetime distribution $P(t)$, namely, the probability of decomposition per unit time, is given by [10]

$$P(t) = k(E)\ \exp[-k(E)t] \tag{2.2}$$

where $k(E)$ is the classical microcanonical unimolecular rate constant. It may be expressed as [12]

$$k(E) = \frac{N^{\ddagger}(E)}{h\rho(E)} \tag{2.3}$$

where $N^{\ddagger}(E)$ is the sum of states at the transition state for decomposition and $\rho(E)$ is the density of states for A*. In accord with the classical/quantum correspondence principle [13,15], the classical and quantum $k(E)$ become equivalent at high energies. However, for E near the unimolecular threshold E_0, the classical $k(E)$ may be significantly larger than the quantum $k(E)$, since classical mechanics allows the transition state to be crossed and products to be performed without the presence of zero-point energy [15].

If the unimolecular dissociation is not random and not in accord with Eq. (2.2) it is thought that classical "bottlenecks" restricting intramolecular vibrational-energy redistribution (IVR) may be manifested in the quantum dynamics [16]. Thus, there is considerable interest in identifying the nature of the unimolecular dynamics of excited molecules. In this section Monte

Carlo sampling schemes are described for exciting A* randomly with a microcanonical ensemble of states and nonrandomly with specific state selection. For pedagogical purposes, selecting a microcanonical ensemble for a normal mode Hamiltonian is described first.

A. Classical Microcanonical Sampling

1. Normal-Mode Hamiltonian

The Hamiltonian for a system of n normal modes is the sum of energies for separable harmonic oscillators and is given by

$$H = E = \sum_{i=1}^{n} E_i = \sum_{i=1}^{n} \frac{P_i^2 + \omega_i^2 Q_i^2}{2} \tag{2.4}$$

To form a microcanonical ensemble random values for the P_i and Q_i are chosen so that there is a uniform distribution in the classical phase space of $H(\mathbf{P}, \mathbf{Q})$ [17]. Two ways are described here to accomplish this. For one method, called microcanonical normal-mode sampling [18], random values are chosen for the mode energies E_i, which are then transformed to random values for P_i and Q_i. In the second method, called orthant sampling [11], random values for P_i and Q_i are sampled directly from the phase space.

Microcanonical normal-mode sampling considers a molecule consisting of n normal modes at a constant total energy E. The probability that normal mode 1 has energy E_1 is proportional to the number of ways the energy $E - E_1$ may be added to the remaining $n - 1$ oscillators. This is given by the density of states for these oscillators [12,19]; thus, the normalized probability that normal mode 1 contains energy E_1 is

$$P(E_1) = (E - E_1)^{n-2}\left(\int_0^E (E - E_1)^{n-2}\, dE_1\right)^{-1} \tag{2.5}$$

After E_1 has been selected according to Eq. (2.5), E_2 is chosen within the energy $E - E_1$ by the probability distribution

$$P(E_2) = (E - E_1 - E_2)^{n-3}\left(\int_0^{E-E_1} (E - E_1 - E_2)^{n-3}\, dE_2\right)^{-1} \tag{2.6}$$

The general expression for $P(E_i)$ becomes

$$P(E_i) = \left(E - \sum_{j=1}^{i-1} E_j - E_i\right)^{n-1-i} \left\{\int \left(E - \sum_{j=1}^{i-1} E_j - E_i\right)^{n-1-i} dE_i\right\}^{-1} \tag{2.7}$$

where the integral is evaluated between the limits 0 to $(E - \Sigma_{j=1}^{i-1} E_j)$. $P(E_i)$ dE_i is the probability of finding E_i in the interval $(E_i, E_i + dE_i)$ once the energies of the $i - 1$ oscillators have been assigned.

Equation (2.7) may be conveniently sampled using the cumulative distribution function (CDF) [1]

$$C(E_i) = \int_0^{E_i} P(E_i)\, dE_i = R_i \tag{2.8}$$

Here and in the following equations, R_i and R represent freshly generated random numbers ranging between 0 and 1. Integration of Eq. (2.8) yields for the individual E_i:

$$E_i = \left(E - \sum_{j=1}^{i-1} E_j\right)(1 - R_i^{1/(n-i)}) \tag{2.9}$$

Random values for the Q_i and P_i are then chosen by giving each normal mode a random phase [20]. The normal-mode coordinates and momenta vary versus time according to $Q_i = A_i \cos(\omega_i t)$ and $P_i = \dot{Q} = -\omega_i A_i \sin(\omega_i t)$, where $A_i = (2E_i)^{1/2}/\omega_i$ is the amplitude and the relationship between ω_i and the vibration frequency v_i is $\omega_i = 2\pi v_i$. Since each time has equal probability during a vibrational period, random Q_i and P_i are given by

$$Q_i = \left[\frac{(2E_i)^{1/2}}{\omega_i}\right] \cos(2\pi R_i) \tag{2.10a}$$

$$P_i = -(2E_i)^{1/2} \sin(2\pi R_i) \tag{2.10b}$$

[The R_i in Eq. (2.10) is different than that used to generate E_i in Eq. (2.9).] For the following discussion, it is useful to note that the probability of a value of Q_i is inversely proportional to the number of Q_i within the time interval dt; i.e. probability $(Q_i) \propto |dQ_i/dt|^{-1} = |P_i|^{-1}$.

Orthant sampling [10] works in the classical phase space of the molecular Hamiltonian $H(\mathbf{P}, \mathbf{Q})$. For a microcanonical ensemble, each phase-space point in the volume element $d\mathbf{P}\, d\mathbf{Q}$ has equal probability [17]. The classical density of states is then proportional to the surface integral of the phase-space shell with $H(\mathbf{P}, \mathbf{Q}) = E$ and is given by [17]

$$\rho(E) = \underset{H=E}{\int \cdots \int} \frac{d\mathbf{P}\, d\mathbf{Q}}{h^n} \tag{2.11}$$

Thus, preparing a microcanonical ensemble involves choosing points at random on the energy shell.

Since Eq. (2.4) defines a $2n$-dimensional ellipsoid, it is straightforward to choose a microcanonical ensemble for the normal-mode Hamiltonian. Random values for the Q_i and P_i satisfying Eq. (2.4) may be chosen by projecting a $2n$-dimensional random unit vector with components x_i on the semiaxes $(2E)^{1/2}$ and $(2E)^{1/2}/\omega_i$:

$$P_i = (2E)^{1/2} x_i, \quad Q_i = \left[\frac{(2E)^{1/2}}{\omega_i}\right] x_{n+i} \tag{2.12}$$

where

$$\sum_{i=1}^{2n} x_i^2 = 1 \tag{2.13}$$

The derivation of the relative probability distribution $P(x_i)$ for the components x_i of the random unit vector has been given previously [10]. The probability distribution is

$$P(x_i) = \left(1 - \sum_{j=1}^{i-1} x_j^2 - x_i^2\right)^{(2n-i-2)/2} \tag{2.14}$$

The most probable value of x_i is 0, for which

$$P^0(x_i) = \left(1 - \sum_{j=1}^{i-1} x_j^2\right)^{(2n-i-2)/2} \tag{2.15}$$

The first $2n - 2$ values of x_i are chosen with the von Neumann rejection technique [21], and a cumulative distribution function is used to select the last two x_i [10]. The von Neumann rejection procedure is performed as follows. Initially $i = 1$ and the sums in Eqs. (2.14) and (2.15) are 0. Randomly generate x_1, $0 < x_1 < 1$. Accept x_1 if

$$\left\{\frac{P(x_1)}{P^0(x_1)}\right\} > R \tag{2.16}$$

Otherwise try a new x_1 and R. [The random number R in Eq. (2.16) and the one that generates x_1 are different.] Continue until an x_1 is accepted. On subsequent steps the procedure is the same except that the sums are nonzero and the sampling range of x_i is

$$0 < x_i < r_i \tag{2.17}$$

with r_i given by

$$r_i = \left(1 - \sum_{j=1}^{i-1} x_j^2\right)^{1/2} \tag{2.18}$$

It is convenient to let $x_i > 0$ throughout the selection procedure, and randomize signs afterward.

Both microcanonical normal-mode and orthant sampling give random sampling of the phase space and prepare microcanonical ensembles. For each, the average energy in a normal-mode is E/n.

2. *Normal-Mode/Rigid-Rotor Hamiltonian*

The classical normal-mode/rigid-rotor Hamiltonian is composed of the vibrational energy E_{vib} of Eq. (2.4) and the classical rotational energy

$$E_{\text{rot}}(\mathbf{j}) = \frac{j_x^2}{2I_x} + \frac{j_y^2}{2I_y} + \frac{j_z^2}{2I_z} \tag{2.19}$$

where $\mathbf{j}$ is the angular momentum vector (with components j_x, j_y, j_z) and I_x, I_y, I_z, are the principal moments of inertia of the molecular system. Initial conditions with $H_{\text{vib}} = E_{\text{vib}}$ and $H_{\text{rot}} = E_{\text{rot}}$ for the Hamiltonians may be selected by forming independent microcanonical ensembles for H_{vib} and H_{rot}. The $\mathbf{Q}$ and $\mathbf{P}$ for H_{vib} may be selected by either microcanonical normal-mode or orthant sampling as described above. Orthant sampling is then used to select j_x, j_y, and j_z for H_{rot}.

To form a microcanonical ensemble for the total Hamiltonian, $H = H_{\text{vib}} + H_{\text{rot}}$, orthant sampling may be used for energy $E = H$. A $(2n + 3)$-dimensional random unit vector is chosen and projected onto the semiaxes for j_x, j_y, and j_z [e.g., the semiaxis for j_x is $(2I_xE)^{1/2}$] as well as the semiaxes for $\mathbf{Q}$ and $\mathbf{P}$. Since rotation has one squared-term in the total energy expression, whereas vibration has two, the average energy in a rotational degree of freedom will be one-half of that in a vibrational degree of freedom.

It is also of interest to sample the normal-mode/rigid-rotor Hamiltonian for fixed total energy E and angular momentum j (the magnitude of $\mathbf{j}$). The sampling involves choosing the components of $\mathbf{j}$, that is, j_x, j_y, and j_z, and the momenta $\mathbf{P}$ and coordinates $\mathbf{Q}$ for the n normal modes. Different projections of $\mathbf{j}$ may have different rotational energies given by Eq. (2.19). The probability of a particular projection of $\mathbf{j}$ is proportional to the density of states of the normal modes with energy $E_{vib} = E - E_{rot}$. Thus, the most probable **j** is the one projected onto the rotational axis with the largest moment of inertia, which minimizes E_{rot} and thus maximizes E_{vib}:

$$P^0(\mathbf{j}) \propto \{E - \min[E_{rot}(\mathbf{j})]\}^{n-1} \tag{2.20}$$

The sampling for fixed E and j is initiated by projecting a three-dimensional random unit vector [Eqs. (2.14)–(2.18), onto j to give a trial **j**:

$$j_x = x_1 j \qquad j_y = x_2 j \qquad j_z = x_3 j \tag{2.21}$$

The relative probability of the chosen **j** is then

$$P(\mathbf{j}) \propto [E - E_{rot}(\mathbf{j})]^{n-1} \tag{2.22}$$

where $E_{rot}(\mathbf{j})$ is determined from Eq. (2.19). The relative probability $P(\mathbf{j})/P^0(\mathbf{j})$ is used with the von Neumann rejection technique of Eq. (2.16) to accept or reject the trial **j**. Once **j** is chosen, either microcanonical normal-mode or orthant sampling may be used to select the **Q** and **P** for the n normal modes with energy $E_{vib} = E - E_{rot}(\mathbf{j})$.

3. *Anharmonic Molecular Hamiltonian*

a. Microcanonical Normal-Mode and Orthant Sampling. Microcanonical normal-mode and orthant sampling may be used to prepare approximate microcanonical ensembles for anharmonic molecular Hamiltonians. As discussed above, these are exact sampling methods for harmonic Hamiltonians.

The first step in applying microcanonical normal-mode sampling is to determine the normal-mode frequencies $\omega = 2\pi\nu$ and normal-mode eigenvector **L** by diagonalizing the mass-weighted Cartesian force constant matrix [22]. Energies for the normal modes are then selected in the same way as described above for the harmonic Hamiltonian. After choosing the normal-mode coordinates **Q** and momenta **P** from Eq. (2.10), the following steps are carried out:

1. The **Q** and **P** are transformed to Cartesian coordinates **q** and momenta **p** for N atoms using the normal-mode eigenvector **L** [22]:

$$\mathbf{q} = \mathbf{q}_0 + \mathbf{M}^{-1/2}\mathbf{L}\mathbf{Q}$$
$$\mathbf{p} = \mathbf{M}^{1/2}\mathbf{L}\mathbf{P} \tag{2.23}$$

where $\mathbf{q}_0$ is a matrix of the equilibrium coordinates and $\mathbf{M}$ is a diagonal matrix whose elements are the atomic masses. Since normal modes are approximate for finite displacements [22], a spurious angular momentum $\mathbf{j}_s$ arises following this transformation [20,23]:

2. The spurious angular momentum is found from

$$\mathbf{j}_s = \sum_{i=1}^{N} \mathbf{r}_i \times m_i \dot{\mathbf{r}}_i \tag{2.24}$$

where m_i is the mass of the ith atom and $\mathbf{r}_i$ its position vector. The desired angular momentum $\mathbf{j}_0$ is added to the molecule by forming the vector

$$\mathbf{j} = \mathbf{j}^0 - \mathbf{j}_s \tag{2.25}$$

and adding the rotational velocity $\boldsymbol{\omega} \times \mathbf{r}_i$ to each of the atoms, where

$$\boldsymbol{\omega} = I^{-1}\mathbf{j} \tag{2.26}$$

and I^{-1} is the inverse of the inertia tensor [5].

3. The actual internal energy E for the Cartesian coordinates and momenta chosen from steps 1 and 2 is calculated using the correct Hamiltonian and compared with the intended energy E^0. If they do not agree within some acceptance criterion, the Cartesian coordinates and momenta are scaled according to

$$q_i' = q_i^0 + (q_i - q_i^0)\left(\frac{E^0}{E}\right)^{1/2}$$
$$p_i' = p_i\left(\frac{E^0}{E}\right)^{1/2} \tag{2.27}$$

Any spurious center of mass translational energy is subtracted from the molecule, and the procedure loops back to step 2.

Orthant sampling follows the same general procedure as described above for harmonic oscillators; however, the sampling is done in Cartesian coordinates and momenta. The steps are as follows:

1. The semiaxes for the Cartesian coordinates and momenta are found at fixed $H(\mathbf{q}, \mathbf{p}) - V_0 = E^0$, where E^0 is the intended energy and V_0 is

the minimum potential energy. This is done by varying the coordinates and momenta to find their maximum and minimum values, q_i^-, q_i^+ and $p_i^- = -p_i^+$.

2. To obtain the correct relationship between the average initial potential $\bar{V}$ and kinetic $\bar{T}$ energies, the $p_i^- = p_i^+$ are uniformly scaled by a parameter p_{scale} [24]. The $\bar{V}$ and $\bar{T}$ may be determined from trajectories integrated for long periods of time. Harmonic systems obey the virial theorem $\bar{V} = \bar{T}$ [25], but this is not necessarily the case for anharmonic systems.
3. The initial momenta and coordinates are then chosen from

$$p_i = x_i(p_i^-, p_i^+)p_{\text{scale}}$$
$$q_i = q_i^0 + x_{3n+i}[(q_i^-, q_i^+) - q_i^0] \tag{2.28}$$

where the x_i are components of a $6N$-dimensional random unit vector, Eqs. (2.14–2.18). Any center-of-mass translation is subtracted from the molecule.

The remaining steps for orthant sampling are the same as steps 2 and 3 described above for microcanonical normal-mode sampling.

b. Accurate Microcanonical Sampling. Classical exact microcanonical sampling for a system with n degrees of freedom is equivalent to choosing a uniform distribution of points in the $2n$-dimensional molecular phase space bounded by hypersurfaces of constant energy H and $H + dH$ [10]. The most straightforward approach is to select in a purely random fashion the coordinates and momenta in a hyperrectangular box (bounded by values allowed by total energy; see discussion above for orthant sampling), and retain phase-space points only for which the total energy lies between H and $H + dH$. It is somehow desirable to have no overall translation and/or rotation in the system, and alternatives to Cartesian coordinates may lie in internal coordinates for small systems [10] or normal-mode coordinates, from which overall translation and rotation are naturally subtracted. In any event, the likelihood of finding random points in the whole molecular phase space that satisfy the energy condition is infinitely small, and such a simple method is computationally not viable.

A much more efficient method consists of choosing all the coordinates and momenta except one as discussed above, and the remaining one from total energy conservation. However, the remaining variable, for instance, P_n, is not sampled uniformly between H and $H + dH$. The proper weighting for selecting P_n is seen from the following transformation for the volume element $d\mathbf{Q}\, d\mathbf{P}$ in Eq. (2.11) [10]:

$$dQ_1 \cdots dQ_n \, dP_1 \cdots dP_n = dQ_1 \cdots dQ_n \, dP_1 \cdots dP_{n-1} \, dH \left| \frac{\partial H}{\partial P_n} \right|^{-1} \quad (2.29)$$

(Here $\mathbf{Q}$ and $\mathbf{P}$ are used to represent general coordinates and momenta, since $d\mathbf{Q}\; d\mathbf{P}$ is independent of the coordinate system [17].) The relative weighting factor for selecting P_n is then

$$P(P_n) \propto \left| \frac{\partial H}{\partial P_n} \right|^{-1} \quad (2.30)$$

which is $|P_n|^{-1}$ for a Hamiltonian with a diagonal kinetic-energy term. The sampled points are then selected using rejection techniques, such as the von Neumann technique of Eq. (2.16). If a coordinate Q_n was chosen from energy conservation instead of a momentum, its relative probability would naturally be

$$P(Q_n) \propto \left| \frac{\partial H}{\partial Q_n} \right|^{-1} \quad (2.31)$$

Other algorithms have been developed for speeding up the uniform sampling of phase-space points. For example, the efficient microcanonical sampling (EMS) series of schemes [26] exploits the possibility of sampling independently the spatial coordinates and momenta, simply by weighting the sampled geometries by their associated momentum space density in the overall procedure. The latter density only depends on the total available kinetic energy, found by subtracting out the potential energy of the given geometry from the total energy:

$$\rho(E, \mathbf{Q}) \propto [E - V(\mathbf{Q})]^{(3N-2)/2} \quad (2.32)$$

Sampled geometrical configurations are accepted or rejected according to this relative probability, using the von Neumann rejection technique. Once a given geometry is accepted, momenta are just chosen randomly so as to yield the proper kinetic energy. Equation (2.32) has been derived by simply integrating the momentum space volume for a given total energy E (and potential energy of the chosen geometry). Similar expressions for excluding overall translation or for $E - j$ resolved densities of momentum states have been reported, and applied to triatomic systems [26]. Note that for the latter system, geometries can be sampled by randomly choosing two bond lengths r_1 and r_2, and an angle α. But because this sampling in phase space is not uniform, a weighting function of the form $r_1^2\, r_2^2 \sin\alpha$ must be incorporated into the relative probability for choosing configurations. This illustrates the importance of using accurate sampling probabilities or weighting functions for uniformly sampling the molecular phase space when not using the simple Cartesian coordinate system.

B. Nonrandom Sampling

1. *Quasi-Classical Distribution*

In the previous sections random sampling procedures were described for preparing a classical microcanonical ensemble of states. However, for comparison with experiment and to study the rate of intramolecular vibrational energy redistribution (IVR) [12,27] within a molecule, it is important to sample the states (i.e., phase space) of the molecule nonrandomly. Here, both normal and local mode [28] quasi-classical [2–4] sampling schemes are described for nonrandom excitation.

In normal-mode sampling, specific energies E_i or quanta v_i are assigned to each normal-mode:

$$E_i = (v_i + \tfrac{1}{2})\hbar\omega_i \tag{2.33}$$

Random values for the coordinates Q_i and momenta P_i of the normal modes are then chosen from Eq. (2.10). Cartesian coordinates and momenta are then selected as described above in Eqs. (2.23–2.27).

In local-mode sampling an individual local mode such as a CH bond in benzene is excited [28]. This type of trajectory calculation has been performed to determine the population of a local-mode state $|v\rangle$ versus time, from which the absorption linewidth of the overtone state may be determined [29]. Good agreement has been found with both experimental and quantum mechanically calculated overtone linewidths for benzene [30] and linear alkanes [31].

The first step in local-mode sampling, to calculate an adsorption linewidth, is to choose Cartesian coordinates and momenta, which correspond to zero-point energy (i.e., $v_i = 0$) in the molecule. This step is the same as described above for normal-mode sampling. Since the normal-mode zero-point energy is added with random phases, the kinetic and potential energies differ for each bond in the molecule after this step. Similarly, the energy in a particular bond will vary after this step between initial conditions. A particular bond is then excited to a local mode state $|v\rangle$ by adding more energy to the bond so that the Einstein–Brillouin–Keller (EBK) [32] semiclassical quantization condition

$$\oint p_r \, dr = (v + \tfrac{1}{2})h \tag{2.34}$$

is realized for a bond energy E_r of

$$E_r = \frac{p_r^2}{2\mu} + V(r) \tag{2.35}$$

where p_r is the bond's radial momentum, $V(r)$ the potential energy, and μ the reduced mass for the two atoms constituting the bond. If $V(r)$ is a Morse function [33], E_r equals

$$E(v) = (v + \tfrac{1}{2})h\nu_e - (v + \tfrac{1}{2})^2 h\nu_e x_e \tag{2.36}$$

where ν_e is the harmonic frequency and x_e the anharmonic correction. The energy is added by either extending or compressing (chosen randomly) the bond, which are the most probable values for the bond length classically [See discussion following Eq. (2.10) and Fig. 1] and quantum mechanically for large values of v (see Fig. 1).

2. *Quantum Distribution*

For the preceding examples classical distribution functions are sampled in choosing initial values for the coordinated and momenta. However, for some simulations it is important to sample a quantum mechanical distribution in selecting coordinates and momentum. The classical probability distribution for a vibrational mode peaks at the classical turning points (Fig. 1). In contrast, for the ground vibrational state, the quantum distribution peaks at the equilibrium value for the vibrational mode. For all vibrational states, the quantum distribution "tunnels" into classically forbidden regions. Including these quantum effects in choosing initial conditions has been found to be critical in simulating photodissociation of molecules that are initially in the ground vibrational state on the ground electronic state [34–37]. Some examples are the photodissociation of triatomics [35] and of the van der Waals complexes Ar–HCl [38] and Ar–H_2O [39].

The quantum probability distribution for choosing initial conditions has been represented by the Husimi [40,41] and Wigner [34,42] functions. For simulations in which only one vibrational mode is sampled quantum mechanically, the actual quantum probability for the coordinate has been used [38]. Here, sampling of the Wigner distribution is illustrated for a two-mode system in its ground vibrational state.

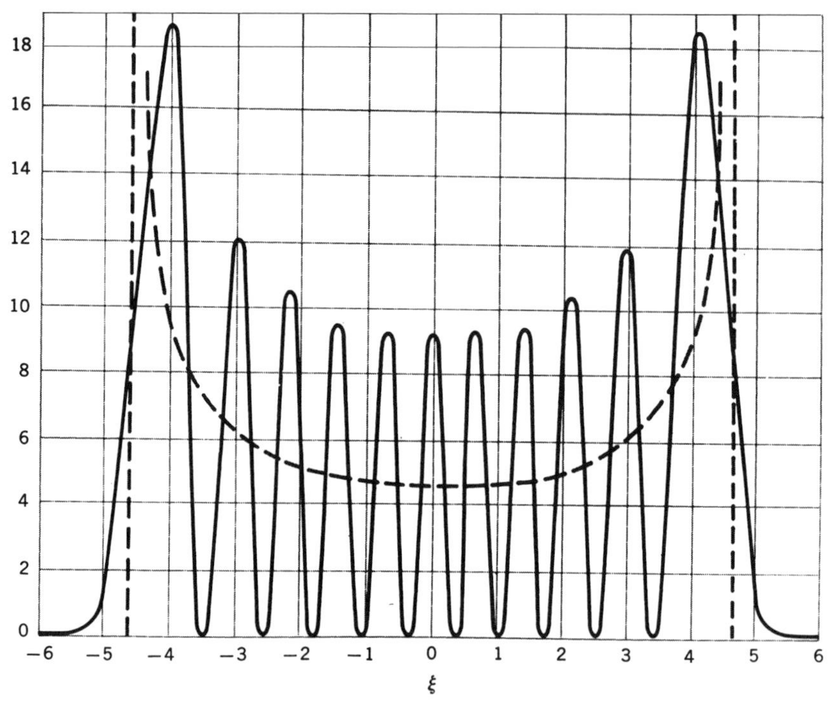

Figure 1. Position probability distribution for the state $n = 10$ of a harmonic oscillator (solid curve) and a classical oscillator of the same total energy (dashed curve). (From L. I. Shiff, *Quantum Mechanics*, 3rd ed., McGraw-Hill, New York, 1968).

In purely classical sampling each phase-space point in the volume element $d\mathbf{P}\,d\mathbf{Q}$ has equal probability. The procedure for selecting the **P** and **Q** for a n-mode system is described in Section II.A.3.b. If the momentum P_n for the nth normal-mode is chosen by energy conservation, the relative probability or weighting factor for the point is $|P_n|^{-1}$. On the other hand, if Q_n is chosen by energy conservation, the weighting factor is $|\partial H/\partial Q_n|^{-1}$. To include quantum effects in the selection of the **P** and **Q**, the Wigner distribution $P_w(\mathbf{P},\mathbf{Q})$ is included in the weighting factor.

For the ground vibrational state of a two-mode system, the Wigner distribution is [36,37]

$$P_w(\mathbf{P}, \mathbf{Q}) = (\pi\hbar)^{-2} \prod_{i=1}^{2} e^{-\omega_i{}^2 Q_i{}^2/\alpha_i} e^{-P_i{}^2/\alpha_i} \tag{2.37}$$

where α_i is related to the mode's fundamental frequency. Initial values for Q_1, P_1, and Q_2 are selected randomly within a range of accessible values, including values for Q_1 and Q_2 in the classically forbidden region. P_2 is selected by energy conservation. The relative probability for the selected Q_1, P_1, Q_2, P_2, phase point is then

$$P_w(\mathbf{Q}, \mathbf{P})|P_2|^{-1} \tag{2.38}$$

which may be sampled by the rejection method [Eq. (2.16)].

III. QUASI-CLASSICAL MODEL FOR BIMOLECULAR REACTIONS

In principle, the cross section for the reaction between A and B to form products:

$$A + B \rightarrow \text{products}$$

may be measured as a function of the A + B relative velocity v_{rel} and the vibrational–rotational energy levels of A and B [43]. Thus, for a reaction between an atom B and diatomic molecule A the reactive cross section may be expressed as $\sigma_R = \sigma_R(v_{\text{rel}}, v_A, J_A)$, where v_A and J_A are respectively the diatomic's vibrational and rotational quantum numbers. If specific values of v_A and J_A are not selected and, instead, there is a distribution of values, for example, the Boltzmann distribution specified by temperature T_A, the reactive cross section becomes

$$\sigma_r(v_{\text{rel}}; T_A) = \sum_{v_A} \sum_{J_A} \sigma_r(v_{\text{rel}}, v_A, J_A) P(v_A; T_A) P(J_A; T_A) \tag{3.1}$$

where $P(v_A; T_A)$ and $P(J_A; T_A)$ are the normalized Boltzmann distributions for v_A and J_A at temperature T_A.

Multiplying either of the above cross sections $\sigma_r(v_{\text{rel}}, v_A, J_A)$ or $\sigma_r(v_{\text{rel}}; T_A)$ by v_{rel} gives the bimolecular rate constant for a fixed relative velocity; for instance

$$k(v_{\text{rel}}; T_A) = v_{rel} \sigma_r(v_{\text{rel}}; T_A) \tag{3.2}$$

Integrating the rate constant in Eq. (3.2) over the Boltzmann relative velocity distribution $P(v_{\text{rel}}; T)$ for temperature $T = T_A$ gives the thermal bimolecular rate constant:

$$k(T) = \int_0^\infty v_{\text{rel}} \sigma_{\text{rel}}(v_{\text{rel}}; T) P(v_{\text{rel}}; T)\, dv_{\text{rel}} \tag{3.3}$$

Regardless of the form of the reaction cross section, whether for a specific vibrational–rotational level or a Boltzmann distribution of levels, the classical mechanical expression for the reaction cross section is

$$\sigma_r = \int_0^{b_{max}} P_r(b) 2\pi b \, db \tag{3.4}$$

where b is the collision impact parameter, b_{max} is the largest impact parameter that leads to reaction, and $P_r(b)$ is the probability of reaction versus b (i.e., the opacity function).

One may determine σ_r from Eq. (3.4) by integrating over $P_r(b)$ or from the average $P_r(b)$, which is given by

$$\begin{aligned} \langle P_r(b) \rangle &= \int_0^{b_{max}} P_r(b) 2\pi b \, db \bigg/ \int_0^{b_{max}} 2\pi b \, db \\ &= \int_0^{b_{max}} P_r(b) 2\pi b \, db / \pi b_{max}^2 \end{aligned} \tag{3.5}$$

Comparison of Eqs. (3.4) and (3.5) shows that

$$\sigma_r = \langle P_r(b) \rangle \pi b_{max}^2 \tag{3.6}$$

The average reaction probability $\langle P_r(b) \rangle$ is evaluated from trajectories with b chosen randomly according to the distribution function

$$P(b) \, db = \frac{2\pi b \, db}{\pi b_{max}^2} \tag{3.7}$$

Random values of b between 0 and b_{max} may be sampled with the CDF

$$R_i = \int_0^b P(b) \, db$$

to give

$$b = (R_i)^{1/2} b_{max} \tag{3.8}$$

With b chosen randomly between 0 and b_{max}, the average reaction probability is $\langle P_r(b) \rangle = N_r/N$, where N is the total number of trajectories and N_r, the number of trajectories that are reactive. Thus, the reaction cross section is

$$\sigma_r = \frac{N_r}{N} \pi b_{max}^2 \tag{3.9}$$

The rate constant $k(T)$ in Eq. (3.3) can be determined from the average reaction cross section. To show this, insert the Boltzmann distribution for $P(v_{rel}\,;T)$ into Eq. (3.3) to give

$$k(T) = \left(\frac{\mu}{2\pi kT}\right)^{3/2} 4\pi \int_0^\infty \sigma_r(v_{rel}) v_{rel}^3\, e^{-m v_{rel}^2/2kT}\, dv_{rel} \tag{3.10}$$

Changing the integration variable to $E_{rel} = \mu v_{rel}^2/2$ gives

$$k(T) = \left(\frac{8kT}{\pi\mu}\right)^{1/2} \int_0^\infty \sigma_r(E_{rel}) \frac{E_{rel}}{(kT)^2}\, e^{-E_{rel}/kT}\, dE_{rel} \tag{3.11}$$

The integral in Eq. (3.11) represents the average reaction cross section for temperature T:

$$\langle \sigma_R(E_{rel}) \rangle = \int_0^\infty \sigma_r(E_{rel}) P(E_{rel})\, dE_{rel} \tag{3.12}$$

where $P(E_{rel})$, normalized to unity for E_{rel} of 0 to ∞, is

$$P(E_{rel}) = \frac{E_{rel}}{(kT)^2}\, e^{-E_{rel}/kT} \tag{3.13}$$

Thus, $k(T)$ in Eq. (3.3) may be written as

$$k(T) = \left(\frac{8kT}{\pi\mu}\right)^{1/2} \langle \sigma_r(E_{rel}) \rangle \tag{3.14}$$

where the first term on the right-hand side is the average thermal velocity. Energy E_{rel} in Eq. (3.13) may be randomly sampled by the rejection method, Eq. (2.16), or by the CDF

$$E_{rel} = -kT\, \ln(R_1 R_2) \tag{3.15}$$

where R_1 and R_2 are two independent uniform random numbers. Choosing both the impact parameter and relative translational energy randomly in accord with Eqs. (3.8) and (3.15), respectively, gives

$$k(T) = \left(\frac{8kT}{\pi\mu}\right)^{1/2} \frac{N_r}{N} \pi b_{\max}^2 \tag{3.16}$$

The preceding presentation describes how the collision impact parameter and the relative translational energy are sampled to calculate reaction cross sections and rate constants. In the following, Monte Carlo sampling of the reactant's Cartesian coordinates and momenta is described for atom + diatom collisions and polyatomic + polyatomic collisions. Initial energies are chosen for the reactants, which corresponds to quantum mechanical vibrational–rotational energy levels. This is the quasi-classical model [2–4].

A. Atom + Diatom

Cartesian coordinates and momenta are selected for the diatomic A by first choosing Cartesian coordinates and momenta that correspond to vibrational and rotational quantum numbers v_A and J_A for A, and then randomly rotating these coordinates about the A center of mass. The first step is to transform J_A and v_A into rotational angular momentum and vibrational–energy. The rotational angular momentum j_A of A is related to its rotational quantum number J_A via

$$j_A = \sqrt{J_A(J_A + 1)}\hbar \tag{3.17}$$

The vibration–rotational energy of A is given by

$$E_{vr} = \frac{p_r^2}{2\mu} + \frac{J(J+1)\hbar^2}{2\mu r^2} + V(r) \tag{3.18}$$

where p_r is the relative momentum along the line-of-centers between the two atoms of A, r is the distance between two atoms, and $V(r)$ is the potential energy. E_{vr} is determined from J_A and v_A by satisfying the semiclassical quantization condition of Eq. (2.34), which for the energy in Eq. (3.18) becomes

$$(v_A + \tfrac{1}{2})h = \oint \left\{2\mu\left[E_{vr} - \frac{J(J+1)\hbar^2}{2\mu r^2} - V(r)\right]\right\} dr \tag{3.19}$$

Once E_{vr} is evaluated, values for r and p_r in Eq. (3.18) are sampled by recognizing the probability of a particular r is proportional to the inverse of its conjugate momentum p_r [44]: $P(r) \propto 1/p_r$. [The probability that r is found in the range $r \rightarrow r + dr$ is proportional to the amount of time the system spends in the interval dr; see discussion following Eq. (2.10).] The von Neumann rejection technique, [Eq. (2.16)]

$$\left\{\frac{P(r)}{P^0(r)}\right\} > R \tag{3.20}$$

may then be used to randomly choose an r, p_r pair. A random value for r is first chosen in the interval between the outer and inner turning points r^+ and r^-, and r accepted if Eq. (3.20) is satisfied. The momentum p_r is zero for the most probable r (i.e., at the turning point) and a small finite value of p_r is used to calculate $P^0(r)$ in Eq. (3.20) for a "most probable" r.

The preceding values for j_A, p_r, and r are transformed to Cartesian coordinates and momenta. Since the diatomic will be randomly rotated, without loss of generality, the diatom is initially placed along the x axis with angular velocity about the z axis. The specific steps are as follows:

1. Cartesian coordinates x_1 and x_2 are chosen for the two atoms of the diatomic (identified as 1 and 2) with the origin at the center of mass.
2. The radial momentum $p_r = \mu\dot{r}$ is then added with the center-of-mass at rest; i.e.

$$\dot{x}_1 = \frac{m_2\dot{r}}{m_1 + m_2}$$
$$\dot{x}_2 = \frac{-m_1\dot{r}}{m_1 + m_2} \tag{3.21}$$

   The Cartesian momenta are then $p_{x1} = m_1\dot{x}_1$ and $p_{x2} = m_2\dot{x}_2$.
3. The angular momentum is added about the z axis to give

$$j_A = I\omega_z \tag{3.22}$$

   where $I = \mu r^2$ is the moment of inertia and ω_z is the angular velocity.
4. The angular velocity is then added using the relationship $\dot{\mathbf{r}}_i = \boldsymbol{\omega} \times \mathbf{r}_i$, where $\mathbf{r}_i$ is the vectorial distance of the atom from the center of mass. With the diatom placed along the x- axis and $\boldsymbol{\omega} = \boldsymbol{\omega}_z$, the resulting Cartesian velocities are only for the y-coordinate and equal

$$\dot{y}_i = \omega_z x_i, \qquad i = 1, 2 \tag{3.23}$$

5. The preceding Cartesian coordinates selected for the diatomic are then randomly rotated through Euler's angles [45] to give a random orientation:

$$\mathbf{q} = ROT(\theta, \phi\ \chi)\mathbf{q}^0 \qquad \dot{\mathbf{q}} = ROT(\theta, \phi, \chi)\dot{\mathbf{q}}^0 \tag{3.24}$$

where $\mathbf{q}^0$ is a vector of the Cartesian coordinates selected above and $ROT(\theta, \phi, \chi)$ is the Euler rotation matrix [45]. The angles θ, ϕ, χ are chosen randomly according to

$$\cos\theta = 2R_1 - 1 \qquad \phi = 2\pi R_2 \qquad \chi = 2\pi R_3 \tag{3.25}$$

where R_1, R_2, and R_3 are three different random numbers.

6. Since A has a random orientation in a space-fixed coordinate frame, the B atom may be placed in the y, z plane without loss of generality. The x, y, z coordinates of B are then

$$x = 0 \qquad y = b \qquad z = (s^2 - b^2)^{1/2} \tag{3.26}$$

where s is the initial separation between B and the A center of mass, and b is the impact parameter.

7. The A + B relative velocity v_{rel} is now added along the z axis with the restraint that the A + B center of mass remain at rest. The space-fixed Cartesian momenta are then

$$\mathbf{P} = \mathbf{M}(\dot{\mathbf{q}} - \dot{\mathbf{q}}_{rel}) \tag{3.27}$$

The elements of the relative velocity $\dot{\mathbf{q}}_{rel}$ are zero for the x and y components and equal $[m_A/(m_A + m_B)]v_{rel}$ for the z component of each atom of B, and equal to $-[m_B/(m_A + m_B)]v_{rel}$ for the z component of each atom of A. The parameters v_{rel}, s, and b are depicted in Figure 2 for an atom + diatom collision.

The cross section $\sigma_r(v_{rel}, v_A, J_A)$ is found from Eq. (3.6) by calculating trajectories for fixed v_A, J_A, and v_{rel}. On the other hand $\sigma_r(v_{rel}; T_A)$ in Eq. (3.1) is determined if v_A and J_A are selected from their Boltzmann distributions

$$P(v_A) \propto \exp\left[-\frac{(v_A + \frac{1}{2})\hbar\omega_A}{kT}\right] \tag{3.28}$$

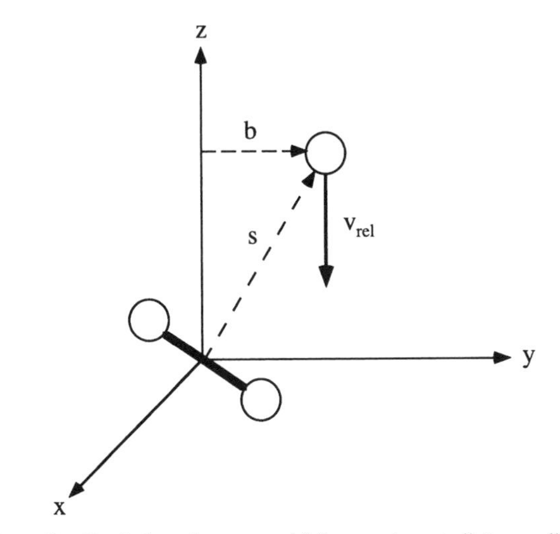

Figure 2. Depiction of v_{rel}, s, and b for an atom + diatom collision.

and

$$P(J_A) \propto (2J_A + 1) \exp\left[-\frac{J_A(J_A + 1)\hbar^2}{2I_A kT} \right] \tag{3.29}$$

These distributions may be sampled by the rejection method [Eq. (2.16)] or by the CDFs

$$\sum_{v_A=0}^{v_A} \frac{e^{-(v_A+1/2)\hbar\omega_A/kT}}{Q_{vib}} = R \tag{3.30}$$

and

$$\sum_{J_A=0}^{J_A} \frac{(2J_A + 1)e^{-J_A(J_A+1)\hbar^2/2I_A kT}}{Q_{rot}} = R \tag{3.31}$$

where the R's are two different random numbers and the Q terms are the partition functions. For most cases, the sum over J_A may be replaced by an integral. If v_{rel} is also sampled randomly, in accord with the $P(E_{rel})$ distribution in Eq. (3.13), the thermal rate constant is found from Eq. (3.16).

Equations (3.28)–(3.31) are based on the harmonic-oscillator/rigid-rotor model. The nonseparable vibrational–rotational v_A, J_A levels, with anharmonicity and vibration–rotation coupling included, may be calculated from Eq. (3.19). The Boltzmann terms for the energy levels, with the $(2J_A + 1)$

degeneracy included, may be summed to form the vibration–rotation partition function. The Boltzmann distributions for these levels may be sampled by the rejection method [Eq. (2.16)] or the appropriate CDF.

Finally, in selecting initial conditions for bimolecular reactions, it is important that the coordinates and momenta for the atoms have random classical phases as given, for example, by Eqs. (2.10) and (3.20). If this is done, the trajectory results will be independent of the initial separation between the reactants, if they no longer interact [46].

B. Polyatomic + Polyatomic

The cross section and rate constant expressions for an A + B reaction, where both reactants are polyatomics, are the same as those above for an atom + diatom reaction, except for polyatomics there are more vibrational and rotational quantum numbers to consider. If the polyatomics are symmetric tops with rotational quantum numbers J and K, the state specific cross section becomes $\sigma_r\,(v_{\rm rel}, \mathbf{v}_A, J_A, K_A, \mathbf{v}_B, J_B, K_B)$, where $\mathbf{v}_A$ represents the vibrational quantum numbers for A. If the polyatomic reactants have Boltzmann distributions of vibrational–rotational energies, the reactive cross section becomes a function of $v_{\rm rel}$ and $T = T_A = T_B$ [i.e., $\sigma_r(v_{\rm rel}\,;\,T)$] and is determined by summing over the quantum numbers as is done in Eq. (3.1). Monte Carlo sampling of the impact parameter gives the reactive cross section in Eq. (3.9), while also sampling $v_{\rm rel}$ gives the thermal rate constant expression in Eq. (3.16).

1. *Normal-Mode/Rigid-Rotor Sampling*

Choosing initial Cartesian coordinates for the polyatomic reactants follows the procedures outlined above for an atom + diatom collision and for normal-mode sampling. If the cross section is calculated as a function of the rotational quantum numbers J and K, the components of the angular momentum are found from

$$
\begin{aligned}
j &= \sqrt{J(J+1)}\hbar \\
j_z &= K\hbar \\
j_x &= (j^2 - j_z^2)^{1/2} \sin 2\pi R \\
j_y &= (j^2 - j_z^2)^{1/2} \cos 2\pi R
\end{aligned}
\tag{3.32}
$$

where R is a random number. When calculating the thermally average cross section $\sigma_r\,(v_{\rm rel}\,;\,T)$, it is sufficiently accurate to sample j and j_z from their classical Boltzmann distributions [47]

$$P(j_z) = \exp\left(\frac{-j_z^2}{2I_z kT}\right) \qquad 0 \le j_z \le \infty \tag{3.33}$$

$$P(j) = j \exp\left(\frac{-j^2}{2I_x kT}\right) \qquad j_z \le j \le \infty \tag{3.34}$$

The von Neumann rejection method [Eq. (2.16] is used to sample j_z from $P(j_z)$, whereas j is sampled by the CDF formula [47]:

$$j = [j_z^2 - 2I_x kT \ln(1 - R_1)]^{1/2} \tag{3.35}$$

The component j_x and j_y of j are found from Eq. (3.32). The vibrational quantum numbers of the reactants, $\mathbf{v}_A$ and $\mathbf{v}_B$, are fixed when calculating the state-specific cross section. However, to calculate σ_r $(v_{rel}; T)$, each vibrational quantum number is sampled from its quantum probability distribution [i.e., Eq. (3.28)].

Cartesian coordinates and momenta for a polyatomic reactant are found from the energies of its normal modes [Eq. (2.33)] and the components of its angular momentum. The procedure is given by Eqn. (210) and steps 1–3 for microcanonical normal-mode sampling in Section II.A.3.a, and is applied to both reactants. Each reactant is randomly rotated through its Euler angles, as described by Eqs. (3.24) and (3.25), and the impact parameter, center-of-mass separation, and relative velocity added as described by Eqs. (3.26) and (3.27).

2. *Sampling of Semiclassically Quantized Levels*

The preceding normal-mode/rigid-rotor sampling assumes the vibrational–rotational levels for the polyatomic reactant are well described by separable normal modes and separability between rotation and vibration. However, if anharmonicities and mode–mode and vibration–rotation couplings are important, it may become necessary to go beyond this approximation and use the Einstein–Brillouin–Keller (EBK) semiclassical quantization conditions [32]

$$J_i = \left(k_i + \frac{m_i}{4}\right)\hbar = \frac{1}{2\pi} \oint_{C_i} \sum_l p_l \, dq_l \tag{3.36}$$

to determine the actions J_i and conjugate angles θ_i for the n modes of vibration and rotation [48,49]. In Eq. (3.36), the C_i denote topologically distinct paths (i.e., a torus for a two-dimensional model) [50], the l labels

Cartesian coordinates and momenta, the k_i are quantum numbers, and m_i is an even integer known as the *Maslov index*, [51] 2, for vibrational motion. Most successful applications of Eq. (3.36) require quasi-periodic motion [50]. There is a unique topologically independent path for each mode, and each vibrational–rotational energy level is specified by a set of quantum numbers k_i.

Initial conditions for rotating–vibrating triatomic molecules have been selected with this semiclassical quantization procedure [52–54]. The first step is to use the Eckardt conditions [55,56] to transform the Cartesian coordinates and momenta to normal-mode and rotational coordinates and momenta for a body-fixed rotating-axis frame. In the most general approach, a Fourier expansion and the fast Fourier transform (FFT) method [57,58] are used to perform the EBK semiclassical quantization and find the action-angle variables J_i and θ_i for the normal-mode and rotational degrees of freedom. The angle is chosen randomly between 0 and 2π, as in Eq. (2.10), and the J_i and θ_i are back-transformed to the Cartesian coordinates and momenta used in the numerical integrations. As stated above, this method assumes quasi-periodic motion and has been applied only to triatomics. The semiclassical quantization becomes more intricate when there are resonance conditions [59] between the modes.

The adiabatic switching semiclassical quantization method [60–62] may also be used to choose initial conditions for polyatomic reactants. This approach does not require an explicit determination of the topologically independent paths C_i and actions J_i for Eq. (3.36) and, in principle, may be more easily applied to larger polyatomics than the EBK semiclassical quantization approach described above. However, what is required is a separable zero-order Hamiltonian H_0 that gives rise to the same "kind" of intramolecular motion as does the complete Hamiltonian [63,64].

Adiabatic switching is based on the Ehrenfest adiabatic theorem [65–67], which states that classical actions and quantum numbers are preserved in adiabatic, slow processes. It is assumed that the actual Hamiltonian H may be written as a sum of a separable zero-order Hamiltonian H_0 and a nonseparable part ΔH:

$$H = H_0 + \Delta H \tag{3.37}$$

The first steps in adiabatic switching are to choose mode energies for H_0, which give the desired quantum numbers, and then transform the energies to momenta and coordinates as is done for the normal-mode/rigid-rotor Hamiltonian. Since H_0 is separable, choosing the mode energies simply requires satisfying the EBK semiclassical quantization condition [Eqs. (2.34) and (3.36)] for each degree of freedom. The next step is to solve the classical trajectory that results for the time-dependent Hamiltonian $H(t)$ with the separable part ΔH slowly switched on:

$$H(t) = H_0 + S(t)\,\Delta H \tag{3.38}$$

where $S(t)$ smoothly changes from 0 to 1 as t increases from 0 to some long time T. The Ehrenfest adiabatic theorem conjectures that the quantum numbers k_i are conserved and the resulting trajectory at $t = T$ corresponds to the semiclassical quantization condition with k_i quanta in the ith mode for the complete Hamiltonian H. Thus, a properly quantized initial condition has been chosen for H. By choosing random sets of coordinates and momenta for H_0, as is done in Eq. (2.10), and then performing adiabatic switching for each of these sets, an ensemble of random initial conditions for the proper quantum state is selected for H.

C. Gas–Surface Collisions

To simulate collisions between a gaseous molecule and a surface, Cartesian coordinates and momenta are first chosen from the energies for the molecule and surface following procedures described above. The distribution of energies for a surface at temperature T_s may be chosen by either classical or quasi-classical normal-mode sampling. From classical statistical mechanics [17], the normalized probability that the ith normal mode has energy E_i is

$$P(E_i) = \frac{\exp(-E_i/kT_s)}{kT_s} \tag{3.39}$$

which may be efficiently sampled by its cumulative distribution function

$$E_i = -kT_s(1 - R) \tag{3.40}$$

For quasi-classical normal-mode sampling, the energies for the normal modes are sampled from their quantum Boltzmann energy distributions [i.e., Eq. (3.28)]. The total surface energy for the initial condition is then ΣE_i. Quasi-classical normal-mode sampling may be used to choose initial conditions for the gaseous molecule, which may have a fixed vibrational–rotational energy specified by the quantum numbers $\mathbf{v}$, J, and K or a Boltzmann distribution of energies given by Eqs. (3.28) and (3.32–3.34). The procedure for transforming the normal-mode energy of the solid and the normal-mode/rigid-rotor energy of the molecule to Cartesian coordinates and momenta is given by Eq. (2.10) and steps 1–3 for microcanonical normal-mode sampling in Section II.A.3.a. A random orientation of the gaseous molecule is chosen by randomly rotating its axis through Euler's angles [Eq. (3.24)].

To calculate a reactive cross section or rate constant for gaseous molecule–surface collisions, the five coordinates b, θ, ϕ_1, ϕ_2, and d, shown in Figure 3, are used to define the position of the center of mass of the gaseous molecule and the orientation of its velocity vector with respect to the surface plane [68,69]. The two coordinates, b and ϕ_1, are used to choose an aiming point on the surface plane. The distance from the surface reactive site b (i.e., the impact parameter) is chosen randomly between 0 and b_{max} using Eq. (3.8). The angle ϕ_1 is randomly chosen from a uniform distribution between 0 and 2π. The angle θ between the incoming H atom's velocity vector and the surface plane is chosen randomly between 0 and θ_{max}. For θ larger than θ_{max}, there are no reactive trajectories. Since the probability of θ is proportional to $\sin \theta$, θ may be chosen randomly from its CDF, which gives

$$\theta = \cos^{-1}[1 - R(1 - \cos \theta_{max})] \tag{3.41}$$

where R is a fresh random number. The angle ϕ_2, for the gaseous molecule, is chosen randomly between 0 and 2π. A sufficiently large initial separation d is chosen so that there is no interaction between the gaseous molecule and the surface.

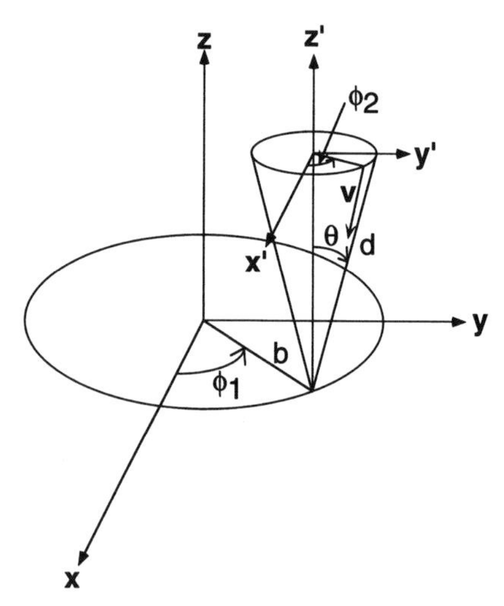

Figure 3. Definition of the coordinates used for sampling trajectory initial conditions for a gas–surface collision. (From Song et al. [69].)

The association cross section for a fixed velocity of the gaseous molecule and a surface temperature T_s is given by

$$\sigma(v_{rel}, T_s) = \tfrac{1}{2}\pi b_{max}^2(1 - \cos\theta_{max})P_r(v_{rel}, T_s) \tag{3.42}$$

where $P_r(v_{rel}, T)$ is the reaction probability with b, θ, ϕ_1, and ϕ_2 chosen randomly as described above. The factor of $\frac{1}{2}$ accounts for only one-half of the gaseous molecule in a thermal distribution moving toward the surface. If the velocity of the gaseous molecule is chosen randomly from its Maxwell–Boltzmann distribution, [e.g., (3.13)], one determines the rate constant given by

$$k(T) = \frac{1}{2}\left(\frac{8kT}{\pi m}\right)\frac{1}{2}\,\pi b_{max}^2(1 - \cos\theta_{max})P_r(T) \tag{3.43}$$

where m is the mass of the gaseous molecule.

IV. BARRIER SAMPLING

The microcanonical and canonical variational transition-state theories are based on the assumption that trajectories cross the transition state (TS) only once in forming products(s) or reactants(s) [70,71]. The correction κ to the transition-state theory rate constant is determined by initializing trajectories at the TS and sampling their coordinates and momenta from the appropriate statistical distribution [72–76]. The value for κ is the number of trajectories that form product(s) divided by the number of crossings of the TS in the reactant(s) $\rightarrow$ product(s), direction. Transition state theories assume this ratio is unity.

Microcanonical transition-state theory (TST) assumes that all vibrational–rotational levels for the degrees of freedom orthogonal to the reaction coordinate have equal probabilities of being populated [12]. The quasi-classical normal-mode/rigid-rotor model described above may be used to choose Cartesian coordinates and momenta for these energy levels. Assuming a symmetric top system, the TS energy $E^\ddagger$ is written as

$$E^\ddagger = E^\ddagger_{vJK} + E^\ddagger_n \tag{4.1}$$

where the reaction coordinate translational energy is given by $E^\ddagger_n = (P^\ddagger_n)^2/2$. The first step is to randomly select one of the TS energy levels [77]. The remaining energy $E^\ddagger - E^\ddagger_{vJK}$ is placed in reaction coordinate translation. The normal-mode coordinates and momenta are chosen from Eq. (2.10) and the rotational angular momenta from Eq. (3.32), and along with the reaction coordinate momentum, are transformed to Cartesian coordinates and momenta as described in steps 1–3 for microcanonical normal-mode sampling in Section II.A.3.a.

The trajectories for this statistical ensemble will give the TS recrossing correction κ for the microcanonical unimolecular rate constant $k(\mathrm{E})$ in Eq. (1.5). By following the trajectories to the product asymptotic limit, product vibrational, rotational, and translational energy distributions may also be determined [78].

Canonical TST assumes there is a Boltzmann distribution of the energy levels at the TS [70,75]. The canonical TST rate constant is given by

$$k(T) = \frac{kT}{h}\frac{Q^\ddagger}{Q} e^{-E_0/kT} \tag{4.2}$$

where $Q^\ddagger$ and Q are the partition functions for the transition state and reactants, respectively, and E_0 is the potential-energy difference between the TS and reactants. The normal-mode/rigid-rotor quasiclassical model may be used to sample the energy levels at the TS and calculate the correction κ for canonical TST. The energies for the vibrational and rotational modes are sampled from Eqs. (3.28) and (3.32–3.34), respectively. The reaction coordinate translational energy $E_n^\ddagger$ is treated classically, and values for $E_n^\ddagger$ may be sampled with the CDF formula of Eq. (3.40). The reaction coordinate momentum is then $P_n^\ddagger = (2E_n^\ddagger)^{1/2}$. The procedure for transforming the normal-mode energies, rotational angular momentum, and reaction coordinate momentum into Cartesian coordinates and momenta is the same as in Eq. (2.10) and steps 1–3 for microcanonical normal-mode sampling in Section II.A.3.a.

REFERENCES

1. D. L. Bunker, *Methods Comp. Phys.* **10**, 287 (1971).
2. R. N. Porter and L. M. Raff, in *Modern Theoretical Chemistry*, W. H. Miller, ed., Plenum Press, New York, 1976, p. 1.
3. D. G. Truhlar and J. T. Muckerman, in *Atom-Molecule Collision Theory*, R. B. Bernstein, ed., Plenum Press, New York, 1979, Chapter 16.
4. L. M. Raff and D. L. Thompson, in *Theory of Chemical Reaction Dynamics*, Vol. III, M. Baer, ed., Chemical Rubber, Boca Raton, FL, 1985.
5. H. Goldstein, *Classical Mechanics*, Addison-Wesley, London, 1950.
6. K. Bolton, W. L. Hase, and G. H. Peslherbe, in *Multidimensional Molecular Dynamics Methods*, D. L. Thompson, ed., World Scientific Publishing, London in press.

7. E. B. Wilson, Jr., J. C. Decius, and P. C. Cross, *Molecular Vibrations*, McGraw-Hill, New York, 1995, p. 307.
8. M. Jacob and G. C. Wick, *Ann. Phys.* **7**, 404 (1959).
9. W. L. Hase, R. J. Duchovic, X. Hu, A. Komornicki, K. F. Lim, D.-H. Lu, G. H. Peslherbe, K. N. Swamy, S. R. Vande Linde, A. Varandas, H. Wang, and R. J. Wolf, *Quantum Chemistry Program Exchange*, **16**, 671 (1996).
10. D. L. Bunker, *J. Chem. Phys.* **37**, 393 (1962); D. L. Bunker, *J. Chem. Phys.* **40**, 1946 (1964).
11. D. L. Bunker and W. L. Hase, *J. Chem. Phys.* **54**, 4621 (1973).
12. T. Baer and W. L. Hase, *Unimolecular Reaction Dynamics. Theory and Experiments*, Oxford, Univ. Press, New York, 1996).
13. M. C. Gutzwiller, *Chaos in Classical and Quantum Mechanics*, Springer-Verlag, New York, 1990.
14. W. L. Hase, in *Modern Theoretical Chemistry*, Vol. 2, W. H. Miller, ed., Plenum Press, New York, 1976, p. 121.
15. W. L. Hase and D. G. Buckowski, *J. Comp. Chem.* **3**, 335 (1982).
16. M. J. Davis and R. T. Skodje, in *Advances in Classical Trajectory Methods*, Vol. 1, W. L. Hase, ed., JAI Press, London, 1992, p. 77.
17. D. A. McQuarrie, *Statistical Thermodynamics*, Harper & Row, New York, 1973, p. 117.
W. L. Hase and D. G. Buckowski, *Chem. Phys. Lett.* **74**, 284 (1980).
19. W. Forst, *Theory of Unimolecular Reactions*, Academic Press, New York, 1973.
20. S. Chapman and D. L. Bunker, *J. Chem. Phys.* **62**, 2890 (1975).
21. J. M. Hammersley and D. C. Handscomb, *Monte Carlo Methods*, Methuen, London, 1964, p. 36.
22. S. Califano, *Vibrational States*, Wiley, New York, 1976, p. 23.
23. C. S. Sloane and W. L. Hase, *J. Chem. Phys.* **66**, 1523 (1977).
24. W. L. Hase, R. J. Wolf, and C. S. Sloane, *J. Chem. Phys.* **71**, 2911 (1979).
25. I. N. Levine, *Quantum Chemistry*, Allan & Bacon, Boston, 1983.
26. E. S. Severin, B. C. Freasier, N. D. Hamer, D. L. Jolly, and S. Nordholm, *Chem. Phys. Lett.* **57**, 117 (1978); G. Nyman, K. Rynefors, and L. Holmlid, *J. Chem. Phys.* **88**, 3571 (1988); G. Nyman, S. Nordholm, and H. W. Schranz, *J. Chem. Phys.* **93**, 6767 (1990); H. W. Schranz, S. Nordholm, and G. Nyman, *J. Chem. Phys.* **94**, 1487 (1991).
27. T. Uzer, *Phys. Rep.* **199**, 73 (1991).
28. M. S. Child and L. Halonen, *Adv. Chem. Phys.* **57**, 1 (1984).
29. E. J. Heller and M. J. Davis, *J. Phys. Chem.* **84**, 1999 (1980); D.-H Lu and W. L. Hase, *J. Phys. Chem.* **92**, 3217 (1988).
30. R. E. Wyatt, C. Iung, and C. Leforestier, *J. Chem. Phys.* **97**, 3477 (1992).
31. M. Topaler and N. Makri, *J. Chem. Phys.* **97**, 9001 (1992).
32. A. Einstein, *Ver. Deut. Phys. Ges.* **19**, 82 (1917); L. Brillouin, *J. Phys.* **7**, 353 (1926); J. Keller, *Ann. Phys.* **4**, 180 (1958).
33. L. Pauling and E. B. Wilson Jr., *Introduction to Quantum Mechanics*, McGraw-Hill, New York, 1935, p. 271; L. D. Landau and E. M. Lifshitz, *Quantum Mechanics*, 2nd ed., Addison-Wesley, New York, 1965, p. 71.
34. S. Goursaud, M. Sizun, and F. Fiquet-Fayard, *J. Chem. Phys.* **65**, 5453 (1976); E. J. Heller, *J. Chem. Phys.* **68**, 2066 (1978); H.-W. Lee and M. O. Scully, *J. Chem. Phys.* **77**, 4604 (1982).

35. R. Schinke, *J. Phys. Chem.* **92**, 3195 (1988).
36. R. Schinke, *Photodissociation Dynamics*, Cambridge Univ. Press, 1993, p. 98.
37. M. D. Pattengill, *Chem. Phys.* **68**, 73 (1981).
38. A. Garcia-Vela, R. B. Gerber, and J. Valentini, *J. Chem. Phys.* **97**, 3297 (1992).
39. K. M. Christoffel and J. M. Bowman, *J. Chem. Phys.* **104**, 8348 (1996).
40. K. Husimi, *Proc. Math. Soc. Jpn.* **22**, 264 (1940).
41. K. Takahashi, *Prog. Theor. Phys. Suppl.* **98**, 109 (1989).
42. E. Wigner, *Phys. Rev.* **40**, 749 (1932).
43. R. D. Levine and R. B. Bernstein, *Molecular Reaction Dynamics and Chemical Reactivity*, Oxford Univ. Press, New York, 1987.
44. R. N. Porter, L. M. Raff, and W. H. Miller, *J. Chem. Phys.* **63**, 2214 (1975).
45. Reference 7, p. 285.
46. J. M. Bowman, A. Kupperman, and G. C. Schatz, *Chem. Phys. Lett.* **19**, 21 (1973).
47. D. L. Bunker and E. A. Goring-Simpson, *Faraday Discuss. Chem. Soc.* **55**, 93 (1973).
48. M. S. Child, *Semiclassical Mechanics with Molecular Applications*, Oxford Univ. Press, New York, 1991, p. 68.
49. M. C. Gutzwiller, *Chaos in Classical and Quantum Mechanics*, Springer-Verlag, New York, 1990, p. 32.
50. Reference 48, p. 147.
51. V. P. Maslov and M. V. Fedoriuk, *Semi-Classical Approximation in Quantum Mechanics*, Reidel, Boston, 1981.
52. G. C. Schatz, *Comp. Phys. Commun.* **51**, 235 (1988).
53. B. G. Sumpter and G. S. Ezra, *Chem. Phys. Lett.* **142**, 142 (1987).
54. C. W. Eaker and D. W. Schwenke, *J. Chem. Phys.* **103**, 6984 (1995).
55. Reference 5, p. 273.
56. Reference 22, p. 17.
57. C. W. Eaker, G. C. Schatz, N. DeLeon, and E. J. Heller, *J. Chem. Phys.* **81**, 5193 (1984).
58. C. C. Martens and G. S. Ezra, *J. Chem. Phys.* **83**, 2900 (1985).
59. Reference 42, p. 159.
60. E. A. Solo'ev, *Sov. Phys. JETP* **48**, 635 (1978).
61. B. R. Johnson, *J. Chem. Phys.* **83**, 1204 (1985).
62. R. T. Skodje, F. Borondo, and W. P. Reinhardt, *J. Chem. Phys.* **82**, 4611 (1985).
63. W. P. Reinhardt, *Lasers, Molecules and Methods*, J. O. Hirschfelder, R. E. Wyatt, and R. D. Coalson, eds., Wiley, New York, 1989, Chapter 23.
64. J. Huang, J. J. Valentini, and J. T. Muckerman, *J. Chem. Phys.* **102**, 5695 (1995).
65. P. Ehrenfest, *Phil. Mag.* **33**, 500 (1917).
66. M. Born and V. Fock, *Zeit. Physik.* **51**, 165 (1928).
67. W. P. Reinhardt, in *Classical Dynamics in Atomic and Molecular Physics*, T. Grozdanov, P. Grujic, and P. Krstic, eds., World Scientific, New York, 1989, p. 26.
68. I. Noorbatcha, L. M. Raff, and D. L. Thompson, *J. Chem. Phys.* **81**, 3715 (1984); B. M. Rice, I. Noorbatcha, D. L. Thompson, and L. M. Raff, *J. Chem. Phys.* **86**, 1602 (1987); M. D. Perry and L. M. Raff, *J. Phys. Chem.* **98**, 4375 (1994).
69. K. Song, P. de Sainte Claire, W. L. Hase, and K. C. Hass, *Phys. Rev. B* **52**, 2949 (1995).

70. W. H. Miller, *Acc. Chem. Res.* **9**, 306 (1976).
71. D. G. Truhlar and B. C. Garrett, *Acc. Chem. Res.* **13**, 440 (1980).
72. J. C. Keck, *Adv. Chem. Phys.* **13**, 85 (1967)
73. J. B. Anderson, *J. Chem. Phys.* **58**, 4684 (1973).
74. W. H. Miller, *J. Chem. Phys.* **61**, 1823 (1974).
75. I. W. M. Smith, *J. Chem. Soc. Faraday Trans. 2* **77**, 747 (1981); R. J. Frost and I. W. M. Smith, *Chem. Phys.* **111**, 389 (1987).
76. D. G. Truhlar, A. D. Isaacson, and B. C. Garrett, in *Theory of Chemical Reaction Dynamics*, Vol. 4, M. Baer, ed. CRC, Boca Raton, Fl., 1985, p. 65; D. G. Truhlar and B. C. Garrett, *Faraday Discuss. Chem. Soc.* **84**, 464 (1987).
77. C. Doubleday, Jr., K. Bolton, G. H. Peslherbe, and W. L. Hase, *J. Am. Chem. Soc.* **811**, 9922 (1996).
78. W. Chen, W. L. Hase, and H. B. Schlegel, *Chem. Phys. Lett.* **228**, 436 (1994).

MONTE CARLO APPROACHES TO THE PROTEIN FOLDING PROBLEM

JEFFREY SKOLNICK

Department of Molecular Biology, The Scripps Research Institute, La Jolla, CA 92037

ANDRZEJ KOLINSKI

Department of Chemistry, University of Warsaw, Warsaw, Poland

CONTENTS

Advances in Chemical Physics, Volume 105, Monte Carlo Methods in Chemical Physics, edited by David M. Ferguson, J. Ilja Siepmann, and Donald G. Truhlar. Series Editors I. Prigogine and Stuart A. Rice.
ISBN 0-471-19630-4

I. INTRODUCTION

Under the appropriate solvent and temperature conditions, globular proteins adopt a well-defined three-dimensional structure [1–3] that is capable of performing a specific biochemical or structural task not only in a living cell [4] but also in vitro. This unique three-dimensional structure is encoded in the protein amino acid sequence. The question of how to relate the sequence to the native structure is commonly referred to as the protein folding problem [5,6]. It is widely believed that proteins obey the "thermodynamic hypothesis" that says that the native conformation of a protein state corresponds to a global free-energy minimum, which is dictated by the various interactions present in a system comprised of proteins and solvent [7,8]. Unfortunately, due to the complexity of the interactions within proteins and between proteins and the surrounding solvent, the task of finding this free-energy minimum in the myriad of multiple minima in the free-energy landscape [9,10] is extremely difficult.

While molecular dynamics (MD) simulations have proven to be very powerful for studying numerous aspects of protein dynamics and structure [11–13], this technique cannot yet access the millisecond-to-second timescales required for folding even a small protein. To address this timescale gap, one has to simplify the protein model by reducing the number of degrees of freedom. Such approaches assume that the basic physics could be reproduced in model systems that employ "united atoms" and effective solvent models. On the basis of recent work, it has become apparent that the crux of the solution to the protein folding problem does not lie in whether a reduced protein model is used, but rather in the development of potentials that can recognize a native-like state from a misfolded state, and techniques that can explore the relevant regions of conformational space [14–20]. In this review chapter, we describe approaches that address these difficulties, and focus in particular on Monte Carlo based approaches to conformational sampling.

II. PROTEIN REPRESENTATION, FORCE FIELD, AND SAMPLING PROTOCOLS

Various levels of simplification of the protein's structure and energy surface have been assumed. As a result, different aspects of the protein folding problem can be addressed by computer simulations. Somewhat arbitrarily, we divide protein models into three classes: simple, exact models; protein-like models; and realistic protein models. Roughly speaking, the level of structural detail taken into account increases in parallel with the classification scheme. At the same time, the resulting simulations become more and more expensive, and analysis of the particular model is less comprehensive. In what follows, we start from the description of simple, exact models that can be studied in great detail. Then, we outline studies of somewhat more complex models of protein-like systems. Finally, realistic protein models are described. Since this issue is devoted to the application of Monte Carlo methods in chemical physics, we focus our attention on studies that employed Monte Carlo techniques as the conformational sampling protocol. However, for completeness, models studied by means of different sampling methods, but which are amenable to investigation by a Monte Carlo–based methodology, will be also mentioned.

A. Sampling Protocols

1. *General Considerations*

Exploration of the conformational space of protein models could be done using different computational techniques. These include MD [21], Brownian dynamics [22,23], Monte Carlo methods [24–27], and other simulation or optimization techniques such as genetic algorithms [25,28–31].

2. *Monte Carlo Sampling Methods*

To address the conformational sampling problem, reduced protein models use techniques that allow for a faster and more efficient exploration of protein conformational space than can be provided by MD. Here, various Monte Carlo techniques are commonly used [32,33]. In a well-designed Monte Carlo sampling method, relatively large conformational changes occur in a single iteration of the simulation algorithm; therefore, the numerous local minima of the energy surface are more easily surmounted.

The well-known Metropolis Monte Carlo (MMC) procedure randomly samples conformational space according to the Boltzmann distribution of (distinguishable) conformations [34]:

$$P_i = \exp\left(\frac{-E_i}{kT}\right) \tag{2.1a}$$

In order to generate this distribution in asymmetric MMC, the transition probability $P_{i,j}$ from an "old" conformation i to a "new" conformation j is controlled by the energy difference $\Delta E_{ij} = E_j - E_i$ via

$$p_{i,j} = \min\left\{1, \exp\left(\frac{-\Delta E_{ij}}{kT}\right)\right\} \tag{2.1b}$$

Obviously, this technique is very sensitive to the presence of energy barriers. To ensure adequate sampling, typically a collection of elemental moves involving end moves and collective motions of two to four bonds are performed randomly. In addition, small distance motions of a large, randomly selected part of the chain are employed. The key to a successful dynamic Monte Carlo protocol is to include a sufficiently large move set so that no element of structure is artificially frozen in space.

To enhance the sampling efficiency, Hao and Scheraga have employed the entropy-sampling Monte Carlo method (ESMC) in their study of simplified protein models [35–37]. ESMC was originally proposed by Lee [38] in the context of a simple Ising model, which is closely related to the multicanonical MC technique of Berg and Neuhaus [39]. Since the formulation of Hao and Scheraga [35–37] is the most straightforward, we briefly review their approach; additional details are found in the chapter "Entropy Sampling Monte Carlo for Polypeptides and Proteins." Unlike MMC, ESMC generates an artificial distribution of states that is controlled by the conformational entropy as a function of the energy of a particular conformation E_i:

$$P_i^{\text{ESMC}} = \exp\left(\frac{-S(E_i)}{k}\right) \tag{2.2a}$$

The transition probability can be formally written as

$$p_{i,j}^{\text{ESMC}} = \min\left\{1, \exp\left(\frac{-\Delta S_{i,j}}{k}\right)\right\} \tag{2.2b}$$

where $\Delta S_{i,j}$ is the entropy difference between energy levels i and j, respectively. At the beginning of the simulation, the entropy is unknown;

however, from a density-of-states energy histogram, $H(E)$, an estimate, $J(E)$, for the entropy $S(E)$ can be iteratively generated. After a sufficient number of runs, all states are sampled with the same frequency. Then, the histogram of $H(E)$ becomes flat, and the curve of $J(E)$ + constant approaches the true $S(E)/E$ curve. ESMC offers the advantage that one can discern whether the simulation has converged over the sampled energy range, and, if so, one has directly determined the free-energy of the system (within a constant). However, a potentially serious problem with ESMC is its computational cost. Kolinski et al. [40] have described a means of increasing the rate of convergence of the simulation. The basic idea is to first run a standard dynamic Monte Carlo folding simulation to provide a library of "seed" structures, and then, periodically, randomly select one of these conformations and introduce it into the conformational pool used to construct the $S(E)/E$ curve. In practice, this technique works quite well.

3. *Use of Dynamic Monte Carlo to Simulate Protein Dynamics*

As demonstrated by Orwoll and Stockmayer [41], a proper sequence of small, random conformational changes of lattice polymers constitutes a solution of a stochastic equation of motion. Of course, the same move sets could be generalized to off-lattice models. However, in all cases, the problem of ergodicity has to be taken into consideration. This problem could be especially dangerous for low-coordination-number lattices. With a high-coordination-number lattice and correspondingly large number of allowed local conformational transitions, the risk of a serious ergodicity problem becomes negligible [42]. A properly designed scheme for local conformational transitions allows simulations of the long-time dynamics of model polymeric systems [43–45], with results consistent with MD [46].

The design of an efficient model of Monte Carlo dynamics for off-lattice models is a nontrivial task. When using all-atom models with the standard semiempirical force fields, the local conformational transitions usually introduce very small changes in chain geometry and are usually uncorrelated with the nature of the energy surface. Consequently, the efficiency of such MC algorithms is not much better than standard MD. An effective set of local moves has to be able to pass over (or rather to neglect) the intervening narrow-width, local energy maxima (which may be of considerable height). Such "smart" moves have to take into account and exploit specific correlations of rotational degrees of freedom [medium-range phi–psi (ϕ–ψ) correlations] over some portion of the polypeptide backbone. One such algorithm has been recently elaborated on by Elofsson et al. [31]. An even more radical approach to local moves has been proposed by Knapp et al. [25,26,30], who assumed a rigid-body representation of peptide bond plates and developed an algorithm that allows for the collective motion of a few peptide units. The moves can interpolate between deep energy minima for phi–psi correlated potentials. When the tertiary interactions are neglected, the model reproduces Rouse dynamics [47,48] of denatured polypeptide chains, that is, the correct long-time dynamics of a polymer in the free draining limit [25,26,30]. Of course, long-range interactions can be introduced, and the model could be employed in folding studies [49]. Efficient local continuous-space deformations of polypeptide chains also have been investigated by Baysal and Meirovitch [50]. These developments are expected to extend the applicability of Monte Carlo off-lattice models in computational studies of protein dynamics and protein folding processes.

Lattice Monte Carlo dynamics of proteins could be studied at various levels of generalization. The dynamics of simple lattice models and equivalent continuous protein-like models led to a reasonable approximation of the main features of the denatured state and gives some insight into protein folding pathways [51–53]. More detailed models can provide additional information on the effect of short-range interactions on polypeptide dynamics at various temperatures, and consequently at various distances from the native state. It has been shown that a high-coordination-lattice discretization of protein conformational space [16,17,20,40,42] leads to Rouse-like dynamics [48] in the denatured state [54]. It was possible to investigate the effect of the short- and long-range cooperativity on the dynamic properties of denatured proteins [55]. Also, the dynamics near the transition state of medium-resolution protein models have been analyzed [16,42,56]. These experiments indicate that high-resolution models could be a powerful tool for model studies of the long-time dynamics of real proteins. The approach is an alternative to the off-lattice models of protein long-time dynamics.

In summary, the recent developments allow Monte Carlo simulations of protein dynamics in their denatured state, in the intermediate states and in the folded state. In the latter case, where the fine details are of major interest, standard MD techniques are usually superior to the Monte Carlo reduced model approaches.

B. Simple, Exact Models

Simple, exact models [51] consist of simple lattice polymers or heteropolymers where each amino acid in the hypothetical protein is represented by a bead occupying a single lattice point. Because of their simplicity, these models have generated considerable attention [57,58].

Short homopolymeric chains restricted to a simple square lattice and to a simple cubic lattice have been studied by Chan and Dill [59–61]. For very small systems, all compact conformations could be enumerated. Not surprisingly, significant intrachain entropy loss occurs on collapse. Expanded conformations could be sampled by a simple Monte Carlo scheme employing various local and global conformational transitions [62]. Chan and Dill suggested that some short-range backbone correlations, similar to protein secondary structure, may be induced solely due to chain compactness [59,60]. This postulate was later questioned [55,63–66]. Also, an earlier computer study of the collapse transition in long flexible lattice homopolymers did not indicate any local ordering of the chain segments [44]. Interestingly, it has been shown that the collapse transition of semiflexible polymers leads to substantial and cooperative local ordering in the compact state [43,67,68].

Much could be learned from computer studies of simple, exact heteropolymeric models of proteins. A classic example is the HP model introduced by Dill and co-workers [61,69,70]. In this model, a simple cubic lattice chain consists of two kinds of beads, H (P), corresponding to nonpolar hydrophobic (polar) residues. Nonbonded nearest neighbor contacts of two hydrophobic (HH) residues contribute a constant negative energy. HP and PP contacts are assumed to be inert. A similar model (AB model) has been investigated by Shakhnovich and co-workers [71,72]. In this case, all residues are attractive; however, the energy of attractive interactions was lower for pairs of the same type (AA or BB) than for different residue pairs (AB). Most simulations were done on 27-mer systems with the target structure being a $3 \times 3 \times 3$ compact cube.

Both models have similar properties and mimic some general features of globular proteins. Not all sequences undergo a collapse transition to a well-defined compact state (neglecting the existence of mirror-image structures that emerge as a result of lack of chiral energy terms in these simple models). This is in accord with the idea that only a small fraction of possible amino acid sequences form well defined globular structures [70,72–76]. Moreover, only a small fraction of HP or AB sequences are "good folders." In cases of such "good folders," the collapse transition is very cooperative. Various sequences of simple, exact models exhibit different folding pathways [76–78]. The folding pathways may suggest (on a very general level) how real proteins reduce their conformational space during the folding process [79–82].

Recent studies of larger-size, simple cubic lattice heteropolymers [83] show that with increasing protein size, the folding pathway becomes more complex. In these models, the initial hydrophobic collapse stage has been followed by a finer structure fixation involving the rearrangement of the surface residues. Some signatures of such a two-stage folding mechanism also have been observed in 27-mer AB- or HP-type systems [81,84,85]. These larger-model systems permit a more meaningful albeit, due to lattice restrictions, still very biased modeling of protein secondary structure.

Longer cubic lattice heteropolymeric chains may also mimic some aspects of multidomain protein folding [86].

Camacho and Thirumalai [87] investigated the effect of strong covalent interactions (disulfide bonds) on the folding pathway of a minimal lattice model. They found that the folding rate is fastest when a proper balance is achieved between effective hydrophobic interactions and the strength of disulfide bonds. This rationalizes how weak denaturing conditions can accelerate the folding process of real crosslinked proteins.

Finally, it should be pointed out that simple, exact models could be used to test new sampling methodologies [88–90] or optimization techniques [84,91]. A good example is the work by Hao and Scheraga [92], where pairwise contact energy parameters have been optimized to maximize the stability of the native state with respect to other conformations for a 27-mer chain composed of 10 different residues types [92].

The tolerance of the protein folding process to noise in the interaction parameters and sequence modifications [93] was also studied using a $3 \times 3 \times 4$ target structure [94]. The conclusions were in agreement with the qualitative picture of real protein folding dynamics and thermodynamics [93]. Such studies are important because they may provide some insights into how specific the protein sequence has to be in order to maintain the structural uniqueness of the native conformation.

It is also possible to design a simple, exact model with side chains [95]. As would be expected, the presence of side chains increases the entropy difference between folded and unfolded states. The possible effects of somewhat more complex and cooperative interactions have also been investigated [96].

Extensive studies of simple, exact models have provided new insights into general principles governing protein folding dynamics, thermodynamics, and the relationship between protein sequence and protein structure. The importance of hydrophobic collapse in the initial stage of the protein folding process is clearly visible in a majority of the preceding studies. As in real proteins, foldable sequences in simple, exact models exhibit specific amino acid patterns of hydrophobic and hydrophilic residues. The sequences of these models control not only the stability of the globular state but also the dynamics of folding. Very likely these and other important generalizations of computer studies stimulated the development of new theoretical descriptions of the protein folding process [58,85,97–102].

C. Protein-Like Models

The simple, exact models described in the previous section employ nearest-neighbor interactions as the only amino acid sequence specific contribution to the conformational energy. Consequently, the dominance of nonpolar tertiary interactions in protein folding is implicitly assumed in these models. Short-range (i.e., local down the chain) interactions are a priori neglected, or more precisely, they implicitly arise from inherent restrictions imposed by the use of a very low-coordination-number lattice. In this section, we describe a class of more complex, protein-like models that tries to address the interplay between secondary and tertiary interactions, a feature characteristic of real proteins. Thus, they account for a wider variety of interactions that control protein folding dynamics, thermodynamics, and the structural uniqueness of the native state. They also attempt to more realistically mimic aspects of the geometry of the major secondary structure motifs that occur in proteins. However, these are not faithful models of real proteins because they lack the ability to reproduce the detailed chain geometry and packing of supersecondary structural elements. In this respect, such protein-like models are similar to simple, exact models, but here the secondary structure and interaction scheme are better defined. We separately review lattice and continuous space models of such protein-like systems.

1. *Lattice Models of Protein-Like Systems*

In their pioneering investigations, Go and co-workers [103–105] investigated the dynamics and thermodynamics of square lattice chains as highly idealized protein-like systems. Three types of interactions were considered: short-range secondary structure conformational propensities, long-range contact potentials (similar to that employed in simple, exact models), and hydrophobic potentials. The sampling scheme consisted of various types of conformational transitions controlled by the Metropolis Monte Carlo algorithm. The results of these and other studies [106,107] showed that long-range interactions consistent with the assumed target structure increase folding cooperativity. In contrast, short-range (but also native-like) interactions decrease folding cooperativity, but accelerate the folding process. All interactions inconsistent with the target structure decrease the folding rate. Similar conclusions were drawn from geometrically more complex models with a similar interaction scheme [108].

Systematic diamond lattice, Monte Carlo studies of various types of interactions and their effects on folding thermodynamics, structure and stability, and folding mechanism have been done by Kolinski et al. [43,67,68] and Skolnick et al. [109–117] The interaction scheme was gradually modified, when necessary, so as to reproduce various features of protein folding. In the simplest model, they investigated the interplay between short- and long-range interactions for various homopolymeric chain lengths. Short-range interactions were modeled as a preference for the *trans* conformation. Long-range interactions (i.e., those that are local in space, but far apart down the chain) were simulated by an energetic preference for contacts between nonbonded chain units. Flexible polymers (having the same a priori probability of all rotational isomeric states) of any finite length undergo a continuous transition to a compact, disordered globular state [44,67]. This is typical for all flexible polymer systems [118,119]. Semi-flexible polymers exhibited a qualitatively different behavior. When the ratio of short-range interactions (the stiffness parameter) to long-range interactions exceeds some critical (but moderate) value, the collapse transition for moderate length chains (50–200 units) becomes discontinuous (pseudo-first-order). There is a cooperative increase of the apparent chain stiffness on collapse, and the globular state exhibits almost perfect orientational ordering [67]. Under the same conditions, longer polymers form several ordered domains [68]. This closely resembles the formation of multiple domains in real proteins. It could be expected that for sufficiently long chains, the collapse transition is again continuous. In all cases, the compact globular state of homopolymers was not unique [43].

Introducing neutral (flexible and neutral with respect to the long-range contacts) turn regions leads to a unique, minimal, four-member (β-barrel globular state. Furthermore, the conformational transition is of the all-or-none type [109,111]. When robust patterns of hydrophobic and hydrophilic residues were introduced and/or when the native conformation of the turns was favored, in all folding simulations, it was possible to reproduce the nontrivial, six-member, Greek key, β-barrel topology [110,112]. The folding scenario could be described as on-site assembly, where already folded fragments of native structure serve as a folding scaffold for the remaining portions of the model chain [110,113].

In a similar way, the minimal requirements for the folding of all typical topologies of four-helix bundles were elucidated [114-116]. In all these studies, the folding process was accelerated when an orientational coupling of long-range interactions had been implemented to model hydrogen bond interactions. The effects of hydrogen bonds and other polar interactions incorporated into a very simple protein-like model have also been investigated by O'Toole et al. [120]. Finally, the diamond lattice protein model has also been used to investigate the possible differences between *in vivo* and *in vitro* folding of typical protein motifs [117].

The diamond lattice representation of a protein, while reasonable for idealized models of β and α proteins, cannot reproduce the chain geometry of more complex motifs such as α/β proteins. To remove this fundamental limitation, the chess-knight model [54] of the protein backbone was developed. In contrast to a diamond lattice model, it allows all possible protein folding motifs to be represented at low resolution [121]. The force field requirements for a unique native state are essentially the same in this finer lattice representation as for diamond lattice models.

The aforementioned studies indicated the crucial importance of the proper balance between short-range versus long-range interactions in order to obtain the requisite cooperativity of the folding process, as well as a unique native state. They also show that the sequence information required for a given structure is rather robust. In addition, a sequential, on site, mechanism of folding was preferred in all cases. Qualitatively, these findings are in agreement with conclusions from simple, exact models and even more so with those of Go and co-workers [104]; however, they are more detailed because of the use of more complex models.

Brower et al. [122] investigated the chain-length dependence of the time required to find the global energy minimum in the framework of various high-coordination-number lattice models and a simple interaction scheme. It has been found that the folding time when simple simulated annealing is used increases exponentially with the chain length. This points out the necessity of more efficient sampling algorithms for the conformational search in nontrivial protein models. In this respect, Finkelstein et al. [123–125] have proposed a very interesting Monte Carlo model of protein assembly, where larger polypeptide fragments are moved around to find an optimal structure.

2. *Continuous Models of Protein-Like Systems*

Valuable insights into the thermodynamics and mechanisms of protein assembly come from Monte Carlo and Brownian dynamics studies of simple, continuous protein-like models. For example, Rey and Skolnick [126] studied a series of reduced models of four-helix bundles. They investigated the role of the balance between short- and long-range interactions in the folding pathways, and the stability and structural uniqueness of the four-helix bundle folding motif. Monte Carlo simulations for various types of models enabled a qualitative assessment of the role of side chains in protein structure fixation. The results have been generalized into a phase diagram describing various regimes of polypeptide behavior. Possible implications for real protein folding and the design of artificial proteins have also been discussed. In related work, Rey and Skolnick [127] demonstrated that the dynamics and folding pathways of various simple folding motifs studied via Brownian dynamics and Monte Carlo dynamics are qualitatively the same. Moreover, the general dynamical characteristics of these continuous models were essentially the same as for related lattice models studied by means of lattice Monte Carlo [127].

Continuous models of minimal β sheets have been investigated by Monte Carlo simulated annealing [128] and by Brownian dynamics [23,52, 53,129]. The design of the β-barrel geometry and the essential features of the force field are complementary to the diamond lattice models of Skolnick et al. that were studied much earlier [109,111]]. Again, the behavior of lattice and continuous models are qualitatively the same [23,52,53,129]. More recently, the coil-to-β-sheet transition and its cooperativity and relation to existing theories has been studied via Monte Carlo simulations on a reduced continuous model [130].

D. Models of Real Proteins

1. Continuous Models

To make protein folding simulations computationally tractable, the protein model must be simplified [131]. Among the earliest examples of such models is the original approach of Levitt and Warshel [22]. There the polypeptide main chain was reduced to its α-carbon trace, and side chains modeled as single united atoms. Long-range interactions were counted only for pairs of side chains and approximated by a Lennard-Jones potential. A constant value of the planar angle between the two consecutive pseudobonds of the α-carbon trace has been assumed. This is a simplification since real proteins exhibit a bimodal distribution of this angle. The local degrees of freedom were the dihedral angles of the α-carbon trace. The torsional potentials associated with these degrees of freedom were derived from a conformational analysis of several "representative" dipeptides. The model has been used in folding simulations of bovine pancreatic trypsin inhibitor. Simulations were done by means of Brownian dynamics; however, the Monte Carlo method could be used as well as for conformational sampling. Similar models have been developed and studied by Kuntz et al. [132], Robson and Osguthorpe [133], and Hagler and Honig [134]. The work of Hagler and Honig showed that similar quality structures of BPTI could be obtained using a sequence code reduced to just two types of amino acids.

Another example of a continuous reduced model is due to Wilson and Doniach [24]. They assumed a fixed planar conformation of peptide bonds, and the main chain was allowed to change its conformation by rotation around the phi (ϕ) and psi (ψ) angles. A single united atom representation of each side group also has been assumed. The importance of their work lies in the systematic application of a knowledge-based force field, an idea applied previously to other aspects of the protein folding problem [135,136]. Short-range interactions were simulated by restricting the allowed values of the phi–psi angles to those commonly observed conformations in known folded protein structures. For long-range (tertiary) interactions, they derived distance dependent pairwise potentials based on a statistical analysis of the regularities in known globular protein structures. Conformational sampling was done using simulated thermal annealing within the framework of a Metropolis-type Monte Carlo scheme. They performed folding simulations of crambin both with and without assumed knowledge of the native secondary structure. Overall, the accuracy of the predicted structures was very low. However, the predicted secondary structure was to a large extent in agreement with that of the native protein and elements of a protein-like hydrophobic core were formed. Also, the native-like pattern of cystine crosslinks was observed.

The accuracy of a reduced protein representation could be improved by taking into account some internal degrees of freedom of the side chains [137]. For example, larger side chains could be represented by two united atoms [138]. Alternatively, an all-atom representation of the main chain could be employed with a reduced representation of the side chains [28]. Using this kind of representation and more elaborate statistical potentials, the structure of short peptides such as melittin, pancreatic polypeptide inhibitor, apamin [28], PPT, and PTHrP [138] have been predicted with an accuracy ranging from 1.7-Å root-mean-square deviation, (RMSD) (measured for the α-carbon positions) for the small single helix of melittin to 4.5-Å RMSD for larger peptides.

Reduced continuous-space models were also employed in studies of the various aspects of protein [27,139] or polypeptide [49,140,141] folding.

2. Discretized, Lattice Models

Lattice models have proved to be extremely useful in studies of simple, exact models and somewhat more complex models of protein-like systems. Similarly, the conformational space of a real protein could be discretized and explored in a very efficient way by various versions of Monte Carlo sampling techniques. Depending on the assumed level of discretization and model of force field, various levels of accuracy can be achieved. For example, simple lattice models of real proteins were studied by Ueda et al. [108], Krigbaum and Lin [142], Dashevskii [143], Covell [144], Covell and Jernigan [145], Hinds and Levitt [146], and others.

As mentioned before, the studies of protein-like models by Go et al. [103,104,107] provided a plausible explanation for the origin of protein folding cooperativity and the role played by short- and long-range interactions in both folding and stability. Monte Carlo studies of a lattice model of lysozyme [108] provided a very nice demonstration of these findings applied to a real protein. The target structure of this 129-residue protein was represented by 116 simple cubic lattice main-chain units plus an additional 15 lattice points for some larger side chains. Thus, there was no one-to-one correspondence between the model and the real protein structural units, but the overall geometry and dense packing of the native structure was reproduced with reasonable accuracy.

Krigbaum and Lin [142] studied a bcc (body-centered cubic) lattice model of PTI. Their simulations demonstrated that some overall features of the native fold (at least for very small structures) could be reproduced using only a one-body, centrosymmetric potential that describes the burial preference of various amino acids.

Covell and Jernigan [145] enumerated all possible compact conformations of several small proteins restricted to an fcc (face-centered cubic) lattice (one residue-per-lattice site). Using pairwise interactions, they found that the native-like conformation could be always found within 2% of the lowest-energy lattice structures. Very much in this spirit, Hinds and Levitt [146] used a diamond lattice to search compact chain conformations where the lattice defines a chain tracing, but each lattice vertex need not correspond to a single residue. Subsequently, Covell [144] investigated simple cubic lattice chains. However, now a more elaborate interaction scheme was used, which consisted of pairwise interactions, surface exposure terms and a packing density regularizing term. These potentials were derived from a statistical analysis of a protein structural database. During Monte Carlo simulations, which employed various local and global moves, the model chains rapidly collapsed to compact states having some features of native proteins.

Kolinski and Skolnick [42,147] developed a series of high-coordination lattice protein models that are capable of describing protein structures at various levels of resolution. The simplest of these, the chess-knight model [54], assumes an α-carbon representation of the polypeptide chain. The model enabled Monte Carlo folding pathway simulations of complex real protein structures, including plastocyanin and two TIM barrels. In these simulations [148,149], the model force field consisted of short-range interactions, biased toward the target native secondary structure, and pairwise interactions between side chains defined according to the Miyazawa–Jernigan [139] hydrophobicity scale.

One problem with low-resolution lattice models is their spatial anisotropy [15]. That is, a given secondary structural element may be represented by substantially different lattice structures as the element assumes various orientations on the lattice. In the hybrid 210-lattice model of protein backbone, this disadvantage has, to a large extent, been removed, mostly due to the larger set of basis vectors (56) employed in building the α-carbon trace. This lattice representation has about a 1.0-Å coordinate RMSD from the corresponding native α-carbon coordinates. The relatively accurate main-chain representation enabled the construction of a side-chain rotamer library. Side chains are represented as single spheres and their various positions, with respect to the main chain, simulated various rotamers of each amino acid. The model force field has been derived from the appropriate statistics of high-resolution crystallographic structures. Short-range interactions consisted of several types of terms that reflect secondary propensities of various amino acids. A somewhat similar potential has been recently investigated by DeWitte and Shakhnovich [150]. Tertiary interactions are described by amino acid–specific burial terms, pair terms, and several kinds of multibody, knowledge-based contributions that act to regularize protein structure. These consisted of a cooperative model of a hydrogen bond network (employing geometric criteria similar to that of Levitt and Greer [151]), and several types of amino acid pair–specific and four-body packing correlations based on side-chain contact patterns seen in globular proteins [152]. This force field was accurate enough to enable the reproducible de novo folding of several simple proteins, starting from a random expanded conformation [16,18].

A finer 310-hybrid lattice model employs a 90-vector basis set for α-carbon trace representation. This model can reproduce the geometry of native proteins with an average α-carbon RMSD of 0.6–0.7 Å. The general principles of the force field were similar to that of the 210-hybrid lattice model. However, the more accurate geometric representation enabled a better knowledge-based modeling of short-range interactions [20] and the more straightforward design of hydrogen bonding [17,40]. As a result, the folding of more complex protein structures became possible [19,40,42,56,153–155]. Single secondary-structure elements, such as long helices, assembled during the Monte Carlo simulations with an accuracy in the range of 1.0–2.0 Å RMSD. The model also predicted some simple protein structures whose coordinate RMSD from native ranged from 2.0 to 5.0 Å, depending on protein size.

3. *Complementary Insights from Various Models*

The preceding outline shows that very different continuous and discretized models of proteins and protein-like systems have been used to address closely related problems of protein physics. Which are better for the study of protein folding dynamics, thermodynamics or folding patterns? The answer depends on the particular aspects of the protein folding problem to be addressed. Certainly, the fine geometric details of protein conformation could be better reproduced by continuous models. Thus, when dealing with small peptides, a properly designed continuous, reduced protein model should be more useful than a lattice model. This is especially true for short-range correlations that are much easier to account for in continuous models. What about larger systems, such as proteins? When small-scale conformational transitions are of interest, a continuous-space model is again the natural choice. However, when dealing with large-scale conformational transitions, especially in simulations of folding from the denatured state, discretized, lattice models offer several distinct advantages. First, because of the lattice structure itself, the conformational space is a priori partitioned, and one avoids many local, and perhaps irrelevant, energy barriers. Second, local conformational transitions could be predefined and used during simulations in a very effective way. Moreover, the conformational energy computation could be accelerated by precalculating many components just once and storing the results in a table. Because of the last two attributes of the lattice models, lattice Monte Carlo simulations could be a couple of orders of magnitude faster than simulations in an otherwise equivalent, continuous-space model [42,127]. However, when designing a lattice model, it is always necessary to carefully analyze the possible effects of the lattice anisotropy and the discrete structure of conformational space [15]. Recently, much progress has been achieved in various methods of discretization [156,157] of protein conformational space [42] and in the analysis of the fidelity of such discretizations [14,156,158–160].

Reduced models of proteins and protein-like models utilize various levels of generalization. On one side there are the simple, exact models that can be analyzed in great detail. The simple cubic lattice representation of polypeptide chains reflects the very fundamental role of the close, crystal-like packing of protein structures. It has been shown that the folding transition of cubic lattice copolymers responds in a reasonable way to sequence variation. For some protein-like sequences, unique compact states represented the minimum of conformational energy. If so, it is very likely that simple, exact models properly mimic the most general physics of hydrophobic collapse and packing complementarity of globular proteins. With a proper model of chain dynamics, the most fundamental features of protein folding pathways also could be studied [41].

However, in spite of their many virtues, it appears that the simple, exact models do not sufficiently account for a rather important feature of polypeptides and proteins, specifically, the complex interplay between short- and long-range interactions associated with the presence of a main chain and side chains. The importance of this aspect of protein structure and dynamics has been addressed in an excellent short review by Honig and Cohen [161]. On the most trivial level, it should be pointed out that proteins are rather stiff polymers. In the range of chain length and conformational stiffness typical of proteins, semiflexible lattice homopolymers undergo a first-order collapse transition to a highly ordered, globular state [67]. This is perhaps the most fundamental aspect of protein folding physics that has to be demonstrated by any protein-like system. Furthermore, the "decoration" of protein-like models with sequence specific short- and long-range interactions (and also with some more complex protein-like geometrical correlations) leads to a structural fixation of compact states characteristic of real proteins. Finally, if the tertiary structure of a real protein is to be predicted, the model must be capable of representing the geometry of the native state at an acceptable level of resolution.

What, then, is the relationship between models of various levels of generalization? They are complementary for several reasons. First, they address various aspects of the protein folding problem. The general insights from studies of the simple, exact models and protein-like models provide guidelines for designing reduced force fields in more detailed models. On the other hand, conformational space cannot be explored with high accuracy when many structural details are taken into account. In such cases, only a limited number of folding simulations could be performed. Consequently, it is more difficult to derive a quantitative description of protein folding dynamics and thermodynamics. However, one can address numerous important aspects of protein physics using these more complex models. When appropriate, the predictions of such models also can be tested experimentally.

III. PROTEIN FOLDING THERMODYNAMICS

A. Nature of the Interactions

A key question is what types of interactions are important for protein stability. Protein backbones contain polar groups that either form hydrogen bonds with water or with other backbone or side-chain atoms; the free energy cost for the absence of hydrogen bonds is substantial [162]. Thus, hydrogen bonds play an important structural regularizing role in proteins [42]. In addition, individual residues exhibit differential secondary structural preferences. The secondary structure found in protein fragments, although small, is nonetheless present and can be significant [163]. Such intrinsic secondary structural preferences act both to assist in the early states of folding and to reduce the configurational entropy of compact states [110]. Another important interaction is the hydrophobic effect that acts to sequester hydrophobic molecules within the interior of the protein [8]. Furthermore, amino acid pair–specific interactions and higher-order tertiary and quaternary interactions may be responsible for the selection of the native fold or the destabilization of an alternative structure [164]. Such interactions may be electrostatic in origin or arise from the complex interplay of potentials of mean force. Thus, there are a variety of interactions present in a globular protein, and the native structure is the result of a balance of such terms.

B. Extent of Frustration in Native Conformations

The thermodynamic hypothesis [1,7] that native proteins are in a global minimum of free energy is now generally accepted. Since the native state has a relatively low conformational entropy, they are in a minimum of conformational energy; however, the larger corresponding entropy of solvent molecules also has to be kept in mind [165]. Of course, this does not mean that all interactions in a globular protein stabilize the native state. There is always competition between some interactions. For example, some pentapeptide fragments [166] can adopt different secondary structures in various proteins. This means that the observed native structure is a compromise between short-range secondary propensities and tertiary interactions. A similar analysis could be performed for other interactions in proteins. Hydrophobic residues are not always buried, and polar residues could be found in otherwise mainly hydrophobic core proteins. Sometimes, such "defects" are essential for structural uniqueness and can cause a "compensation" effect—the destabilization of one part of a protein increases the stability in another part. This leads us to very important question: What are the molecular reasons for protein structural uniqueness and the cooperativity of the folding process?

C. Origin of Cooperativity and Two-State Thermodynamic Behavior in Protein Folding

Typically, the in vitro folding of a single domain globular protein resembles a first-order phase transition in the sense that the thermodynamic properties undergo an abrupt change, and the population of intermediates at equilibrium is very low. In other words, the process is cooperative and is well described by a two-state model [8]. The first attempts to explain protein folding cooperativity focused on the formation of secondary structure. Theoretical and experimental analysis of coil–helix transitions indeed proved that the process is cooperative [167]. However, the helix–coil transition is always continuous [168], and thus it cannot explain the two-state behavior of the protein folding transition.

Simulations of protein-like models with target short- and long-range interactions [103,104,107,108,110] demonstrated that the abrupt transition is due to strong native-like tertiary interactions. Native-like short-range interactions increased the stability of the globular state; however, they appeared to decrease folding cooperativity. The dominant role of the long-range interactions for folding cooperativity also has been demonstrated in studies of simple, exact models. Shakhnovich et al. [57,71–73,169,170], Bryngelson et al. [85,99,100], and others [65,75,78,84,171,172] pointed out the importance of the entire energy landscape for the protein folding cooperativity. In this context, a cooperative all-or-none transition also could be expected for systems with many competing interactions provided that there is a well-pronounced energy minimum corresponding to the native state [65,75,78,84,172].

An interesting realization of such an energy landscape is provided by semiflexible polymer models. As was suggested by mean field considerations [173,174], semiflexible polymers should undergo a pseudo-first-order phase transition. This has been quantitatively illustrated in Monte Carlo studies of homopolymeric lattice systems by Kolinski et al. [43,67,68]. Thus, a cooperative collapse transition may also result from the interplay between nonspecific short-range and uniform long-range interactions.

A very convincing explanation of protein folding cooperativity has been given recently by Hao and Scheraga [35]. They used the entropy-driven Monte Carlo (ESMC) method [38] to investigate the collapse transition in the chess-knight model [54] of a minimal β-barrel protein (see the chapter "Entropy Sampling Monte Carlo for Polypeptides and Proteins"). Their model force field consisted of short-range interactions, a qualitative model of hydrogen bonding and pairwise side-chain interactions. The side chains lacked internal degrees of freedom. Computer experiments demonstrated that some amino acid sequences exhibit an abrupt first-order transition from a random coil state to a well-defined globular state. The cooperativity of the folding process emerged from a characteristic depletion of the system's conformational entropy as a function of conformational energy as one moves away from the native state. More random sequences exhibited a continuous collapse transition [36]. Sequence-specific long-range interactions were responsible for the protein-like, all-or-none folding, while the short-range interactions only contributed to the stability of the globular state [37].

A similar study on a somewhat more complex 310-hybrid lattice model of an idealized Greek-key fold has been performed by Kolinski et al. [40,42]. The model force field consisted of knowledge-based, short-range interactions that quantitatively reproduced secondary propensities of various sequences, a cooperative model of hydrogen bonds, and several (also knowledge-based) types of tertiary interactions. The model employs a single united-atom multiple-rotamer representation of the side chains. A first-order transition has been obtained only when the explicit cooperativity of the side-chain interactions was taken into account. These cooperative terms had the form of four-body potentials preferring protein-like side-chain packing arrangements. The transition-state of such a cooperative model has an overall conformation very similar to that of the native state with most of the native secondary structure (in a sense of loose interactions of expanded β strands); however, the average volume is larger. The folding transition was associated with the fine-tuning of side-chain packing accompanied by small rearrangements of the entire structure. Such "side-chain fixation" increased the relative contribution of the multibody interactions and produced a unique native state. Since the cooperative interactions contribute very little to the transition-state energy, they acted by providing a sufficient energy gap from native for the all-or-none transition to emerge. Interestingly, the native state still had a substantial energetic degeneracy (and thereby a substantial entropy) with a large number of conformations having essentially the same hydrophobic core packing. However, these simply reflect small structural differences that arise mostly in the loop and turn regions. The first-order transition was due to the change of the conformational entropy as a function of energy. The entropy changed most rapidly near the transition-state. The conformational entropy of the side chains may play a very important role in the late stages of the folding process [40,42]. This picture of the transition state is consistent with many experimental facts and with the theoretical postulate that the molten globule is a general transition state in protein assembly [169,175–184].

All these studies seem to indicate that the cooperativity of the folding process is due to the specific pattern of tertiary interactions and/or the specific interplay between short- and long-range interactions. This may appear to be a trivial statement, but detailed analysis of the results from the simple, exact model, protein-like models and reduced models of real proteins show several specific requirements for the protein folding cooperativity. It is very encouraging that over the entire spectrum of theoretical model studies of the protein folding process, these requirements essentially overlap [35–37,51,83,92,95]. Naturally, various models place different stress on the specific interactions that may control protein folding and structural uniqueness.

IV. TERTIARY STRUCTURE PREDICTION

Knowledge of protein structure is important for understanding their biological functions. However, the number of known protein sequences is many times larger than the number of solved three-dimensional structures. Thus, the ability to predict protein structure from amino acid sequence is one of the most challenging goals of contemporary molecular biology. Considerable progress toward a solution of the protein folding problem has been made recently, although a complete solution is not yet in hand. One area of such progress has been through the use of threading approaches that are designed to match a sequence to the most compatible structure in a library of known protein structures [185]. Here, we limit ourselves to methods based on the direct search of protein conformational space, and focus mainly on Monte Carlo and closely related methods.

A. Peptides and Cyclic Peptides

At first glance, it may appear that the problem of structure prediction in small peptides should be much simpler than in globular proteins. To some extent this is true. On other hand, a minimum-energy conformation of a peptide may differ very little from the manifold of competing conformations. Thus, the accuracy requirements are greater for the force field. In this respect, Scheraga and co-workers [186] have developed a semiempirical force field (ECEPP) [187] that is particularly well suited for studies of oligopeptides, where the short-range interactions may dominate the structural properties. Using various minimization [9,10,29,91,188–190] and simulation techniques [191,192], the conformational space of (Met)enkephalin [191,193], the cyclic decapeptide gramicidin S, cyclic hexaglycine, and the 20-residue membrane-bound fragment of melittin [194], alamethicin, and other peptides and small proteins [195–198] have been investigated in detail. Possible effects of the solvent environment also have been analyzed. Good-quality structure predictions of various coiled-coil and peptide crystal structures also have been achieved [186,199,200]. Although not all of these works have employed a Monte Carlo search methodology, we list them to provide an overview of the present status of the oligopeptide structure prediction field.

Using a statistical potential and a genetic algorithm as the optimization procedure, Sun has achieved a relatively high resolution (1.66-Å) RMSD for the 20-residue membrane fragment of melittin) prediction of a few oligopeptides [28]. However, as demonstrated by Hansmann and Okamoto [201], peptide conformational space could be effectively explored by a Monte Carlo multicanonical algorithm. Okamoto [202] employed the ECEPP potential and a simulated annealing Monte Carlo procedure to estimate the helix-forming tendencies of several nonpolar amino acids within the context of homooligopeptides. Such an approach can perhaps be used to obtain better semiempirical short-range interaction potentials for use in protein simulations. Finally, a very interesting Monte Carlo minimization procedure has been employed by Meirovitch et al. [203–205] for the determination of peptide conformations in solution from averaged NMR (nuclear magnetic resonance) spectra.

This brief overview shows that for particular applications, Monte Carlo methods and other minimization methods could be complementary to standard molecular dynamics methods (see some reviews on MD techniques and applications to proteins and peptides) [11,12,206].

B. Proteins

1. *De Novo Approaches*

Early attempts at de novo predictions of protein structure focused on a few small proteins. In spite of the rather simple test cases, the obtained structures were of low resolution [22,24,132]. It is interesting that the simple cubic lattice simulations of Covell [144] led to a similar accuracy. Also, a very simple lattice representation has been employed in the work of Hinds and Levitt [146]. These results suggest that the proper design of the interaction scheme is no less important than the accuracy of the geometrical representation [207,208]. The more elaborate Monte Carlo model of Wallqvist and Ullner [138] allowed for the prediction of the 36-residue PPT structure with an accuracy of 4.5-Å RMSD. This is the same accuracy as obtained by Sun [28] via a completely different search procedure.

An interesting hierarchical Monte Carlo procedure for the prediction of protein structures has been proposed recently by Rose et al. [209]. Their model employs an all-atom representation of the main chain and a crude representation of the side groups interacting via a simple contact potential. The method seems to be quite accurate in the prediction of protein secondary and supersecondary structure; however, the overall global accuracy of the folded structures is rather low.

Kolinski, Skolnick, and co-workers [16–18,42,56,147,153,155,210] have developed high-coordination-number lattice models of protein conformation and dynamics. The method has been tested in de novo structure predictions of several small proteins, including protein A fragments, two designed four-helix bundle proteins, a monomeric version of ROP, crambin and several coiled-coil motifs (see Section V.A). The Monte Carlo folding procedure employed a simulated annealing protocol and a knowledge-based force field derived from protein structural regularities [152,211–213]. For globular proteins, the accuracy of these predictions (measured as the coordinate RMSD from the α-carbon trace of known structures) ranged from 2.25 Å for the B domain of protein A and 4.0 Å for crambin to about 5 Å for the 120-residue ROP monomer. The obtained lattice folds were consistent with an all-atom representation, and detailed atomic models could always be reconstructed at virtually identical accuracy as the corresponding lattice models [18,155,210]. This high-coordination-number lattice model also has been used in de novo computer-aided design (CAD) [19,40] and in the redesign of protein topology [153,154]. However, the method fails for larger proteins of complex topology. In these cases, fragments of the protein under consideration usually form native-like motifs; however, topological errors (wrong strand reversals, etc.) lead to globally misfolded structures. The knowledge-based force field of these models seems be qualitatively correct (it recognizes most native folds in threading experiments), but the native energy minimum is not very deep. Therefore, the applicability of the model to structure prediction could be substantially extended with the help of some theoretically or experimentally derived restraints.

2. *Prediction Using Known Secondary Structure and Correct Tertiary Restraints*

One possible way of improving tertiary structure prediction is to use either known or predicted secondary structural information. In that regard, using an off-lattice model and exact knowledge of the secondary structure, Friesner et al. have successfully folded two four-helix bundle proteins, cytochrome b562 and myohemerythrin [214]. Furthermore, assuming known native secondary structure, Dandekar and Argos [215,216] have reported encouraging results for simple helical and β proteins using a genetic algorithm to search conformational space. Mumenthaler and Braun [217] have developed a self-correcting distance geometry method that also assumes known secondary structure. They successfully identified the native topology of six of eight helical proteins. There also have been a number of studies that incorporate the known native secondary structure and a limited number of known and correct long-range restraints to predict the global fold of a globular protein. In particular, Smith-Brown, Kominos, and Levy [218] have modeled the protein as a chain of glycine residues and used Monte Carlo as the search protocol. Restraints are encoded via a biharmonic potential, with folding forced to proceed sequentially via successive implementation of the restraints. A number of proteins were examined. By way of example, flavodoxin, a 138-residue α/β protein, was folded to a structure whose backbone RMSD from native was 3.18 Å for 147 restraints. Another effort to predict the global fold of a protein from a limited number of distance restraints is due to Aszodi et al. [219]. This approach is very much in the spirit of the work of Mumenthaler and Braun and is based on distance geometry, where a set of experimental tertiary distance restraints are supplemented by a set of predicted interresidue distances. Here, these distances are obtained from patterns of conserved hydrophobic amino acids that have been extracted on the basis of multiple sequence alignments. In general, they find that to assemble structures below 5-Å RMSD, on average, typically more than $N/4$ restraints are required, where N is the number of residues. Even then, the Aszodi et al. [219] method has problems selecting out the correct fold from competing alternatives. An advantage of the Aszodi et al. approach is that it is very rapid, with a typical calculation taking on the order of minutes on a typical contemporary workstation.

Turning to lattice-based methods that sample conformations via Monte Carlo, the MONSSTER (*mo*deling of *n*ew *s*tructures from *s*econdary and *te*rtiary *r*estraints) method for folding of proteins using a small number of tertiary distance restraints and loosely defined knowledge of the secondary structure of regular fragments has been developed [220]. The method incorporates potentials reflecting statistical preferences for secondary structure, side-chain burial and pair interactions, and a hydrogen bond potential based on the α-carbon coordinates. Using this algorithm, several globular proteins have been folded to moderate resolution native-like, compact states. For example, flavodoxin, with 35 restraints, has been folded to structures whose average RMSD is 4.28 Å. Plastocyanin with 25 long-range restraints adopts conformations whose average RMSD is 5.44 Å. These results compare very favorably with the Aszodi et al. and Smith-Brown et al. approaches [218,219]. In general, preliminary indications are that helical proteins can be folded with roughly $N/7$ restraints, while β and $\alpha\beta$ proteins require about $N/4$ restraints, where N is the number of residues. Of course, for any particular case, the accuracy depends on the location of the restraints. This point requires additional investigation.

3. *Method for Prediction of Tertiary Structure from Predicted Secondary Structure and Tertiary Restraints*

Recently, a procedure has been developed that does not rely on a priori knowledge of the native conformation to fold a number of single-domain proteins [221,222]. The procedure can be derived into two parts: restraint derivation and structure assembly using the MONSSTER algorithm. With respect to restraint derivation, using multiple sequence alignment, protein secondary structure is predicted using the PHD method [223]. Regions where the chain reverses global direction (U-turns) are predicted using the method of Kolinski et al. [224] and override PHD predictions because they have been shown to be highly accurate. Then, the multiple sequence alignment and the secondary structure prediction for the topological elements between U-turns are combined to predict side-chain contacts. Such contact map prediction is carried out in two stages. First, an analysis of correlated mutations is carried out to identify pairs of topological elements of secondary structure (regions of the chain between U-turns) that are in contact [225]. The rationale is that in this way it is possible to restrict the predictions to rigid elements of the core, for which the assumption of closeness in space as reflected in the covariance of their mutational behavior is, in principle, more valid. Then, inverse folding [226] is used to select compatible fragments in contact, thereby enriching the number and identity of predicted side-chain contacts. Parenthetically, we note that for topological elements that are known to touch, this procedure produces contacts of which 67% are correct within ± 1 residue. This is of comparable accuracy to the situation where the contacts are predicted. Of course, they do not employ any information about the native structure in the prediction protocol. The final outcome of the prediction protocol is a set of noisy secondary and tertiary restraints.

The predicted secondary and tertiary restraints are inputted to the

MONSSTER method described above. However, now restraint information is predicted rather than being extracted from the native structure. Implementation of the restraints is carried out so as to take their low resolution into account. Low-energy structures are then searched for by simulated annealing, followed by isothermal refinement. Those structures having the lowest average and minimum energies are assigned as native.

The above protocol has been applied to the set of 17 proteins. All are extrinsic to the set of proteins employed in the derivation of the potentials. The average RMSD of the lowest-energy set of structures ranges from 2.8 Å for small helical proteins to roughly 6 Å for β and mixed-motif proteins. In all cases, the global topology is recovered. It is very important to emphasize that all structural predictions use the identical parameter set and folding protocol. Furthermore, they also considered one case, 1ctf, where both the Rost–Sander prediction and the U-turn prediction protocol were poor. Because of the failure to identify the core regions of the molecule, the correlated mutation analysis yielded only two tertiary contacts (i.e., the protocol failed). More generally, the range of validity of this protocol remains to be established. Nevertheless, it is very encouraging that in the test case where the algorithm was grossly incorrect, the result yielded by the protocol was that the molecule is not foldable. It is very important to have an algorithm that indicates when it cannot be used; this preliminary result provides some encouragement that this might be the case.

V. QUATERNARY STRUCTURE PREDICTION

A. Coiled Coil, De Novo Simulations

Theoretical studies of coiled-coil systems are surprisingly scarce. Nevertheless, coiled coils represent an important test bed for quaternary structure prediction methods. Their quaternary structure consists simply of the side-by-side association of α helices wrapped around each other with a slight supertwist [227]. In the 1980s, Skolnick and Holtzer generalized the Zimm Bragg helix–coil theory [167] to include interhelical interactions [228]. The resulting phenomenological theory was able to predict the quaternary structure of tropomyosin and its fragments over a broad range of temperature and pH. However, these studies were mainly limited to dimeric coiled coils and required that the native structure be assumed. More detailed atomic modeling of coiled coils commenced in 1991 when Nilges and Brunger developed an automated algorithm for the prediction of coiled-coil structure, based on the assumption that the helices are parallel and in register [200,229]. They applied this approach to predict a structure of the GCN4 leucine zipper, whose backbone atom RMSD from the subsequently solved crystal structure is 1.2 Å. They concluded that given the correct registration, the best packing of the hydrophobic residues in the core dictates the detailed geometry. This observation was first made by Crick in 1953. At about the same time, Novotny et al. [230] estimated the stability of GCN4, fos, and jun leucine zippers from molecular mechanics calculations. Their conclusions suggest that Leu in the *d* position of the canonical coiled-coil heptet makes a major contribution to the stability of dimers, whereas residues in the *a* positions are far less important. These conclusions are supported by the studies of Zhang and Hermans on a model leucine zipper [231]. Most recently, Harbury and Kim have developed a very simple and fast algorithm for predicting structures with ideal coiled-coil geometry. Unfortunately, it will not work if the coiled coil geometry is distorted [232]. To date, the only de novo simulations of coiled-coil assembly that starts from a pair of random chains and results in structures that are at the level of experimental resolution are those of Vieth et al. [17,155]. Starting from random denatured states and using their high-coordination-number lattice protein model [17], they produced parallel, in-register folding conformations on lattice. Full atomic models are built and refined using a molecular dynamics annealing protocol. This produced structures whose backbone RSMD from the crystal structure is 0.81 Å.

B. Method for Prediction of Equilibrium Constants of Multimeric Helical Proteins

Harbury et al. [233] have examined the shift in equilibrium among coiled-coil dimers, trimers, and tetramers associated with varying the identity of the residues of the hydrophobic residues in the coiled-coil heptad repeat of the GCN4 leucine zipper. Vieth and co-workers have developed a method to calculate the equilibrium constant among a manifold of assumed conformations (using a hybrid Monte Carlo–transfer matrix approach) [234]. When applied to the eight mutants studied by Harbury et al., their calculations are in agreement with experiment over the entire concentration range for five of eight sequences, and over a portion of the concentration-range, they are in agreement for an additional two sequences. They find that local, intrinsic secondary structure preferences and side-chain entropy act to favor lower order multimers, whereas tertiary and quaternary interactions act to favor higher-order multimers. The model also correctly predicted the state of association of a number of GNC4 leucine zipper fragments [235], as well as the quaternary structure of the DNA binding proteins fos and jun. Finally, they examined the coil-ser sequence designed by Eisenberg, DeGrado, and co-workers. In agreement with the experiment, an antiparallel, three-helix bundle is predicted to be the native conformation [236].

VI. SIMULATION OF PEPTIDE INSERTION INTO MEMBRANES

A. Small Proteins and Peptides

An off-lattice Monte Carlo dynamics simulation has been used to investigate the behavior of filamentous bacteriophage coat proteins pf1 and fd and a number of peptides such as m2δ, melittin, and magainin2 in a model membrane environment [237,238]. The objective is to predict the orientation, location and conformation of the proteins with respect to a model lipid bilayer. The protein is represented by a chain of balls with centers at the α carbons and whose radii are residue-dependent. Here, a continuous-space model is used; that is, the α-carbon positions are not restricted to lattice sites. The membrane is treated as a z-coordinate-dependent effective potential representing the environments of water, lipid–water interface and lipid. The hydrophobic interaction energy depends not only on amino acid identity but also on its position in space. In this set of simulations, the membrane is modeled as an effective medium. The simulations begin with a random conformation in the water portion of the box, and sampling occurs using Monte Carlo dynamics. These simulations predict that despite the very low sequence similarity between the major coat proteins of Pf1 and fd bacteriophages, their structures in a membrane environment are very similar. This suggests that the hydrophobic effect exerts an important influence on membrane protein structure. Focusing on fd by way of example, the simulations predict that an amphipathic helix runs from residues 5 to 19, and that a transbilayer helix runs from residues 25 to 41. This is to be compared with the experiment that indicated that the N-terminal amphipathic helix runs from residues 7 to 20, and the transbilayer helix is located between residues 23 and 42. Thus, good agreement with experiment is found [239,240]. Similar agreement with experiment was found for the other systems studied.

B. Translocation Across Model Membranes

A legitimate question concerns the limitations of a model where the membrane is treated as an effective medium. To explore this issue, a series of simulations were done where each lipid molecule in the bilayer was represented by a dumbbell, and the effect of curvature on the transport of a structureless polymer was explored [241]. Interestingly, when the membrane is highly curved, there is almost irreversible transport from the outside to the inside of the spherical vesicle. This arises from entropic effects. Basically, because the interior leaflet is of lower density, after penetrating into the membrane, the polymer prefers to remain in this region. In contrast, this effect vanishes when the membrane is treated as a structureless, hydrophobic medium. Thus, there may well be situations where the structure of the membrane exerts important effects.

The translocation of a polymer driven by an external field bias of strength ε across a model membrane has also been investigated using Monte Carlo methods [242]. Again, each lipid molecule is modeled as a dumbbell. It is found that below a characteristic field strength, ε^*, the membrane is practically impermeable. In the high-field limit of $\varepsilon > \varepsilon^*$, the permeability is independent of the field strength. At low fields, the permeability decreases according to $\exp(-\varepsilon^*/\varepsilon)$. The translocation mechanism can be understood as a Kramers process describing the escape of a particle over potential barriers.

VII. PERSPECTIVES

A. Where Are We Now?

Computer studies of simple, exact models [51,57,78], protein-like models, and more realistic protein models have greatly increased our understanding of the protein folding process and the stability of protein structures. For example, Levinthal's paradox [243–245] seems to be rationalized by the simulation results. It is now more or less clear how proteins partition their conformational space, thereby sampling only a very small fraction of all possible states [58]. Studies of simple, exact models have demonstrated that one such way of reducing the available conformational space is by hydrophobic collapse. Also, simple, exact models and protein-like models provide a plausible explanation of why some sequences fold rapidly, whereas others fold very slowly or not at all. Different folding scenarios can be now rationalized by analogy to a spin-glass [85,99,100,246].

From studies on somewhat more complex models, we have learned about the interplay between the short- and long-range interactions [35–37,40,42,103,104,107] and their possible role in both the thermodynamics and kinetics of protein folding. The formation of secondary structure nuclei can also drastically reduce available conformational space, providing a somewhat different (although in principle equivalent) explanation of Levinthal's paradox. Actually, an opposite paradox, termed here the *proximity paradox*, should be addressed: Why does a confined (by chain connectivity) molecular system consisting of thousands of atoms (whose degrees of conformational freedom are locally strongly coupled) require times ranging from milliseconds to seconds to adopt its native state, whereas folding into an "almost correct" structure from the denatured state is very rapid. The type of collective molecular mechanism responsible for the "side-chain fixation" or molten globule–native state transition is not yet known. Simulation studies [131], physical experiments on natural proteins [247], and the investigation of artificial proteins [177,248–252] have started to address this issue. Recent studies seem to indicate that the satisfactory solution of this problem will require the analysis of the various types of physical interactions in proteins, including explicit multibody approximations to tertiary interactions.

Protein structure prediction [253,254] is one of the major goals of computational molecular biology. Up to now, homology based and threading methods have been the most successful [185,226,255–264]. However, due to the increasing ratio of the number of known protein sequences to the number of solved protein structures, the development of de novo (or related) methods would be extremely valuable. To date, only limited, nevertheless encouraging, progress has been achieved in such direct approaches [42,254]. Purely de novo predictions are now possible only for peptides [186] and very small and structurally simple proteins [42,200,229]. It appears that the most promising are the methods that employ knowledge-based factorization of protein interaction and structural regularities [28,135,139,152,207,211,212,214,265,266]. Over the short term, hierarchic techniques that exploit evolutionary information and fragmentary experimental data should become a standard tool for low-resolution protein structure prediction [222,267].

B. Shortcomings of the Present Force Fields, Possible Refinements, and Reformulations

The work of Novotny et al. [268] and others [269,270] has shown that the classic, detailed atomic force fields are locally, but not necessary globally, correct. It is unlikely that for these force fields the global minimum of conformational energy corresponds to the native state. On other hand, in many cases, even an extremely simplified interaction scheme can match a sequence to its native structure in a threading experiment. In other words, when a sequence is fitted to various protein-like structures, the demands on the force field are much less. In such cases, the numerous interactions that make a polypeptide chain a protein become to a large extent irrelevant, and only sequence specific components have to be correctly accounted for in threading experiments. The conclusion is rather simple—an efficient force field for simulation of the protein folding process and, consequently, for the de novo protein structure prediction must discriminate against a majority of conformations that never occur in folded proteins. Otherwise, a conformational search procedure would mostly explore rather irrelevant portions of the model conformational space. In the early attempts to build reduced models of proteins, this important fact was not sufficiently appreciated.

Within the context of the above, let us try to formulate some necessary requirements for the design of a moderate resolution reduced model (based on united-atom representation and knowledge-based potentials) of real proteins and its force field:

1. The resolution of the model has to be at least on the level of 2 Å. Otherwise, packing complementarity of the side groups would be lost.
2. Short-range interactions (knowledge-based or semiemperical) must account for the polypeptide chain's conformational stiffness [20,42]. In other words, protein-like correlations enforced by the potential should extend over several residues. This would considerably narrow the available conformational space.
3. At least for main-chain units, a highly directional model of hydrogen bond interactions (perhaps with an explicit cooperative term) has to be designed [16,17,19,40,42,55]. Globular proteins have a regular network of hydrogen bonds. The energy difference between a peptide group hydrogen-bonded to water molecules and the same group hydrogen-bonded to another peptide group is minor. However, the free energy price for having non saturated hydrogen bonds in a protein interior is large. Thus, the hydrogen bond network plays a very important structure-regularizing role, eliminating the majority of otherwise possible compact conformations of polypeptide chains.
4. Side chains (even when modeled as single-interaction spheres) should have some conformational freedom that reflects their internal mobility in real proteins. Pairwise interaction potentials should be as specific as possible, and in the absence of explicit solvent, a burial potential that reflects the hydrophobic effect may be necessary [17,42].
5. Somehow, higher-order than pair, multibody interactions have to be built into the model. Since unrestrained MD simulations with all atom potentials lead to liquid-like (instead of native-like) packing of side chains [269], it is rather unlikely that a protein-like pattern of side-group contacts would emerge in a reduced model context without tertiary structure-regularizing interactions. Such side-chain multibody interactions provide a bias toward a protein-like packing. The possible role of multibody correlations for protein structure modeling and recognition is starting to receive some attention [16,17,40,42,56,153,154,213,271,272].

Of course, the specific realization of these proposed minimal requirements may differ in various models. Some other features may be necessary.

The practical realization of the preceding set of requirements allowed the design of protein models that were capable of folding a subset of small, relatively simple proteins [16,42,56,155,210]. Very likely, further progress will depend on the development, refinement and consistent implementation of more complete knowledge-based force fields.

C. Promises of New Sampling Methods

Several recent developments in sampling methodology have increased the accuracy and reliability of Monte Carlo simulations of protein folding processes. In standard Metropolis-type algorithms, sampling efficiency can be immensely improved by the application of "smart" local conformational updates. This is true for both off-lattice continuous-space models [25,30,31] and for lattice models of protein chains [17,20,42]. These local moves, in contrast to purely random conformational updates, are designed in such a way that the model system visits only the local conformational energy minima; thus, it ignores all conformations that would be rejected as a result of steric overlaps, large distortions of valence angles, or other factors. Local minimization techniques [189] could be combined with the Metropolis sampling scheme.

Recently developed high-coordination-number lattice models of protein dynamics and structure have several methodological advantages. Such models have an acceptable resolution, no worse than that of most off-lattice reduced models. However, due to the use of a lattice, the simulation algorithms are much faster than those otherwise equivalent continuous-space models. It also should be noted that the resolution of recently developed reduced models is sufficient for hierarchical simulations. In these cases, large-scale relaxation can be performed via Monte Carlo dynamics on a reduced model and the intervening fragments of trajectory could be simulated via the MD technique. It is easy to do fast projections of all-atom models onto reduced models. Conversely, rebuilding all-atom structures from a reduced representation could also be done rapidly with reliable accuracy.

A novel approach to protein conformation is the entropy-sampling Monte Carlo method (ESMC), which is described in detail in another contribution to this volume. The method provides a complete thermodynamic description of protein models, but it is computationally quite expensive. However, because of the underlying data-parallel structure of ESMC algorithms, computations could be done on massively parallel computers essentially without the communication overhead typical for the majority of other simulation techniques. This technique will undoubtedly be applied to numerous systems in the near future.

VIII. CONCLUSIONS

Over the last decade, Monte Carlo–based approaches have proved to be a powerful tool for addressing various aspects of the protein folding problem. To date, their greatest impact has been in the application to reduced protein models designed either to explore general questions of protein folding or to predict low to moderate resolution native structures. Because dynamic Monte Carlo can effectively simulate dynamic processes, it has provided numerous insights into the mechanism by which proteins fold. Both dynamic and entropy-sampling Monte Carlo have permitted the investigation of the panoply of interactions responsible for the unique native structure characteristic of globular proteins. More recently, these methods have been applied with increased success to the prediction of protein tertiary and quaternary structure. This is especially true when knowledge-based, statistical potentials are combined with evolutionary information provided by multiple sequence alignments. Such combined approaches are very likely to yield additional progress in the protein folding problem in the near future.

This review has emphasized the application of reduced models to problems of protein folding. Such models combined with Monte Carlo sampling also could be used in the study of other biological processes, such as the mechanism of virus assembly [273]. Furthermore, Monte Carlo-driven, reduced protein models could provide effective reaction coordinates, which are then sampled by molecular dynamic simulations of atomic models. This would combine the best of both approaches: the ability to simulate large-scale, long-time processes and the ability to examine such processes at atomic detail. Thus, by being employed in novel contexts, Monte Carlo–based sampling approaches are likely to continue to provide numerous valuable insights into the behavior of biological systems.

REFERENCES

1. C. B. Anfinsen and H. A. Scheraga, *Adv. Prot. Chem.* **29**, 205–300 (1975).
2. T. E. Creighton, *Proteins: Structures and Molecular Properties*, Freeman, New York, 1993.
3. R. L. Jernigan, *Curr. Opin. Struct. Biol.* **2**, 248–256 (1992).
4. M. J. Gething and J. Sambrook, *Nature* **355**, 33–45 (1992).
5. O. B. Ptitsyn, *J. Prot. Chem.* **6**, 273–293 (1987).
6. T. E. Creighton, *Biochem. J.* **270**, 131–146 (1990).
7. C. B. Anfinsen, *Science* **181**, 223–230 (1973).
8. P. L. Privalov and S. J. Gill, *Adv. Prot. Chem.* **39**, 191–235 (1988).
9. L. Piela, J. Kostrowicki, and H. A. Scheraga, *J. Phys. Chem.* **93**, 3339–3346 (1989).

10. D. R. Ripoll, L. Piela, M. Velasquez, and H. A. Scheraga, *Proteins* **10**, 188–198 (1991).
11. C. L. I. Brooks, M. Karplus, and B. M. Pettitt, *Proteins: A Theoretical Perspective of Dynamic Structure and Thermodynamics*, Wiley, New York, 1988.
12. M. Karplus and G. A. Petsko, *Nature* **347**, 631–639 (1990).
13. J. A. McCammon, *Rep. Prog. Phys.* **47**, 1–46 (1984).
14. B. H. Park and M. Levitt, *J. Mol. Biol.* **249**, 493–507 (1995).
15. A. Godzik, A. Kolinski, and J. Skolnick, *J. Comp. Chem.* **14**, 1194–1202 (1993).
16. A. Kolinski, A. Godzik, and J. Skolnick, *J. Chem. Phys.*, **98**, 7420–7433 (1993).
17. A. Kolinski and J. Skolnick, *Proteins* **18**, 338–352 (1994).
18. J. Skolnick, A. Kolinski, C. Brooks III, A. Godzik, and A. Rey, *Curr. Biol.* **3**, 414–423 (1993).
19. A. Kolinski, W. Galazka, and J. Skolnick, *J. Chem. Phys.* **103**, 10286–10297 (1995).
20. A. Kolinski, M. Milik, J. Rycombel, and J. Skolnick, *J. Chem. Phys.* **103**, 4312–4323 (1995).
21. Y. Zhou, C. K. Hall, and M. Karplus, *Phys. Rev. Lett.* **77**, 2822–2825 (1996).
22. M. Levitt and A. Warshel, *Nature* **253**, 694–698 (1975).
23. J. D. Honeycutt and D. Thirumalai, *Proc. Natl. Acad. Sci. USA* **87**, 3526–3529 (1990).
24. C. Wilson and S. Doniach, *Proteins* **6**, 193–209 (1989).
25. E. W. Knapp, *J. Comp. Chem.* **13**, 793–798 (1992).
26. D. Hoffmann and E. W. Knapp, *Phys. Rev. E* **53**, 4221–4224 (1996).
27. B. Lee, N. Kurochkina, and H. S. Kang, *FASEB J.* **10**, 119–125 (1996).
28. S. Sun, *Prot. Sci.* **2**, 762–785 (1993).
29. A. A. Rabow and H. A. Scheraga, *Prot. Sci.* **5**, 1800–1815 (1996).
30. E. W. Knapp and A. Irgens-Defregger, *J. Comp. Chem.* **14**, 19–29 (1993).
31. A. Elofsson, S. M. Le Grand, and D. Eisenberg, *Proteins* **23**, 73–82 (1996).
32. K. Binder, *Monte Carlo Methods in Statistical Physics*, Springer-Verlag, Berlin, 1986.
33. K. Binder, paper, Institut Für Physik, Johannes Gutenberg-Universität, Mainz, 1991.
34. N. Metropolis, A. W. Rosenbluth, M. N. Rosenbluth, A. H. Teller, and E. Teller, *J. Chem. Phys.* **51**, 1087–1092 (1953).
35. M.-H. Hao and H. A. Scheraga, *J. Phys. Chem.* **98**, 4940–4948 (1994).
36. M.-H. Hao and H. A. Scheraga, *J. Phys. Chem.* **98**, 9882–9893 (1994).
37. M.-H. Hao and H. A. Scheraga, *J. Chem. Phys.* **102**, 1334–1348 (1995).
38. J. Lee, *Phys. Rev. Lett.* **71**, 211–214 (1993).
39. B. A. Berg and T. Neuhaus, *Phys. Rev. Lett.* **68**, 9–12 (1991).
40. A. Kolinski, W. Galazka, and J. Skolnick, *Proteins* **26**, 271–287 (1996).
41. R. A. Orwoll and W. H. Stockmayer, *Adv. Chem. Phys.* **15**, 305–324 (1969).
42. A. Kolinski and J. Skolnick, *Lattice Models of Protein Folding, Dynamics and Thermodynamics*, Landes, Austin, TX, 1996.
43. A. Kolinski, J. Skolnick, and R. Yaris, *Biopolymers* **26**, 937–962 (1987).
44. A. Kolinski, J. Skolnick, and R. Yaris, *Macromolecules* **20**, 438–440 (1987).
45. A. Kolinski, J. Skolnick, and R. Yaris, *J. Chem. Phys.* **86**, 7164–7173 (1987).
46. A. Kolinski and J. Skolnick, *J. Phys. Chem.* **97**, 3450 (1993).
47. P. E. J. Rouse, *J. Chem. Phys.* **21**, 1272–1278 (1953).

48. R. Zwanzig, *J. Chem. Phys.* **60**, 2717–2720 (1974).
49. D. Hoffmann and E. W. Knapp, *Eur. Biophys. J.* **24**, 387–403 (1996).
50. C. Baysal and H. Meirovitch, *J. Chem. Phys.* **105**, 7868–7871 (1996).
51. K. A. Dill, S. Bromberg, K. Yue, K. M. Fiebig, D. P. Yee, P. D. Thomas, and H. S. Chan, *Prot. Sci.* **4**, 561–602 (1995).
52. Z. Guo and D. Thirumalai, *J. Chem. Phys.* **97**, 525–536 (1992).
53. Z. Guo and D. Thirumalai, *Biopolymers* **36**, 83–102 (1995).
54. A. Kolinski, M. Milik, and J. Skolnick, *J. Chem. Phys.* **94**, 3978–3985 (1991).
55. A. Kolinski and J. Skolnick, *J. Phys. Chem.* **97**, 9412–9426 (1992).
56. A. Kolinski and J. Skolnick, *Proteins* **18**, 353–366 (1994).
57. E. I. Shakhnovich, *Folding & Design* **1**, R50–R54 (1996).
58. M. Karplus and A. Sali, *Curr. Opin. Struct. Biol.* **5**, 58–73 (1995).
59. H. S. Chan and K. A. Dill, *Macromolecules* **22**, 4559–4573 (1989).
60. H. S. Chan and K. A. Dill, *Proc. Natl. Acad. Sci. USA* **87**, 6388–6392 (1990).
61. H. S. Chan and K. A. Dill, *J. Chem. Phys.* **95**, 3775–3787 (1991).
62. H. S. Chan and K. A. Dill, *J. Chem. Phys.* **99**, 2116–2127 (1993).
63. L. M. Gregoret and F. E. Cohen, *J. Mol. Biol.* **219**, 109–122 (1991).
64. M.-H. Hao, S. Rackovsky, A. Liwo, M. R. Pinkus, and H. A. Scheraga, *Proc. Natl. Acad. Sci. USA* **89**, 6614–6618 (1992).
65. N. D. Socci and J. N. Onuchic, *J. Chem. Phys.* **100**, 1519–1528 (1994).
66. G. N. Hunt, L. M. Gregoret, and F. E. Cohen, *J. Mol. Biol.* **241**, 214–225 (1994).
67. A. Kolinski, J. Skolnick, and R. Yaris, *J. Chem. Phys.* **85**, 3585–3597 (1986).
68. A. Kolinski and J. Skolnick, *Proc. Natl. Acad. Sci. USA* **83**, 7267–7271 (1986).
69. K. A. Dill, *Biochemistry* **24**, 1501–1509 (1985).
70. K. F. Lau and K. A. Dill, *Macromolecules* **22**, 3986–3997 (1989).
71. E. Shakhnovich, G. Farztdinov, and A. M. Gutin, *Phys. Rev. Lett.* **67**, 1665–1668 (1991).
72. E. I. Shakhnovich and A. M. Gutin, *Proc. Natl. Acad. Sci. USA* **90**, 7195–7199 (1993).
73. E. I. Shakhnovich and A. M. Gutin, *Protein Eng.* **6**, 793–800 (1993).
74. M. E. Kellman, *J. Chem. Phys.* **105**, 2500–2508 (1996).
75. A. Sali, E. Shakhnovich, and M. Karplus, *J. Mol. Biol.* **235**, 1614–1636 (1994).
76. N. D. Socci, J. N. Onuchic, and P. G. Wolynes, *J. Chem. Phys.* **104**, 5860–5868 (1996).
77. V. I. Abkevich, A. M. Gutin, and E. I. Shakhnovich, *J. Chem. Phys.* **101**, 6052–6062 (1994).
78. A. Sali, E. Shakhnovich, and M. Karplus, *Nature* **369**, 248–251 (1994).
79. P. E. Leopold, M. Montal, and J. N. Onuchic, *Proc. Natl. Acad. Sci. USA* **89**, 8721–8725 (1992).
80. L. A. Mirny, V. Abkevich, and E. I. Shakhnovich, *Folding & Design* **1**, 103–116 (1996).
81. H. S. Chan and K. A. Dill, *J. Chem. Phys.* **100**, 9238–9257 (1994).
82. H. S. Chan and K. A. Dill, *Proteins* **24**, 335–344 (1996).
83. A. R. Dinner, A. Sali, and M. Karplus, *Proc. Natl. Acad. Sci. USA* **93**, 8356–8361 (1996).
84. N. D. Socci and J. N. Onuchic, *J. Chem. Phys.* **103** 4732–4744 (1995).
85. J. D. Bryngelson, J. N. Onuchic, N. D. Socci, and P. G. Wolynes, *Proteins* **21**, 167–195 (1995).

86. V. I. Abkevich, A. M. Gutin, and E. I. Shakhnovich, *Prot. Sci.* **4**, 1167–1177 (1995).
87. C. J. Camacho and D. Thirumalai, *Proteins*, **22**, 27–40 (1995).
88. E. M. O'Toole and A. Z. Panagiotopoulos, *J. Chem. Phys.* **97**, 8644–8652 (1992).
89. T. C. Beutler and K. A. Dill, *Prot. Sci.* **5**, 2037–2043 (1996).
90. K. Yue, K. M. Fiebig, P. D. Thomas, H. S. Chan, E. I. Shakhnovich, and K. A. Dill, *Proc. Natl. Acad. Sci. USA* **92**, 325–329 (1995).
91. M.-H. Hao and H. A. Scheraga, *J. Phys. Chem.* **100**, 14540–14548 (1996).
92. M.-H. Hao and H. A. Scheraga, *Proc. Natl. Acad. Sci. USA* **93**, 4984–4989 (1996).
93. V. I. Abkevich, A. M. Gutin, and E. I. Shakhnovich, *Folding & Design* **1**, 221–230 (1996).
94. A. F. P. de Araujo and T. C. Pochapsky, *Folding & Design* **1**, 299–314 (1996).
95. S. Bromberg and K. A. Dill, *Prot. Sci.* **3**, 997–1009 (1994).
96. S.-I. Segawa and T. Kawai, *Biopolymers* **25**, 1815–1835 (1986).
97. M. Karplus and E. Shakhnovich, in *Protein Folding*, T. E. Creighton, ed., Freeman, New York, 1992, pp. 127–196.
98. H. Frauenfelder and P. G. Wolynes, *Phys. Today* **47**, 58–64 (1994).
99. J. D. Bryngelson and P. G. Wolynes, *Biopolymers* **30**, 117–188 (1990).
100. J. D. Bryngelson, *J. Chem. Phys.* **100**, 6038–6045 (1994).
101. J. N. Onuchic, P. G. Wolynes, Z. Luthey-Schulten, and N. D. Socci, *Proc. Natl. Acad. Sci. USA* **92**, 2626–3630 (1995).
102. A. V. Finkelstein, A. Y. Badretdinov, and A. M. Gutin, *Proteins* **23**, 142–150 (1996).
103. N. Go and H. Taketomi, *Proc. Natl. Acad. Sci. USA* **75**, 559–563 (1978).
104. N. Go, H. Abe, H. Mizuno, and H. Taketomi, eds., *Protein Folding*, Elsevier/North Holland, Amsterdam, 1980.
105. H. Taketomi, F. Kano, and N. Go, *Biopolymers* **27**, 527–559 (1988).
106. H. Taketomi, Y. Ueda, and N. Go, *Int. J. Pept. Prot. Res.* **7**, 445–449 (1988).
107. N. Go, *Ann. Rev. Biophys. Bioeng.* **12**, 183–210 (1983).
108. Y. Ueda, H. Taketomi, and N. Go, *Biopolymers* **17**, 1531–1548 (1978).
109. J. Skolnick, A. Kolinski, and R. Yaris, *Proc. Natl. Acad. Sci. USA* **85**, 5057 (1988).
110. J. Skolnick and A. Kolinski, *Annu. Rev. Phys. Chem.* **40**, 207–235 (1989).
111. J. Skolnick, A. Kolinski, and R. Yaris, *Biopolymers* **28**, 1059–1095 (1989).
112. J. Skolnick, A. Kolinski, and R. Yaris, *Proc. Natl. Acad. Sci. USA* **86**, 1229–1233 (1989).
113. J. Skolnick and A. Kolinski, *J. Mol. Biol.* **212**, 787–817 (1989).
114. A. Sikorski and J. Skolnick, *Biopolymers* **28**, 1097–1113 (1989).
115. A. Sikorski and J. Skolnick, *Proc. Natl. Acad. Sci. USA* **86**, 2668–2672 (1989).
116. A. Sikorski and J. Skolnick, *J. Mol. Biol.* **212**, 819–836 (1990).
117. A. Sikorski and J. Skolnick, *J. Mol. Biol.* **215**, 183–198 (1990).
118. A. Kolinski and A. Sikorski, *J. Polym. Sci. Polym. Lett. Ed.* **20**, 177–179 (1982).
119. A. Kolinski and A. Sikorski, *J. Polym. Sci. Polym. Chem. Ed.* **22**, 97–106 (1984).
120. E. O'Toole, R. Venkataramani, and A. Z. Panagiotopoulous, *AIChE J.* **41**, 954–958 (1995).
121. J. Skolnick and A. Kolinski, *J. Mol. Biol.* **221**, 499–531 (1991).
122. R. C. Brower, G. Vasmatiz, M. Silverman, and C. Delisi, *Biopolymers* **33**, 329–334 (1993).
123. A. V. Finkelstein and B. A. Reva, *Nature* **351**, 497–499 (1991).

124. O. V. Galzitskaya and A. V. Finkelstein, *Prot Eng.* **8**, 883–892 (1994).
125. A. V. Finkelstein and B. A. Reva, *Prot. Eng.* **9**, 387–397 (1996).
126. A. Rey and J. Skolnick, *Proteins* **16**, 8–28 (1993).
127. A. Rey and J. Skolnick, *Chem. Phys.* **158**, 199–219 (1991).
128. D. G. Garrett, K. Kastella and D. M. Ferguson, *J. Am. Chem. Soc.* **114**, 6555–6556 (1992).
129. J. D. Honeycutt and D. Thirumalai, *Biopolymers* **32**, 695–709 (1992).
130. F. Kano and H. Maeda, *Mol. Simul.* **16**, 261–274 (1996).
131. J. Skolnick and A. Kolinski, in *Encyclopedia of Computational Chemistry*, P. von Rauge Schleyer and P. Kollman, eds., Wiley, New York, (in press) 1998.
132. I. D. Kuntz, G. M. Crippen, P. A. Kollman, and D. Kimelman, *J. Mol. Biol.*, **106**, 983–994 (1976).
133. B. Robson and D. J. Osguthorpe, *J. Mol. Biol.* **132**, 19–51 (1979).
134. A. T. Hagler and B. Honig, *Proc. Natl. Acad. Sci. USA* **75**, 554–558 (1978).
135. S. Tanaka and H. A. Scheraga, *Macromolecules* **9**, 945–950 (1975).
136. G. M. Crippen and V. N. Viswanadhan, *Int. J. Pept. Prot. Res.* **25**, 487–509 (1985).
137. F. Fogolari, G. Esposito, P. Viglino, and S. Cattarinussi, *Biphys. J.* **70**, 1183–1197 (1996).
138. A. Wallqvist and M. Ullner, *Proteins* **18**, 267–289 (1994).
139. S. Miyazawa and R. L. Jernigan, *Macromolecules* **18**, 534–552 (1985).
140. J. A. McCammon, S. H. Northrup, M. Karplus, and R. M. Levy, *Biopolymers* **19**, 2033–2045 (1980).
141. D. M. Ferguson and D. G. Garret, in *Adaptation of Simulated Annealing to Chemical Optimization Problems*, Vol. 15, by J. H. Kalivas, ed., Elsevier, New York, 1995.
142. W. R. Krigbaum and S. F. Lin, *Macromolecules* **15**, 1135–1145 (1982).
143. V. G. Dashevskii, *Molekulyarnaya Biologiya* (*Engl. transl.*), **14**, 105–117 (1980).
144. D. G. Covell, *Proteins* **14**, 409–420 (1992).
145. D. G. Covell and R. L. Jernigan, *Biochemistry* **29**, 3287–3294 (1990).
146. D. A. Hinds and M. Levitt, *Proc. Natl. Acad. Sci. USA* **89**, 2536–2540 (1992).
147. J. Skolnick and A. Kolinski, in *Computer Simulations of Biomolecular Systems. Theoretical and Experimental Studies*, W. F. van Gunsteren, P. K. Weiner, and A. J. Wilkinson, eds., ESCOM Science. Publishers B. V., Leiden, 1997.
148. J. Skolnick and A. Kolinski, *Science* **250**, 1121–1125 (1990).
149. A. Godzik, J. Skolnick, and A. Kolinski, *Proc. Natl. Acad. Sci. USA* **89**, 2629–2633 (1992).
150. R. S. DeWitte and E. I. Shakhnovich, *Prot. Sci.* **3**, 1570–1581 (1994).
151. M. Levitt and J. Greer, *J. Mol. Biol.* **114**, 181–293 (1977).
152. A. Godzik, J. Skolnick, and A. Kolinski, *Prot. Eng.* **6**, 801–810 (1993).
153. K. A. Olszewski, A. Kolinski, and J. Skolnick, *Prot. Eng.* **9**, 5–14 (1996).
154. K. A. Olszewski, A. Kolinski, and J. Skolnick, *Proteins* **25**, 286–299 (1996).
155. M. Vieth, A. Kolinski, C. L. Brooks III, and J. Skolnick, *J. Mol. Biol.* 237, 361–367 (1994).
156. D. S. Rykunov, B. A. Reva, and A. V. Finkelstein, *Proteins* **22**, 100–109 (1995).
157. A. Monge, E. J. P. Lathrop, J. R. Gunn, P. S. Shenkin, and R. A. Friesner, *J. Mol. Biol.* **247**, 995–1012 (1995).
158. B. A. Reva, D. S. Rykunov, A. J. Olson, and A. V. Finkelstein, *J. Comp. Biol.* **2**, 527–535 (1995).

159. B. A. Reva, A. V. Finkelstein, M. F. Sanner, and A. J. Olson, *Proteins* **25**, 379–388 (1996).
160. B. A. Reva, A. V. Finkelstein, D. S. Rykunov, and A. J. Olson, *Proteins* **26**, 1–8 (1996).
161. B. Honig and F. E. Cohen, *Folding & Design* **1**, R17–R20 (1996).
162. G. D. Rose and R. Wolfenden, *Annu. Rev, Biophys. Biomol. Struct.* **22**, 381–415 (1993).
163. J. H. Dyson and P. E. Wright, *Curr. Biol.* **3**, 60–65 (1993).
164. K. Lumb, J, and P. Kim, S., *Biochemistry* **34**, 8642–8648 (1995).
165. R. A. Abagyan, *FEBS Lett.* **325**, 17–22 (1993).
166. P. Argos, *J. Mol. Biol.* **197**, 331–348 (1987).
167. B. H. Zimm and J. K. Bragg, *J. Chem. Phys.* **31**, 526–535 (1959).
168. D. Poland and H. A. Scheraga, *Theory of Helix-Coil Transitions in Biopolymers*, Academic Press, New York, 1970.
169. E. I. Shakhnovich and A. V. Finkelstein, *Biopolymers* **28**, 1667–1680 (1989).
170. E. I. Shakhnovich, *Phys. Rev. Lett.* **72**, 3907–3910 (1994).
171. I. Bahar and R. L. Jernigan, *Biophys. J.* **66**, 467–481 (1994).
172. C. J. Camacho and D. Thirumalai, *Proc. Natl. Acad. Sci. USA* **90**, 6369–6372 (1993).
173. C. B. Post and B. H. Zimm, *Biopolymers* **18**, 1487–1501 (1979).
174. I. C. Sanchez, *Macromolecules* **12**, 980–988 (1979).
175. S. F. Betz, D. P. Raleigh, and W. F. DeGrado, *Curr. Opin. Struct. Biol.* **3**, 601–610 (1993).
176. C. L. I. Brooks, *Curr. Opin. Struct. Biol.* **3**, 92–98 (1993).
177. T. Handel and W. F. DeGrado, *Biophys. J.* **61**, A265 (1992).
178. K. Kuwajima, *Proteins* **6**, 87–103 (1989).
179. O. B. Ptitsyn, R. H. Pain, G. V. Semisotnov, E. Zerovnik, and O. I. Razgulyaev, *FEBS Lett.* **262**, 20–24 (1990).
180. D. P. Raleigh and W. F. DeGrado, *J. Am. Chem. Soc.* **114**, 10079–10081 (1992).
181. J. Skolnick, A. Kolinski, and A. Godzik, *Proc. Natl. Acad. Sci. USA* **90**, 2099–2100 (1993).
182. A. V. Finkelstein and E. I. Shakhnovich, *Biopolymers* **29**, 1681–1694 (1989).
183. R. L. Baldwin and H. Roder, *Curr. Biol.* **1**, 219–220 (1991).
184. O. B. Ptitsyn, *Curr. Opin. Struct. Biol.* **5**, 74–78 (1995).
185. A. Godzik, A. Kolinski, and J. Skolnick, *J. Comp. Aided Mol. Design* **7**, 397–438 (1993).
186. M. Vásquez, G. Nemethy, and H. A. Scheraga, *Chem. Rev.* **94**, 2183–2239 (1994).
187. F. A. Momany, R. F. McGuire, A. W. Burgess, and H. A. Scheraga, *J. Phys. Chem.* **79**, 2361–2381 (1975).
188. M. H. Lambert and H. A. Scheraga, *J. Comp. Chem.* **10**, 817–831 (1989).
189. Z. Li and H. A. Scheraga, *Proc. Natl. Acad. Sci. USA* **84**, 6611–6615 (1987).
190. H. Meirovitch, M. Vasquez, and H. A. Scheraga, *Biopolymers* **27**, 1189–1204 (1988).
191. M. Vásquez and H. A. Scheraga, *Biopolymers* **24**, 1437–1447 (1985).
192. H. A. Scheraga, M.-H. Hao, and J. Kostrowicki, in *Methods in Protein Structure Analysis*, M. Z. Atassi and E. Appela, eds., Plenum Press, New York, 1995.
193. M. Vasquez, E. Meirovitch, and H. Meirovitch, *J. Phys. Chem.* **98**, 9380–9382 (1994).
194. D. R. Ripoll and H. A. Scheraga, *Biopolymers* **30**, 165–176 (1990).
195. A. Liwo, M. R. Pincus, R. J. Wawak, S. Rackovsky, and H. A. Scheraga, *Prot. Sci.* **2**, 762–785 (1993).
196. A. Liwo, M. R. Pincus, R. J. Wawak, S. Rackovsky, and H. A. Scheraga, *Prot. Sci.* **2**,

2725–1731 (1993).

197. A. Liwo, S. Oldziej, J. Ciarkowski, G. Kupryszewski, M. R. Pinkus, R. J. Wawak, S. Rackovsky, and H. A. Scheraga, *J. Prot. Chem.* **13**, 375–380 (1994).
198. B. Cheng, A. Nayeem, and H. A. Scheraga, *J. Comp. Chem.* **17**, 1453–1480 (1996).
199. R. J. Wawak, K. D. Gibson, A. Liwo, and H. A. Scheraga, *Proc. Natl. Acad. Sci. USA* **93**, 1743–1746 (1996).
200. M. Nilges and A. T. Brunger, *Proteins* **15**, 133–146 (1993).
201. U. H. E. Hansmann and Y. Okamoto, *J. Comp. Chem.* **14**, 1333–1338 (1993).
202. Y. Okamoto, *Proteins* **19**, 14–23 (1994).
203. H. Meirovitch and E. Meirovitch, *J. Phys. Chem.* **99**, 4847–4854 (1995).
204. E. Meirovitch and H. Meirovitch, *Biopolymers* **38**, 68–88 (1996).
205. H. Meirovitch and E. Meirovitch, *J. Phys. Chem.* **100**, 5123–5133 (1996).
206. W. F. van Gunsteren and P. K. Weiner, *Computer Simulations of Biomolecular Systems. Theoretical and Experimental Applications*, ESCOM Science Publishers B.V., Leiden, 1989.
207. G. M. Crippen and M. E. Snow, *Biopolymers* **29**, 1479–1489 (1990).
208. G. M. Crippen, *Biochemistry* **30**, 4232–4237 (1991).
209. R. Srinivasan and G. D. Rose, *Proteins* **22**, 81–99 (1995).
210. M. Vieth, A. Kolinski, C. L. Brooks, III, and J. Skolnick, *J. Mol. Biol.* **251**, 448–467 (1995).
211. A. Godzik, A. Kolinski, and J. Skolnick, *Prot. Sci.* **4**, 2107–2117 (1995).
212. A. Godzik, *Current Biol.* **4**, 363–366 (1996).
213. M. Milik, A. Kolinski, and J. Skolnick, *Prot. Eng.* **8**, 225–236 (1995).
214. R. Friesner and J. R. Gunn, *Annu. Rev. Biophys. Biomol. Struct.* **25**, 315–342 (1996).
215. B. Persson and P. Argos, *J, Mol. Biol.* **237**, 182–192 (1994).
216. T. Dandekar and P. Argos, *J. Mol. Biol.* **256**, 645–660 (1996).
217. C. Mumenthaler and W. Braun, *Prot. Sci.* **4**, 863–871 (1995).
218. M. J. Smith-Brown, D. Kominos, and R. M. Levy, *Prot. Eng.* **6**, 605–614 (1993).
219. A. Aszodi, M. J. Gradwell, and W. R. Taylor, *J. Mol. Biol.* **251**, 308–326 (1995).
220. J. Skolnick, A. Kolinski, and A. R. Ortiz, *J. Mol. Biol.* **265**, 217–241 (1997).
221. A. Ortiz, A. Kolinski, and J. Skolnick, *J. Mol. Biol.* **277**, 419–448 (1998).
222. A. Ortiz, A. Kolinski, and J. Skolnick, *Proc. Natl. Acad. Sci. USA* **95**, 1020–1025 (1998).
223. B. Rost and C. Sander, *J. Mol. Biol.* **232**, 584–599 (1993).
224. A. Kolinski, J. Skolnick, A. Godzik, and W-P. Hu, *Proteins* **27**, 290–308 (1997).
225. U. Gobel, C. Sander, R. Schneider, and A. Valencia, *Proteins* **18**, 309–317 (1994).
226. A. Godzik, J. Skolnick, and A. Kolinski, *J. Mol. Biol.* **227**, 227–238 (1992).
227. F. H. C. Crick, *Acta Cryst.* **6**, 689–697 (1953).
228. A. Holtzer, M. E. Holtzer, and J. Skolnick, in *Protein Folding. Deciphering the Second Half of the Genetic Code*, Vol. 4, L. M. Gierash, and J. King AAAS, Washington, DC, 1990, pp. 2401–2411.
229. M. Nilges and A. T. Brunger, *Prot. Eng.* **4**, 649–659 (1991).
230. S. R. Krystek Jr., R. E. Bruccoleri, and J. Novotny, *Int. J. Pept. Prot. Res.* **38**, 229–236 (1991).
231. L. Zhang and J. Hermans, *Proteins* **16**, 384–392 (1993).

232. P. Harbury, H., B. Tidor, and P. S. Kim, *Proc. Natl. Acad. Sci. USA* **92**, 8408–8412 (1995).
233. P. B. Harbury, T. Zhang, P. S. Kim, and T. Alber, *Science* **262**, 1401–1407 (1993).
234. M. Vieth, A. Kolinski, and J. Skolnick, *Biochememistry* **35**, 955–967 (1996).
235. K. J. Lumb, C. M. Carr, and P. S. Kim, *Biochemistry* **33**, 7361–7367 (1994).
236. B. Lovejoy, S. Choe, D. K. McRorie, W. F. DeGrado, and D. Eisenberg, *Science* **259**, 1288–1293 (1993).
237. M. Milik and J. Skolnick, *Proteins* **15**, 10–25 (1993).
238. M. Milik and J. Skolnick, *Biophys J.* **69**, 1382–1386 (1995).
239. P. McDonnell, K. Shon, Y. Kim , and S. Opella, *J. Mol. Biol.* **233**, 447–463 (1993).
240. K. J. Shon, Y. Kim, L. A. Colnago, and S. J. Opella, *Science* **252**, 1303–1305 (1991).
241. A. Baumgartner and J. Skolnick, *Phys. Rev. Lett.* **74**, 2142–2145 (1995).
242. A. Baumgartner and J. Skolnick, *J. Phys. Chem.* **98**, 10655–10658 (1994).
243. C. Levinthal, *J. Chim. Phys.* **65**, 44–45 (1968).
244. R. Zwanzig, A. Szabo, and B. Bagchi, *Proc. Natl. Acad. Sci. USA* **89**, 20–22 (1992).
245. P. G. Wolynes, in *Protein Folds. A Distance-Based Approach*, H. Bohr and S. Brunak, ed., CRC Press, New York, 1996, pp. 3–17.
246. P. G. Wolynes, in *Physics of Biomaterials: Fluctuations, Selfassembly and Evolution*, T. Riste and D. Sherrington, eds., Kluwer Academic Publishers, Netherlands, 1996, pp. 235–248.
247. D. Eliezer, P. A. Jennings, P. E. Wright, S. Doniach, K. O. Hodgson, and H. Tsuruta, *Science* **270**, 487–488 (1995).
248. W. F. DeGrado, Z. R. Wasserman, and J. D. Lear, *Science* **243**, 622–628 (1989).
249. T. M. Handel, S. A. Williams, and W. F. DeGrado, *Science* **261**, 879–885 (1993).
250. D. P. Raleigh, S. F. Betz, and W. F. DeGrado, *J. Am. Chem. Soc.* **117**, 7558–7559 (1995).
251. C. Sander, G. Vriend, F. Bazan, A. Horovitz, H. Nakamura, L. Ribas, A. V. Finkelstein, A. Lockhart, R. Merkl, L. J. Perry, *Proteins* **12**, 105–110 (1992).
252. M. H. Hecht, J. S. Richardson, D. C. Richardson, and R. C. Ogden, *Science* **249**, 884–891 (1990).
253. G. Bohm, *Biophys. Chem.* **59**, 1–32 (1996).
254. M. Levitt, *Curr. Opin. Struct. Biol.* **1**, 224–229 (1991).
255. J. W. Ponder and F. M. Richards, *J. Mol. Biol.* **193**, 775–791 (1987).
256. H. Flockner, M. Braxenthaler, P. Lackner, M. Jaritz, M. Ortner, and M. J. Sippl, *Proteins* **23**, 376–386 (1996).
257. D. T. Jones, W. R. Taylor, and J. M. Thornton, *Nature* **358**, 86–89 (1992).
258. R. Luethy, J. U. Bowie, and D. Eisenberg, *Nature* **356**, 83–85 (1992).
259. S. H. Bryant and C. E. Lawrence, *Proteins* **16**, 92–112 (1993).
260. S. J. Wodak and M. J. Rooman, *Curr. Opin. Struct. Biol.* **3**, 247–259 (1993).
261. J. M. Thornton, T. P. Flores, D. T. Jones, and M. B. Swindells, *Nature* **354**, 105–106 (1991).
262. P. D. Thomas and K. A. Dill, *J. Mol. Biol.* **257**, 457–469 (1996).
263. M. J. Sippl and S. Weitckus, *Proteins* **13**, 258–271 (1992).
264. C. Chothia and A. Finkelstein, *Annu. Rev. Biochem.* **59**, 1007–1039 (1990).
265. J. Richardson and D. C. Richardson, in *Prediction of Protein Structure and the Principles of Protein Conformation*, G. D. Fasman, ed., Plenum Press, New York, 1989, pp. 1–99.

266. V. N. Maiorov and G. M. Crippen, *Proteins* **20**, 167–173 (1994).
267. A. R. Ortiz, W.-P. Hu, A. Kolinski, and J. Skolnick, *J. Mol. Graph.* (in press).
268. J. Novotny, R. Brucolleri, and M. Karplus, *J. Mol. Biol.* **177**, 787–818 (1984).
269. A. Elofsson and L. Nilsson, *J. Mol. Biol.* **223**, 766–780 (1993).
270. S. DeBolt and J. Skolnick, *Prot. Eng.* **9**, 637–655 (1996).
271. R. K. Singh, A. Tropsha, and I. I. Vaisman, *J. Comput. Biol.* **3**, 213–221 (1996).
272. A. Tropsha, R. K. Singh, I. I. Vaisman, and W. Zheng, in *Pacific Symposium on Biocomputing'96*, L. Hunter and T. E. Klein, eds., World Scientific, Singapore, 1996, pp. 614–623.
273. D. C. Rapaport, J. E. Johnson, and J. Skolnick, *Biophys. J.*, submitted.

ENTROPY SAMPLING MONTE CARLO FOR POLYPEPTIDES AND PROTEINS

HAROLD A. SCHERAGA AND MING-HONG HAO

Baker Laboratory of Chemistry, Cornell University, Ithaca, NY 14853-1301

CONTENTS

I. INTRODUCTION

Monte Carlo (MC) methods have been applied in a variety of ways in simulation studies of polypeptides and globular proteins [1,2]. Over the years, our laboratory has developed and implemented a number of MC methods for treating different problems concerning polypeptides and globular proteins. These MC techniques can be divided into two classes according to their objectives: those for finding the lowest-energy conformation of a polypeptide and those for studying the statistical mechanics of protein folding.

In searching for the lowest-energy conformation of a polypeptide, the major technical requirement for an MC algorithm is to find the lowest-energy conformation as quickly as possible. This means that the ideal MC method for finding the global-energy minimum of a polypeptide should be one that samples the conformational space sparsely and pinpoints the target region of the conformational space quickly. The Monte-Carlo–plus

Advances in Chemical Physics, Volume 105, Monte Carlo Methods in Chemical Physics, edited by David M. Ferguson, J. Ilja Siepmann, and Donald G. Truhlar. Series Editors I. Prigogine and Stuart A. Rice.
ISBN 0-471-19630-4

minimization (MCM) method is an example of such an algorithm [3,4]. The conventional MC method which searches the whole conformational space is not efficient in finding the global energy minimum of a polypeptide. The MCM method overcomes this deficiency by sorting through only the local energy minima of a polypeptide by energy minimization within the framework of the conventional MC algorithm. This approach is very efficient in finding the lowest-energy conformations of polypeptides. In a test of the MCM method [4], by allowing random changes of more than one dihedral angle at a time and by making more frequent changes in the backbone dihedral angles than in the side-chain dihedral angles, the procedure found the *exact* lowest-energy conformation of the pentapeptide Met-enkephalin from all 18 random starting conformations. This method has been found to be superior to simulated annealing (SA) for seeking the global energy minimum of Met-enkephalin [5]. Later claims to the contrary by Okamoto et al. [6], repeated by Wilson and Cui [7], were without foundation because Okamoto et al. [6] found *regions* near the global minimum in only 25% of their SA runs, and none of them corresponded to the *exact* global minimum. The basic idea of the MCM method has been extended in several algorithms for searching for the global energy minimum of polypeptides, including the electrostatically driven Monte Carlo [8–10] method which also adopts some ideas from the self-consistent electrostatic field procedure [11]; the MC search method, which uses an additional compact-volume constraint [12]; and the alternative heating and cooling MC minimization algorithm of von Freyberg and Braun [13]. The various approaches in this area have been reviewed previously [14,15].

The main focus of this chapter is the statistical–mechanical study of polypeptide and protein folding by MC methods. The major technical requirements for MC methods in statistical–mechanical studies of proteins are to sample the complete relevant conformational space of a protein reliably and reproducibly, not to be trapped in local-energy minima or substates, and to produce a high-quality conformational sample with a sampling uncertainty as small as possible within a given amount of computational time. Aside from the accuracy of the potential function, these are the necessary and sufficient conditions for an accurate determination of the statistical–mechanical properties of proteins. Because the conformational space of an even modest-size polypeptide is huge and complicated, purely random sampling could hardly reveal the essential information about a protein. Therefore, importance-sampling techniques are indispensable for effective MC simulations of polypeptides and globular proteins.

The basic importance-sampling MC algorithm is the Metropolis method [16]. Here, the conformational space is sampled by a set of MC moves through a Markov chain process. A trial move from a conformation (or state) m to a conformation (or state) n is accepted on the basis of the probability $p_{(m \to n)} = \min(1, \alpha_{nm} \rho_n / \alpha_{mn} \rho_m)$, where α_{mn} is the a priori transition probability from conformation m to n, and ρ_m is the probability density of conformation m. Energetic importance sampling is realized in the conventional MC method by choosing the probability density ρ_m as the Boltzmann probability: $\rho_m = \exp(-E_m/kT)$, where E_m is the energy of conformation m, T is the temperature, and k is the Boltzmann constant. However, the probability density factor ρ_m can be varied in different ways in order to obtain an efficient MC sampling algorithm. In parallel, the factor α_{mn} may also be varied by selecting proper move sets to achieve better sampling.

In statistical–mechanical studies of polypeptides and globular proteins, a quantity of fundamental importance is the microscopic entropy, $S(E)$, which is proportional to the logarithm of the number density of conformations at energy level E. For either proteins or polypeptides of modest lengths, $S(E)$ is a quasi-continuous function, which is usually an increasing function of E in relevant energy regions. If the molecular system has a unique lowest-energy structure, then the $S(E)$ function of the lowest-energy state also has the smallest value in all relevant energy ranges. By simple analogy to the energy importance-sampling MC method, one may construct an entropy importance-sampling MC algorithm by choosing the probability density ρ_m in the Metropolis algorithm as $\rho_m = \exp[-S(E_m)/k]$. Such an algorithm would sample the low-entropy conformations in a manner as the conventional MC method samples the *energetically* important conformations. This algorithm is called the entropy-sampling Monte Carlo (ESMC) method [17–19]. It has a broader meaning than the simple analogy described above. This chapter presents an in-depth analysis of the ESMC method in its applications to polypeptides and globular proteins, focusing on its advantages over conventional MC methods, on its existing problems, and on the approaches to treat these problems.

Although the ESMC method is an effective algorithm for sampling the conformations of small polypeptides, the most important application of the method perhaps lies in its treatment of globular proteins. To reveal the connection between the ESMC method and the protein folding problem, we first discuss the microcanonical approach to the problem of protein folding.

II. THE MICROCANONICAL APPROACH TO PROTEIN FOLDING

The physical process of protein folding involves a phase transition from a statistical coiled state to a uniquely compact native state. A powerful approach to define these systems is to make use of the microcanonical entropy function [18–27]. The entropy function $S(E)$ is related to the density of states, $\Omega(E)$, by the relationship

$$S(E) = k \ln \Omega(E) \tag{2.1}$$

From the entropy function, one can directly define the Helmholtz free energy F as a function of energy and temperature:

$$F(E, T) = E - TS(E) \tag{2.2}$$

These two functions provide a microcanonical characterization of the folding behavior of a given protein. Figure 1 illustrates the characteristics of the $S(E)$ function of the two basic types of protein models, denoted as A and

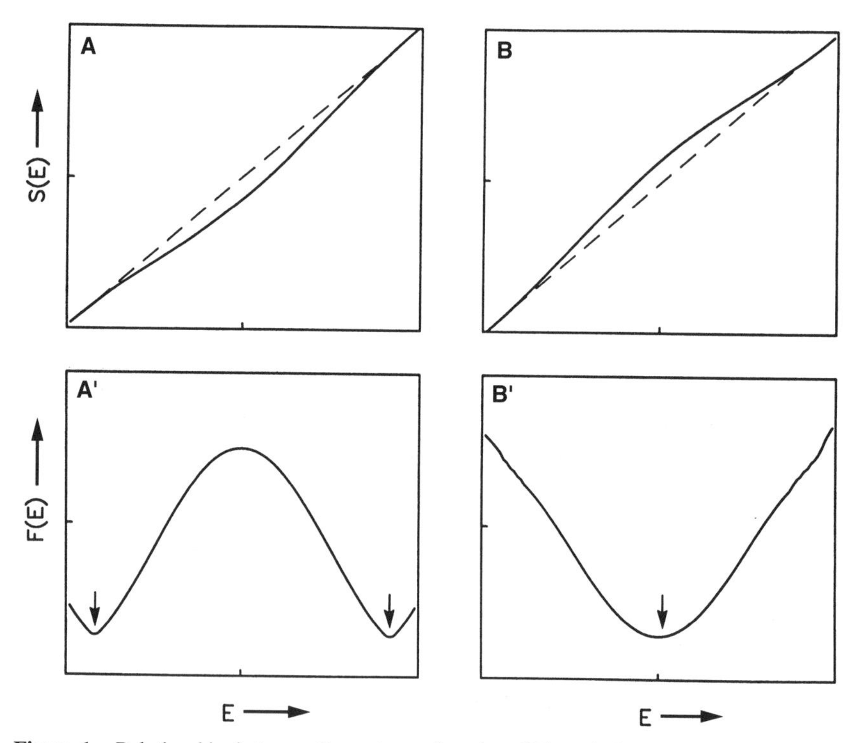

Figure 1. Relationship between the entropy function $S(E)$ and the Helmholtz free-energy function $F(E)$ at a given temperature. Panels A and A' show a protein model with two-state folding character; B and B' show a protein model with a gradual folding process.

B, respectively. Type A protein models involve a two-state folding transition, they are characterized by an entropy function with a concave segment as shown by Figure 1*A*. The corresponding free energy function (Fig. 1*A*′) has two minima that define the native and nonnative states of the protein, respectively. Type B polypeptide models are characterized by a convex $S(E)$ function as shown by Figure 1*B*; the free energy function of such models (Fig. 1*B*′) has only one minimum at all temperatures. Polypeptides with the type B $S(E)$ function do not have a unique phase transition in their folding, and, for polypeptides of the size of proteins, type B models mostly fold to random globules. The $S(E)$ and $F(E)$ functions also provide a definition of the transition state in protein folding; the transition state corresponds to the free-energy maximum in Figure 1*A*′. The type A model plays a central role for understanding the statistical–mechanical basis of protein folding to a unique native structure; more complicated folding behavior can be treated as a modification of that of the type A protein model.

Gō [20] is one of the first authors to invoke the $S(E)$ function for defining the characteristics of protein folding. He used the concave segment of the $S(E)$ function as a quantitative definition of the all-or-none folding and unfolding transition of proteins. More recently, researchers have begun to use the energy landscape concept to understand the folding behavior of proteins [28–30]. It is proposed that the energy landscapes of polypeptides have two types of behavior; for random heteropolymers, the energy landscape consists of numerous funnels leading to globally different low-energy states, but for proteins with a unique folding transition to the native state, the funnels in the energy landscape are all globally connected to the single native state. At present, the energy landscape models for proteins are only qualitative illustrations; it is not possible to determine directly the complete energy landscape for protein models. However, the entropy function $S(E)$ is a one-dimensional mapping of the multidimensional energy landscape, and $S(E)$ can be determined quantitatively by the MC methods discussed later in this chapter. In foldable protein models whose energy landscape is characterized by a dominant tunnel to the native state, the corresponding $S(E)$ function is characterized by a concave segment between the native and the nonnative states.

From the $S(E)$ function, all thermodynamic properties of protein models can be evaluated. The canonical average energy of the system as a function of temperature can be calculated as

$$\langle E \rangle_T = \frac{\sum_{E'} E' \exp[S(E')/k - E'/kT]}{\sum_{E'} \exp[S(E')/k - E'/kT]} \tag{2.3}$$

The canonical free energy is calculated as

$$\langle F \rangle_T = -kT \ln\left\{\sum_E \exp[S(E)/k - E/kT]\right\} \tag{2.4}$$

and the canonical entropy is

$$\langle S \rangle_T = \frac{\langle E \rangle_T - \langle F \rangle_T}{T} \tag{2.5}$$

The thermal average of other intensive properties, say, M, can be calculated as

$$\langle M \rangle_T = \frac{\sum_E \hat{M}(E) \exp[S(E)/k - E/kT]}{\sum_E \exp[S(E)/k - E/kT]} \tag{2.6}$$

where $\hat{M}(E)$ is the average value of the intensive property M inside the energy bin E. Therefore, once the entropy function $S(E)$ of a protein model is determined, the statistical–mechanical problem of the system is solved.

The heart of the microcanonical approach to protein folding is an accurate determination of the $S(E)$ function, or the density of states, for protein models of interest. Many years ago, Miyazawa and Jernigan [21] determined the $S(E)$ function for studying the equilibrium folding and unfolding of the protein BPTI. These authors used a biased MC sampling technique, a modified Rosenbluth and Rosenbluth procedure [31–33] to generate the conformations of the protein. With this technique, they generated a total of 230,000 conformations spanning the denatured states to the completely folded state, from which an approximate $S(E)$ function for BPTI was obtained. The resulting $S(E)$ function exhibits a strong concave segment, indicating a two-state folding transition of the BPTI model. This result is exactly what one would expect on the basis of the native-specified type of force field used in that study; such force fields usually lead to a strong two-state folding behavior in model proteins [34,35]. However, from a computational point of view, the sampling approach of Miyazawa and Jernigan is not satisfactory because, for such a large system, only a modest number of conformations was generated and no convergence test could be made.

The reliability of a microcanonical approach to the protein folding problem depends on the accuracy of the resulting $S(E)$ function. To obtain a sufficiently accurate $S(E)$ for complicated protein systems by MC methods, a number of basic problems have to be addressed rigorously. These problems include (1) how to ensure that all relevant energy regions of the conformational space are sampled adequately and sufficiently, (2) how to reduce the sampling uncertainty or variances in MC simulations, and (3) how to check the convergence and correctness of the computed $S(E)$. The ESMC method provides a way to address these problems directly; it also provides an effective algorithm for sampling conformations of polypeptides and globular proteins.

III. THE ENTROPY SAMPLING MONTE CARLO METHOD

The sampling problems in MC determinations of the thermodynamic functions of proteins were encountered early in MC studies of simple physical systems and, accordingly, effective approaches have evolved over the years. One of the earliest works in this area was that of Torrie and Valleau [36, 37]. Their problem was to determine the free-energy difference between a state of interest and a reference state, say, high- and low-temperature states. The relative free-energy difference can be expressed as a canonical average over the reference system (denoted by the subscript 0): $\Delta A/kT = -\ln\langle \exp(-E/kT + E/kT_0)\rangle_0$, where E is the potential function of the system, and T and T_0 are the two temperatures of interest. In order to determine the average in the brackets $\langle\ \rangle$ accurately, the MC sampling probability between and including the two states must be sufficiently large to ensure good statistics. The standard Metropolis MC formulation with the Boltzmann probability does not satisfy such a requirement because the sampling probability in such MC simulations normally peaks in only one region and vanishes quickly in other regions. Torrie and Valleau [36,37] proposed the use of an artificial probability function $\rho(E) = W(E)$ to replace the Boltzmann factor in the Metropolis MC simulation. The bridge function $W(E)$ is chosen in such a way that the sampling probabilities over all the interested conformational ranges are uniformly large. The canonical distribution can be recovered from a reweighting of the sampled conformational distribution with the bridge function. The free energy difference can be calculated as

$$\Delta A/kT = -\ln\left\langle \exp\frac{-E/kT + E/kT_0}{W(E)}\right\rangle_W + \ln\left\langle\frac{1}{W(E)}\right\rangle_W$$

where the subscript W indicates averaging over the conformational sample obtained with the modified Metropolis algorithm. In the work of Torrie and Valleau, the probability function $W(E)$ was constructed by trial-and error; the authors felt [37] that a trustworthy a priori estimate of the correct value of $W(E)$ could not be made because this would require prior knowledge of the conformational distribution.

The preceding problem was solved by Berg et al. in their multicanonical MC method [38,39]. Berg et al. were concerned with efficient simulation of the first-order phase transition and the spin-glass transition in spin systems. It is known that ordinary MC simulations of first-order phase transitions suffer from a critical slowdown phenomenon due to the free-energy barrier between the two phases in the transition, which causes the tunneling time between the two phases to increase exponentially with the size of the system. The free-energy barrier, and therefore the critical slowdown, can be removed with the approach of Torrie and Valleau, [36,37] specifically, by carrying out the Monte Carlo simulation with a probability function that has a uniformly flat distribution between the two states of the transition. Berg et al. [38,39] used a probability function of the form $\rho(E) = \exp[-\beta(E)E + \alpha(E)]$, which is a linear combination of a series of Boltzmann functions at different temperatures, hence the term *multi-canonical MC*. To find the correct values for the functions $\beta(E)$ and $\alpha(E)$ that lead to a flat density distribution in the MC sampling, an iterative procedure was proposed. Starting with an initial estimate (or an arbitrary one) for the functions, an MC simulation is carried out with the initial estimated probability function, and the energy histogram $H(E)$ is collected from the simulation. Then the function $\beta(E)$ is updated by the formula

$$\beta(E) = \beta(E)_0 + \frac{\ln H(E') - \ln H(E)}{E' - E} \tag{3.1}$$

where E and E' are the energies of the two neighboring bins in the energy histogram. The function $\alpha(E)$ is treated as a free parameter that can be determined after $\beta(E)$ is fixed. For simple systems, a few iterations are sufficient to determine the correct $\beta(E)$ and $\alpha(E)$. In comparison to the approach of Torrie and Valleau, the multicanonical method needs no trial-and-error; all required quantities are defined explicitly and can be obtained automatically by the iterative procedure. However, the basic objective of the two methods is identical: finding a probability function $\rho(E)$ with which the sampling probability of the Metropolis MC method is uniformly flat in the range of conformational space of interest.

A direct connection between the MC probability function and the microcanonical entropy of the system is realized in the ESMC algorithm [17]. The condition for a uniform (or flat) probability distribution of energy in a Metropolis-type MC algorithm can be expressed as

$$P(E) = \Omega(E)\rho(E) = \text{constant} \tag{3.2}$$

In the algorithms discussed above [36–39], different choices have been used for $\rho(E)$. However, from the relationship between the entropy function $S(E)$ and the density of states $\Omega(E)$, it can be seen that, to satisfy Eq. (3.2), a simple choice for the function $\rho(E)$ is

$$\rho(E) = \Omega(E)^{-1} = e^{-S(E)} \tag{3.3}$$

with the constant k set to unity for simplicity. On the basis of the preceding relationships, Lee [17] proposed the entropy-sampling Monte Carlo method, in which the conventional Boltzmann probability function $\exp(-E/T)$ in the Metropolis MC algorithm is replaced by the entropy function $\exp[-S(E)]$. With this formulation, the sampled probability of energy states in an MC simulation is exactly constant. When Eq. (3.3) is inserted into the Metropolis MC algorithm [16], the criterion for accepting or rejecting a new conformation in an ESMC simulation becomes

$$p = \min[1, e^{-S(E)+S(E)_0}] \tag{3.4a}$$

where $S(E)$ and $S(E)_0$ are the entropies of the new and the old conformation, respectively. Since the exact entropy of the system is unknown, the following iterative procedure is used instead [17]. Starting with an arbitrary function $J(E)$, carry out a Metropolis MC simulation with the acceptance criterion:

$$p = \min[1, e^{-J(E)+J(E)_0}] \tag{3.4b}$$

where $J(E)$ is a trial entropy function. The energy space of the system is divided into bins. From an ESMC simulation, the histogram of the energy density $H(E)$ is collected. After each simulation, the function $J(E)$ is updated by the formula

$$J(E)_{\text{new}} = J(E)_{\text{old}} + \ln H(E) \tag{3.5}$$

when $H(E) > 0$. This process is repeated until the sampled energy density $H(E)$ becomes constant, at which stage the function $J(E)$ converges to the entropy function $S(E)$ plus a constant. The convergence of the preceding process can be shown as follows [18]. From the definition of $J(E)$, it is clear that, the larger the function $J(E)$, the smaller is the sampling probability of the state of energy E; therefore, the histogram for that state obtained in the simulation will be smaller, and the increase of the new $J(E)$ after the updating will be smaller. As a result, the moving direction of $P(E)$ in a series of updatings is toward the uniformity of the function $P(E)$. Once the correct entropy function is obtained, a single ESMC simulation run can sample all energy states, including the lowest-energy one.

The ESMC algorithm resolves some basic problems in accurately determining the entropy functions for protein models by MC simulations. First, the ESMC technique samples all energy states with equal probability; therefore, the presence of local energy minima will no longer be a problem in an ESMC simulation. All energy states, including the lowest-energy state, can be covered in an ESMC simulation in the process of iteratively updating the entropy function. Second, in an ESMC simulation with the correct entropy function, the sampling probabilities of all energy regions of interest are equal, which avoids wasting computational time in sampling some energy states excessively while sampling other states only sparsely or not at all. Therefore, the ESMC method can increase the sampling efficiency in a Monte Carlo determination of the entropy function. Finally, when the correct entropy function $S(E)$ is approached closely after a number of updatings, the energy histogram should satisfy the relationship

$$H(E_i) \propto \Omega(E_i) \exp[-S(E_i)] \tag{3.6}$$

i.e. the histogram is constant over all energy ranges. This relationship provides a simple and rigorous check on the correctness of the $S(E)$ function.

Berg *et al.* [40] discussed the equivalence of the ESMC algorithm and the multicanonical MC method in terms of the uniform sampling and iterative updating of the probability function. Hao and Scheraga [41] pointed out that the ESMC method is not merely a reformulation of the previous methods [36–39]; it not only leads to a simpler algorithm, but more importantly, it rigorously establishes the relationship between the microcanonical entropy of a protein model and the uniform sampling probability in an MC algorithm. There are clear conceptual advances from the umbrella sampling MC method of Torrie and Valleau [36,37] to the multicanonical MC [38,39] methods, and from the latter to the ESMC algorithm. The multicanonical ensemble corresponds to an artificial state; and the parameters $\beta(E)$ and $\alpha(E)$ as well as the multicanonical weighting functions are mere computational devices that are to be eliminated in computing physical meaningful properties of the system by the reweighting process. In comparison, the weighting function $S(E)$ in the ESMC algorithm can serve the same computational purpose just as well, and, in addition, $S(E)$ has the important physical meaning as the microcanonical entropy, which provides the basis for the microcanonical analysis of the system. After the ESMC method was introduced, a hybrid MC procedure was also proposed [42] in which the entropy function is first calculated by the ESMC procedure $S(E)_{\kappa+1} = S(E)_\kappa$

$+ \ln H(E)$, then the multicanonical parameters are calculated by $\beta(E) = \beta_0 + [S(E') - S(E)]/(E' - E)$, $\alpha(E) = \alpha(E') + [\beta(E') - \beta(E)]E'$, and finally the weighting function is expressed as $\exp[-\beta(E)E - \alpha(E)]$. There appears to be no merit in using these cumbersome processes to calculate the multicanonical parameters since the weighting function can be expressed in a straightforward form as $e^{-S(E)}$.

Before exploring the application of the ESMC method to polypeptides and globular proteins, it is useful to discuss some differences between the statistical mechanics of small molecules and that of globular proteins. In simple homogeneous physical systems such as a collection of interacting particles or spins, there exist some scaling laws that permit a rather accurate extrapolation of the simulation results from smaller systems to larger systems. However, proteins are different from simple discrete physical systems in that a protein is a single-chain molecule with highly heterogeneous interactions. Proteins are also different from a simple enlargement of oligopeptides in that protein sequences are highly optimized (through biological evolution) such that the native structures of proteins exhibit strong cooperative interactions that could not occur in oligopeptides. Because of these complications, there appears to be no simple scaling law from small polypeptides to proteins. The difference between small polypeptides and globular proteins is readily reflected in their conformational transitions. The folding of globular proteins has the characteristics of a two-state phase transition; however, there is no phase transition in oligopeptides. The conformational distribution of small polypeptides shifts gradually with temperature, and, even at quite low temperature, there is still a considerable conformational fluctuation and dispersion in their energies. The helix–coil transition in homopolypeptides is also a gradual process, different from protein folding.

There is a subtle difference in the way that the ESMC method is used for studying small polypeptides and globular proteins. For small and/or simple polypeptides, the exact determination of the density of the lowest-energy state is not critically important because the conformational distribution is not dominated by the lowest-energy conformation in the relevant temperature range. A roughly estimated entropy function is sufficient for simulation studies of small polypeptides with the ESMC method; the estimated entropy is a mere computational device that is eliminated later by the reweighting process, so whether the estimated function is actually correct does not have a direct physical consequence. The emphasis on the ESMC method as it applies to oligopeptides is as an efficient algorithm for searching the conformational space and obtaining the average properties. On the other hand, for protein models, the emphasis on the ESMC method is for accurately determining the microcanonical entropy of the protein models.

An exact knowledge of the microcanonical entropy, or the density of the states, of a protein model in both the native and nonnative states is crucial for a precise characterization of the folding process of the model, such as whether the folding is first-order or gradual, whether the model can fold uniquely to the native structure, whether there is a discontinuity in the order parameter of the conformation in the folding transition, and so on. Once the microscopic entropy function is accurately determined, the statistical mechanics of the protein folding problem is solved. The accurate determination of the entropy function of protein models by the ESMC method requires a proper treatment of the computational problems discussed in Section IV.

IV. THE QUASI-ERGODICITY PROBLEM IN ESMC SIMULATIONS

The ESMC simulation of protein models encounters the quasi-ergodicity problem [43] in which, even though all conformations of a system can be accessed by the conformational moves used in the algorithm on the basis of structural considerations, certain parts of the conformational space cannot be sampled in the actual ESMC simulation because of entropy barriers. Those inaccessible conformational states may have low energy or low entropy, so they are important to the thermodynamics of protein folding. Quasi-ergodicity, a well-known problem in standard MC simulations, arises because of the energy barriers when the temperature of a simulation is below the glass-transition temperature. The quasi-ergodicity phenomenon in ESMC simulation is a new problem occurring in a complicated system such as heteropolymers and protein models. Such a problem has not been reported in ESMC (or multicanonical MC) simulations of either simple spin systems [17,38,39] or small polypeptides [42,44,45].

The following example illustrates the general feature of the quasi-ergodicity problem in an ESMC simulation. The protein model used here is a 45-residue cubic–lattice chain. The force field for this model consists of contact-type interactions; the energy parameters between different residues have been optimized so that the native structure is the lowest-energy conformation for this model. The standard move set on a cubic–lattice chain [46–48], including the end-bond rotations, two-bond corner flips, and three-bond crankshift motions, was used in the ESMC simulations. This move set is believed to produce reasonably correct dynamics for the coiled state of the model chain. Figure 2 shows the energy histograms of four consecutive ESMC simulations on this cubic lattice chain with an estimated entropy function which covers all the energy states based on data from previous MC simulations. After each simulation shown in Figure 2, the

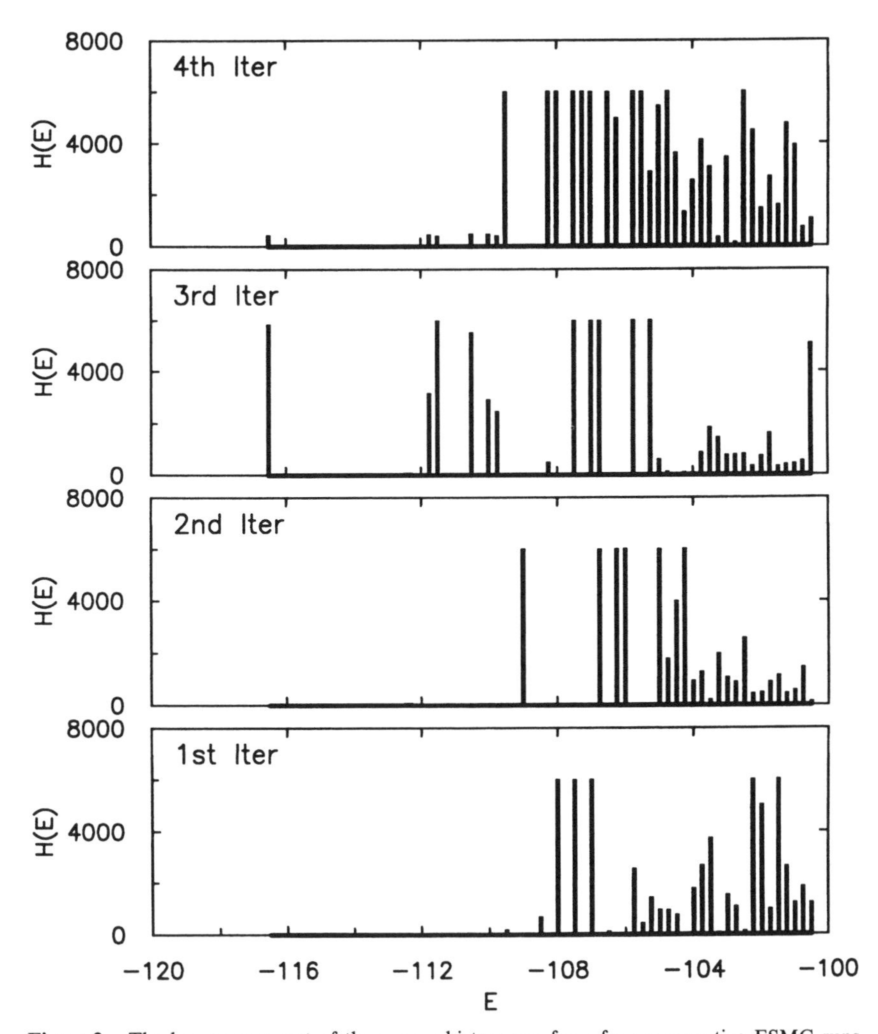

Figure 2. The low-energy part of the energy histograms from four consecutive ESMC runs with the entropy function updated after each simulation. The model is a 45-mer cubic lattice chain, with a standard move set for the cubic lattice chain used in the MC simulation. The lowest-energy state is indicated by the leftmost point of the thick lines.

entropy function was updated according to the ESMC procedure described above. For clarity, only the lower parts of the histogram in the energy spectra are shown in the figure while the rest of the histogram is omitted even though the ESMC simulations were carried out over the whole energy range. It can be seen from Figure 2 that, in each simulation, only a fraction of the low-energy states are sampled. When the entropy of those sampled states is raised in the updating procedure following a simulation, then, in the next simulation, say from the first iteration (Iter) to the second iteration, some of the previously sampled states disappear, whereas some new states appear in the histogram from the ESMC sampling. Clearly, in each given simulation, some energy states are not sampled, even though these states are accessible with a different entropy function. This is the quasi-ergodicity problem in the ESMC simulation. This problem cannot be alleviated by increasing the number of MC moves in an MC run, and the behavior can be repeated over an infinite number of iterations. When such a behavior occurs, the variances in part of the sampled energy density are very large, the entropy function does not converge to a definite value in the iteration process, and an accurate estimate of the relative entropy function cannot be obtained for the protein model.

The quasi-ergodicity in ESMC simulations of protein models is caused by the rugged energy landscape of protein models and the polymeric nature of the compact native structures. First, when the free-energy landscape of a protein model is rugged, there are many separated energy valleys; the $S(E)$ function of such a system is a rapidly varying function near the lower end of the energy spectrum. From a computational point of view, it is difficult to obtain a good estimate for a strongly varying entropy function near the bottom of the free-energy valleys. Second, the trial $S(E)$ function used by the ESMC algorithm contains not only the intrinsic variations in the $S(E)$ function of the protein system but also random variations because of the incorrect estimate of the $S(E)$ function from sampling variances. The combined variances of the $S(E)$ function from the two sources can create many real and/or artificial entropy barriers. Finally, because of the polymeric nature of the protein structure, it is often difficult to find the proper moves, either local or global ones, in which a local energy-minimum conformation can be moved to another minimum in one MC step. All these factors contribute to making it difficult to overcome the entropic barriers of protein models and lead to the quasi-ergodicity phenomenon in ESMC simulations.

The severity of the quasi-ergodicity problem in ESMC simulations depends on the nature of the free-energy surface. Simple systems such as fluids or spin systems are easier to simulate than polymer systems are because, in the former systems, a move from one energy minimum to another does not require a global structural rearrangement as occurs in long-chain molecules. Similarly, short polypeptides are easier to simulate by the ESMC method than the long chains of proteins. For polypeptide models of the typical size of proteins, the free-energy landscape is a determining factor. If the free-energy landscape of a protein model is smooth, the entropy function of the system will also be a smooth function within the relevant ranges, and hence ESMC simulations will encounter fewer entropy barriers. Ideal two-state systems with the native states having a smooth free-energy surface can also be handled easily by the ESMC method. For long protein models, even if the free-energy surface of the system has a good nature, the elemental move sets in MC simulations still play an important part in determining the outcome of ESMC simulations. A main area in which one can improve the effectiveness of ESMC simulations on protein models is to design good move sets. The basic guide for designing effective move sets for ESMC simulations is that, with one elemental move, the protein conformation can be changed from one local energy minimum to another. This will overcome the quasi-ergodicity problem in ESMC simulations of protein models. Two of the effective move sets that have been implemented in the ESMC simulations are the conformational biased MC [18] method and the jump-walking technique [19]. With these techniques, it was possible to improve the convergence of the ESMC simulations of the protein models significantly.

V. EFFECTIVE SAMPLING TECHNIQUES FOR ESMC SIMULATIONS

One of the useful moves for ESMC simulations of protein models is the conformation-biased MC (CBMC) method [49,50]. This method can be considered as an extension of the modified Rosenbluth and Rosenbluth chain growth method [31–33]. In the Rosenbluth and Rosenbluth method, one complete chain is generated at a time, and all generated chains are independent. In the CBMC method, instead of growing a complete chain each time, only part of the chain is discarded, and then the chain is grown back residue by residue to generate a new conformation. In the regrowth process, an energetic bias is applied as each residue is added, according to the interactions of the residue with all those already on the chain; hence the term *conformation-biased Monte Carlo*. After a new conformation is generated by the chain partial regrowth process, the decision as to whether to accept it is made according to the Metropolis criterion.

The chain regrowth move is implemented in the ESMC simulation in the following way [18]. A residue of the chain molecule is first selected randomly, a decision is also made randomly about which part of the chain is discarded starting at the selected residue. The discarded part of chain is then grown back. In growing the mth residue on the chain, all possible local chain states (LCS) of the mth residue are scanned, and each LCS is given a weight factor proportional to its Boltzmann factor. One of the allowed

LCSs is selected randomly according to its Boltzmann probability:

$$w_{i,m} = \frac{e^{-\beta E_i}}{\sum_j e^{-\beta E_j}} \tag{5.1}$$

where E_i is the interaction energy of residue m in the ith LCS with all residues that are already on the chain, β is an inverse temperature parameter, and the sum is evaluated over all possible LCSs for the added residue. The term $w_{i,m}$ carries the significance of the probability of growing residue m on the chain. The probability of growing residue m in the old conformation is also calculated in a similar way, but the selected LCS for the residue is always the state in which residue m exists in the old chain conformation. In this way, the old conformation will be recovered after calculating growth probabilities for all the residues in the discarded part of the chain. After the new conformation is completed according to this procedure, the probability of acceptance of the new conformation with respect to the old is calculated according to the ESMC criterion, but with a correction based on the different generation probabilities for new and old conformations:

$$P(x \to x') = \min\left[1, \frac{e^{-S[E(x')]}}{e^{-S[E(x)]}} \frac{\prod_m w_{i,m}}{\prod_m w_{i',m}}\right] \tag{5.2}$$

where $w_{i,m}$ and $w_{i',m}$ are the probabilities for growing the mth residue in the new and old conformations, respectively.

The preceding conformation-biased chain generation step is introduced stochastically into the regular conformational updating procedure based on local perturbations. In general, one regrowth of a part of the chain could require a time equivalent to several hundreds of local chain moves. However, the conformation biased chain generation procedure is effective in sampling new conformational regions and, with a proper choice of the inverse temperature, can generate low-energy compact conformations. A conformational move by the chain regrowth method can also produce a conformation independent of the old one, thus reducing the correlation among conformations that occurs when only local perturbations are made. Therefore, the CBMC method can help to reduce the quasi-ergodicity problem in ESMC simulations. Figure 3 shows an example of an ESMC simulation on a 38-residue (210) lattice-chain protein model [18]. The elemental moves for this model in the ESMC simulation include (1) the end-bond flip, (2) two consecutive-bond spike-like flips, (3) four consecutive-bond rearrangements, (4) eight consecutive-bond rearrangements, and (5) a chain partial regrowth attempt for every four swaps of

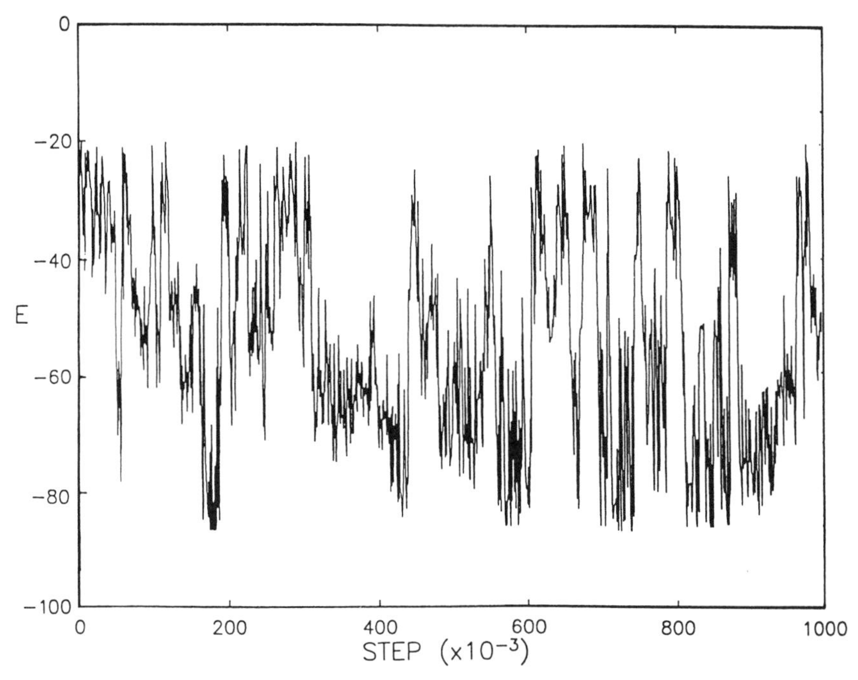

Figure 3. The energy trajectory in an ESMC simulation. The native structure has the lowest energy of −87.2 (in units of kT); the upper cutoff energy level in the sampling is −20. (Adapted from Hao and Scheraga [18].)

randomly selected local moves over the whole chain. The complete set of all these moves is counted as one chain updating in the ESMC simulation. The plot in Figure 3 shows the energy trajectory of an ESMC simulation with the correct entropy function that was determined by the iterative entropy updating procedure. The global-energy minimum of the model is −87.2. It can be seen from the figure that the algorithm is quite efficient in sampling all the energy regions; the mean time for traveling from the high-energy state to the lowest-energy state takes about 2.5×10^3 chain updatings. The ESMC procedure samples all the energy states (confined to $E < -20$ in Fig. 3) uniformly. The uniform sampling behavior is in contrast to conventional MC simulations in which the population of the sampled states commonly exhibits a normal distribution. Figure 4 shows a snapshot of a coiled conformation and the lowest-energy structure of this model [18].

The impact of the biased chain regrowth method on the efficiency of MC simulations depends significantly on the nature of the polymer models. Specifically, the ability of the CBMC method to generate unique low-energy

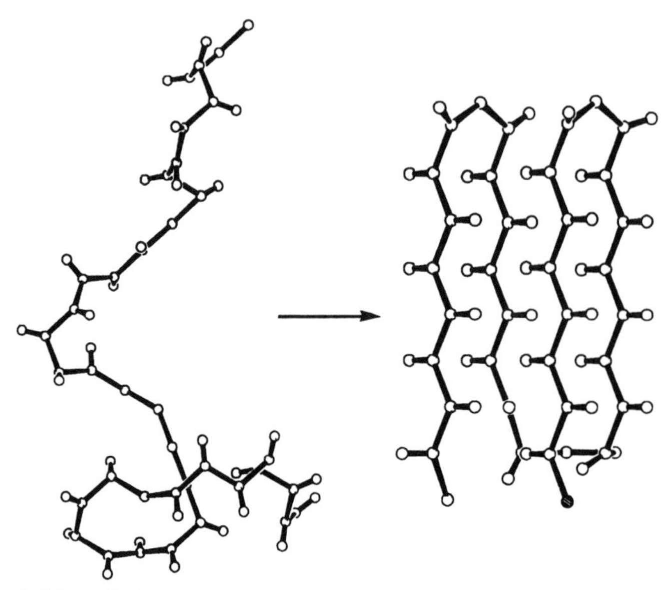

Figure 4. Folding of a 38-mer (210) lattice protein model from the coiled state to the native state. The left side is a snapshot of the unfolded conformation; the right is the lowest-energy conformation. (Adapted from Hao and Scheraga [18].)

compact conformations depends on the localized interacting environment at each step of the chain growth. The stronger the local energies contribute to the overall energy of the molecule, the higher the probability that the algorithm can generate low-energy conformations. This is because the chance of selecting a good growth direction at each step increases with the importance of the localized interactions. For an identical chain length, the chain regrowth method works better for stiffer chain models than softer ones. In a way, it appears that the efficiency of the conformational biased chain regrowth method is correlated with the persistence length of the polymer model. When the chain length is sufficiently long, growing one residue at a step becomes less effective for generating good conformations with low energies. To deal with such problems, one can consider the interactions of several more residues ahead of the current residue in growing each residue. Such algorithms have been considered in the scanning method developed by Meirovich [51,52]. However, the computational cost of scanning increases rapidly with the number of residues whose conformations are scanned; thus, we would expect that there is a limit to the chain length of protein models below which the conformational biased chain regrowth approach is effective for generating low-energy conformations.

Another useful technique for overcoming the quasi-ergodicity problem in ESMC simulations is the jump–walking (JW) method. This technique was originally proposed for the standard MC procedure [53,54]. In applying this method to ESMC simulations, the basic idea is to break a single long MC run into many short runs, each of which starts with an independent new conformation [19]. This clearly provides a way to jump across the entropy barrier in the ESMC simulations. The starting new conformations in the ESMC simulation are not generated completely randomly, but rather are drawn from a conformational pool that is generated and saved in a previous ESMC run or in an independent standard MC run. When the probabilities of occurrence of the conformations in the pool are uniform over all energy ranges, the conformations drawn randomly from the pool will be nearly uniform over all energy levels. Then, in an identical amount of computational time, the simulation with a number of short runs starting from new and independent conformations produces a more uniform sampling over the whole energy range than a simulation in a single long run, and the convergence of the ESMC simulations can be greatly improved.

The JW technique is implemented in the following way [19]. First, starting from a normal ESMC simulation, randomly selected conformations in the simulation are saved to form a conformational pool. Then, in the following simulations, the procedure jumps occasionally, with a certain probability, to the conformational pool saved previously to choose a new conformation with which to continue the regular ESMC run. The jump is treated as a particular kind of conformational move. Suppose that the estimated entropy functions for the current simulation and for the previous simulation from which the conformational pool was generated are $S_{\text{I}}(E)$ and $S_{\text{II}}(E)$, respectively. In a jump move, the conformation x at the current step is switched to the conformation x' randomly chosen from the pool, and then the regular simulation procedure is resumed. If the transition between conformations x and x' were carried out by a *direct* move from the current simulation, the forward and backward transition probabilities, $\pi(x'; x)$ and $\pi(x; x')$, would be determined by

$$\frac{\pi(x'; x)}{\pi(x; x')} = \frac{e^{-S_{\text{I}}[E(x')]}}{e^{-S_{\text{I}}[E(x)]}} \equiv e^{-\Delta S_{\text{I}}(x'; x)} \tag{5.3}$$

where $\Delta S_{\text{I}}(x'; x) \equiv S_{\text{I}}[E(x')] - S_{\text{I}}[E(x)]$. However, the new conformation chosen through the jump is not selected completely randomly, but rather is determined by the distribution of conformations in the pool. The transitions between x and x' through the jump process are, therefore, affected by the relative statistical probabilities of the two conformations in distribution II. To correct for the effect of the unequal probabilities of conformations x and

x' in distribution II, the forward and backward transition probabilities in a jump have to be determined by Eq. (5.4) instead of Eq. (5.3):

$$\frac{\pi(x'; x)}{\pi(x; x')} = \frac{e^{-\Delta S_{\mathrm{I}}(x'; x)}}{e^{-\Delta S_{\mathrm{II}}(x'; x)}} \tag{5.4}$$

where $\Delta S_{\mathrm{II}}(x'; x) \equiv S_{\mathrm{II}}[E(x')] - S_{\mathrm{II}}[E(x)]$. The probability of acceptance for the jump is finally expressed as

$$P_j(x \to x') = \min[1, e^{[-\Delta S_{\mathrm{I}}(x'; x) + \Delta S_{\mathrm{II}}(x'; x)]}] \tag{5.5}$$

The JW procedure was used in an ESMC simulation study of 38- and 54-residue (210) lattice polypeptide models [19,22] described above. Figure 5 shows the computed entropy function and its standard deviations obtained by the ESMC method with the JW procedure (including also the CBMC technique). The standard deviations of the computed entropy func-

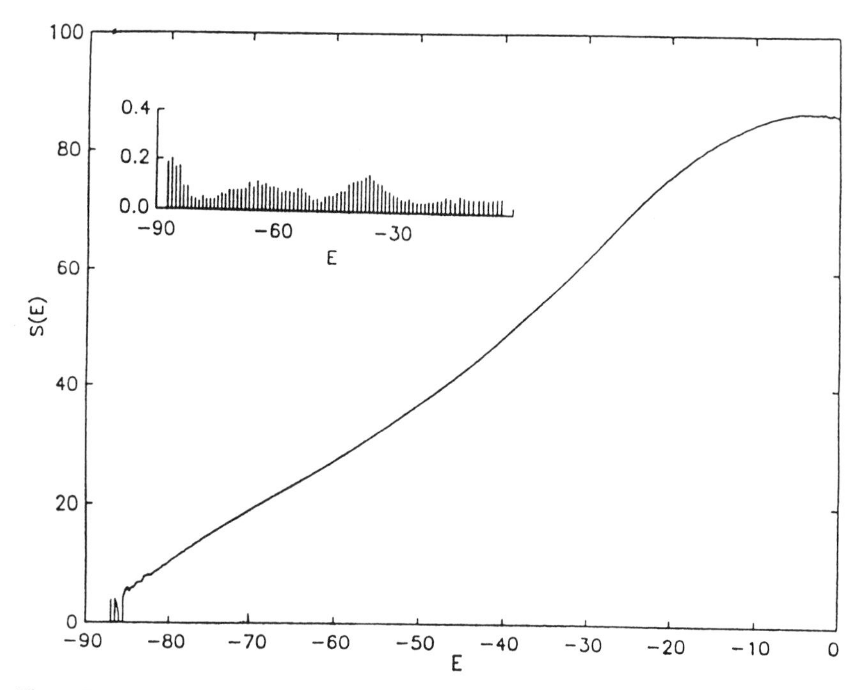

Figure 5. Relative entropy of a 38-mer (210) lattice protein model determined by the ESMC procedure with enhanced sampling procedures (including both the CBMC and jump–walking techniques). The inset shows the standard deviations of the computed entropy function.

tion are defined as the variances of the computed entropy function in the last five updatings in the ESMC algorithm. As shown in the figure, with the JW procedure [19], the maximum standard deviation of the computed entropy is only about 0.2 entropy unit. In comparison, without the JW procedure [18], the maximum standard deviations of the computed entropy function are about 1.3 entropy units, even though the average entropy function is basically identical in the two cases. Therefore, application of the JW procedure reduces the variances of the calculated entropy function significantly, and the convergence rate of the ESMC simulations is also increased (e.g., the iteration process can be finished with few entropy updates). Kolinski and Skolnick [55] studied a more complicated polypeptide model of designed sequences using the ESMC method. They also found that the application of a version of the JW procedure is necessary to achieve a reasonable convergence of the ESMC simulations.

VI. APPLICATIONS OF THE ESMC METHOD TO POLYPEPTIDES AND PROTEINS

The ESMC method in the form of the multicanonical MC has been used to simulate conformations of oligopeptides by Hansmann and Okamoto [42,44,45]. These studies were aimed at determining the canonical properties of the molecules and searching for the low-energy conformations. Therefore, only approximate multicanonical functions, for example, with no convergence check on these functions, were used in the conformational sampling. It was reported that, for the small polypeptides studied, about 10–40% of the total computation was used in obtaining the multicanonical functions. With the computed multicanonical weighting functions, only one MC simulation was required to sample all energy states in the range of interest. In a study of the 5-residue polypeptide Met-enkephalin [44], it was found that, in one simulation with the ESMC algorithm and with the approximately computed multicanonical function, the lowest-energy conformation was sampled 6 times. It was therefore concluded [44] that the ESMC method is more efficient than the MC-simulated annealing method in finding the global energy-minimum conformation for the small polypeptide studied [6]. The same method was also applied to study the thermodynamics of the helix–coil transition of homooligopeptides of $(\text{Gly})_{10}$, $(\text{Ala})_{10}$, and Val_{10} [42]. It was found that lowest-energy conformation of $(\text{Ala})_{10}$ is an α helix, but that $(\text{Gly})_{10}$ and Val_{10} have no unique lowest-energy conformation because there are several conformations with comparable low energies. However, the computed multicanonical functions for the homooligopeptides were not reported; therefore, it is not clear whether the accuracy of the computed multicanonical function has any effect on the simulation results. No effort was made to analyze the multicanonical function itself, which is related to the microcanonical entropy of the polypeptides and can provide information about the nature of the conformational transitions of these polypeptides.

Hao and Scheraga studied protein models with the ESMC method [18,19,22,23,25–27]. The focus of these studies was the statistical mechanics of protein folding, including the types of folding transitions in protein models, the factors that affect the folding behavior of protein models, and the interactions that lead protein models to fold to unique native structures. In several studies of the (210) lattice-chain model [18,19,22] proposed by Kolinski and Skolnick [56,57], it was found that the ESMC method is a very effective approach for these systems: the simulations could be carried out to full convergence in a reasonable amount of computational time, and the entropy functions of these protein models could be determined to a sufficiently high degree of accuracy as judged by the small standard deviations of the calculated quantities. Protein models with different sequences, and under varying interactions, were examined. With the calculated entropy functions, it was possible to define the character of protein models with different folding behavior quantitatively. It was shown that foldable protein models have an entropy function with a concave segment but the entropy function of unfoldable protein models is convex [19]. A concave segment in the entropy function of foldable protein models implies that, in the folding process, the energy of the system decreases faster than the entropy, a signal of the cooperativity of the interactions in the native structure of the model [18]. Thus, the results from ESMC simulations provide unique insights into the interactions that lead proteins to fold to their unique native structures [22]. When a reasonably good estimate for the entropy function of a protein model is obtained by the ESMC method, the foldability of the model can be determined; therefore, in later studies [25–27], the entropy function computed by the ESMC method was used as a monitor for optimization of the interaction energy parameters for folding protein models.

The ESMC method was also used to study the statistical mechanics of the folding transition of fine-grained lattice-chain models for the proteins BPTI [23] and crambin [25]. With optimized energy parameters, it was shown that the folding of these two protein models has a two-state character because their entropy functions have a concave segment. However, the force fields for the real proteins were less optimized than those for the simpler lattice chains based on the evidence that the lowest-energy states of the real protein models have a strong glassy behavior over which the entropy function could not be computed accurately. Kolinski and Skolnick [55] also studied the folding transition of fine-grained lattice protein models of designed sequences with the ESMC method. The objective of their study was to determine the origin of the cooperative folding transition in these models. The folding behavior of the protein models was determined by the microcanonical entropy function. They found that those models that include an explicit cooperative side-chain packing term exhibit a well defined all-or-none transition from a denatured, statistical coil state to a high-density native-like state, but that models lacking such a term exhibit a conformational transition that is essentially continuous. As with the protein models studied by Hao and Scheraga [23,25], but different from the simpler lattice-chain models studied earlier [18,19,22], the energy landscapes of the protein models studied by Kolinski and Skolnick appear to be so complicated that, even after considerable computations, the computed entropy functions were still not completely converged in a narrow range of the lowest-energy states.

Kidera [58] introduced the scaled collective-variable algorithm of Noguti and Gō [59,60] into the ESMC method for simulating atom-level structures of polypeptides. In this method, the conformational moves are made according to the normal modes of the torsion-angle variables of a polypeptide, and the acceptance criterion is determined by the ESMC algorithm. He applied this method to a 9-residue peptide that contained a disulfide bond between the first and the last residues [58]. This is a constrained polypeptide, and the scaled collective variables appeared to be very efficient for sampling the conformational space of the molecule. The energy density, sampled by a single canonical MC simulation at 1000 K, supplemented with a function extrapolation, was sufficient to produce a roughly correct entropy function for the system, as judged by the basically flat distribution of energy states in the follow-up ESMC simulation. With such a computed entropy function, a long ESMC simulation was carried out from which the average properties at different energy levels were obtained. Then, after reweighting, the canonical averages at different temperatures were obtained. It was found that, whereas the ESMC simulation can sample the lowest-energy state, the conventional MC simulations with the scaled collective variables at low temperatures were trapped in local-energy minima. This indicates that, while the canonical MC cannot surmount the energy barriers, the ESMC algorithm can overcome the energy barriers in this molecular system. Kidera's procedure is most effective for systems in which a correct estimate of the microscopic entropy function can be easily obtained. When an extensive iterative process has to be carried out in order to find the correct entropy function (as for longer polypeptides and globular proteins), the computational demand of determining the collective variables would be expected to be very high in this approach. The convergence of the algorithm for more complicated systems has not yet been studied.

VII. COMPARISONS WITH RELATED MC METHODS

In determining the density of states of protein models by MC methods, a number of other approaches are also useful. For systems for which it is feasible, exhaustive enumeration of all the conformations of the system by computer provides a viable and exact solution to the problem [61–66]. A number of powerful techniques have been developed, some for exhaustive enumeration of all conformations of relatively short lattice chains [61,63]; others, for enumeration of protein conformations that are confined to a compact space [62,64–66]. However, in general protein models, the number of conformations is beyond the power of computer enumeration; for these systems, MC methods are the main effective techniques.

The conventional Metropolis MC method is a basic technique for simulating the folding transition of protein models [67,68] as well as for extracting information about the density of states of proteins [69,70]. The sampled energy histogram from a conventional Metropolis MC simulation reflects the canonical distribution of states at the simulation temperature. Such information is very useful for identifying the nature of the folding transition, and the approach has been used in MC studies of simple cubic lattice protein models [69,70]. The density of states of protein models is accessible from a conventional MC simulation by the relationship between the sampled energy density $h(E_i, T_j)$ for the energy level E_i at temperature T_j and the Boltzmann probability:

$$\frac{h(E_i, T_j)}{\sum_{E_{i'}} h(E_{i'}, T_j)} = \frac{\Omega(E_i) \exp(-E_i/kT)}{\sum_{E_{i'}} \Omega(E_{i'}) \exp(-E_{i'}/kT_j)} \tag{7.1}$$

Therefore, the entropy function can be determined from the Metropolis sampling probability by the relationship

$$S(E_i) = \frac{E_i}{T_j} + k \ln h(E_i, T_j) - c_j \tag{7.2}$$

where c_j is an undetermined constant [because the sum in the denominator on the right-hand side of Eq. (7.1) is generally unknown]. In the ESMC approach, the determined density of states can be refined systematically in the iterative process; however, the conventional MC method lacks such a refinement mechanism, and the energy histogram obtained from a single MC simulation contains a large sampling uncertainty.

A powerful procedure for reducing the errors in the computed density of states from conventional MC simulations is the Monte Carlo histogram

(MCH) method [71–73]. The MCH method provides a way of extracting the maximum amount of information from a number of conventional MC simulations. There are several formulations of the MCH method; a simpler one is that proposed by Poteau et al. [72]. In this approach, the uncertainty, Δp, of the sampled energy density from a single MC run is expressed as

$$\frac{\Delta p}{p} = \frac{1}{h(E_i, T_j)^{1/2}} \tag{7.3}$$

If a number of conventional MC simulations have been carried out at different temperatures, the $S(E)$ function has the same relationship with the sampled energy density in each simulation as shown by Eq. (7.2). In general, however, the Boltzmann distribution of the states peaks at only one particular energy region and is negligibly small in other energy regions at a given temperature; therefore, the sampled energy densities at different temperatures provide reliable information only in each different energy range. In order to determine the microcanonical entropy accurately for all the relevant energy ranges, the MCH method synthesizes all the pieces of information together in such a way that the $S(E_i)$ function is fitted to all the histograms $h(E_i, T_j)$ while minimizing the uncertainties of the histograms by the following least-squares formula [72]:

$$\min\left\{\sum_{ij} h(E_i, T_j)\left[S(E_i) + c_j - \frac{E_i}{T_j} - k \ln h(E_i, T_j)\right]^2\right\} \tag{7.4}$$

The constants c_j were determined by minimizing this expression with respect to c_j.

The effective range of the MCH method is limited to cases in which the conventional MC method can sample all the conformational states reliably at the proper temperatures. The MCH method has been used to study the folding thermodynamics of 27-mer cubic lattice chains whose native state is a maximally compact cube [24], and to characterize the statistical–mechanical properties of longer lattice-chain protein models with an optimized force field [25,26] (i.e., interresidue interaction parameters). However, because the conventional MC method suffers from the multiple-minima problem in simulating polypeptides and globular proteins, the MCH method is not reliable for determining the energy density of states of long random polypeptides or protein models with an unoptimized force field. Another problem with the MCH method is error propagation. Because the entropies of different energy states determined by the MCH method are all correlated through the least-squares formulation, the errors at the different energy levels can accumulate and propagate over the entire entropy function. This is in contrast to the ESMC method, where the entropy of each energy state is determined independently.

The MCH and ESMC methods have different strengths and weaknesses. The standard MC simulation can be considered as an ESMC simulation with a straight-line approximation to the entropy function in the following sense. The ESMC algorithm uses the acceptance/rejection criterion of $p = \min[1, e^{-\Delta S(E)}]$; the standard MC method uses the criterion $p = \min[1, e^{-\Delta E/T}]$; therefore, the latter is equivalent to the use of the function E/T to approximate $S(E)$. The efficiency of an MC algorithm is related to the ability to sample all the relevant energy states reliably. For a system with a smooth free-energy landscape, and that has a first-order transition, the ESMC method is more effective than the MCH method because the ESMC simulation follows a more faithful description of the free-energy surface in carrying out the MC simulation. On the other hand, even though the straight-line approximation to the $S(E)$ function in the standard MC algorithm is not always the most efficient method, it is a solid approximation without introducing artificial variances to the $S(E)$ function. In cases in which the potential surface is rugged, the ESMC algorithm can create many artificial entropy barriers and encounter serious quasi-ergodicity problems, but the standard MC simulation does not have such problems and could be more stable and effective in sampling the energy states. A practical approach is that, when an ESMC simulation seems to have problems, one should use the standard MC method with the MCH approach. After one obtains a reasonably good estimate of the $S(E)$ function, one can then try an ESMC simulation to check whether the estimated $S(E)$ is correct by using the fact that the ESMC method provides a straightforward check on the correctness of the computed $S(E)$ function.

Finally, an approach that is different from all the methods discussed above is the Monte Carlo chain generation method based on the modified Rosenbluth–Rosenbluth algorithm [31–33]. Unlike the conventional MC methods that sample protein conformations by following a Markov chain process, the Rosenbluth–Rosenbluth method first generates an ensemble of independent chains, and then calculates the various statistical–mechanical properties. Such an approach has been used by Miyazawa and Jernigan in determining the density of states of the small protein BPTI [21]. There are two major problems in applying the original Rosenbluth–Rosenbluth method to longer polypeptide chains. First, there is the chain attrition problem; because of the excluded volume effects, the number of successfully generated chains diminishes exponentially with chain length. Second, the number of conformations in the native state of a protein is negligibly small compared to the number of nonnative high-energy conformations; therefore, the chance of generating the native conformations by a random generation process is vanishingly small. Meirovitch addressed these problems with the scanning method [51,52]. In this method [51,52], when growing the *n*th residue onto the chain, all possible conformations of a segment of residues starting from the *n*th residue are scanned, from which one feasible position is chosen for the *n*th residue. Furthermore, energetic biases are introduced to increase the weight of the conformation with low energy to be selected in the chain growth process. The scanning method holds the promise of being capable of generating a sufficiently large number of independent conformations that cover all relevant energy ranges, from which the density of states of a protein model can be determined. The scanning method has been used to study the free energy and stability of polypeptides [74,75]. A potential problem in applying the scanning method to proteins is that, in order to obtain a good sample of conformations in the low-energy states, the number of future residues scanned in the chain growth process needs to be large [75]. However, increasing the number of future residues to be scanned dramatically increases the computational demand. The study by O'Toole and Panagiotopoulos [76] of a lattice protein model with a procedure that is essentially identical to the scanning method provides an estimate of the effectiveness and limitation of the chain growth technique in studying protein folding.

VIII. CONCLUSIONS

Our main purpose here has been to review a new MC technique, entropy-sampling Monte Carlo, for simulating polypeptides and globular proteins. First, we discussed the microcanonical approach in studying the statistical mechanics of protein folding. The microcanonical entropy $S(E)$ provides a one-dimensional, quantitative definition of the free-energy surface of protein systems; and it is a convenient and powerful way to determine the nature of a protein folding transition. Second, we described how the entropy function is related to the ESMC algorithm, which, in turn, leads to an efficient method for accurately computing the entropy function. The ESMC method, however, is a general Monte Carlo technique characterized by sampling all energy states with equal probability, thereby avoiding the local energy-minimum problem. Third, we discussed in detail the major problem with ESMC simulations of protein models, the quasi-ergodicity problem arising from the entropy barriers, which is more serious in polymer systems with a rugged free-energy landscape. If this problem is not dealt with properly, ESMC simulations of protein models are not only extremely computationally demanding, but also could produce poorer results than conventional MC methods. Fourth, we described our approaches to use more efficient move sets for overcoming this problem. Two efficient conformational sampling techniques, the conformational biased chain growth method and the jumping–walking procedure, were shown to be useful for enhancing the efficiency of ESMC simulations of model proteins.

Fifth, we evaluated the applications of the ESMC method in simulating oligopeptides and protein models. For small and/or simple polypeptides, the exact determination of the density of states is not critically important; a roughly correct entropy function is sufficient for an ESMC simulation. In these systems, the ESMC method has been shown to be more efficient than conventional MC methods. For models of globular proteins, the emphasis on the ESMC method is to determine the microcanonical entropy, or the density of states, accurately. The reason for this emphasis is not only because the information is fundamentally important for characterizing the nature of the protein model but also because a reasonably accurate entropy function is the basis for an effective and productive simulation of a protein model with the ESMC method. Finally, we compared the relative strengths and weaknesses of the ESMC method and other related MC approaches. The Monte Carlo histogram approach based on the conventional MC method and the ESMC method can be used in combination to produce an accurate estimate of the entropy functions of protein models. Overall, the ESMC method provides a new, simple, and efficient technique for simulation studies of polypeptides and globular proteins, but one should be aware that the method may entail problems that do not occur in conventional Monte Carlo methods.

ACKNOWLEDGMENTS

This work was supported by grants from the National Institutes of Health (GM-14312) and the National Science Foundation (MCB95-13167). Support was also received from the Association for International Cancer Research and the Cornell Biotechnology Center.

REFERENCES

1. J. Skolnick and A. Kolinski, *Annu. Rev. Phys. Chem.* **40**, 207 (1989).
2. R. A. Friesner and J. R. Gunn, *Annu. Rev. Biophys. Biomol. Struct.* **25**, 315 (1996).
3. Z. Li and H. A. Scheraga, *Proc. Natl. Acad. Sci. USA* **84**, 6611 (1987).
4. Z. Li and H. A. Scheraga, *J. Mol. Struct.* (*Theochem.*) **179**, 333 (1988).
5. A. Nayeem, J. Vila, and H. A. Scheraga, *J. Comp. Chem.* **12**, 594 (1991).
6. Y. Okamoto, T. Kikuchi, and H. Kawai, *Chem. Lett.*, 1275 (1992).
7. S.R. Wilson and W. Cui, in *The Protein Folding Problem and Tertiary Structure Prediction*, K. M. Merz, Jr. and S. M. LeGrand, eds., Birkhauser, Boston, 1994, pp. 43–70.
8. D. R. Ripoll and H. A. Scheraga, *Biopolymers* **27**, 1283 (1988).
9. D. R. Ripoll and H. A. Scheraga, *Biopolymers* **30**, 165 (1990).

10. D. R. Ripoll, L. Piela, M. Vásquez, and H. A. Scheraga, *Proteins* **10**, 188 (1991).
11. L. Piela and H. A. Scheraga, *Biopolymers* **26**, S33 (1987).
12. A. Liwo, M. R. Pincus, R. J. Wawak, S. Rackovsky, and H. A. Scheraga, *Protein Sci.* **2**, 1715 (1993).
13. B. von Freyberg and W. Braun, *J. Comp. Chem.* **12**, 1065 (1991).
14. H. A. Scheraga, *Chem. Scripta* **29A**, 3 (1989).
15. H. A. Scheraga, in *Reviews in Computational Chemistry*, Vol. 3, K. B. Lipkowitz and D. B. Boyd, eds., VCH Publishers, New York, 1992, p. 73.
16. N. Metropolis, A. W. Rosenbluth, M. N. Rosenbluth, A. H. Teller, and E. Teller, *J. Chem. Phys.* **21**, 1087 (1953).
17. J. Lee, *Phys. Rev. Lett.* **71**, 211 (1993); Erratum **71**, 2353 (1993).
18. M.-H. Hao and H. A. Scheraga, *J. Phys. Chem.* **98**, 4940 (1994).
19. M.-H. Hao and H. A. Scheraga, *J. Phys. Chem.* **98**, 9882 (1994).
20. N. Gō, *Int. J. Pept. Protein Res.* **7**, 313 (1975).
21. S. Miyazawa and R. L. Jernigan, *Biopolymers* **21**, 1333 (1982).
22. M.-H. Hao and H. A. Scheraga, *J. Chem. Phys.* **102**, 1334 (1995).
23. M.-H. Hao and H. A. Scheraga, *Supercomputing'95, Proc. 1995 ACM*/IEEE Supercomputer Conference, San Diego, CA, Dec. 3–8, 1995. Association for Computing Machinery, New York, p. 57.
24. N. D. Socci and J. N. Onuchic, *J. Chem. Phys.* **103**, 4732 (1995).
25. M.-H. Hao and H. A. Scheraga, *Proc. Natl. Acad. Sci. USA* **93**, 4984 (1996).
26. M.-H. Hao and H. A. Scheraga, *J. Phys. Chem.* **100**, 14540 (1996).
27. M.-H. Hao and H. A. Scheraga, *Physica A* **244**, 124 (1997).
28. P. E. Leopold, M. Montal, and J. N. Onuchic, *Proc. Natl. Acad. Sci. USA* **89**, 8721 (1992).
29. J. N. Onuchic, P. G. Wolynes, Z. Luthey-Schulten, and N. D. Socci, *Proc. Natl. Acad. Sci. USA* **92**, 3626 (1995).
30. J. D. Bryngelson, J. N. Onuchic, N. D.Socci, and P. G. Wolynes, *Proteins* **21**, 167 (1995).
31. M. N. Rosenbluth and A. W. Rosenbluth, *J. Chem. Phys.* **23**, 356 (1955).
32. F. T. Wall, S. Windwer, and P. J. Gans, *J. Chem. Phys.* **37**, 1461 (1962).
33. S. Premilat and J. Hermans, *J. Chem. Phys.* **59**, 2602 (1973).
34. N. Gō and H. Taketomi, *Proc. Natl. Acad. Sci. USA* **75**, 559 (1978).
35. Y. Ueda, H. Taketomi, and N. Gō, *Biopolymers* **17**, 1531 (1978).
36. G. M. Torrie and J. P. Valleau, *Chem. Phys. Lett.* **28**, 578 (1974).
37. G. M. Torrie and J. P. Valleau, *J. Comp. Phys.* **23**, 187 (1977).
38. B. A. Berg and T. Neuhaus, *Phys. Rev. Lett.* **68**, 9 (1992).
39. B. A. Berg and T. Celik, *Phys. Rev. Lett.* **69**, 2292 (1992).
40. B. A. Berg, U. H. E. Hansmann, and Y. Okamoto, *J. Phys. Chem.* **99**, 2236 (1995).
41. M.-H. Hao and H. A. Scheraga, *J. Phys. Chem.* **99**, 2238 (1995).
42. Y. Okamoto and U. H. E. Hansmann, *J. Phys. Chem.* **99**, 11276 (1995).
43. M.-H. Hao and H. A. Scheraga, unpublished work.
44. U. H. E. Hansmann and Y. Okamoto, *J. Comp. Chem.* **14**, 1333 (1993).
45. U. H. E. Hansmann and Y. Okamoto, *Physica A* **212**, 415 (1994).
46. P. H. Verdier and W. H. Stockmayer, *J. Chem. Phys.* **36**, 227 (1962).

47. L. Monnerie and F. Geny, *J. Chem. Phys.* **50**, 1691 (1969).
48. H. J. Hilhorst and J. M. Deutch, *J. Chem. Phys.* **63**, 5153 (1975).
49. J. I. Siepmann and D. Frenkel, *Mol. Phys.* **75**, 59 (1992).
50. J. J. de Pablo, M. Laso, and U. W. Suter, *J. Chem. Phys.* **96**, 2395 (1992).
51. H. Meirovitch, *J. Chem. Phys.* **89**, 2514 (1988).
52. H. Meirovitch, *Int. J. Mod. Phys. C* **1**, 119 (1990).
53. D. D. Frantz, D. L. Freeman, and J. D. Doll, *J. Chem. Phys.* **93**, 2769 (1990).
54. C. J. Tsai and K. D. Jordan, *J. Chem. Phys.* **99**, 6957 (1993).
55. A. Kolinski and J. Skolnick, *Proteins* **26**, 271 (1996).
56. A. Kolinski, M. Milik, and J. Skolnick, *J. Chem. Phys.* **94**, 3978 (1991).
57. J. Skolnick and A. Kolinski, *J. Mol. Biol.* **221**, 499 (1991).
58. A. Kidera, *Proc. Natl. Acad. Sci. USA* **92**, 9886 (1995).
59. T. Noguti and N. Gō, *Biopolymers* **24**, 527 (1985).
60. N. Gō and T. Noguti, *Chem. Scripta* **29A**, 151 (1989).
61. H. S. Chan and K. A. Dill, *J. Chem. Phys.* **90**, 492 (1989).
62. H. S. Chan and K. A. Dill, *Macromolecules* **22**, 4559 (1989).
63. H. S. Chan and K. A. Dill, *J. Chem. Phys.* **92**, 3118 (1990).
64. D. G. Covell and R. L. Jernigan, *Biochemistry* **29**, 3287 (1990).
65. E. Shakhnovich and A. Gutin, *J. Chem. Phys.* **93**, 5967 (1990).
66. H. Li, R. Helling, C. Tang, and N. Wingreen, *Science* **273**, 666 (1996).
67. A. Kolinski and J. Skolnick, *Proteins* **18**, 353 (1994).
68. J. Skolnick, A. Kolinski, and A. R. Ortiz, *J. Mol. Biol.* **265**, 217 (1997).
69. V. I. Abkevich, A. M. Gutin, and E. I. Shakhnovich, *J. Chem. Phys.* **101**, 6052 (1994).
70. A. M. Gutin, V. I. Abkevich, and E. I. Shakhnovich, *Biochemistry* **34**, 3066 (1995).
71. A. M. Ferrenberg and R. H. Swendsen, *Phys. Rev. Lett.* **63**, 1195 (1989).
72. R. Poteau, F. Spiegelmann, and P. Labastie, *Z. Physik D* **30**, 57 (1994).
73. D. M. Ferguson and D. G. Garrett, in *Adaption of Simulated Annealing to Chemical Optimization Problems*, J. H. Kalivas, ed., Elsevier Science B. V., 1995, pp. 369–394.
74. H. Meirovitch, M. Vásquez, and H. A. Scheraga, *Biopolymers* **27**, 1189 (1988).
75. H. Meirovitch, M. Vásquez and H. A. Scheraga, *J. Chem. Phys.* **92**, 1248 (1990).
76. E. M. O'Toole and A. Z. Panagiotopoulos, *J. Chem. Phys.* **97**, 8644 (1992).

MACROSTATE DISSECTION OF THERMODYNAMIC MONTE CARLO INTEGRALS

BRUCE W. CHURCH, ALEX ULITSKY, AND DAVID SHALLOWAY

Biophysics Program, Section of Biochemistry, Molecular and Cell Biology, Cornell University, Ithaca, NY 14853

CONTENTS

Advances in Chemical Physics, Volume 105, Monte Carlo Methods in Chemical Physics, edited by David M. Ferguson, J. Ilja Siepmann, and Donald G. Truhlar. Series Editors I. Prigogine and Stuart A. Rice.
ISBN 0-471-19630-4

I. INTRODUCTION

Increasing the efficiency of multidimensional Monte Carlo integration is the key to solving a number of fundamentally important scientific problems. For example, the solution to the protein folding problem, the task of predicting the thermodynamically stable shape of an amino acid chain at temperature T, is given by the mean conformation $\langle R \rangle(\beta)$, where R represents all the conformational degrees of freedom ($R \equiv \{\vec{r}_n : n = 1 \cdots N_a\}$), $\beta \equiv (k_B T)^{-1}$ is the inverse temperature and k_B is Boltzmann's constant. If, as is believed to often be the case, the folded conformation is in thermodynamic equilibrium, the different conformations will have occupation probabilities determined in classical approximation by the Gibbs–Boltzmann distribution

$$p_B(\beta; R) \equiv \frac{e^{-\beta V(R)}}{Z_c(\beta)} \tag{1.1}$$

where $V(R)$ is the potential energy function and

$$Z_c(\beta) \equiv \int e^{-\beta V(R)} \, dR \tag{1.2}$$

is the conformational partition function.

Then

$$\langle R \rangle(\beta) = \int R p_B(\beta; R) \, dR \tag{1.3}$$

where Eq. (1.3) can, in principle, be computed by Monte Carlo integration. It is only the slow numerical convergence that makes the protein folding problem so difficult.

Equation (1.3) is a particular case of the general expression for a (classical) thermodynamic average

$$\langle f \rangle(\beta) \equiv \int f(R) p_B \, dR \tag{1.4}$$

where f is the variable being averaged. Important cases include

$f = R$	average position	(1.5a)
$= \Lambda^2 \equiv (R - \langle R \rangle)(R - \langle R \rangle)$	fluctuation tensor	(1.5b)
$= V$	average potential energy	(1.5c)
$= k_B \beta^2 (V - \langle V \rangle)^2$	spatial contribution to specific heat	(1.5d)

The convergence of all Monte Carlo integrals of this type slows dramatically as the number of degrees of freedom, N, increases, particularly when the integrand is multimodal. The problem is particularly severe for protein folding since N is often $\gtrsim O(10^4)$, and $V(R)$, and hence p_B, is extremely jagged. Unless care is taken, most computational effort will be expended in sampling excluded volume or high-potential-energy regions of conformation space where p_B is very small due to unfavorable van der Waals interactions. The extended regions of conformation space where p_B has significant support roughly correspond to metastable macroscopic states (*macrostates*) of the physical system [1]. The most difficult task is the initial (approximate) localization of these regions.

Historically, attention has focused on using local minima for this purpose. The potential-energy landscapes of small molecules in vacuum have only a few local minima separated by large energy barriers [2], and each of these corresponds to a macrostate at low temperature. Their catchment regions define a temperature-independent inherent structure [3,4] that dissects conformation space into subregions. But studies of atomic microclusters show that the number of local minima grows exponentially with the number of atoms [5] and similar behavior is expressed for polypeptides and proteins; therefore exhaustive enumeration of all local minima is impossible in larger problems. Furthermore, there is a one-to-one correspondence between local minima and macrostates only at very low temperatures—at physical temperatures many of the energy barriers that separate local minima are insignificant compared to the thermal energy $k_B T$ and many local minima can be contained within each macrostate [6]. That is, the temperature-independent inherent structure defined by the local minima, while useful for $T \approx 0$, needs to be replaced by a temperature-dependent description.

Experimental evidence indicates that the local minima of protein energy landscapes have a complex, hierarchical structure [7], and computational methods have been developed for clustering local minima into hierarchies using various metrics for measuring the distance between them. Both spatial (e.g., root-mean-square distance in conformation space between local minimizers [8,9]) and energetic (i.e., height of the potential-energy barrier separating two minimizers [10–13]) metrics have been used. The energetic approaches can identify the level of clustering that is relevant at a given temperature, but it is not known whether clusters defined in this way correspond to localized subregions of conformation space. Even with small systems (e.g., 19-atom microclusters [12] or tetrapeptides [11,13]), only a tiny fraction of the local minima can be identified and clustered, but thermodynamic and kinetic cluster properties have been calculated assuming that the sampled subset is representative. Regardless of whether such an assumption is justified, it is not evident how such "bottom–up" approaches can be scaled up to larger systems.

The alternative is to use a "top–down" approach which begins with a single macrostate including all conformation space and recursively dissects it into a hierarchical tree of smaller (in both size and energy) macrostates until the local minima are reached. In this type of approach the exponential explosion of tree complexity appears gradually, and can be systematically managed by intelligent pruning procedures. *Macrostate annealing* (also called *packet annealing*) is an approach of this type. It was originally developed [14–17] in the context of statistical physics. However, it can also be viewed as an efficient way of organizing Monte Carlo integrations over integrands that are dominated by nonnegative distributions as in Eq. (1.1). The method uses thermal annealing to identify concentrations or *packets* of probability density within the Gibbs–Boltzmann distribution. Each of these is approximated by a Gaussian *characteristic packet* that roughly identifies a macrostate. Since the excluded volume is determined by the relationship between the potential energy barriers and the temperature, both the macrostate boundaries and the number of macrostates depend on temperature. Once this dissection has been performed, the complete integral can be computed by sampling each macrostate region separately with its own customized *integration kernel* determined by the corresponding characteristic packet. This can greatly enhance efficiency.

In Section II we show how characteristic packets and macrostate annealing can be used to efficiently compute thermodynamic integrals over complicated potentials, and in Section III we show how the same ideas can be used to efficiently compute intermacrostate transition rates. In Section IV we show how this approach can be adapted for analyzing macromolecular integrals and demonstrate it on the pentapeptide Met-enkephalin. We summarize and discuss the next steps to be taken in Section V. For simplicity we consider the evaluation of the conformational partition function Z_c, but the method also applies to averages such as Eqs. (1.5). Although we focus on proteins, it should be possible to apply a similar approach to problems involving Monte Carlo integrations over other nonnegative probability distributions.

II. MACROSTATE ANNEALING: FORMALISM

A. Gaussian Integration Kernels

The Monte Carlo integration kernel $\mu(R)$ is the normalized probability distribution used to sample conformation space. For example, Z_c [Eq. (1.2)] is numerically integrated as

$$Z_c(\beta) = \int e^{-\beta V(R)} \mu^{-1}(R)[\mu(R)\, dR] \tag{2.1a}$$

$$\approx N_{MC}^{-1} \sum_{i=1}^{N_{MC}} e^{-\beta V(R^{(i)})} \mu^{-1}(R^{(i)}) \tag{2.1b}$$

where the $R^{(i)}$ are values of R randomly selected with distribution $\mu(R)$. The expected variance of the integral for a given N_{MC} is [18]

$$\sigma_{MC}^2[\mu] = \frac{Z_c^2}{N_{MC}} \int \left[\frac{p_B}{\sqrt{\mu(R)} - \sqrt{\mu(R)}} \right]^2 dR \tag{2.2}$$

We see that the convergence rate is determined by the match between the integrand and $\mu(R)$, and will be fastest when

$$\mu \approx p_B$$

Kernels that have constant support over simple geometric regions such as parallelepipeds and ellipsoids can be used for simple bounded problems, but Gaussian kernels are more useful for localized distributions over extended spaces. These distributions

$$\mu^G(R^0, \Lambda^0; R) \equiv \det{}^{-1}(\sqrt{2\pi}\,\Lambda^0)\; e^{-(R-R^0)(\Lambda^0)^{-2}(R-R^0)/2} \tag{2.3}$$

which are normalized to satisfy

$$\int \mu^G(R)\, dR = 1$$

are simply parameterized by a centroid vector R^0 and fluctuation tensor $(\Lambda^0)^2$. Random vectors $R^{(i)}$ having this distribution are efficiently generated numerically by [19]

$$R^{(i)} = R^0 + GU^{(i)} \tag{2.4}$$

where $GG^{\mathrm{T}} = (\Lambda^0)^2$ is the Cholesky factorization of $(\Lambda^0)^2$ and $U^{(i)}$ is a vector of normally distributed random numbers with unit variance.

The problem is that the appropriate kernel can be heard to find and that a single Gaussian kernel is often not adequate.

B. Macrostate Trajectory Diagrams

For example, consider the two-dimensional potential and corresponding Gibbs–Boltzmann distributions shown in Figure 1. At high temperature where $k_B T_{\mathrm{hi}}$ exceeds the largest potential-energy barrier, $\exp(-\beta_{\mathrm{hi}} V)$ is roughly unimodal and covers much of the space, so it is not difficult to find an appropriate Gaussian kernel (e.g., the corresponding distribution in Fig. 1*c*) for efficient integration. However, at low-temperature T_{lo}, finding the regions where the distribution has support is a needles-in-the-haystack problem that (in more complicated multidimensional problems) can be difficult to solve by direct search. We can find these regions by thermal annealing in which we start from an extended kernel at T_{hi} and track the kernel parametes as T is lowered to T_{lo}. During this process we will pass through intermediate temperatures (e.g., T_{med} in Fig. 1), where the distribution will divide into multiple macrostates that should be integrated using individualized Gaussian kernels. The number of macrostates and kernels required will depend on the temperature. These will be hierarchically related—large, high-temperature *parents* will bifurcate into smaller, low-temperature *children* until there is a macrostate corresponding to each local minimum. The procedure uncovers an intrinsic *variable-scale coarse-grained hierarchical decomposition* of the conformation space. Once this fundamental structure has been discovered, the thermodynamic properties of the system can be efficiently computed by Monte Carlo integration at every temperature. This information can be summarized in *macrostate trajectory diagrams* like the conformational free-energy trajectory diagram shown in Figure 1*e*. At each temperature every branch identifies a macrostate of known size and thermodynamic properties. A complete diagram would provide a complete hierarchical description of the potential–energy landscape down to the level of the local minima, but at excessive computational cost. However, partial diagrams may be adequate for finding the dominating macrostates at much less expense.

The same general features will hold for macrostate trajectory diagrams

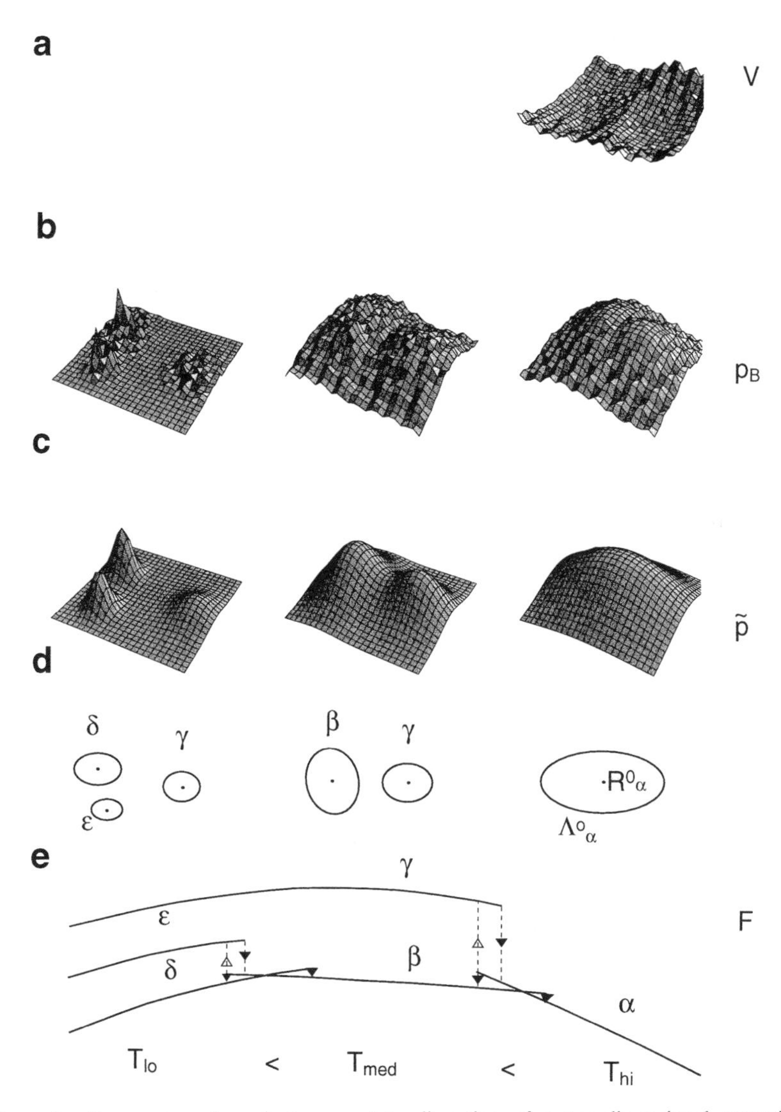

Figure 1. Temperature-dependent macrostate dissection of a two-dimensional potential-energy landscape. (*a*) Potential V as a function of two coordinates. (*b*) Gibbs–Boltzmann distribution p_B at low (left), medium (middle), and high (right) temperatures. (*c*) Corresponding $\tilde{p}$ at each temperature constructed from solutions to the characteristic packet equations. (*d*) Characteristic packet solution parameters R^0 and Λ^0 for each macrostate (labeled with indices α, β, δ, γ, and ε). (*e*) Trajectory diagram of macrostate conformational free energies F_α as a function of temperature. (Reproduced from Church et al. [17] with permission obtained.)

of complex systems like proteins, although the diagrams will be much more complex. If a protein folds to a unique conformation, only a single macrostate will have significant probability at biological temperature. The protein folding problem is thereby subsumed within a more general analysis.

C. The Unimodal Case

A single Gaussian kernel is appropriate when p_B is unimodal at T_{hi} as in Figure 1. In this case, we could substitute

$$\mu(R) \to \mu^G(R^0, \Lambda^0; R)$$

into Eq. (2.2) and attempt to find optimal kernel parameters by minimization with respect to R^0 and Λ^0. However, because of the μ^G in the denominator, the resulting integral for the expected variance would converge poorly and be difficult to evaluate numerically as $R \to \infty$. Instead, we use an approximate criterion and simply match the kernel parameters to the first and second spatial moments of p_B. That is, we choose[20]

$$R_i^0 = \langle R_i \rangle \tag{2.5a}$$

$$(\Lambda^0)_{ij}^2 = \langle (R - R^0)_i (R - R^0)_j \rangle \tag{2.5b}$$

(Throughout, subscripts refer to vector and tensor indices.) Of course, the averages in Eqs. (2.5) are not known in advance but must themselves be calculated, and the kernel updated, as the computation proceeds.

D. The Multimodal Case

The kernel defined by Eqs. (2.5) will be inefficient when p_B is multimodal. In this situation we must dissect the conformation space into separate macrostate regions $\alpha \in \{a, b, c, \ldots\}$ and find Gaussian kernels that match p_B in each region. This can be accomplished using *window functions* and *characteristic packets*, concepts that have previously been introduced in the context of global optimization [14,15], potential-energy landscape analysis [16], and stochastic dynamics [1]. Here we present a more direct, but heuristic, discussion.

1. *Window Function Dissection*

A macrostate region α can be described by a *window function* w_α having the properties

$$w_\alpha(R) \approx \begin{cases} 1 & \text{if } R \text{ is inside } \alpha \\ 0 & \text{otherwise} \end{cases} \tag{2.6}$$

In a "hard" dissection, the w_α exactly equal one or zero and the macrostate regions are distinct. However, we allow for "soft" dissections [16,21] in which the w_α satisfy only the weaker conditions

$$0 < w_\alpha(\beta; R) < 1 \qquad\qquad \text{(2.7a)}$$

$$\forall R$$

$$\sum_\alpha w_\alpha(\beta; R) = 1 \qquad\qquad \text{(2.7b)}$$

where the sum is over all macrostates. Soft window functions typically overlap at the boundaries of the macrostate regions. Nonetheless, once they have been determined, Eq. (2.7b) implies that we can decompose Eq. (2.1a) as

$$Z_c(\beta) = \sum_\alpha \int e^{-\beta V(R)} \mu_\alpha^{-1}(R) w_\alpha(\beta; R)[\mu_\alpha(R)\, dR] \qquad \text{(2.8)}$$

where $\mu_\alpha(R)$ is the optimal Gaussian kernel for macrostate α:

$$\mu_\alpha(R) \equiv \mu^G(R_\alpha^0, \Lambda_\alpha^0; R)$$

In principle, optimal w_α and μ_α can be determined by minimizing the summed variance of the Monte Carlo integrations used to evaluate Eq. (2.8), but this would be even more complicated than minimizing the variance in the unimodal case. Instead we use a two-step approximation procedure.

2. *Characteristic Packet Equations*

The first step is to approximate p_B by $\tilde{p}$, a sum of Gaussians

$$p_B \approx \tilde{p} \approx \sum_\alpha (\phi_\alpha^0)^2 \qquad \text{(2.9)}$$

where the Gaussian characteristic packets

$$\phi_\alpha^0(\beta; R) \equiv e^{-\beta/2[V_\alpha^0 + (R - R_\alpha^0)K_\alpha^0(R - R_\alpha^0)]} / \sqrt{Z_c}$$

are parametrized by scalar, vector, and tensor *characteristic packet parameters* $V_\alpha^0(\beta)$, $R_\alpha^0(\beta)$, and $K_\alpha^0(\beta) \equiv (\Lambda_\alpha^0)^{-2}/(2\beta)$. (We use either K_α^0 or Λ_α^0 to characterize the size of Gaussian packets and kernels as is convenient.) The expected variance of Monte Carlo integration of Z_c using $\tilde{p}$ as the kernel is obtained from Eq. (2.2) with the substitution $\mu \to \tilde{p}$. Using Eq. (2.9) to approximate $p_B/\sqrt{\tilde{p}} \approx \sqrt{p_B}$, we get

$$\sigma^2_{MC}[\tilde{p}] \approx \left(\frac{Z_c^2}{N_{MC}}\right)\chi(p_B, \tilde{p}) \tag{2.10}$$

where

$$\chi(p_1, p_2) = \tfrac{1}{2}\int (\sqrt{p_1} - \sqrt{p_2})^2 \, dR$$

where χ is one-half the *Hellinger distance* [22] between normalized distributions p_1 and p_2 and varies between 0 (when they are identical) and 1 (when they have no overlap). Equation (2.10) will not be accurate when $\tilde{p}$ is far from p_B, but will be useful in the vicinity of a solution of Eq. (2.9). This suggests that $\sigma^2_{MC}[\tilde{p}]$ and $\chi(p_B, \tilde{p})$ will have similar minimizers in the space of characteristic packet parameters, and thus, that $\chi(p_B, \tilde{p})$ is a good objective function for determining these parameters.

As long as the characteristic packets do not overlap significantly (a condition that is self-consistently satisfied)

$$\sqrt{\tilde{p}} \approx \sum_\alpha \phi^0_\alpha$$

and the minimization conditions

$$\frac{\partial\chi[p_B, \tilde{p}]}{\partial V^0_\alpha} = 0, \qquad \frac{\partial\chi[p_B, \tilde{p}]}{\partial R^0_\alpha} = 0, \qquad \frac{\partial\chi[p_B, \tilde{p}]}{\partial K^0_\alpha} = 0$$

imply the *characteristic packet equations* [1,16]

$$\int \sqrt{p_B}\,\phi^0_\alpha \, dR = \int (\phi^0_\alpha)^2 \, dR \tag{2.11a}$$

$$\frac{\int \sqrt{p_B}\, R_i \phi^0_\alpha \, dR}{\int \sqrt{p_B}\,\phi^0_\alpha \, dR} = \frac{\int R_i (\phi^0_\alpha)^2 \, dR}{\int (\phi^0_\alpha)^2 \, dR} = (R^0_\alpha)_i \tag{2.11b}$$

$$\begin{aligned}\frac{\int \sqrt{p_B}(R - R^0_\alpha)_i (R - R^0_\alpha)_j \phi^0_\alpha \, dR}{\int \sqrt{p_B}\,\phi^0_\alpha \, dR} &= \frac{\int (R - R^0_\alpha)_i (R - R^0_\alpha)_j (\phi^0_\alpha)^2 \, dR}{\int (\phi^0_\alpha)^2 \, dR} = (2\beta K^0_\alpha)^{-1}_{ij} \\ &= (\Lambda^0_\alpha)^2_{ij}\end{aligned} \tag{2.11c}$$

Each solution of these coupled equations corresponds to a macrostate. Equations (2.11b) and (2.11c) are somewhat similar to the unimodal Eqs. (2.5), but there are important differences. Whereas Eqs. (2.5) determine R^0 and Λ^0 explicitly, Eqs. (2.11b) and (2.11c) are coupled and determine them implicitly. They have solutions when the expectation values of R and R^2 in the geometric mean distribution $\sqrt{p_B(\phi_\alpha^0)^2}$ equal their expectation values in $(\phi_\alpha^0)^2$ itself. Moreover, because they are nonlinear, they will have a varying number of solutions depending on the temperature. This corresponds to the varying number of macrostates that are needed to dissect p_B (e.g., as in Fig. 1).

Equations (2.11b) and (2.11c) can be solved by numerical iteration [16] to find a single solution in the vicinity of initial conditions ${}^{(0)}R_\alpha^0$, ${}^{(0)}K_\alpha^0$. Convergence can be assessed using $\chi[({}^{(n)}\hat{\phi}_\alpha^0)^2, ({}^{(n+1)}\hat{\phi}_\alpha^0)^2]$ as a dimensionless indicator of change, where ${}^{(n)}\phi_\alpha^0$ is the characteristic packet at the nth iteration. However, finding appropriate initial conditions is a nontrivial problem whose solution is postponed to Section II.E. Once R_α^0 and Λ_α^0 have been computed, Eq. (2.11a) explicitly determines V_α^0 as

$$V_\alpha^0 = -2\beta \log\left[\det{}^{-(1/2)}\left(\frac{\pi}{\beta K_\alpha^0}\right)\int e^{-\beta/2V(R)}\, e^{-\beta/2(\mathrm{R}-\mathrm{R}_\alpha^0)K_\alpha^0(\mathrm{R}-\mathrm{R}_\alpha^0)}\, dR\right]$$

3. *Characteristic Window Functions*

Once the characteristic packets have been determined (as described below), the second step is to define *characteristic window functions* [1,16]

$$w_\alpha^0(\beta;\, R) \equiv \frac{\phi_\alpha^0(\beta;\, R)}{\sum_\gamma \phi_\gamma^0(\beta;\, R)}$$

The w_α^0 dissect conformation space into subregions according to Eq. (2.6) and satisfy the window conditions of Eqs. (2.7). Thus they are reasonable approximations to the w_α. Although the precise location and shape of the edges of the w_α^0 are arbitrary in this approximation, this is numerically unimportant since p_B is very small in the boundary regions. An example of the relationship of the characteristic packets and window functions to p_B is illustrated in Figure 2.

4. *Macrostate Integrals*

The kernel for integrations within each macrostate region can be determined by the unimodal approximation Eqs. (2.5) with the substitution $\exp(-\beta V) \to w_\alpha^0 \exp(-\beta V)$. Alternatively, it is often adequate to use the

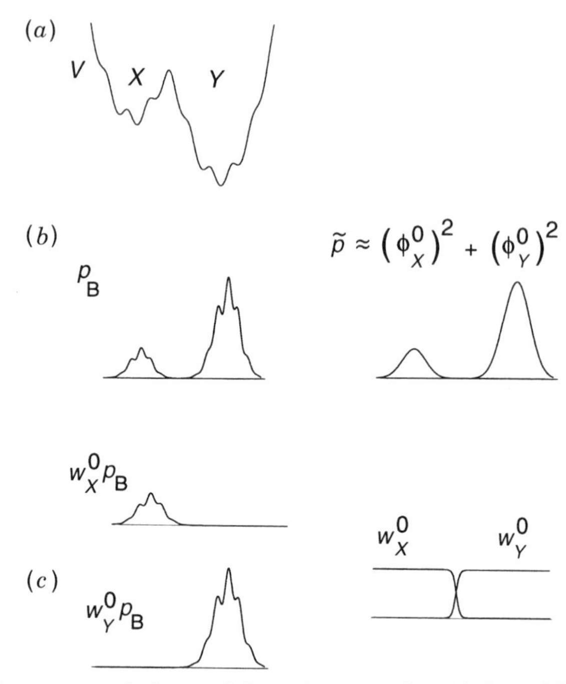

Figure 2. Macrostates and characteristic packet expansion. (a) Potential $V(R)$ that can support multiple macrostates. (*b*) The equilibrium distribution p_B at a temperature where only two macrostates are supported and the corresponding characteristic packet expansion $\tilde{p}$. (*c*) The dissection of p_B into macrostate distributions using the characteristic window functions.

normalized squared characteristic packet as kernel:

$$\mu_\alpha \approx \frac{(\phi_\alpha^0)^2}{\int (\phi_\alpha^0)^2 \, dR} \equiv (\hat{\phi}_\alpha^0)^2$$

The goodness of fit of these kernels to the window-weighted probability distribution can be assessed by $\chi(p_{B,\alpha}, \mu_\alpha)$, where

$$p_{B,\alpha} \equiv e^{-\beta V} \frac{w_\alpha^0}{Z_\alpha}$$

is the renormalized macrostate equilibrium probability distribution and

$$Z_\alpha = \int e^{-\beta V} w_\alpha \, dR$$

is the macrostate conformational partition function. If $\chi(p_{B,\alpha}, \mu_\alpha)$ is small, Monte Carlo integrations over the macrostate will be efficient.

Macrostate thermodynamic parameters are computed as integrals over $p_{B,\alpha}$. The macrostate conformational free energy F_α is

$$F_\alpha = -\beta^{-1} \log\left[\frac{Z_\alpha}{(h_R)^N}\right] \tag{2.12}$$

where h_R is an arbitrary constant with the dimensions of a conformation space coordinate. The macrostate average of a variable f is

$$\langle f \rangle_\alpha = \int f p_{B,\alpha}\, dR$$

Macrostate integrals and averages can be combined to give properties of the complete system:

$$Z_c = \sum_\alpha Z_\alpha$$

and

$$\langle f \rangle = \sum_\alpha p_\alpha \langle f \rangle_\alpha$$

where

$$p_\alpha(\beta) \equiv \frac{Z_\alpha(\beta)}{Z_c(\beta)}$$

is the *macrostate probability*.

E. Computing Macrostate Trajectory Diagrams

The solution to the problem of finding initial conditions for solving the characteristic packet equations is to use an annealing procedure. The unique high-temperature characteristic packet can be found by solving Eqs. (2.11) using almost any initial conditions. This packet can be updated as the temperature is lowered through a sequence of discrete values by using the characteristic packet at the previous temperature as an initial condition for numerical iteration. By this means we can track the (in principle, continuous) vector and tensor functions $R_\alpha^0(\beta)$, $K_\alpha^0(\beta)$ down in discrete temperature steps. This process will be disrupted whenever the temperature is lowered below a *branch temperature* at which the characteristic packet solution will bifurcate into two (or possibly more) child solutions. For example, this occurs between temperatures T_{hi} and T_{med} in Figure 1. Somewhere in this temperature range the iterative solver will be trapped by one of the children. A key point is that the resultant discontinuity can be distinguished from numerical fluctuation, and thereby detected, by using the Hellinger distance between sequentially computed characteristic packets as a measure. Numerical fluctuations of R_α^0 and K_α^0 will have $\chi \ll 1$, while discontinuities resulting from a bifurcation in the packet solution space will have $\chi \lesssim 1$. Once one child of a bifurcation has been detected, it is easy to find the sibling by comparing the known child to the parent and using initial conditions corresponding to the location and distribution of the missing probability mass [16]. This procedure can be continued recursively as temperature is lowered. (For physically relevant examples, see Ref. 16 for trajectory diagrams of six-, seven-, and eight-atom microclusters.)

In more complicated cases, computational cost will force us to "prune" the tree by terminating trajectories that have little chance of leading to dominating macrostates (having high p_α) at low temperatures. The success of the macrostate annealing approach in large systems will depend on our ability to gauge the odds during an annealing run and thereby to construct effective *branch selection algorithms.* A significant amount of information is available for guiding these choices. Attributes that may be important include the macrostate energy, entropy, free energy, and heat capacity—all of which can be calculated as functions of temperature. Which will be optimal will depend on the structure of the potential–energy landscape being considered [17,23]. We do not know how to determine this a priori. However, by empirically examining the trajectory diagrams of multiple systems, it may be possible to identify general characteristics of entire classes of problems that can be exploited to construct effective algorithms. The studies described in Section IV are a first step in this direction.

III. APPLICATION TO TRANSITION RATES

Intermacrostate transition rates are often computed using reaction path-based formulations such as transition-state theory [2,24]. However, Monte Carlo integration can be used in a procedure that focuses instead on the *transition region*—a localized subregion of conformational space positioned between the macrostates [1,25]. As in equilibrium computations, the key to efficiency is to identify an appropriate integration kernel.

A. Kinetic Rates and the Smoluchowski Equation

The diffusive relaxation of the probability density $p(R, t)$ of an ensemble of systems towards the Gibbs–Boltzmann equilibrium distribution is described by the Smoluchowski equation

$$\frac{\partial p}{\partial t} = D\nabla[e^{-\beta V}\nabla(e^{\beta V}p)] \tag{3.1}$$

where D is the diffusion constant. By introducing Ψ, the *reduced distribution*

$$\Psi \equiv p(R, t)/\sqrt{p_{\mathrm{B}}(\beta; R)}$$

this is transformed to the *reduced Smoluchowski equation*

$$\frac{\partial \Psi}{\partial t} = -D[e^{\beta/2V}(\nabla^2\, e^{-\beta/2V}) - \nabla^2]\Psi \equiv -\mathscr{H}\Psi \tag{3.2}$$

governed by the hermitian operator $\mathscr{H}$ (see Ref. 1 for review). Equation (3.2) is an imaginary-time Schrödinger equation and its solutions can be expanded in terms of the eigenfunctions of $\mathscr{H}$. The lowest-lying eigenfunction is

$$\psi_0(\beta; R) = \sqrt{p_{\mathrm{B}}(\beta; R)}$$

which always has eigenvalue

$$\gamma_0 = 0$$

as demanded by stability of the Gibbs–Boltzmann distribution.

When the temperature is low enough so that the energy barriers separate conformation space into multiple macrostate regions, the relaxation process can be described by a phenomenological master equation

$$\frac{dp_\alpha}{dt}(\beta; t) = \sum_\gamma \Gamma_{\alpha\gamma}(\beta)p_\gamma(\beta; t)$$

where $p_\alpha(\beta; t) = \int p(R, t)w_\alpha(\beta; R)\, dR$. The intermacrostate transition rates $\Gamma_{\alpha\gamma}(\beta)$ are determined by the eigenvalues of $\mathscr{H}$ [1,21,25–27]. In a system having two macrostates, the forward (k_{f}) and backward (k_{b}) transition rates are determined by the lowest nonzero eigenvalue γ_1:

$$k_{\text{f}} = p_X \gamma_1, \qquad k_{\text{b}} = p_Y \gamma_1$$

where p_X and p_Y are the macrostate equilibrium probabilities [25]. As in quantum mechanics, γ_1 is

$$\gamma_1 = \frac{\langle \psi_1 | \mathscr{H} | \psi_1 \rangle}{\langle \psi_1 | \psi_1 \rangle}$$

where bracket notation denotes integration over R. The corresponding eigenfunction ψ_1 can be determined by variational minimization:

$$\left. \frac{\delta}{\delta \psi_1} \right|_{\langle \psi_0 | \psi_1 \rangle = 0} \gamma_1 = 0 \tag{3.3}$$

B. Solving the Variational Equation

Iterative solution of Eq. (3.3) requires many Monte Carlo integrations, and it is important to use appropriate kernels. On physical grounds, we expect that these will be concentrated in the boundary region which lies between the two macrostates. To see how this transition region can be identified, consider a one-dimensional, two-macrostate system like that shown in Figure 3. The ground-state eigenfunction ψ_0 has no nodes and, as in quantum mechanics, ψ_1 has one node located at the point of maximum energy lying between macrostate regions X and Y. This conformation corresponds to the "transition state" of transition state theory. (In multiple dimensions ψ_1 has a nodal surface, and the transition state is a saddle point of ψ_0.) Away from the node, ψ_1 is approximately proportional to ψ_0 but with different proportionality constants of opposite sign and magnitude, which are determined by the eigenfunction orthonormality conditions. We express the ratio of the eigenfunctions in terms of a *transition function* $\sigma(\beta; R)$:

$$\frac{\psi_1(\beta; R)}{\psi_0(\beta; R)} = \frac{\sigma(\beta; R) - \langle \psi_0 | \sigma | \psi_0 \rangle}{\sqrt{\langle \psi_0 | \sigma^2 | \psi_0 \rangle - \langle \psi_0 | \sigma | \psi_0 \rangle^2}} \tag{3.4}$$

This form ensures that

$$\langle \psi_0 | \psi_1 \rangle = 0, \qquad \langle \psi_1 | \psi_1 \rangle = 1$$

for any σ.

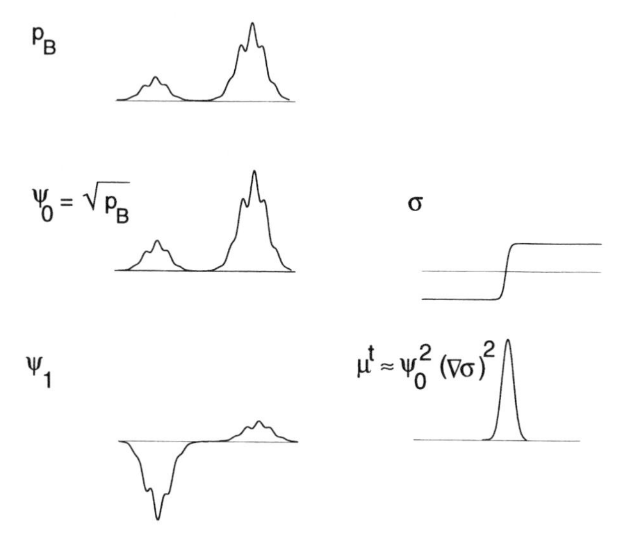

Figure 3. Functions for transition-rate calculations. Shown for a two-macrostate, one-dimensional system characterized by p_B are the ground-state (ψ_0) and first excited-state (ψ_1) eigenfunctions of $\mathscr{H}$, the transition function σ and the transition kernel μ^t.

Using Eq. (3.4), Eq. (3.3) becomes the unconstrained variational condition

$$\frac{\delta}{\delta\sigma} \frac{D\langle\psi_0|(\nabla\sigma)^2|\psi_0\rangle}{\langle\psi_0|\sigma^2|\psi_0\rangle - \langle\psi_0|\sigma|\psi_0\rangle^2} = 0 \tag{3.5}$$

As long as the variation is constrained within reasonable limits around a form such as that shown in Figure 3, the denominator will not change much and the numerator will dominate the variation. The numerator depends strongly on $(\nabla\sigma)^2$, so σ should be parametrized in a way that permits good control of its behavior in the nodal region. At the same time, to account for the constant proportionality of ψ_0 and ψ_1 away from the nodal surface, we want $\lim_{|R|\to\infty}|\sigma(R)| \to$ (constant). As long as the integrand of the numerator has support over only a relatively small region (i.e., the usual case in which the saddle in the potential-energy landscape is narrow compared to the distance between the macrostates), we can, without much loss of accuracy, simplify the parametrization by using a σ that has a planar nodal surface. (This will be tangent to the nodal surface of the exact σ in the transition region.) A useful form of this type is

$$\sigma \approx s = \mathrm{erf}[\eta(\beta)R + \xi(\beta)] \tag{3.6}$$

where erf is the error function and the vector η and scalar ξ determine the position of the nodal *transition plane*

$$\eta R + \xi = 0$$

where η is a vector normal to the plane (i.e., in the *transition direction*) and $|\eta|$ is inversely proportional to the width of the transition region; ξ determines the position of the plane along the transition direction. Substituting Eq. (3.6) into Eq. (3.5) gives variational conditions

$$\frac{\partial}{\partial \xi} \gamma_1(\eta, \xi) = 0 \tag{3.7a}$$

$$\frac{\partial}{\partial \eta} \gamma_1(\eta, \xi) = 0. \tag{3.7b}$$

where

$$\gamma_1(\eta, \xi) = -4v^2(\eta, \xi) H_{XY}(\eta, \xi) \tag{3.8a}$$

$$H_{XY}(\eta, \xi) \equiv -\frac{D|\eta|^2}{\pi} \int \psi_0^2 \, e^{-2(\eta R + \xi)^2} \, dR \tag{3.8b}$$

$$v^{-2}(\eta, \xi) \equiv \int \psi_0^2 \, \mathrm{erf}^2(\eta R + \xi) \, dR - \left[\int \psi_0^2 \, \mathrm{erf}(\eta R + \xi) \, dR \right]^2 \tag{3.8c}$$

Equations (3.7) can be solved iteratively using Monte Carlo integrations to evaluate the numerator and denominator [25]. Since the main contributions to the v^{-2} integrals come from the central macrostate regions where $\psi_0^2 = p_B$ is maximal, they can be performed using kernels μ_X and μ_Y, which are determined as described in Section II.C. However, integration of H_{XY} requires a special *transition kernel*

$$\mu_{XY}^t(\beta; R) \equiv \mu^G(R_{XY}^t, \Lambda_{XY}^t; R).$$

The $\exp[-2(\eta R + \xi)^2]$ term in the integrand of Eq. (3.8b) has support only near the transition plane whereas the ψ_0^2 term restricts support in the orthogonal subspace; therefore μ_{XY}^t need cover only a limited region. In many cases, this integrand will be unimodal and the procedure described in Section II.C can be used with the substitution $p_B \rightarrow \psi_0^2 \exp[-2(\eta R + \xi)^2]$. When it is not (i.e., when multiple distinct flux channels participate in the transition), the characteristic packet equations can be used to obtain a multiple-Gaussian dissection. The finite transition region identified by the transition kernel extends the single-conformation transition state of transition-state theory, and it can be shown that the transition region shrinks to the transition state in the $T \to 0$ limit [25]. The term Λ_{XY}^t provides the important additional information that is needed to compute the influence of the surrounding conformations at finite temperatures.

The variational method has been tested by computing temperature-dependent transition rates in one- and two-dimensional model systems and in the six-atom Lennard-Jones microcluster [25]. The variational transition rates were similar to those computed from Brownian dynamics mean first-passage times but were obtained at $\sim 1\%$ the cost. This reflects the enhanced efficiency that results from focusing computational effort in the transition region. In contrast, standard dynamical trajectory methods expend most computational effort wandering in the intramacrostate regions, whereas visits to the important transition region are rare. New trajectory-based methods have recently been developed that can also greatly enhance efficiency [28,29], and it will be interesting to compare them with the variational method.

IV. APPLICATION TO PEPTIDES

Sections II and III provide a framework for dissecting thermodynamic Monte Carlo integrals for problems where Cartesian coordinates are appropriate. But it must be modified when other coordinate systems are needed. Macromolecules are a case in point. Their thermal fluctuations tend to preserve bond lengths and angles, resulting in tortuously curved macrostate regions that cannot be well modeled by Gaussians in Cartesian coordinates. These regions are more compact when described in internal coordinates. However, the fluctuation tensor in these coordinates can be poorly conditioned because changes in some angles (e.g., backbone angles near the middle of the molecule) cause very large conformational changes while others (e.g., side-chain angles) move only a few atoms. This makes the solution of multidimensional characteristic packet equations expensive. This problem can be overcome by using a hybrid system that describes individual conformations with internal coordinates but identifies macrostates using distance coordinates. We describe this procedure using the pentapeptide Met-enkephalin, H_2N–Tyr–Gly–Gly–Phe–Met–COOH, in vacuo as an example. It has 75 atoms, 24 torsion angles, 74 bond lengths, and 121 bond angles (Figure 4), and is a neural hormone and a natural ligand for opiate receptors [30]. While the physiological properties of Met-enkephalin are strongly influenced by solvent and receptor interactions, we use the simpler

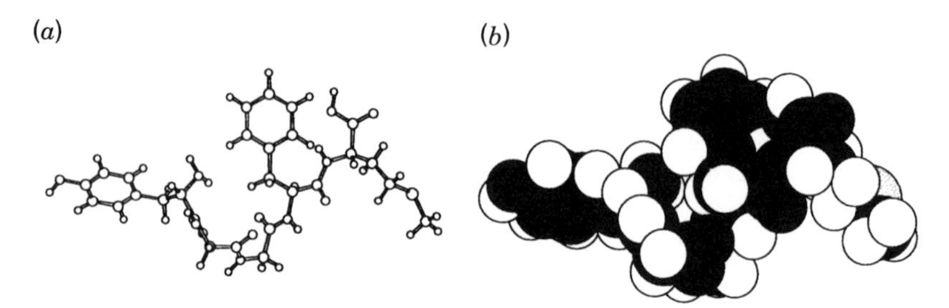

Figure 4. The pentapeptide Met-enkephalin in (*a*) ball-and-stick and (*b*) space-filling representations.

unsolvated system for demonstration because it has been analyzed in a number of theoretical studies [31–34] and a putative global potential energy minimum for the ECEPP/3 potential [35] is known [31].

A. Transformation to Internal Coordinates

The conversion between Cartesian and internal coordinates for a system with N_a atoms requires only $O(N_a)$ operations. The $3N_a$ Cartesian coordinates R are replaced with $3N_a - 6$ internal coordinates Q that uniquely describe the polypeptide conformation and six external coordinates X that specify overall position and rotational orientation. For a noncyclic chain, Q consists of the $N_b = N_a - 1$ bond lengths $B \equiv \{b_i\}$, N_Θ bond angles $\Theta \equiv \{\Theta_i\}$, and N_Ω torsion angles $\Omega \equiv \{\Omega_i\}$; $N_\Theta + N_\Omega = 2N_a - 5$. The Jacobian of the transformation has a simple form that depends only on bond lengths and bond angles, not on the torsion angles or atomic masses [36,37]. This can also be computed in only $O(N_a)$ operations and, since the evaluation of $V(R)$ is $O(N_a^2)$, does not significantly increase computational complexity even when bond lengths and angles vary. However, because relatively stiff covalent linkages hold bond lengths and angles close to their strain-free values, thermodynamic integrals are accurately approximated even when these variables are fixed [37]. This greatly reduces the dimensionality of the problem (a factor of 10 is not unusual) and only torsion angles remain as independent variables. In this approximation $R(B, \Theta, \Omega)$ reduces to $R(\Omega)$ and the Jacobian is a conformation-independent constant that factors out of all thermodynamic averages.

B. Multivariate Wrapped Gaussian Distribution

To describe correlated peptide conformation fluctuations in internal coordinates, the multivariate Gaussian distribution μ^G used for characteristic packets and integration kernels in Cartesian space must be adapted to the periodic and multiply connected torsion angle space. Although the Cartesian Gaussian distribution and conditions such as Eqs. (2.5) can be used in angular coordinates when the angular fluctuation tensor $(\Xi^0)^2$ is small (on the scale set by π^2), such conventional quasi-harmonic procedures poorly model angular distributions and so lead to inefficient computation when fluctuations are large and periodicity is important. To remedy this we introduce the *multivariate wrapped Gaussian* (MWG) distribution $\mu^{WG}(\Omega^0, \Xi^0; \Omega)$ [38], which is constructed by "wrapping" the nonperiodic Gaussian $\mu^G(\Omega^0, \Xi^0; \Omega)$ onto the torus $\mathscr{T} \equiv (-\pi \le \Omega_i < \pi) \otimes \cdots \otimes (-\pi \le \Omega_{N_\Omega} < \pi)$. This gives

$$\mu^{WG}(\Omega^0, \Xi^0; \Omega) = \det^{-1}(\sqrt{2\pi}\,\Xi^0) \times \sum_{M \in \mathbf{Z}^{N_\Omega}} e^{-(\Omega - \Omega^0 - 2\pi M)(\Xi^0)(\Omega - \Omega^0 - 2\pi M)/2} \tag{4.1}$$

where vector Ω^0 specifies the center of the distribution and M is a vector on the N_Ω-dimensional integer lattice $\mathbf{Z}^{N_\Omega}$. This is normalized so that

$$\int_{-\pi}^{\pi} \mu^{WG}(\Omega^0, \Xi^0; \Omega)\, d\Omega = 1$$

When fluctuations are small (i.e., the eigenvalues of $(\Xi^0)^2$ are $\ll \pi^2$), only the $M = 0$ term will contribute to Eq. (4.1). Then μ^{WG} will approximate the Cartesian space kernel [Eq. (2.3)] with the replacements

$$R \to \Omega, \qquad R^0 \to \Omega^0, \qquad \Lambda \to \Xi, \qquad \Lambda^0 \to \Xi^0 \tag{4.2}$$

When fluctuations are large [i.e., the eigenvalues of $(\Xi^0)^2$ are $\gtrsim \pi^2$], the unwrapped distribution extends beyond $(-\pi, \pi)$ and wraps back onto $\mathscr{T}$; the infinite sum ensures periodicity.

Although Eq. (4.1) involves an infinite sum, a set of random torsion space vectors $\{\Omega\}$ having distribution μ^{WG} is easily generated by first generating a set of unwrapped vectors $\{\Omega^U\}$ having the Cartesian distribution $\mu^G[\Omega^0, \Xi^0; \Omega]$ [as in Eq. (2.4)] and mapping each Ω^U onto $\mathscr{T}$ by

$$\Omega_i = (\Omega_i^U + \pi) \bmod 2\pi, \qquad i = 1, \ldots, N_\Omega$$

The remaining task is to compute the parameters Ω_α^0 and Ξ_α^0 for macrostate α. We assume that the conformation space has already been dissected with window functions so that we can restrict consideration to unimodal distributions. Figure 5 shows the performance of the conventional quasi-harmonic torsion-angle approximation [i.e., the Cartesian formulas of Eqs. (2.5) with the substitutions of Eqs. (4.2)] for the $\phi - \psi$ angles of terminally

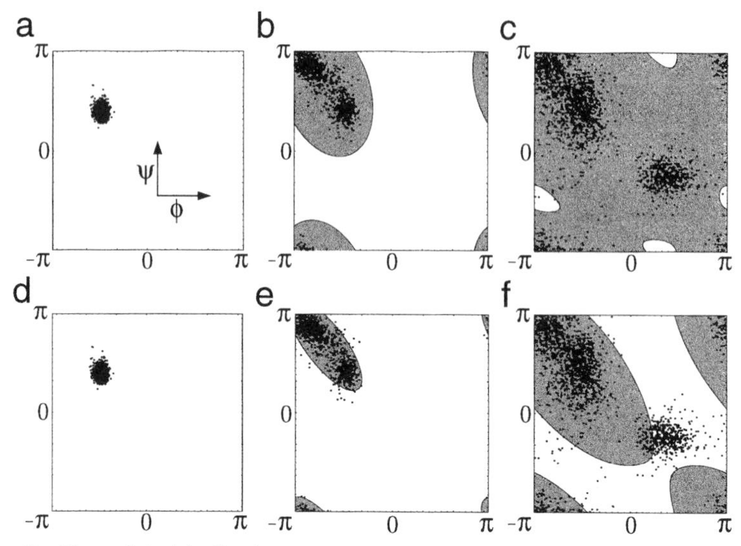

Figure 5. Thermal ϕ–ψ (radians) scatterplots for terminally blocked valine at $T = 0.2$ kcal/mol (*a*, *d*), 0.6 kcal/mol (*b*, *e*), and 2.2 kcal/mol (*c*, *f*) and the corresponding descriptions produced by the angular linear quasi-harmonic (*a*–*c*) and multivariate wrapped Gaussian (*d*–*f*) models. The shaded regions are bounded by the contours where $p(\phi, \psi)/p(\phi^0, \psi^0) = e^{-2}$. In contrast to the quasi-harmonic method, the MWG method accurately models the data set even at high temperatures. (Adapted from Church and Shalloway [38].)

blocked valine at three temperatures. We see that this approximation is satisfactory for low-temperature compact distributions (e.g., panel *a*), but is very inaccurate for the extended distributions that are generated at high temperatures (e.g., panels *b* and *c*). This is not surprising given that these formulas disregard periodicity. To rectify this, we use the periodic procedure summarized below [38].

The mean conformation of the macrostate is computed as

$$(\Omega_\alpha^0)_j = \arg\langle e^{i\Omega_j}\rangle_\alpha \tag{4.3}$$

By analogy to Eq. (2.5b), it might be thought that $(\Xi_\alpha^0)^2_{ij}$ could be computed by matching expectation values of exponential moments of the form $\langle e^{i[\Omega_i - (\Omega_\alpha^0)_i]}\, e^{i[\Omega_j - (\Omega_\alpha^0)_i]}\rangle_\alpha$. However, the resultant tensor is not always positive semidefinite and can have negative eigenvalues that will cause μ^{WG} to diverge. We require an expression for $(\Xi_\alpha^0)^2$ that satisfies the following conditions:

1. $(\Xi_\alpha^0)^2$ is positive semidefinite.

2. For small fluctuations (i.e., all eigenvalues $\ll \pi^2$), $(\Xi_\alpha^0)^2_{ij} \approx \langle [\Omega_i - (\Omega_\alpha^0)_i][\Omega_j - (\Omega_\Omega^0)_j] \rangle_\alpha$.
3. For uncorrelated fluctuations (i.e., $(\Xi_\alpha^0)^2$ diagonal) of any magnitude, $(\Xi_\alpha^0)^2_{ii} = -2 \log|\langle \exp(i\Omega_i) \rangle_\alpha|$.

These conditions are satisfied if we fix $(\Xi_\alpha^0)^2$ by

$$\left[\frac{\sinh(\Xi_\alpha^0)^2}{4}\right]_{jk} = \frac{\left\langle \sin\left(\frac{\Omega_j - (\Omega_\alpha^0)_j}{2}\right) \sin\left(\frac{\Omega_k - (\Omega_\alpha^0)_k}{2}\right)\right\rangle_\alpha}{\langle e^{i[\Omega_j = (\Omega_\alpha^0)_j]} \rangle_\alpha^{1/4} \langle e^{i[\Omega_k - (\Omega_\alpha^0)_k]} \rangle_\alpha^{1/4}} \tag{4.4}$$

The MWG kernel is compared with that resulting from the conventional quasi-harmonic torsion-angle approximation in Figure 5. Both kernels are similar when modeling low-temperature small fluctuations (Figs. 5*a*, *d*) but the MWG kernel is much more compact and accurate when high-temperature large fluctuations must be modeled (Figs. 5*b*, *c*, *e*, *f*). Figure 6, which plots the spectrum of eigenvalues of $(\Xi_\alpha^0)^2$ computed for Met-enkephalin at 2.7 kcal/mol by the two methods, demonstrates that the MWG and torsion-angle quasi-harmonic approximations can report significantly different distribution properties. The shift of the spectrum to much smaller eigenvalues indicates that the MWG characterizations (panel *a*) are much more compact than the quasi-harmonic ones (panel *b*) for a preponderance of modes.

C. Identifying Macrostates in Interatomic Distance Space

While torsion angles are convenient for describing conformations and their fluctuations, it is difficult to use them to identify macrostate bifurcations because, as mentioned earlier, $(\Xi_\alpha^0)^2$ can be poorly conditioned.

A more efficient approach is to use the interatomic distances to specify window functions. For example, consider the conformational transition associated with opening and closing of the two "flaps" that surround the active site of HIV (human immunodeficiency virus) protease [39]. This transition involves simultaneous changes in many torsion angles, but we can discriminate between the open and closed macrostates just by specifying a single *indicator distance* r_{ij} between atoms i and j, one near the tip of each flap, and defining hard window functions $w_{\text{open}}(r_{ij}) = w^{ij}_{>}(r_{ij})$ and $w_{\text{closed}}(r_{ij}) = w^{ij}_{<}(r_{ij})$, where

$$w^{ij}_{>}(r_{ij}) \equiv \theta(r_{ij} - r^t_{ij}) \tag{4.5a}$$

$$w^{ij}_{<}(r_{ij}) \equiv \theta(-r_{ij} + r^t_{ij}) \tag{4.5b}$$

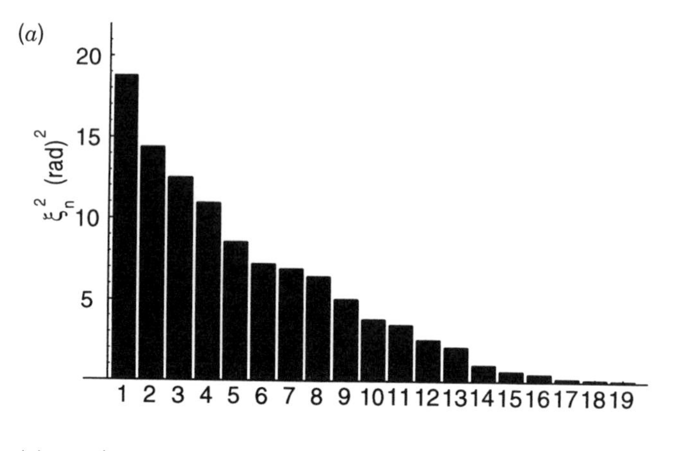

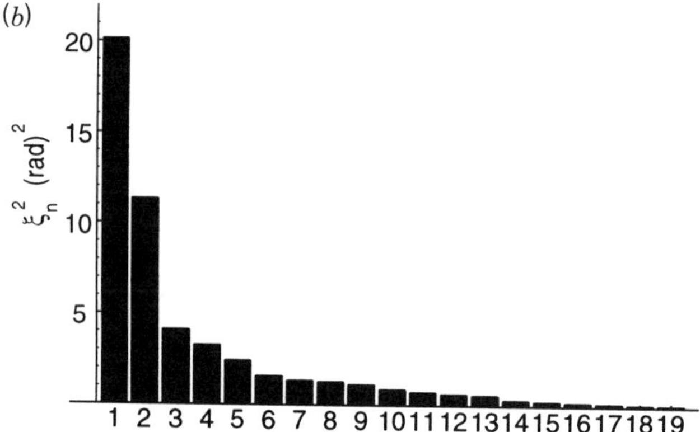

Figure 6. Comparison of eigenvalue spectra for Met-enkephlin at $T = 2.7$ kcal/mol for (*a*) quasi-harmonic estimate of fluctuation kernel and (*b*) MWG estimate of fluctuation kernel. The MWG approximation produces a much more compact distribution than the linear quasi-harmonic calculation. (Adapted from Church and Shalloway [38].)

$\theta(x) = (x + |x|)/2x$ is the unit step function, and r_{ij}^t is a transition-state value between the open and closed states. While a single indicator distance cannot completely describe the conformational transition, it is sufficient for dissecting the macrostate probability mass into two children. In general, many indicator distances can be used, and each will dissect conformation space somewhat differently. However, as long as reasonable choices are made, we expect that the differences will be unimportant since the dissections will differ only in regions where p_B is low.

Multiple macrostates can be distinguished by increasing the number of indicator distances. For example, the open macrostate defined above can be subdivided into children by using a second indicator distance r_{kl} and dissecting $w_{\text{open}} = w^{ij}_{>}$ into $w_{\text{open},1} = w^{ij}_{>} w^{kl}_{>}$ and $w_{\text{open},2} = w^{ij}_{>} w^{kl}_{<}$. Such subdissection can be continued until window functions for all descendents are expressed as products of the form

$$w_{\alpha} \approx \prod_{ij \in \mathscr{R}_{\alpha}} w^{ij}_{b_{\alpha}}(r_{ij}), \qquad b_{\alpha} \in \{<, >\}$$

that depend on a macrostate-specific set $\mathscr{R}_{\alpha}$ of indicator distances.

The success of this binary-tree approach requires that indicator distances can be found to distinguish between each pair of siblings. This is possible because of the very large number of available distances: A molecule containing N_a atoms has $N_a(N_a - 1)/2$ interatomic distances but only $3N_a - 6$ conformational degrees of freedom. Thus there are roughly $N_a/6$ distances for each degree of freedom. Because of this extensive redundancy, when N_a is large, we are likely to find at least one indicator distance that can parameterize each intermacrostate transition. Because of the growth of redundancy with N_a, in large problems it may not be necessary to consider each interatomic distance but only a smaller *representative subset* [40].

1. *Distance Probability Distributions and Effective Potentials*

To identify bifurcations using the indicator distances, we project p_{B} onto the curvilinear r_{ij} coordinates and look for effective energy barriers in the projected distribution. To maintain physical relevance, the projection must yield a one-dimensional Smoluchowski equation when the same transformation is applied to $p(\Omega, t)$ under the adiabatic assumption that the relaxation of p toward p_{B} is much faster in the directions transverse to r_{ij} than along r_{ij}. This assumption will be sensible when there is an effective energy barrier in this coordinate; thus, the assumption can be self-consistently verified. Care is required because $r_{ij}(\Omega)$ is a nonlinear function of Ω and we do not get an equation of the form of Eq. (3.1) by projecting onto r_{ij} itself. However, we do get such an equation by projecting onto the transformed coordinate $s(r_{ij})$ defined by [41]

$$s(r) = \int_0^r \left| \frac{d\Omega}{dr} \right| (r')\, dr' \tag{4.6}$$

where

$$\left| \frac{d\Omega}{dr} \right| (r) = \left[\frac{\int p_{\text{B}}(\beta; \Omega)\, |\nabla r_{ij}(\Omega)|^2\, \delta[r_{ij}(\Omega) - r]\, d\Omega}{\int p_{\text{B}}(\beta; \Omega)\, \delta[r_{ij}(\Omega) - r]\, d\Omega} \right]^{-1/2} \tag{4.7}$$

Defining the projected probability distribution

$$p_s(\beta;\, s,\, t) = \frac{1}{|\,d\Omega/dr\,|\,[r(s)]} \int p(\Omega,\, t)\, \delta[r_{ij}(\Omega) - r(s)]\; d\Omega$$

the projected equilibrium distribution

$$p_{B,\,s}(\beta;\, s) = \frac{1}{|\,d\Omega/dr\,|\,[r(s)]} \int p_B(\beta;\, \Omega)\, \delta[r_{ij}(\Omega) - r(s)]\; d\Omega$$

and the potential of mean force V_s,

$$V_s(\beta;\, s) = -\beta^{-1} \log p_{B,\,s}(\beta;\, s)$$

we get the projected Smoluchowski equation

$$\frac{1}{D}\frac{\partial}{\partial t} p_s = \frac{d}{ds}\left[e^{-\beta V_s} \frac{d}{ds}\left(e^{\beta V_s} p_s\right)\right]$$

The $|\nabla r_{ij}|^2$ that appears in Eq. (4.7) can be calculated analytically in $O(N_\Omega)$ operations for each r_{ij} considered. In practice, $|d\Omega/dr|$ is approximately constant for atoms that are separated by more than four bonds and can be ignored in these cases.

The projected distributions $p_{B,\,s}$ are computed by accumulating histograms over each pairwise distance r_{ij}. This adds little cost because the pairwise distances are already needed for evaluating $V[R(\Omega)]$. The term $p_{B,\,s}$ for an indicator distance from Met-enkephalin is shown in Figure 7 at temperatures above and below a bifurcation.

2. *Branch Recognition Temperature*

Adapting the procedure of Section II.E, we use macrostate annealing in the mixed torsion angle–distance coordinate system to compute macromolecule macrostate trajectory diagrams. The characteristic packet equations can be applied to $p_{B,\,s}$ to identify bifurcations, but it is unnecessarily expensive to use these equations to test for branching after every temperature step. Instead, we use a pre-test to anticipate bifurcations during cooling. Empirically, we have found that we can anticipate a bifurcation in r_{ij} if $p_{B,\,s}[\beta;\, s(r_{ij})]$ has two nodes with a minimizer r_{ij}^t between them that satisfy the following criteria:

1. The total probability contained in each node is at least 1% of the total probability.

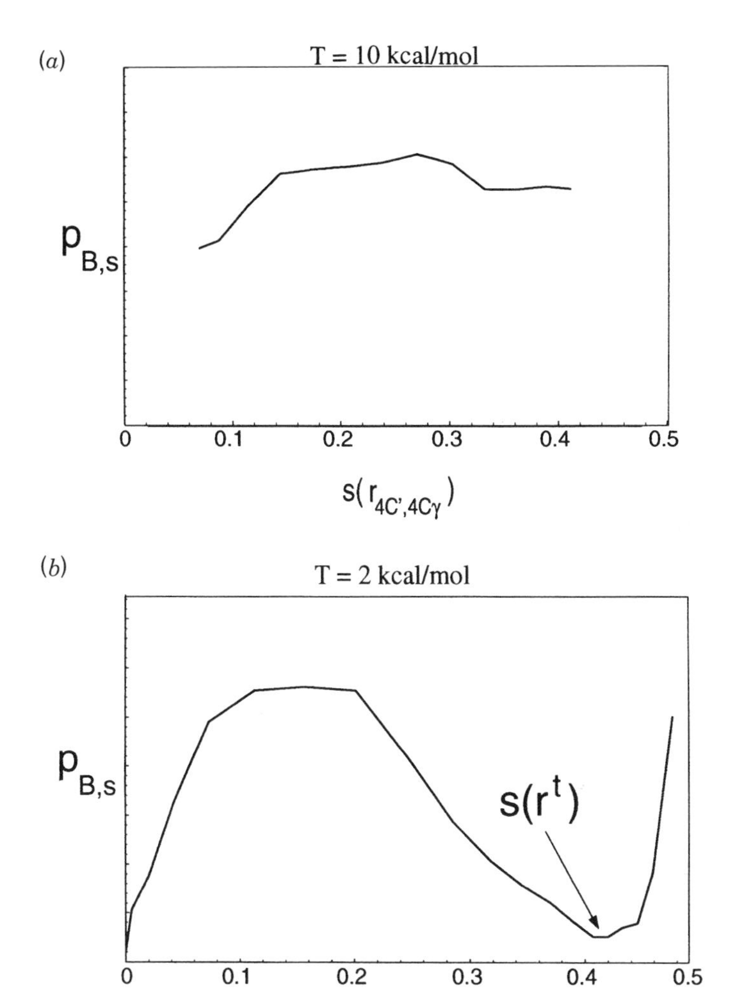

Figure 7. Projected probability distribution $p_{B,s}$ for the distance between the phenylalanine C′ and C_γ atoms of Met-enkephalin measured at high (10 kcal/mol) and low (2 kcal/mol) temperatures. The distribution is plotted over the transformed distance $s(r)$ defined by Eq. (4.6). At high temperature (*a*) the distribution is continuous; at low temperature (*b*) an effective barrier at $s(r^t)$ has emerged, causing a bimodal distribution.

2. $p_{B,s}[\beta; s(r_{ij}^t)] < 0.15 \times \max(p_B)$ in the smaller node.

The first condition prevents noise from inducing a false signal. If the recognition criteria are satisfied for multiple indicator distances at once, the one with the minimum ratio of $p_{B,s}[\beta; s(r_{ij}^t)]$ to the smaller node peak is chosen as the indicator for the bifurcation. We call the temperature T_{ij}^R at which these conditions are first satisfied during cooling the *branch recognition temperature.* Precise packet branching temperatures can then be determined by applying the characteristic packet equations to $V_s[s(r_{ij})]$. However, when rough dissections are adequate, this refinement is not needed and the T_{ij}^R can be used as approximate bifurcation temperatures. In this approximation hard window functions are constructed using Eqs. (4.5)

D. Enhanced Metropolis Monte Carlo Integration

Macromolecules can pose a special problem—even after coarse-grained dissection the macrostate probability distributions at high temperatures are seldom well modeled by Gaussian kernels, even in torsion-angle space. For example, we examined arbitrarily selected macrostates of Met-enkephalin and found that $\chi(p_{B,\alpha}, \mu_\alpha^{WG})$, a useful goodness-of-fit measure, was typically small (on the scale 0–1) at very low temperatures (e.g., $\chi \approx 0.2$ at $T = 0.01$ kcal/mol) but large at moderate and high temperatures (e.g., $\chi \approx 0.9$ at $T = 1$ kcal/mol). This indicates that Monte Carlo integration, even with the MWG kernel μ_α^{WG} selected by Eqs. (4.3) and (4.4), is likely to be inefficient.

We overcome this problem by using Metropolis Monte Carlo (MMC) sampling [42] within each macrostate. In contrast with Monte Carlo integration, which uses a fixed kernel to select each sample point $\Omega^{(i)}$, in MMC a *trial* sample conformation $\Omega^{(i)'}$ is selected with a *transition distribution* centered about the previous accepted sample conformation $\Omega^{(i-1)}$. We accept $\Omega^{(i)'}$ with probability p_{MMC}, which is determined as in standard MMC procedure except that the potential is modified to account for the restriction of sampling to the macrostate region defined by w_α:

$$p_{MMC} = \begin{cases} e^{-\beta[V_\alpha(\Omega^{(i)'}) - V_\alpha(\Omega^{(i-1)})]}, & V_\alpha(\Omega^{(i)'}) - V_\alpha(\Omega^{(i-1)}) > 1 \\ 1, & V_\alpha(\Omega^{(i)'}) - V_\alpha(\Omega^{i-1)}) \leq 1 \end{cases}$$

where

$$V_\alpha \equiv V/\beta^{-1} \log(w_\alpha) \tag{4.8}$$

As the number of accepted sample points, N_{MMC}, becomes large, this procedure samples conformation space with probability proportional to $w_\alpha p_B$.

Therefore, macrostate thermodynamic averages can be approximated by

$$\langle f \rangle_\alpha \approx N_{\mathrm{MMC}}^{-1} \sum_{i}^{N_{\mathrm{MMC}}} f(\Omega^{(i)}) \tag{4.9}$$

The convergence rate hinges on the choice of transition distribution. Vanderbilt and Louie [43] have suggested using, in Cartesian coordinates, a Gaussian transition distribution with a fluctuation tensor proportional to $(\Lambda_\alpha^0)^2$. Replacing Gaussians with MWGs for torsion-angle coordinates, we use $\mu^{\mathrm{WG}}(0, c\Xi_\alpha^0; \Delta\Omega)$ as the transition distribution, where

$$\Delta\Omega \equiv \Omega^{(i)'} - \Omega^{(i-1)}$$

and c, the *relative jump length*, is a scalar selected to give an empirically optimal [18,19] acceptance ratio of ~ 0.3–0.5. During annealing, we use the fluctuation tensor computed at the previous temperature $(\Xi_\alpha^0)^2(\beta_{n-1})$ to estimate the current fluctuation tensor according to

$$(\Xi_\alpha^0)^2(\beta_n) \approx \frac{\beta_n}{\beta_{n-1}} (\Xi_\alpha^0)^2(\beta_{n-1})$$

(This estimate would be exact if the potential were quadratic.) The trial sample conformations are generated by

$$\Delta\Omega = cGU \tag{4.10}$$

where G is the Cholesky factor of $\beta_n/\beta_{n-1}(\Xi_\alpha^0)^2(\beta_{n-1})$ and U is vector of normally distributed random numbers with unit variance. The modified MMC procedure converged at least 10 times faster than Monte Carlo sampling in numerical tests of Met-enkephalin.

We tested whether sampling efficiency could be further enhanced by using "swapping" [44] and "jump-walking" [45], two improvements on the classical MMC method that combine sampling at two different temperatures to generate search trajectories that can more easily cross barriers.

In swapping, MMC sampling is performed concurrently at high $[T = (k_{\mathrm{B}}\beta_{\mathrm{hi}})^{-1}]$ and low $[T = (k_{\mathrm{B}}\beta_{\mathrm{lo}})^{-1}]$ temperatures. A swap is attempted every N_{Sw} steps in which conformations Ω_{hi} and Ω_{lo} from the high- and low-temperature trajectories are exchanged with probability p_{Sw}:

$$p_{\mathrm{Sw}} = \begin{cases} e^{-\beta_{\mathrm{lo}}[V_\alpha(\Omega_{\mathrm{hi}}) - V_\alpha(\Omega_{\mathrm{lo}})] - \beta_{\mathrm{hi}}(\Omega_{\mathrm{lo}}) - V_\alpha(\Omega_{\mathrm{hi}})]} & \\ \qquad \beta_{\mathrm{lo}}[V_\alpha(\Omega_{\mathrm{hi}}) - V_\alpha(\Omega_{\mathrm{lo}})] + \beta_{\mathrm{hi}}[V_\alpha(\Omega_{\mathrm{lo}}) - V_\alpha(\Omega_{\mathrm{hi}})] > 0 & \\ 1, \quad \beta_{\mathrm{lo}}[V_\alpha(\Omega_{\mathrm{hi}}) - V_\alpha(\Omega_{\mathrm{lo}})] + \beta_{\mathrm{hi}}[V_\alpha(\Omega_{\mathrm{lo}}) - V_\alpha(\Omega_{\mathrm{hi}})] \le 0 & \end{cases} \tag{4.11}$$

where V_α is defined in Eq. (4.8). As N_{MMC} becomes large, the high- and low-temperature accepted sample distributions approach $w_\alpha p_{\mathrm{B}}$ for their respective temperatures. N_{Sw} should be chosen to maximize the rate at which the low-temperature sampling trajectory can cross barriers. We have found in numerical experiments with Met-enkephalin ($N_{\mathrm{Sw}} = 100$) that swapping accelerates the convergence rate about two- to three-fold.

Jump-walking can be used when equilibrated high-temperature sample points are already available (as is the case during macrostate annealing except during the initialization step). At every $N_{\mathrm{J}}^{\mathrm{th}}$ trial, a high-temperature sample conformation Ω_{hi} is selected at random from the high-temperature sample set and used as a low-temperature trial sample point. It is accepted into the low-temperature trajectory according to Eq. (4.11) using the last accepted low-temperature sample conformation as Ω_{lo}. In numerical experiments with Met-enkephalin ($N_{\mathrm{J}} = 100$), jump-walking accelerated the convergence rate about 20- to 100-fold compared to standard MMC.

Based on these results, we used jump-walking to compute Met-enkephalin macrostate trajectory diagrams.

E. Computing Macrostate Entropy and Free Energy

Average macrostate properties such as average potential energy $U_\alpha \equiv \langle V \rangle_\alpha$ can be computed using Eq. (4.9). However, the macrostate conformational free-energy F_α and conformational entropy S_α cannot be computed as averages [46]. Instead, changes in these thermodynamic parameters with temperature are computed by summing finite differences along the macrostate trajectories. Between bifurcation temperatures, $S_\alpha(\beta)$ is a continuous function that satisfies

$$\frac{dS_\alpha(\beta)}{dT} = \frac{C_{v,\alpha}(\beta)}{T} \tag{4.12}$$

where $C_{v,\alpha}$ is the conformational contribution to the macrostate specific heat.

$$C_{v,\alpha} \equiv \frac{dU_\alpha}{dT} = k_{\mathrm{B}}\beta^2(\langle V^2 \rangle_\alpha - \langle V \rangle_\alpha^2)$$

This equation can be integrated (with linear interpolation for $C_{v,\alpha}$) to determine the change in entropy between two annealing temperatures. The conformational free energy can then be determined from [47]

$$F_\alpha = U_\alpha - TS_\alpha \tag{4.13}$$

A different procedure is needed to compute changes in S_α and F_α when a parent macrostate p bifurcates into children $c1$ and $c2$. Conservation of probability implies that

$$e^{-\beta F_p} = e^{-\beta F_{c1}} + e^{-\beta F_{c2}} \tag{4.14}$$

and the relative probabilities of the children, p_{c1}/p_{c2}, determine the free-energy difference according to

$$\frac{p_{c1}}{p_{c2}} = e^{-\beta(F_{c1} - F_{c2})} \tag{4.15}$$

Assuming that the bifurcation was detected using indicator r_{ij}, p_{c1}/p_{c2} can be calculated by integrating $p_{\mathrm{B},s}$ for $s < s(r_{ij}^t)$ and $s > s(r_{ij}^t)$ at T_{ij}^R (see Fig. 7). Equations (4.14) and (4.15) determine F_{c1} and F_{c2} relative to F_p at T_{ij}^R. S_{c1} and S_{c2} can then be calculated from Eq. (4.13).

This will determine all the S_α and F_α up to an arbitrary constant corresponding to h_R in Eq. (2.12).

F. Macromolecular Macrostate Trajectory Diagram

An extensive but partial probability macrostate trajectory diagram for Met-enkephalin was computed using the methods described above (see also Fig. 8). It includes most macrostates with $p_\alpha > 10^{-3}$ but not the very large number of low probability macrostates that are present at low temperature (corresponding to the estimated 10^{11} local minima of the Met-enkephalin potential-energy landscape [31]). The diagram includes much more detail than is required for most numerical analyses to facilitate its use as a model for computational explorations of trajectory diagram properties. It displays some properties that will be common to all trajectory diagrams. At very high temperatures (in this case, $T > 20$ kcal/mol) only one macrostate spans the entire conformation space. Since it is unique, $p_\alpha = 1$. At intermediate temperatures (0.4 kcal/mol $\lesssim T \lesssim$ 20 kcal/mol) the diagram becomes more complex as bifurcations distribute probability among an increasing number of macrostates. At still lower temperatures ($T \lesssim$ 0.4 kcal/mol), the diagram is simplified since macrostate G, which contains the global-energy minimum, gathers most of the probability while the probabilities of the other macrostates approach zero.

G. Branch Selection Strategies

At biological temperature ($\sim$0.6 kcal/mol) $p_G \approx 0.1$, and averages over G and a handful of other macrostates are adequate for computing thermodynamics properties. For simplicity, we focus on finding G, but the same

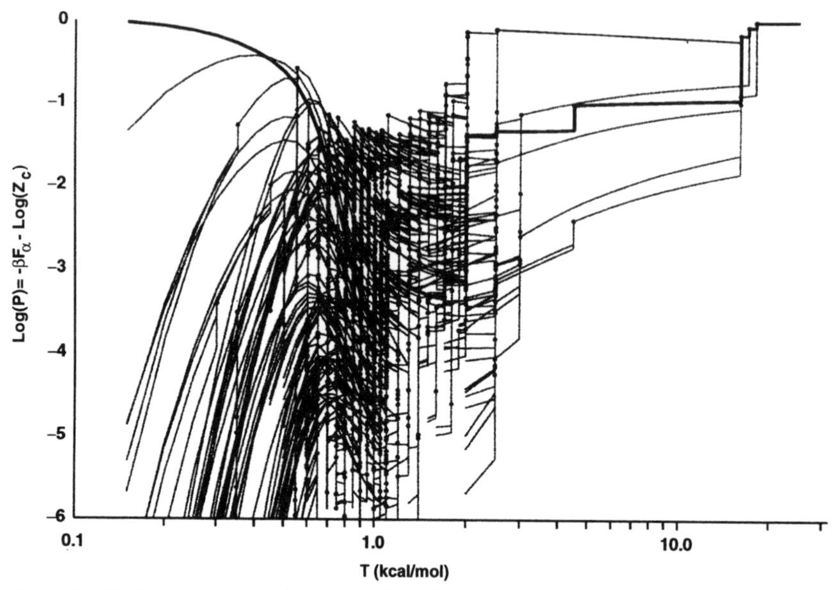

Figure 8. Trajectory diagram for Met-enkephalin plotting the logarithm of the macrostate probability against the logarithm of temperature for a selected set of trajectories. The thick black line is the ground-state branch. The thin black line marks the path selected by choosing the most probable branch at each bifurcation.

approach should find all states having significant probability at low temperature. This is accomplished by tracking the *ground-state branch* (indicated by the thick black line in Fig. 8) from the single high-temperature macrostate down to G at the low temperature of interest. As discussed in Section II.E, our goal is to find branch selection algorithms that enable us to prune the trajectory diagram as much as possible while retaining this branch. The simplest algorithm is a rule that specifies which child trajectories to follow at each bifurcation point. An obvious rule is to select the child having the highest probability. This generates the trajectory drawn with a thin black line that has high probability for $T > 2$ kcal/mol but ends up having low probability for $T < 0.6$ kcal/mol. Evidently this is not a good rule—we learn that for this system the macrostate of lowest free energy does not always include the global minimum. The rule could be generalized to include simultaneous tracking of multiple trajectories by comparing the probabilities of all the current macrostates and, after each bifurcation, keeping the N_{traj} with highest probability. But, the observation that the ground-state branch has $p_G < 10^{-3}$ at $T \approx 1$ kcal/mol indicates

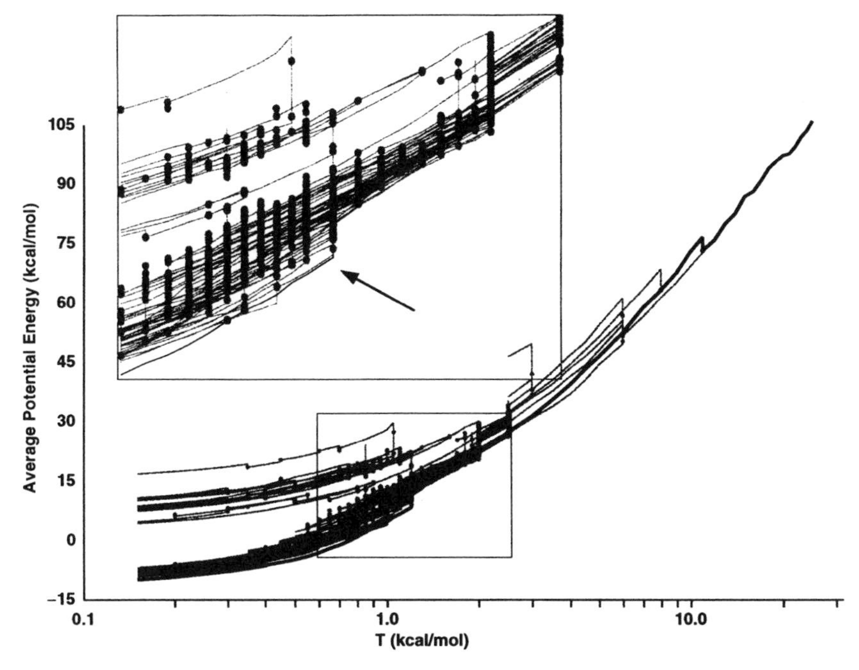

Figure 9. Mean energy trajectory diagram for Met-enkephalin. The thick black line is the ground-state branch.

that N_{traj} would have to be very large for success even in this relatively simple problem. It is likely that N_{traj} would have to be even larger for bigger problems.

However, an interesting property appears when the trajectory diagram is replotted to show the macrostate mean potential energies as a function of temperature (Fig. 9). Now the ground-state branch remains close to the bottom of the trajectory diagram at all temperatures. This indicates that the rule of keeping those macrostates that have the lowest U_α will be much more efficient for finding the global energy minimum of Met-enkephalin. Since the ground-state branch does not have the very lowest U_α at every temperature, it is still necessary to trace multiple trajectories. $N_{traj} = 20$ is adequate in this case.

We do not know if this property holds in general for peptides or proteins, nor how the required N_{traj} scales with the number of degrees of freedom. Thus we do not know if the mean-energy branch selection algorithm will be adequate for large problems. But, if it is not, other branch selection algorithms, based on other combinations of macrostate thermodynamic properties, may be useful.

V. CONCLUSION

Monte Carlo integrals over the Gibbs–Boltzmann distribution appear in many large-scale computational problems, particularly when analyzing macromolecules. We have described a new method for performing such integrals that recursively dissects conformation space into variable-size macrostate regions, thereby uncovering an intrinsic, hierarchical temperature-dependent dissection of conformation space that is induced by the potential-energy landscape. This includes the temperature-independent "inherent structure," which is based on local minima catchment regions [3,4] as the $T \rightarrow 0$ limit. The method has been applied to covalently bonded macromolecules by using a combination of internal and distance coordinates to reduce the difficult multidimensional problem to a collection of simple one-dimensional problems. The resulting macrostate trajectory diagrams provide "roadmaps" of the possible routes to the free-energy global minimum and other low-energy states. The same concepts can be used to compute relaxation rates, and macrostates provide a natural basis for a master equation description and for variational calculation of transition rates.

The method synergistically combines annealing and dissection to enhance numerical control and hence efficiency. For example, simulated annealing will converge to the global minimum as long as the cooling schedule is "slow enough" [48], but there is no easy way to determine what "slow" means during the process. Furthermore, slow convergence in just one region of conformation space can retard the overall cooling process. In contrast, the macrostate dissection provides well-defined convergence criteria for each region separately so that computational effort can be differentially allocated. Conversely, annealing enhances the efficiency with which the dissection can be constructed, since each effective energy barrier is detected and characterized at an appropriate branch recognition temperature. This temperature is low enough so that the activation energy can be computed using only a modestly accurate projected probability distribution yet high enough so that the relative probabilities of the children can be computed by efficient sampling with a single kernel.

Beyond their use in Monte Carlo integration, macrostate trajectory diagrams describe the dynamical relationships between macrostates. Since packets bifurcate when the thermal energy is comparable to the energy of the effective energy barrier, the bifurcation temperatures are roughly proportional to the effective activation energies which govern the inter-macrostate transitions. Thus, we expect that the macrostate trajectory diagram hierarchy will be similar to hierarchies generated by bottom–up methods that cluster local minima based on transition rates or saddle-point energies [10–13]. An important difference is that the top–down procedure opens the possibility of selecting only a computable fraction of the trajectory space by using appropriate predictive criteria.

On a broader level, the topology of the diagrams can be used to infer global characteristics of potential landscapes that control the dynamic relaxation of ensembles. For example, most speculations as to whether protein landscapes have "funneling" or "glassy" properties [49,50] have been based on computational studies performed on lattice models that attain simplicity at the expense of accuracy [51–53]. However, whereas these models may adequately account for important packing constraints [51] they are unlikely to accurately model the effective-energy barriers between macrostates that govern stochastic relaxation. The trajectory diagrams may provide more accurate insights since they are computed using continuous macromolecular potentials. For example, for a system having perfect funneling, the ground-state branch will be at the bottom of the mean potential-energy trajectory diagram at all temperatures. (Examples of this behavior are provided by the trajectory diagrams of six- to eight-atom Lennard-Jones microclusters [16].) Conversely, trajectory crossings can identify kinetic traps and complicated patterns of ensemble probability flow [23].

So far only proof-of-principle demonstrations have been performed, and much remains to be done. Accuracy must be improved by adapting the method for potentials that include solvation. Efficiency can be improved in a number of ways, including the following:

1. *Improved Branch Selection Algorithms.* The central problem in the hierarchical approach is to learn how to predict which high-temperature trajectories will lead to the dominating macrostates at low temperature. We have seen, for the case of Met-enkephalin, that the high-probability selection rule is poor whereas the low-mean-potential-energy rule is surprisingly good. Empirical analysis of the trajectory diagrams of a spectrum of problems is needed to test the generality of these observations and to develop better branch selection strategies.
2. *Smoothed Effective Potentials.* The Monte Carlo integrations at high and intermediate temperatures only need to be accurate enough to enable the branch selection algorithm to function. They can be more efficiently evaluated using smoothed effective potentials that can be

approximated by analytic forms [32,54,55] or lookup tables [14,15,56, 57]. When combined with macrostate annealing, this approach is comparable to renormalization group methods [58] in that it replaces a "flat" integration over the entire space with a hierarchically organized set of integrations over a sequence of varying-size scales.

3. *Adaptive Temperature Jump.* The size of a temperature jump is restricted only by the requirement that the initial conditions for solving the characteristic packet equations (provided by the higher-temperature characteristic packet) be within the attractor region of the lower temperature solution. This will be so until temperature is lowered below a branchpoint when one solution will trap the solver. As long as we have not jumped to a temperature too far below the branchpoint, we can find the other solution by looking for the missing probability mass that is present in the parent but not in the child [16]. An optimal strategy will have to balance the cost of taking unnecessarily short steps against the cost of having to backtrack to correct for an overly long steps that jump past multiple branchpoints.

Clearly, this is a work in progress, not yet a completely developed system.

ACKNOWLEDGMENTS

We thank Jason Gans and Matej Orešič for helpful discussions. This work was supported by grants from the National Institutes of Health (GM48874), the Air Force Office of Scientific Research (F49620), and a postdoctoral fellowship to A. U. from the National Science Foundation.

REFERENCES

1. D. Shalloway, *J. Chem. Phys.* **105**, 9986 (1996).
2. S. Glasstone, K. J. Laidler, and H. E. Eyring, *The Theory of Rate Processes*, McGraw-Hill, New York, 1941.
3. F. H. Stillinger and T. A. Weber, *Phys. Rev. A* **25**, 978 (1982).
4. F. H. Stillinger and T. A. Weber, *Science* **225**, 983 (1984).
5. M. R. Hoare, *Adv. Chem. Phys.* **40**, 49 (1979).
6. R. Elber and M. Karplus, *Science* **235**, 318 (1987).
7. H. Frauenfelder, S. G. Sligar, and P. G. Wolynes, *Science* **254**, 1598 (1991).
8. T. Noguti and N. Go, *Proteins* **5**, 97 (1989).
9. J. M. Troyer and F. E. Cohen, *Proteins* **23**, 97 (1995).
10. R. Czerminski and R. Elber, *Proc. Natl. Acad. Sci. USA* **86**, 6983 (1989).
11. R. Czerminski and R. Elber, *J. Chem. Phys.* **92**, 5580 (1990).
12. R. E. Kunz and R. S. Berry, *J. Chem. Phys.* **103**, 1904 (1995).
13. O. M. Becker and M. Karplus, *J. Chem. Phys.* **106**, 1495 (1997).

14. D. Shalloway, "Packet Annealing: A Deterministic Method for Global Minimization. Application to Molecular Conformation," in *Recent Advances in Global Optimization*, C. A. Floudas and P. M. Pardalos, eds., Princeton Series in Computer Science, Princeton Univ. Press, Princeton, NJ, 1996, pp. 433–477.
15. D. Shalloway, *J. Global Optim.* **2**, 281 (1992).
16. M. Orešič and D. Shalloway, *J. Chem. Phys.* **101**, 9844 (1994).
17. B. W. Church, M. Orešič, and D. Shalloway, "Tracking Metastable States to Free-Energy Global Minima," in *Global Minimization of Nonconvex Energy Functions: Molecular Conformation and Protein Folding: DIMACS Workshop*, P. Pardalos, D. Shalloway, and G. Xue, eds., Vol. 23 of *DIMACS Series in Discrete Mathematics and Theoretical Computer Science*, American Mathematical Society, Providence, RI, 1996, pp. 41–64.
18. M. H. Kalos and P. A. Whitlock, *Monte Carlo Methods*, Vol. I: *Basics*, Wiley, New York, 1986.
19. M. P. Allen and D. J. Tildesley, *Computer Simulations of Liquids*, Oxford Univ. Press, Oxford, 1987.
20. Vanderbilt and Louie (Ref. 43) discuss the use of Gaussian kernels determined by Eq. (2.5) in the closely related situation of Metropolis Monte Carlo–simulated annealing.
21. G. J. Moro, *J. Chem. Phys.* **103**, 7514 (1995).
22. R. Beran, *Ann. Stat.* **5** 445 (1977). (The Hellinger distance satisfies the triangle inequality and is invariant under changes of coordinate.)
23. D. Shalloway, "Variable-Scale Coarse-Graining in Macromolecular Global Optimization," in *Proceedings of IMA Conference on Large Scale Optimization*, L. Biegler, T. Coleman, A. Conn, and F. Santosa, eds., Springer-Verlag, New York, pp. 135–161.
24. P. Hänggi, P. Talkner, and M. Borkovec, *Rev. Mod. Phys.* **62**, 251 (1990).
25. A. Ulitsky and D. Shalloway, manuscript in preparation.
26. B. Widom, *J. Chem. Phys.* **55**, 44 (1971).
27. J. T. Bartis and B. Widom, *J. Chem. Phys.* **60**, 3474 (1974).
28. H. Grubmüller, *Phys. Rev. E* **52**, 2893 (1995).
29. R. Olender and R. Elber, *J. Chem Phys.* **105**, 9299 (1997).
30. J. Hughes,, T. W. Smith, H. W. Kosterlitz, L. A. Fothergill, B. A. Morgan, and H. R. Morris, *Nature* **258**, 577 (1975).
31. Z. Li and H. A. Scheraga, *Proc. Natl. Acad. Sci. USA* **84**, 6611 (1987).
32. J. Kostrowicki and H. A. Scheraga, *J. Phys. Chem.* **96**, 7442 (1992).
33. U. Hansmann and Y. Okamoto, *J. Comp. Chem.* **14**, 1333 (1993).
34. C. D. Maranas, I. P. Androulakas, and C. A. Floudas, "A Deterministic Global Optimization Approach for the Protein Folding Problem," in *Global Minimization of Nonconvex Energy Functions: Molecular Conformation and Protein Folding: DIMACS Workshop*, P. Pardalos, D. Shalloway, and G. Xue, eds., Vol. 23 of *DIMACS Series in Discrete Mathematics and Theoretical Computer Science*, American Mathematical Society, Providence, RI, 1996, pp. 133–150.
35. G. Némethy, K. D. Gibson, K. A. Palmer, C. N. Yoon, G. Paterlini, A. Zagari, S. Rumsey, and H. A. Scheraga, *J. Phys. Chem.* **96**, 6472 (1992).
36. D. R. Herschbach, H. S. Johnston, and D. Rapp, *J. Chem. Phys.* **31**, 1652 (1959).
37. N. Go and H. A. Scheraga, *Macromolecules* **9**, 535 (1976).
38. B. W. Church and D. Shalloway, *Polymer* **37**, 1805 (1996).

39. L. K. Nicholson, T. Yamazaki, D. A. Torchia, S. Grzesiek, A. Bax, J. S. Stahl, J. D. Kaufman, P. T. Wingfield, P. Y. S. Lam, P. K. Jadhav, C. N. Hodge, P. J. Domaille, and C.-H. Chang, *Nat. Struct. Biol.* **2**, 274 (1995).

40. For large N_a, the number of members in such subsets probably needs to grow only like $N_a \log N_a$. For example, for a linear chain the set of pairs $\{(i, i + 2^n): i = 1, \ldots, N_a; n = 1, \ldots, \mathrm{Floor}[\log_2(N_a - i)]\}$ should be an adequate set of indicator distances.

41. D. Shalloway, unpublished.

42. N. Metropolis, A. W. Metropolis, M. Rosenbluth, A. H. Teller, and E. Teller, *J. Chem. Phys.* **21**, 1087 (1953).

43. D. Vanderbilt and S. G. Louie, *J. Comp. Phys.* **56**, 259 (1984).

44. C. J. Geyer, "Markov Chain Monte Carlo Maximum Likelihood," in *Computing Science and Statistics: Proceedings of the 23rd Symposium on the Interface*, E. Keramidas, ed., American Statistical Association, 1991, pp. 156–163.

45. D. D. Frantz, D. L. Freeman, and J. D. Doll, *J. Chem. Phys.* **93**, 2769 (1990).

46. Formally, these can be expressed in terms of $\langle \exp(\beta V) \rangle_\alpha$, but this converges so poorly as to be a computationally useless expression.

47. This relationship holds even though only the conformational contributions over the macrostate regions (and not the momentum contributions) are being considered. U_α would be replaced by a conformational enthalpy if the window functions were allowed to vary with temperature.

48. S. Kirkpatrick, J. C. D. Gelatt, and M. P. Vecchi, *Science* **220**, 671 (1983).

49. P. G. Wolynes, J. N. Onunchi, and D. Thirumalai, *Science* **267**, 1619 (1995).

50. K. S. Dill, *Curr. Opin. Struct. Biol.* **3**, 99 (1993).

51. K. A. Dill, S. Bromberg, K. Yue, K. M. Fiebig, D. P. Yee, P. D. Thomas, and H. S. Chan, *Prot. Sci.* **4**, 561 (1995).

52. C. J. Camacho and D. Thirumalai, *Proc. Natl. Acad. Sci.* **90**, 6369 (1993).

53. E. I. Shakhnovich, *Curr. Opin. Struc. Biol.* **7**, 29 (1997).

54. J. Kostrowicki, L. Piela, B. J. Cherayil, and H. A. Scheraga, *J. Phys. Chem.* **95**, 4113 (1991).

55. J. Ma and J. E. Straub, *J. Chem. Phys.* **101**, 533 (1994).

56. T. Coleman, D. Shalloway, and Z. Wu, *Comp. Optim. Appl.* **2**, 145 (1993).

57. T. Coleman, D. Shalloway, and Z. Wu, *J. Global Optim.* **4**, 171 (1994).

58. K. Wilson, *Rev. Mod. Phys.* **47**, 773 (1975).

SIMULATED ANNEALING—OPTIMAL HISTOGRAM METHODS

DAVID M. FERGUSON

Department of Medicinal Chemistry and Minnesota Supercomputer Institute, University of Minnesota, Minneapolis, MN 55455

DAVID G. GARRETT

Lockheed Martin TDS Eagan, St. Paul, MN 55164

CONTENTS

Advances in Chemical Physics, Volume 105, Monte Carlo Methods in Chemical Physics, edited by David M. Ferguson, J. Ilja Siepmann, and Donald G. Truhlar. Series Editors I. Prigogine and Stuart A. Rice.
ISBN 0-471-19630-4

I. INTRODUCTION

Monte Carlo-simulated annealing has been applied to a wide variety of problems in chemistry. Although the methodology has gained fairly widespread popularity for use in global optimization studies, annealing protocols are also quite valuable for examining the thermodynamics of chemical processes as well. In systems that undergo phase transitions, for instance, simulated annealing provides a route to track key physical properties of the system during a cooling or heating process. The most familiar may be the specific heat, which is often used to gauge transition temperatures in molecular systems; pinpointing important structural events during the process. As with many simulation techniques, however, the utility of simulated annealing for these types of studies is greatly limited by finite system size effects and sampling errors that result in slow convergence of thermodynamic averages. While it is sometimes possible to scale the system sizes to explore problems associated with the former, the latter issue can be more difficult to address since sampling times are inherently limited by computing power. In many cases, simulation length are simply doubled or tripled to determine the effect of the time length on the property of interest. This is not only impractical in some cases, but may also be misleading in others since the overall relaxation time of the property is typically not known. This is particularly true for properties that are derived from fluctuations in the system, such as specific heat, that are notoriously slow to converge.

The reliability of thermodynamic properties calculated during a typical simulated annealing process is therefore difficult to assess. Although this has little impact on optimization studies, it certainly limits the overall potential of annealing protocols for many applications. One in particular is protein folding. As in the majority of molecular conformations problems, the main goal is structure prediction. However, proteins are significantly more complex then their small molecule counterparts and display physical properties more similar to those of heterogeneous polymers systems. Globular proteins are known to undergo a characteristic phase transition during the annealing (cooling) or folding process [1–3]. Although it is generally thought that this event corresponds to a hydrophobic collapse of the aliphatic side chains in the system, the exact details of the folding mechanism remain obscure.

The ideal Monte Carlo experiment should probe the physical and structural properties of the protein during the folding process, while simultaneously converging to the global minimum structure or native state. Although a great deal of effort has already been expended on the development of computational models for this purpose, most have focused on energy functions or structural analysis. Very few have examined the thermodynamics of the proposed models in detail. This is due, in part, to the limitations of calculations of this type as mentioned above. In our own work, however, we have begun to address this problem through an extension of the optimal histogram methodology first reported by Ferrenberg and Swendsen [4–8].

Although single histogram methods have been shown to useful studying critical-point behavior when the histogram is collect at or near the transitions temperature [4], these methods require some "predetermined" knowledge of the transition temperature. Various efforts [9,10,12] have been made to extend the temperature ranges by using multistage or umbrella sampling techniques. The advantage of the method presented in this work lies in the use of multiple temperature measurements in estimating the density of states. This, in turn, allows key thermodynamic properties of the system (i.e., specific heat) to be reliably determined from a single Monte Carlo–simulated annealing simulation, which has significant benefits as system sizes grow in size and complexity.

The approach developed applies variance optimizations (minimization) techniques to calculate a "best estimate" of the density of states for a protein system from an annealing simulation. From this calculation, it is then possible to obtain more reliable estimates of all thermodynamic parameters during the folding process, including specific heat. Here, we fully derive, implement, and apply the optimal histogram equations to examine the physical behavior of a simple protein model. In addition, we introduce structure-based overlap functions to probe the nature of the predicted phase transitions. Care is taken throughout the chapter to outline the methodology to enable fairly straightforward adaptations to other chemical systems.

II. OPTIMAL HISTOGRAM METHODOLOGY

A. Histogram Equations

The basic methodology developed here is a fairly straightforward adaptation of standard simulated annealing protocols used to solve a wide variety of problems in chemistry [13]. As in most applications, the system is slowly cooled over many discrete temperatures, T_i, during which various equilibrium properties are sampled. It is common practice to then plot the results from each individual simulation, i, as a function of T to identify important chemical events as the system cools. The average energy, for example, may show a large decrease over a small range of temperatures at potential phase transitions. Using Monte Carlo–Metropolis sampling, this property is easily estimated at each T_i sampled from the simulation by summation.

$$\langle E \rangle_i = \frac{1}{N} \sum_{j=1}^{N} E_j \tag{2.1}$$

where E_j is the configurational energy and N is the total number of observed steps. Unfortunately, the uncertainties in average properties measured in this way are typically large (i.e., they are slow to converge). If instead, we rewrite the right-hand side (RHS) of Eq. (2.1) in terms of the configurational integrals over the density of states of the system, an alternative scheme can be derived to calculated the average energy. For a canonical ensemble, the average is given by

$$\langle E \rangle_i = \frac{1}{Z(\mathcal{N}, V, T_i)} \int E\, e^{-\beta_i E} \Omega(E)\, dE \tag{2.2}$$

where β_i is the inverse temperature, $1/kT_i$, $\Omega(E)$ is the density of states, $\mathcal{N}$ is the number of particles, and Z is the configurational partition function defined by

$$Z(\mathcal{N}, V, T) = \int e^{-\beta E} \Omega(E)\, dE \tag{2.3}$$

By discretizing the energy of the system into a histogram, $H(E_j)$, each simulation at T_i can be used to generate an independent estimate of $\Omega(E)$. As we shall show, this set of histograms can then be combined over all T_i values sampled in a manner that minimizes the variance in $\Omega(E)$, hence optimizing the estimates of $\langle E \rangle$ from the RHS of Eq. (2.2).

Although this brief introduction captures the general spirit of the approach to follow, a much more detailed description is required to fully appreciate the methodology as well as its implementation. We start by considering a complete annealing run over a set of temperatures, T_i. For each T_i a histogram of energies is generated with bin widths of ΔE_j. In this case, however, j is taken over the number of discrete energies bins, not over N of Eq. (2.1). The probability of finding the system in energy bin ΔE_j centered at energy E_j is given by

$$P_{\beta_i}(E_j) = \frac{e^{-\beta_i E_j} \Omega(E_j)(\Delta E_j)}{Z(\beta_i)} \tag{2.4}$$

The expected value recorded in the histograms bin corresponding to energy E_j is then

$$\overline{H_i(E_j)} = N_i P_{\beta_i}(E_j) \tag{2.5}$$

where N_i is the number of samples on the ith annealing sweep (i.e. at temperature T_i). By using Eqs. (2.4), and (2.5), we can estimate the density of states up to a proportionality constant, e^{f_i}, by

$$W^{(i)}(E_j) \equiv \Omega(E_j)\,\Delta E_j = \frac{H_i(E_j)}{N_i}\,e^{\beta_i E_j - f_i}. \tag{2.6}$$

Here the superscript (i) denotes the fact that the estimate of $W(E)$ is based on the histogram observed at T_i. Clearly, $e^{-f_i} = \sum_j W(E_j)\,e^{-\beta_i E_j}$, and f_i is related to the free energy at T_i by $f_i = \beta_i F(\beta_i)$.

The goal of our calculation is to find an improved estimate of $W(E)$ utilizing all the histograms collected. This achieved by observing that each histogram provides an independent estimator of $W(E)$. By averaging the estimates of $W(E)$ weighted by the inverse of the variance of the estimators we can produce a maximum likelihood or "optimal" estimator for $W(E)$. As can be seen from Eq. (2.6), the variance in $W(E)$ is proportional to the uncertainty in $H_i(E)$, which we will denote as $\delta^2 H_i(E)$. Notice that $\delta^2 H_i(E)$ depends not only on the temperature T_i but also on the energy bin E_j. An estimator for $\delta^2 H_i(E)$ is afforded by the observation that the distribution of $H_i(E)$ is Poison [10,11]. That is, if we were to perform multiple MC sweeps at a fixed temperature, the observed distribution of counts in bin ΔE_j [i.e., $H_i(E_j)$] would have a mean of $\widehat{H_i(E_j)}$ and a variance of

$$\delta^2 H_i(E_j) = g_i\,\overline{H_i(E_j)} \tag{2.7}$$

The extra factor of g_i depends on the correlation of successive MC configurations. We expect $\widehat{H_i(E_j)} \approx \overline{H_i(E_j)}$. The "hat" notation indicates that $\widehat{H_i(E_j)}$ is sample statistic [vs. the population statistic $\overline{H_i(E_j)}$]. If the successive configurations are not correlated, then $g_i = 1$. However, this is rarely the case. When the successive MC configurations are correlated the correct value for g_i is given by

$$g_i = 1 + 2\tau_i \tag{2.8}$$

where τ_i is the observed autocorrelation time in energy of successive configurations. In other words, $\widehat{H_i(E_j)}$ is a biased estimator for $\delta^2 H_i(E_j)$, independent of the number of MC steps used to obtain H_i. The factor of g_i in Eq. (2.7) removes this bias. We will estimate $\widehat{H_i(E_j)}$ by $H_i(E_j)$, where $H_i(E_j)$ is the value observed on a single MC sweep.

By combining Eqs. (2.4), (2.5), and (2.7) we can write the variance in $H_i(E)$ as

$$\delta^2 H_i(E) = \frac{N_i}{1 + 2\tau_i} W(E)\, e^{-\beta_i E + f_i} \tag{2.9}$$

Using Eq. (2.6), we can now obtain the variance in $W^{(i)}(E)$ as follows:

$$\begin{aligned} \delta^2 W^{(i)}(E) &= (e^{\beta_i E - f_i} N_i^{-1})^2 \delta^2 H_i(E) \\ &= e^{\beta_i E - f_i} \frac{(1 + 2\tau_i)}{N_i} W(E) \end{aligned} \tag{2.10}$$

By combining the estimates of $W^{(i)}(E)$ obtained from the several $H_i(E)$, we see that the maximum likelihood estimate of $W(E)$ can be formed:

$$W(E) = \frac{\sum_i \dfrac{W^{(i)}(E)}{\delta^2 W^{(i)}(E)}}{\sum_i \dfrac{1}{\delta^2 W^{(i)}(E)}} \tag{2.11}$$

Substituting for $W^{(i)}(E)$ and $\delta^2 W^{(i)}(E)$ from Eqs. (2.6) and (2.10), respectively, we obtain

$$W(E) = \frac{\sum_i \dfrac{H_i(E)}{1 + 2\tau_i}}{\sum_k \dfrac{N_i}{1 + 2\tau_i}} e^{-\beta_i E + f_i} \tag{2.12}$$

At this point the goal has in effect been reached. Equation (2.12) combined with $e^{-f_i} = \Sigma_j\, W(E_j)\, e^{-\beta_i E_j}$ can be solved self-consistently subject to the constraint that $\Sigma_j\, W(E_j) = 1$. Numerically, this requires an initial guess for f_i, and then $W(E)$ is computed and normalized. The resulting $W(E)$ is used to compute a new value of f_i and the procedure repeats until some convergence criteria is met. The serious draw back to this procedure is the exceedingly slow convergence of $W(E)$. In our past experiences, this calculation was found to require as much or even more computer time than was required to collect the original histograms [7,8]. The histogram equations, however, can be recast in a more computationally efficient form. Following

Ferrenberg and Swendsen, we compute the probability that the system is in energy bin ΔE_j, denoted by $P_\beta(E_j)$. Given that $\Omega(E)$, and hence $W(E)$, are independent of temperature, $P_\beta(E_j)$ can be expressed as

$$P_\beta^{(i)}(E_j) = e^{-(\beta - \beta_i)E_j} \frac{H_i(E_j)}{\sum_j H_i(E_j)\, e^{\beta_i E_j}} \tag{2.13}$$

The estimate of $P_\beta^{(i)}(E)$ contains all of the information about the system found in $\Omega(E)$ and $Z(\beta)$ in sense that the temperature dependence of the average value of any function of E can be obtained from

$$\langle A \rangle = \sum_j A(E_j) P_\beta(E_j) \tag{2.14}$$

For computational convenience we change the normalization by letting

$$P(E, \beta) \equiv e^{-f} P_\beta(E) = W(E)\, e^{-\beta E} \tag{2.15}$$

Substituting for $W(E)$, we have the essential equations of the method:

$$P(E_j, \beta) = \frac{d(E_j)\, e^{-\beta E_j}}{\sum_i C_i(E_j)\, e^{f_i}} \tag{2.16}$$

$$e^{-f_i} = \sum_j P(E_j, \beta_i) \tag{2.17}$$

where

$$d(E_j) \equiv \sum_i \frac{H_i(E_j)}{1 + 2\tau_i}$$

and

$$C_i(E) \equiv \frac{N_i}{1 + 2\tau_i}\, e^{-\beta_i E}$$

Here again these equations must be solved self-consistently. Notice also that Eqs. (2.16) and (2.17) determine f_i only to within an additive constant. This is overcome by setting $f_0 = 0$ (where T_0 is the maximum temperature for which a histogram has been collected). In fact, Eq. (2.16) does not converge significantly faster that the direct computation of $W(E)$. However, this definition does lead a numerically stable set of Newton–Raphson equations.

B. Newton–Raphson Formulation

The Newton–Raphson approach to solving Eqs. (2.16) and (2.17) begins by defining

$$\lambda_i(f_i, f_2, \ldots, f_R) = f_i + \ln \sum_j P(E_j, \beta_i)$$
$$= f_i + \ln \sum_j \frac{d(E_j)\, e^{\beta_i E_j}}{\sum_k C_k(E_j)\, e^{f_k}} \tag{2.18}$$

where R is the number of discrete T_i values used to collect the histograms. The vector $\vec{f} = (f_0, f_1, \ldots, f_{R-1})$ such that $\lambda_i(\vec{f}) = 0\ \forall i$ is the solution that represents the best estimate of f_i, and hence $P(E, \beta)$. That is, if Eqs. (2.16) and (2.17) are satisfied by $\vec{f}$, then $\vec{\lambda}(\vec{f}) = 0$. The Newton–Raphson procedure is carried out by first supplying an initial guess for $\vec{f}^{(0)}$, and then iteratively computing

$$\vec{f}^{(k+1)} = \vec{f}^{(k)} + \delta\vec{f}^{(k)} \tag{2.19}$$

where $\delta\vec{f}^{(k)}$ is the solution to the set of linear equations

$$\left[\frac{\partial\lambda_i(\vec{f} = \vec{f}^{(k)})}{\partial f_j}\right] \delta\vec{f}^{(k)} = -\vec{\lambda}(\vec{f} = \vec{f}^{(k)}) \tag{2.20}$$

Since Eqs. (2.16) and (2.17) determine f_i only to within an additive constant, the same is true of $\vec{\lambda}(\vec{f}) = 0$. This can be overcome by modifying the equation for $\lambda_0(\vec{f})$ to

$$\lambda_0(\vec{f}) = 0 + \ln \sum_j P(E_j, \beta_i) \tag{2.21}$$

This effectively "pins" the value of f_0 to 0.

A simple and reasonably effective convergence criterion is provided by $\varepsilon_k \equiv \max|\vec{\lambda}(\vec{f}^{(k)})|$. Iteration of Eq. (2.19) is continued until ε_k is less than some predetermined value. The rank of $[\partial V_i/\partial f_j]$ depends on the number of temperatures, R. Generally, 30–50 temperatures values are sufficient, so the system in Eq. (2.20) is easily solved using standard methods. The term $[\partial V_i/\partial f_i]$ is obtained in terms of $H_i(E_j)$ by differentiating Eq. (2.18). Computationally this term requires little additional processing as most of the summations in $[\partial V_i/\partial f_j]$ are also required to compute $\vec{\lambda}(\vec{f})$.

Finding the roots of a system of nonlinear simultaneous equations like those in Eq. (2.18) can be problematic. No solution is guaranteed, and convergence and numeric stability can be dependent on the initial solution guess as well as other factors. Experience with Eq. (2.18) indicates that a reasonable initial guess for $\vec{f}$ is provided by setting $f_i = 0 \ \forall i$ on the RHS of Eq. (2.17) to obtain

$$f_i^{(0)} = -\ln\left(\frac{\sum_j d(E_j)\, e^{-\beta E_j}}{\sum_j C_i(E)}\right) \tag{2.22}$$

Using the preceding initial guess, we find that the Newton–Raphson equations generally converge within less than 20 iterations. This means that given about 40 histograms, $P(\beta, E)$ can be found with in a few minutes of computation time on a moderate workstation. The only remaining numerical difficulties that can arise are due to the magnitude of some terms in the various summations that are required. In particular, $\partial V_i/\partial f_j$ has terms of the form $C_i(E_j)\, d(E_j)\, e^{-\beta_i E_j}$, which can lead to under or overflows (i.e., numbers with magnitudes beyond the capability of the computer hardware). This generally arises only at low temperatures, where $-\beta E$ can be large and positive. For protein folding, and other chemically interesting problems this typically occurs far below the phase transition temperature(s) and is easily avoided.

III. PROTEIN MODEL

Although there are a wide variety of models for simulating protein folding, we have chosen a fairly simple force field and tethered-bead model to demonstrate the optimal histogram methodology. The basic model was first described by Honeycutt and Thirumalai to model the folded states of β-barrel structures [14,15] and has since been re-parameterized by us to model α-helical-type structures as well [16]. The specific sequence studied here

(PHB)–(PHB)–(PHL)–(PHL)–(PHB)–(PHB)–(PHL)–(PHL)–(PHB)–
(NEU)–(NEU)–(NEU)–(PHL)–(PHB)–(PHB)–(PHL)–(PHL)–
(PHB)–(PHB)–(PHL)–(PHL)–(PHB)

is designed to fold to a helix-turn-helix motif with a well-defined hydrophobic core and a variable loop section. The three residue types, PHB for hydrophobic, PHL for hydrophilic, and NEU for neutral, are parameterized to represent the dominant forces on peptide structures. This is

TABLE I
Force Field Parameters of Eq. (3.1)[a]

Bonds	$K_l/2$	l_0	
X–X[b]	100.0	1.0	
Angles	$K_\theta/2$	θ_0	
X–X–X	20.0	91.0	
NEU–NEU–X	10.0	104.0	
Nonbonds	A	B	
PHB–PHB	4.0000	4.0000	
PHB–PHL	2.6666	−2.6666	
PHL–PHL	2.6666	−2.6666	
PHB–NEU	4.0000	0.0000	
PHL–NEU	4.0000	0.0000	
NEU–NEU	4.0000	0.0000	
Torsion	$V_n/2$	γ	n
X–PHB–PHB–X	1.0	0	3
	1.2	240	1
X–PHL–PHL–X	1.0	0	3
	1.2	240	1
X–PHB–PHL–X	1.0	0	3
	1.2	240	1
X–NEU–PHB–X	0.2	0	3
X–NEU–PHL–X	0.2	0	3
X–NEU–NEU–X	0.2	0	3

[a] See Ref. 16 for a complete description of the force field. Energy units are in kilocalories per mole.
[b] X stands for any bead type.

accomplished using a fairly standard energy function borrowed from popular molecular mechanics programs:

$$E_{\text{tot}} = \sum_{i,j} K_l \Delta l_{i,j}^2 + \sum_{i,j,k} K_\theta \Delta\theta_{i,j,k}^2 + \sum_{i,j,k,l} \sum_n \frac{V_n}{2} [1 + \cos(n\phi_{i,j,k,l} + \gamma)] + \sum_{i,j} \frac{A}{r_{i,j}^{12}} - \frac{B}{r_{i,j}^6}, \tag{3.1}$$

Although the details of parameterization are given elsewhere [7,16], for completeness, we have listed the force constants, coefficients, and geometric parameters for this expression in Table I. In past studies, this protein model has been shown to produce the desired helix-turn-helix folded state as the global energy minimum [16]. In addition, a great deal of information has been collected about the low-energy states of this model, which should help simplify the analysis. More examples and descriptions of models of this type can be found in the seventh chapter in this volume (by Skolnick and Kolinski).

IV. MONTE CARLO SAMPLING

A. Simulated Annealing

Although the energy histograms can be generated using a number of techniques, we have implemented the methodology as part of a simulated annealing protocol based on Metropolis sampling. This scheme generates acceptances with probability

$$P(\Delta\mathscr{E}) = \begin{cases} e^{-\beta\,\Delta\mathscr{E}} & \Delta\mathscr{E} > 0 \\ 1 & \Delta\mathscr{E} \leq 0 \end{cases} \tag{4.1}$$

where $\Delta\mathscr{E}$ is the change in configurational energy produced by the trial move and β is the inverse temperature factor, $1/kT$. The general approach begins with a randomly generated configuration annealed at infinite temperature (i.e., $\beta_0 = 0$). After equilibrium has been reached, the energy histogram is collected. The temperature is then set to some large value (sufficiently greater than the phase-transition temperature) and the system is annealed again starting with the configuration observed with the lowest energy (often referred to as the best-so-far configuration). Successive sweeps are performed at $T_i = \gamma T_{i-1}$ for $i = 1, \ldots, R$, where γ is a scale factor less than 1.0. This schedule compensates to some degree for the expected slow down in thermodynamic processes at lower temperatures. Here, we use $\gamma = 0.8$ and initialize each annealing sweep with the last accepted configuration. The schedule is further modified as the system approaches the phase transition. Small linear steps are taken to concentrate the sampling around the phase-transition temperature where relaxation times are longest (as discussed in Section IV.D). The linearization can be started at any point in the cooling schedule, but should begin far enough above the phase transition temperature to ensure no physical events during the annealing process are missed. The overall schedule is given by

$$T_i = \begin{cases} \gamma T_{i-1} & \text{when } T_{i-1} > T_{lin} \\ T_{i-1} - \Delta T & \text{otherwise} \end{cases} \tag{4.2}$$

where T_{lin} is the temperature at which linearization begins.

B. Equilibration Criterion

To ensure the system has reached equilibrium before the histograms are collected, a heuristic criterion has been developed to estimate the equilibration times after each temperature change. Prior to collecting the histogram data, several preliminary annealing runs are made. At each temperature step the mean energy in a moving window is computed. In the "demonstration" problem presented below it was observed that after 4000–6000 acceptances, the change in mean energy between windows was approximately zero and much smaller than the variance in the energy. This was nearly independent of the details of the force field and temperature. Thus, we established the criteria of sampling a minimum of 10,000 acceptances to ensure equilibrium has been reached at each T_i.

C. Move Scheme

As in most, if not all, Monte Carlo applications, the equilibration time also depends on the efficiency of the move scheme. While move schemes can be quite complex with a variety of goals in mind, we have chosen a fairly straightforward set of move types that approximate the structural changes commonly associated with proteins. These types include bond length and angle changes, torsional rotations, rotations of subsections with in the protein model (i.e. crankshaft motion), complimentary bending at two beads (i.e., kinking), and compression or elongation within a subsection of the protein model. Despite their simplicity, these move types effectively expand the conformational space around each configuration that can be reached in one move, which results in reduced equilibration times and increased sampling efficiencies.

D. Autocorrelation Time

The sampling efficiency is also related to the autocorrelation or relaxation time of the system, τ. For each temperature, τ_i is the characteristic decay time of the energy autocorrelation function for the system once equilibrium has been reached. The autocorrelation function, $a(t)$, can be estimated from Eq. (4.3) as the annealing sweep progresses:

$$a(t) \equiv \frac{\langle E_{t0} E_{t0+t} \rangle - \langle E_{t0} \rangle^2}{\langle E_{t0} E_{t0} \rangle - \langle E_{t0} \rangle^2} = C\, e^{-t/\tau_i} \tag{4.3}$$

The value of τ_i is then determined by fitting

$$\ln a(t) = -\frac{1}{\tau_i} t + \ln C \tag{4.4}$$

in the linear regime.

The autocorrelation time in question here is not easily related to a physically observable quantity. Time τ_i depends on not only the potential function E_{tot} but also on the simulated annealing move scheme used that defines the neighborhood in configuration space that is reachable in a single move. The "size" of the neighborhood and the shape of the energy surface determines the statistical properties of the energy sequence observed. Although τ_i should be related to the relaxation time for dynamics of the protein, the overall scaling of τ_i is unknown, making direct comparisons to physical measurements difficult and unreliable.

V. STRUCTURE ANALYSIS

Analyzing the configurations visited above and below the phase transition is key to understanding the nature of phase transitions in molecular systems. In protein folding, such understanding is central to the debate surrounding the existence and nature of the *folding pathways*. Several structural measures have been used to facilitate this type of analysis, including density, RMS dihedral deviations, and hydrophobic contacts. Although these measures have proven very useful [2,14,17], taken individually or collectively, they do not completely characterize the structure of the system in the sense that many unrelated conformations have the same measure (the same density, dihedral deviation, etc.).

To improve on this situation, we have developed an *overlap function* applicable to continuous space protein models. This overlap function is motivated by the order parameters found in spin-glass theory and provides a more powerful method of determining the behavior of the system in configuration space. Overlap functions arise naturally in spin-glass applications and are a measure of the distance between two states of the system in configuration space [18,19]. These functions are order parameters for glass-like phase transitions. For an on-lattice protein model, the order parameter is given by the number of beads in each chain that occupy the same lattice position [20,21]. Beads in each chain that occupy the same lattice positions have, in effect, the same *local* conformation, thus the overlap function can be thought of as a measure of the degree of similarity between the conformations.

In a continuous model, the conformation is primarily determined by the hydrophobic contact pattern and the dihedral angles of the molecule. We ignore the bond-angle and bond-length effects as these are nearly fixed with respect the torsion angles and contacts. To apply the overlap function methodology to the continuous case, we must discretize the continuous space and capture information about both the torsion angles and contacts.

Discretization of conformation space can be done using the internal coordinates of the potential function as a reference. First, two vectors $\vec{\chi}(\alpha)$ and $\vec{\Phi}(\alpha)$ are introduced with components constrained to be 0 or 1. To each component of $\vec{\chi}(\alpha)$ we assign a pair of nonadjacent phobic beads i and j. We then define the value of $\chi_{(ij)}$ as

$$\chi_{(ij)} \equiv \begin{cases} 1 & \text{if } E_{\text{nb}}(r_i, r_j) \leq E_{\text{contact}} \\ 0 & \text{otherwise} \end{cases} \tag{5.1}$$

where $E_{\text{contact}} = \frac{1}{2}\min\{E_{\text{nb}}\}$ and E_{nb} is the energy of the nonbonded interactions. Thus, the elements of the vector $\vec{\chi}(\alpha)$ record the phobic contacts within a given state. Although the definition of E_{contact} is somewhat arbitrary here, our results are not sensitive to it.

Similarly for each torsion angle, ϕ_j, two components from the vector $\vec{\Phi}(\alpha)$ are assigned with values $\Phi_j^{(2)}$ defined by

$$\Phi_j^{(2)} \equiv \begin{cases} 00 & \text{if } \phi_j \text{ is } \textit{cis} \\ 01 & \text{if } \phi_j \text{ is } \textit{gauche} \\ 10 & \text{if } \phi_j \text{ is } \textit{trans} \\ 11 & \text{if } \phi_j \text{ is } \textit{gauche} \end{cases} \tag{5.2}$$

The local maxima in the torsion potential are used to define the boundaries in torsion angles between *gauche*$^-$ and *trans*, and *trans* and *gauche*$^+$. The boundary between *cis* and *gauche*$^\pm$ is chosen so that the corresponding energy is the same as the local maximum between *gauche*$^\pm$ and *trans*. Thus, the elements of $\vec{\Phi}(\alpha)$ represent a discretization of torsion space. Since the configurations are effectively determined by the torsion states (*cis*, *gauche*$^\pm$, *trans*) and the phobic contacts, the vectors $\vec{\chi}(\alpha)$ and $\vec{\Phi}(\alpha)$ effectively determine the structures and discretize the continuous spatial coordinates.

Given two conformations, α and β, a distance function can be defined using Eqs. (5.1) and (5.2) by

$$X(\alpha, \beta) \equiv \|\vec{\chi}(\alpha) - \vec{\chi}(\beta)\| + \|\vec{\Phi}(\alpha) - \vec{\Phi}(\beta)\| \tag{5.3}$$

where $\|\cdots\|$ denotes the Hamming distance between the two vectors with components constrained to 0 or 1. The Hamming distance between two such vectors, $\vec{a}$ and $\vec{b}$, is given by $\sum_k (1 - \delta(a_k - b_k))$ (i.e., the number of components in vectors that are different [22]). The distance measure $X(\alpha, \beta)$ has the property that $X(\alpha, \alpha) = 0$ and $X(\alpha, \beta) \geq 1$ for conformations, α and β, which differ in either the number of phobic contacts or torsional angles. To precisely follow the spin-glass order parameter "prescription," we would like to normalize $X(\alpha, \beta)$ and then define $q_{\alpha\beta} = 1 - X_N(\alpha, \beta)$. It is not at all obvious which structures maximize $X(\alpha, \beta)$, and since $X(\alpha, \beta)$ as defined in Eq. (5.3) contains all the information of the more usual normalized order parameter, we will ignore the issues of normalization and use the overlap function $X(\alpha, \beta)$ directly.

To study the physical behavior of the system during an annealing run, the thermodynamic average of the overlap function $X(\alpha, \beta)$ is introduced. After reaching equilibrium at some temperature, T, we generate a set of states, $A = \{\alpha\}$ during an annealing run. Using Eq. (5.3), we define the average overlap with state β and its variance by

$$\langle X_A \rangle_T = \frac{1}{N_A} \sum_{\alpha \in A} X(\alpha, \beta) \tag{5.4}$$

$$\delta^2 \langle X_A \rangle_A = \frac{1}{N_A} \sum_{\alpha \in A} (X(\alpha, \beta) - \langle X_A \rangle_T)^2 \tag{5.5}$$

where N_A is the number of elements in the set A and β is defined as the lowest energy state observed during the run. If the system becomes trapped in a local-energy minimum below the phase-transition temperature, T_m, then $\delta^2 \langle X_A \rangle_T \rightarrow 0$ as $T \rightarrow T_m$, signaling that the system is just fluctuating about a local minimum. However, with $\delta^2 \langle X_A \rangle$ alone we cannot determine whether the local minimum is the ground state, nor can we say anything about the relationship between local minima observed on separate annealing runs.

The relationship between pair of annealing runs can be studied using the measure $\langle X_{A,B} \rangle$ defined by

$$\langle X_{A,B} \rangle_T = \frac{1}{N_A N_B} \sum_{\alpha \in A, \beta \in B} X(\alpha, \beta) \tag{5.6}$$

where A and B represent the set of conformations observed on two separate annealing sweeps at the same temperature and $\langle X_{A,B} \rangle$ is a measure of similarity between the annealing sweeps. If the $\langle X_{A,B} \rangle \rightarrow 0$ as T is reduced, this implies that both annealing runs are falling into the same energy minima. In the case where $\langle X_{A,B} \rangle$ remains greater than zero the annealing runs are falling into different local energy minima with differing conformations.

An interesting case arises when we observe $\delta^2 \langle X_A \rangle \rightarrow 0$ and $\langle X_{A,B} \rangle 0$ as $T \rightarrow T_m$. $\delta^2 \langle X_A \rangle \approx 0$ implies the system is falling into local minima and is becoming trapped at $T = T_m$; $\langle X_{A,B} \rangle > 0$ below T_m implies that they are different local minima. Such behavior is characteristic of a glass-like transition, suggesting that interconversion times between states may be very large below T_m [21–25].

For further details regarding the development and interpretation of the order parameters, Refs. 20 and 26–28 consider on-lattice protein models, while Refs. 21, 23 and 29 consider field theory approaches. References 24, 25, and 30 provide overviews of and general arguments for the spin-glass approach to the protein folding problem.

VI. APPLICATION TO SAMPLE PROBLEM

A. Phase Transition

Table II compares the results of several annealing experiments performed using different cooling schedules and sampling lengths. For each experiment, five independent annealing runs were combined to estimate the uncertainty in the predicted phase-transition temperature, T_m, and the variance in C_v averaged over all $T \in [100,400]$ [31]. Although all simulation protocols produced similar estimates of T_m, the uncertainty is significantly higher using the nonlinear cooling schedule ($T_{\text{lin}} = 0$). This is most likely due to the larger temperature increments applied through this schedule, which limits the number of histograms collected during the run, reducing the number of independent samples of the density of states. As might be expected, the uncertainty also decreases with increasing sampling times (number of acceptances). One exception is noted, however, for the annealing schedule using $\Delta T = 7$ K. In this case, the errors are slightly greater than anticipated in both T_m and C_v. As will be discussed further in Section III.D, this is apparently due to differences in the folding pathways that are populated as the system is cooled.

1. *Comparison to Conventional Methodology*

Table II also compares the results of the histogram methodology to the more conventional approach of estimating C_v from the variance in the configurational energy [32]. This more traditional approach estimates C_v by

$$C_v^{(\text{var})}(T_i) = \frac{\langle (E^{(i)} - \langle E^{(i)} \rangle)^2 \rangle}{T_i^2}$$
$$= \frac{1}{T_i^2} \left[\frac{1}{N_i} \sum_{j=1}^{N_i} (E_j^{(i)})^2 - \left(\frac{1}{N_i} \sum_{j=1}^{N_i} E_j^{(i)} \right)^2 \right] \tag{6.1}$$

where $E_j^{(i)}$ is the sequence of energies observed at temperature T_i during the ith annealing sweep. Although both methods give comparable predictions

TABLE II

Phase-Transition Temperature, T_m, and $\delta^2 C_v$ Averaged over Five Independent Annealing Experiments with Various Cooling Schedules

Cooling Schedule[a]					Histogram Method		Variance Method	
T_{lin} (K)	ΔT (K)	Number of Acceptances	Total Number of Trials	Number of Histograms	T_m [b]	$\delta^2 C_v$	$T_m^{(\text{var})}$ [b]	$\delta^2 C_v^{(\text{var})}$
0	—	45,000	3.8×10^6	40	239.1 ± 27.0	36.4	243.4 ± 0.0	55.6
300	4	15,000	2.2×10^6	76	233.8 ± 22.0	36.7	244.2 ± 48.2	210.0
300	10	45,000	5.3×10^6	53	241.8 ± 10.6	8.1	238.2 ± 41.2	58.9
400	7	75,000	12.3×10^6	75	235.3 ± 12.4	9.6	236.1 ± 45.6	72.7
400	6	100,000	20.0×10^6	86	242.3 ± 5.0	2.4	223.1 ± 20.6	69.3

[a] Cooling schedule terms are defined in Eq. (4.2).

[b] T_m value reported is the mean value of five runs $\pm 2\sigma$.

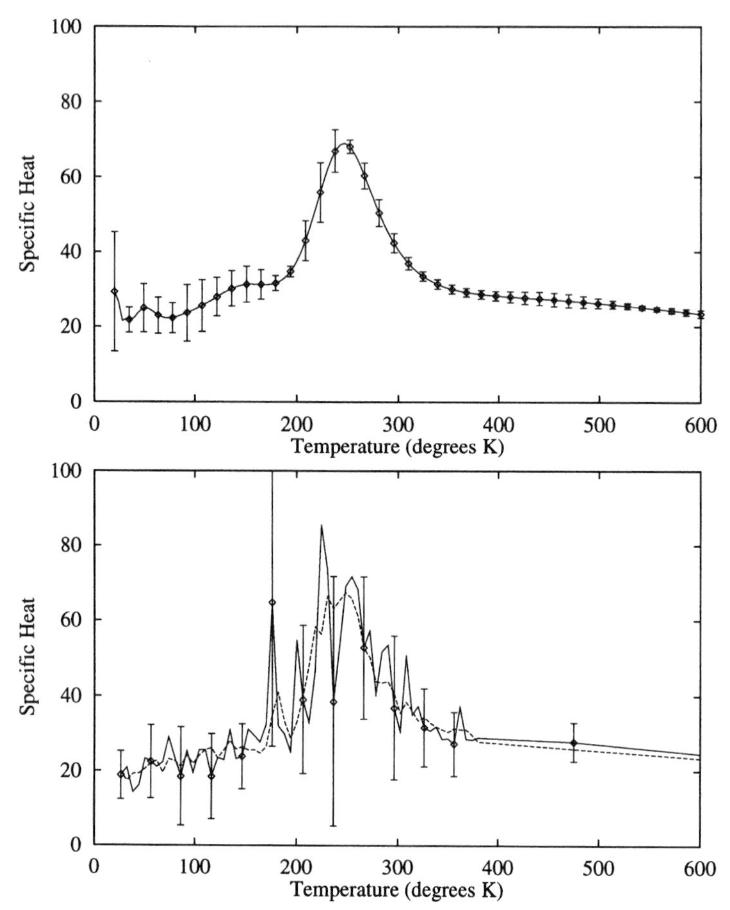

Figure 1. C_v versus T (upper graph) computed using the histogram method. The solid line in the lower graph is the C_v versus T for the same annealing run computed using $\langle(E - \langle E\rangle)^2\rangle T^{-2}$. The dotted line is $\langle(E - \langle E\rangle)^2\rangle T^{-2}$ averaged over five runs. In both graphs, the error bars represent the 2σ point, where σ is variance observed over five runs.

of T_m, the histogram-derived values show a substantial reduction in uncertainty. The reproducibility of $T_m^{(\mathrm{var})}$ and the ability to estimate $\delta T_m^{(\mathrm{var})}$ is greatly affected by the number of discrete temperatures sweeps. Some reduction in the uncertainty is noted as the simulation lengths are increased. However, the mean values show considerable variation, casting some doubt to the validity of the predictions. In addition, the nonlinear protocol shows zero variance in T_m. This artifact is a result of the temperature step size in this particular cooling schedule that is too large to

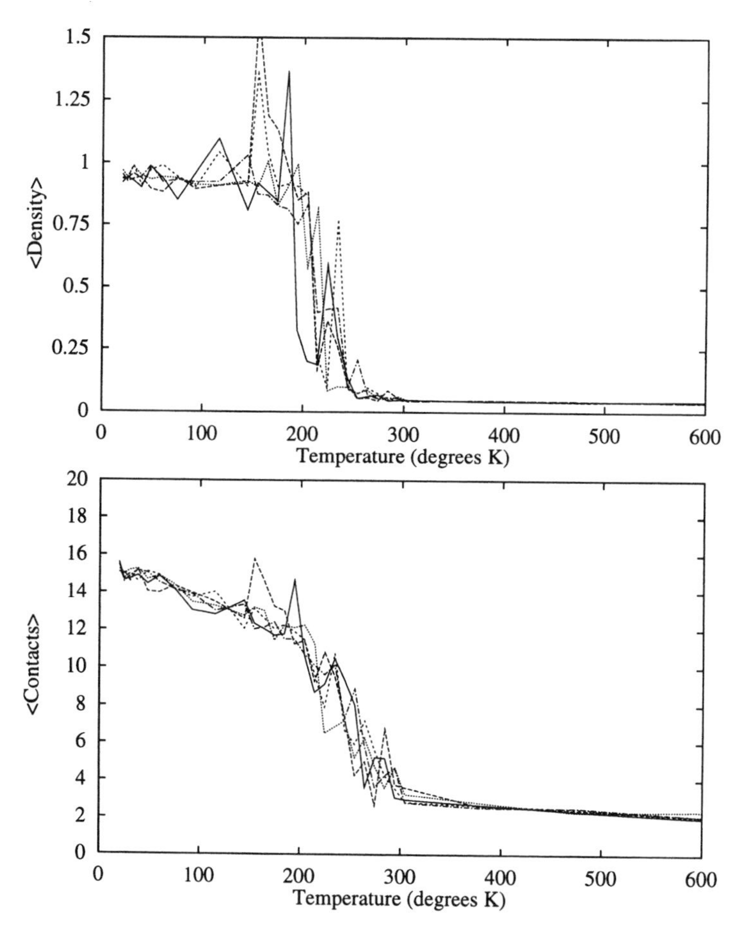

Figure 2. Density, $\langle \rho \rangle$, versus T (upper graph) and average number of phobic contacts, $\langle \text{Contacts} \rangle$, versus T for 5 independent simulated annealing runs.

resolve differences in C_v. Considering the size of the temperature steps applied, a more realistic error in T_m would be ± 70 K.

Figure 1 provides additional insight into convergence of both methods. The upper graph of Figure 1 shows C_v versus T computed with the histogram methodology, while the lower graph shows $C_v^{(\text{var})}$ estimated from the observed variance in E for the same annealing run. In both cases, the error bars are computed from the sample variance observed in five independent runs. Although the $\langle (E - \langle E \rangle)^2 \rangle T^{-2}$ estimator does generally have a peak value near $T_m = 242$ K (as determined by the histogram method), the extreme jagged behavior of $\langle (E - \langle E \rangle)^2 \rangle T^{-2}$ indicates that the variance method has not converged. Figure 1 also reports the average $C_v^{(\text{var})}$ computed over all five independent annealing runs using Eq. (6.1). This curve shows much better convergence and is similar to the histogram result. It should be emphasized, however, that the average $C_v^{(\text{var})}$ incorporates 5 times as many samples to produce a similar-quality result.

B. Physical Properties

Figure 2 shows the density, $\langle \rho \rangle$ and number of hydrophobic contacts $\langle \text{Contacts} \rangle$, of the protein model as a function of T during the annealing run. The measure of protein density used here is defined as $\rho = R_0^3 / \langle R^3 \rangle$, where R is the radius of gyration and R_0 is the radius of gyration of the minimum-energy conformation [2]. $\langle \text{Contacts} \rangle$ is defined as the thermodynamic average of the number of hydrophobic contacts in each configuration. Using $\langle \rho \rangle_T = 0.5$ and $\langle \text{Contacts} \rangle_T = 0.5 \max \{\langle \text{Contacts} \rangle\}$ as a criterion to determine the phase-transition point, we obtain $T_m^{(\rho)} = 210$ K and $T_m^{(\text{contacts})} = 239$ K, respectively. The value obtained from the $\langle \text{Contacts} \rangle$ is consistent with the peek observed in C_v. This strongly suggests that the system is collapsing at T_m around a hydrophobic core of contacts. However, the behavior of the density as a function of temperature suggests that the system remains somewhat extended above 210 K. Taken together, the results indicate the protein model folds through two steps or phases. In effect, T_m represents the collapse of the helices, whereas final folding or compaction occurs at approximately 210 K. A more detailed analysis of the contacts and torsion angles observed during the cooling process for this protein model is available elsewhere [7].

C. Nature of the Phase Transition

The nature of the phase transition can be studied by examining $P_\beta(E)$ as defined by Eq. (2.15). If the phase transition is first-order, we expect $P_{\beta_c}(E)$ to be bimodal at T_m. For temperatures above T_m, $P_{\beta_c}(E)$ should be dominated by "high" energies and below T_m, $P_\beta(E)$ should be dominated by "low" energies. Figure (3*a*) is a graph of log $P_\beta(E)$ for several temperatures (above, at, and below T_m). The $P_\beta(E)$ obtained here is characteristic of a second-order phase transition. At T_m, $P_{\beta_c}(E)$ is essentially flat, indicating a smooth transition from a random coil to a collapse state. This is in contrast to the all-or-nothing nature of a first-order transition. Alternatively, one could argue that finite size effects preclude the existence of a first-order phase transition for this protein model. It is important to point out, however, that a bimodel $P_\beta(E)$ has also been observed in lattice models with 38 beads [33].

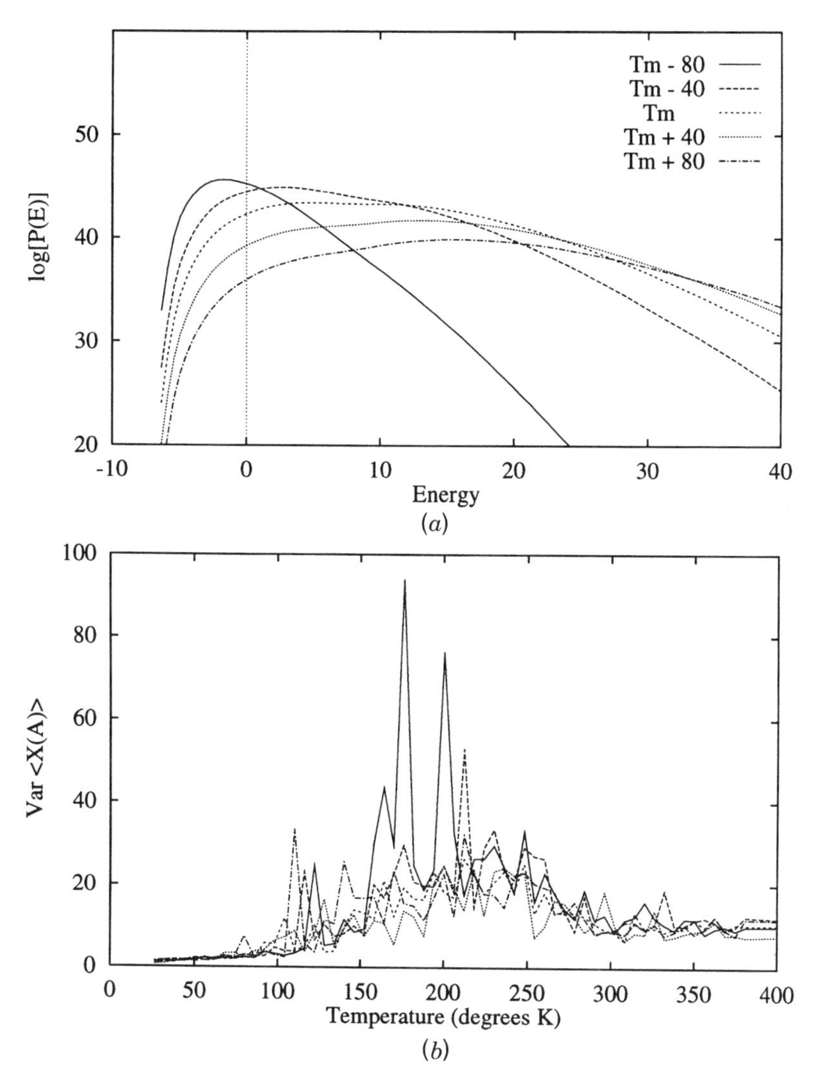

Figure 3. Graphs of (*a*) *P*(*E*) versus *E* at five temperatures; (*b*) $\delta^2\langle X_A \rangle$ versus *T* for five independent annealing runs.

D. Overlap Functions

Figure 3*b* is a plot of $\delta^2\langle X_A \rangle_T$ versus T for the runs with $T_{\text{lin}} = 400$ K, $\Delta T = 6$ K. $\delta^2\langle X_A \rangle_T$ reaches essentially zero for all runs at or before 100 K. Similar results are obtained independent of the cooling schedule, implying that the system is no longer transitioning between local minima below 100 K. The nature of the frozen states can also be explored by examining $\langle X_{AB} \rangle$ as a function of T. This data are shown in Figure 4 for the nonlinear cooling schedule (i.e., $T_{\text{lin}} = 0$, and using 15,000 acceptances). The divergence noted is characteristic of multiple frozen states, suggesting the protein has become trapped in two or more local minima. A more detailed examination of the low temperature configurations reveals four distinct local minima are populated; three α-helical hairpin-like conformers and one highly disordered form. Although this indicates the system may be freezing into a collection of states (including nonhelical structures) near 100 K, this behavior appears to be an artifact of the rapid cooling schedule. Similar ground states (α-helical hairpin-like) were located using the linear cooling schedules. The artifact appears to be due to the size of the temperature increment which may cause rapid quenching through the phase transition.

While all annealing experiments with linear cooling schedules produced the desired α-helical hairpin structure as the frozen state, the overlap functions indicate some variation in local structure occurs below this freezing transition. An analysis of $\langle X_A \rangle$ for the simulation using $T_{\text{lin}} = 400$ K, $\Delta T = 7$ K reveals that two of the annealing runs end in a slightly higher-energy structure than the observed ground state. Although this state is α-helix hairpin-like in conformation, the differences in energy states populated are apparently enough to affect C_v and T_m. Recall this particular run produced higher uncertainties in these values. A more detailed structural analysis of the entire annealing experiment indicates that these two runs fail

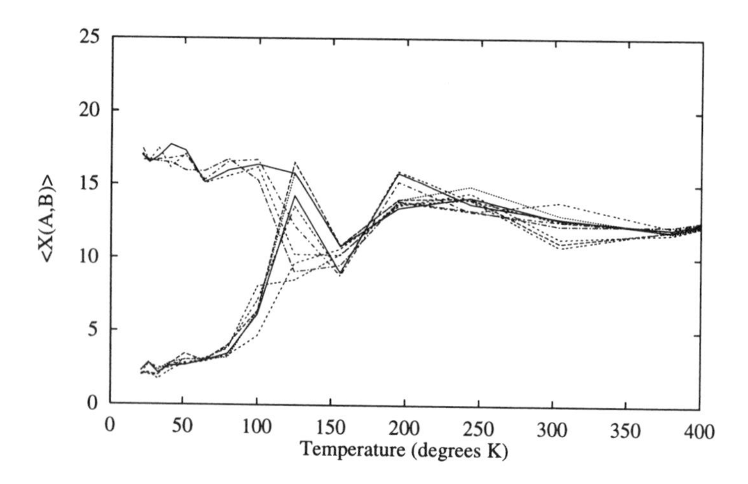

Figure 4. $\langle X_{A,B} \rangle$ versus T for an annealing run with a fast cooling schedule.

to develop significant helical content at or just below T_m as in the other annealing runs (both with this and slower cooling schedules). These "anomalous runs" developed helices at a lower temperature, and both developed a different set of helix–helix contacts from those observed in ground state. This behavior is very similar to the rapid quenching observed during the runs with faster cooling schedules. In particular, a similar effect is seen in the $T_{\text{lin}} = 300$ K, $\Delta T = 4$ K runs. Since the behavior is not observed as the equilibration time is extended to 100,000 acceptances, the higher-energy conformers most likely stem from the cooling schedules and not from in the inherent physical characteristics of the force field or protein model.

VII. CONCLUSIONS

This chapter has described the implementation of optimal histogram methodology to study the physical behavior of a simple protein model. Although some additional complexity is involved is solving the histogram equations, the results indicate that this methodology provides a fairly straightforward extension of standard Monte Carlo–simulated annealing protocols to study the thermodynamic properties of chemical systems. As the results presented above show, the approach provides fairly sensitive estimates of C_v as a function of temperature, especially when compared to the more common method of monitoring the variance in configurational energy. One simulation using the histogram methodology yields results of quality similar to that of multiple runs using the more conventional variance method. It is important to point out that the histogram equations can also be applied to estimate ΔG, ΔH, and ΔS, for the annealing or freezing process. These parameters can be derived from Eq. (2.12) (and normalized using an infinite temperature simulation to determine $V^{-\mathcal{N}}$ of the configurational partition function) to yield absolute estimates as a function of T or compared to provide relative estimates for independent annealing runs. Given the pervasive role Monte Carlo–simulated annealing plays in conformational search and/or global optimization of molecular systems, calculations of this type may provide more detailed information regarding the nature and behavior of force field models as compared to experimental data.

One of the key advances we have described here in implementing the optimal histogram equations is the use of the Newton–Raphson formulation in Eq. (2.18). This approach rapidly converges and, in contrast to solving the histogram equations self-consistently, produces numerically stable solutions independent of the problem size. In practice, this allows the number of temperatures and histograms to be greatly expanded, increasing the number of independent samples of the density of states. More importantly, using $\varepsilon_k \equiv \max|\vec{\lambda}(\vec{f}^{(k)})|$ as a convergence criteria for the Newton–

Raphson equations, we can control the amount of numerical error in the estimates of $P(E, \beta)$ and hence, the error in estimating the thermodynamic properties of the system. The technique effectively expands the general applicability and reliability of the methodology, especially as system sizes grow in complexity and size.

In addition to the histogram methodology, this chapter has also described and applied overlap functions to analyze the structural behavior of the protein model during the cooling process. When applied in concert with the histogram calculations, the results indicate the system folds in two transitions. The first occurs at $T_m \sim 242$ K, where the system undergoes a hydrophobic collapse to what could be referred to as a "molten globule" [34,35]. This collapse is further supported by an analysis of the number of hydrophobic contacts that sharply rise near 240 K. Although there is some evidence that compaction continues to take place slightly below T_m (at ~ 210 K, where the density reaches a maximum value), this process does not apparently support a phase transition. A second transition, however, is evident near 100 K, where the overlap functions reveal the system either freezes into a fixed ground state, or if the cooling rate is to rapid, freezes into a glass-like state with the system trapped in various local minima. Although this freezing could be characterized as a glass transition, it is important to point out that the folding process is thermodynamically reversible when equilibration periods are extended. Overall, these results are consistent with theoretical model predictions based on the "folding funnel view" of protein folding [23,34,35]. The three folding processes described above (collapse, compaction, and freezing) may, in fact, correspond to the three structural transitions proposed to occur in the folding funnel at T_θ, T_{fold}, and T_{glass} [23,34,35]. Such assignments, however, should be viewed with caution since the scaling behavior of the protein model has not been examined. Similar studies of larger system sizes are therefore required to fully delineate the folding processes described here as well as the role these models play in understanding protein folding mechanisms.

REFERENCES

1. H. S. Chan and K. A. Dill, "The Protein Folding Problem," *Phys. Today* **46**, 24–32 (1993).
2. K. A. Dill, "Dominant Forces in Protein Folding," *Biochemistry*, **29**, 7133–7155 (1990); K. A. Dill, "Theory of the Folding and Stability of Globular Proteins," (1985) *Biochemistry* **24**, 1501–1509 (1985).
3. J. M. Troyer and F. E. Cohen, "Simplified Models for Understanding and Predicting Protein Structure," in *Reviews in Computational Chemistry*, Vol. 2, K. B. Lipkowitz and D. B. Boyd, eds., VCH Publishers, New York, 1991.
4. A. M. Ferrenberg and R. H. Swendsen, "New Monte Carlo Technique for Studying Phase Transitions," *Phys. Rev. Lett.* **61**, 2635 (1988).

5. A. M. Ferrenberg and R. H. Swendsen, "Optimized Monte Carlo Data Analysis," *Phys. Rev. Lett.* **63**, 1195 (1988).
6. A. M. Ferrenberg and R. H. Swendsen, "Optimized Monte Carlo Data Analysis," *Comp. Phys.*, 101–104 (Sept./Oct. 1989).
7. D. M. Ferguson and D. G. Garrett, "Simulated Annealing-Optimal Histograms Application to the Protein Folding Problem," in *Adaption of Simulated Annealing to Chemical Optimization Problems*, J. Kalivas, ed., Elsevier Science, 1995.
8. D. G. Garrett, K. Kastella, and D. M. Ferguson, "New Results on Protein Folding from Simulated Annealing," *J. Chem. Soc.* **114**, 6555–6556 (1992).
9. J. P. Valleau and D. N. Card, "Monte Carlo Estimation of the Free Energy by Multistage Sampling," *J. Chem. Phys.* **57**, 5457 (1972).
10. C. H. Bennett, "Efficient Estimation of Free Energy Differences from Monte Carlo Data," *J. Comp. Phys.* **22**, 245 (1976).
11. A. M. Ferrenberg, D. P. Landau, and K. Binder, "Statistical and Systematic Error in Monte Carlo Sampling," *J. Stat. Phys.* **63**, 867–882 (1991).
12. G. Torrie and J. P. Valleau, "Non-Physical Sampling Distributions in Monte-Carlo Free-Energy Estimation—Umbrella Sampling," *J. Comp. Phys.* **23**, 187 (1977).
13. By J. Kalivas, ed., *Adaptation of Simulated Annealing to Chemical Optimization Problems*. Elsevier Science, 1995.
14. J. D. Honeycutt and D. Thirumalai, "Metastability of the Folded States of Globular Proteins," *Proc. Natl. Acad. Sci. USA* **87**, 3526–3529 (1990).
15. J. D. Honeycutt and D. Thirumalai, "The Nature of Folded States of Globular Proteins," *Biopolymers* **32**, 695–709 (1992).
16. D. M. Ferguson, A. Marsh, T. Metzger, D. G. Garrett, and K. Kastella, "Conformational Searches for the Global Minimum of Protein Models," *J. Global Opt.* **4**, 209 (1993).
17. A. Rey and J. Skolnick, "Comparison of lattice Monte Carlo Dynamics and Brownian Dynamics Folding Pathways of α-Helical Hairpins," *Chem. Phys.* **158**, 199–219 (1991).
18. M. Mezard, G. Parisi, and M. A. Virasoro, *Spin Glass Theory and Beyond*, World Scientific Press, New York, 1984.
19. S. F. Edwards and P. W. Anderson, "Theory of Spin Glasses," *J. Phys. Metal Phys.* **5**, 965–974 (1975).
20. E. I. Shakhnovich and A. M. Gutin, "Implications of the Thermodynamics of Protein Folding form Evolution of Primary Sequences," *Nature* **346**, 773–775 (1990).
21. J. D. Bryngelson and P. G. Wolynes, "A Simple Statistical Field Theory of Heteropolymer Collapse with Application to Protein Folding," *Biopolymers* **30**, 177–188 (1990); J. D. Bryngelson and P. G. Wolynes, "Spin Glasses and the Statistical Mechanics of Protein Folding," *Proc. Natl. Acad. Sci. USA* **84**, 7524–7528 (1987); J. D. Bryngelson and P. G. Wolynes, "Intermediates and Barrier Crossing in a Random Energy Model with Applications to Protein Folding," *J. Phys. Chem.* **93**, 6902–6915 (1989).
22. R. Rammal and G. Toulouse, "Ultrametricity of Physicists," *Rev. Mod. Phys.* **58**, 765–788 (1986).
23. J. D. Bryngelson, J. N. Onuchic, N. D. Socci, and P. G. Wolynes, "Funnels, Pathways, and the Energy Landscape of Protein Folding: A Synthesis," *Prot. Struct. Funct. Gen.* **21**, 167–195 (1995).
24. H. Frauenfelder, K. Chu, and R. Philipp, in *Physics from Proteins, Biologically Inspired Physics*, L. Peliti, ed., Plenum Press, New York, 1991.

25. P. G. Wolynes, "Spin Glass Ideas and the Protein Folding Problems," in *Spin Glasses and Biology*, D. L. Stein, ed., World Scientific Press, New York, 1992.
26. E. I. Shaknovich, G. Farztdinov, A. M. Gutin, and M. Karplus, "Protein Folding Bottlenecks: A Lattice Monte Carlo Simulation," *Phys. Rev. Lett.* **67**, 1665–1668 (1991).
27. E. I. Shakhnovich and A. M. Gutin, "Enumeration of All Compact Conformations of Copolymers with Random Sequence Links," *J. Chem. Phys.* **93**, 5967–5971 (1990).
28. E. I. Shakhnovich and A. M. Gutin, "Formation of Unique Structure in Polypeptide Chains Theoretical Investigation with the Aid of a Replica Approach," *Biophys. Chem.* **34**, 187–199 (1989).
29. B. Derrida, "Random-Energy Model: An Exactly Solvable Model of Disordered Systems," *Phys. Rev. B* **24**, 2613–2626 (1981).
30. R. H. Austin and C. M. Chen, "The Spin-Glass Analogy in Protein Dynamics," in *Spin Glasses and Biology*, D. L. Stein, ed., World Scientific Press, New York, 1992.
31. The variance in C_v reported is computed by first computing the observed variance in C_v at a fixed set of temperatures, $T \in [100{,}400]$, over the five runs. These $\delta^2 C_v(T_i)$'s are then averaged to produce a single value for each annealing experiment. This is essentially the same as computing the total variance in C_v over $T \in [100{,}400]$ for each experiment. The choice of the temperature rejoin was made to emphasize the transition rejoin of C_v where the issue of convergence is most critical. The same number of temperature points were used to avoid biasing the $\delta^2 C_v$ values reported.
32. M. P. Allen and D. J. Tildesley, *Computer Simulation of Liquids*, Clarendon Press, Oxford, 1987.
33. M. H. Hao and H. A. Scheraga, "Monte Carlo Simulation of a First-Order Transition for Protein Folding," *J. Phys. Chem.* **98**, 4940–4948 (1994).
34. P. G. Wolynes, J. N. Onuchic, and D. Thirumalai, "Navigating the Folding Routes," *Sciences* **267**, 1619–1620 (1995).
35. J. N. Onuchic, P. G. Wolynes, Z. Luthey-Schulten, and N. D. Socci, "Toward a Outline of the Topography of a Realistic Protein-Folding Funnel," *Proc. Natl. Acad. Sci. USA* **92**, 3626–3630 (1995).

MONTE CARLO METHODS FOR POLYMERIC SYSTEMS

JUAN J. DE PABLO AND FERNANDO A. ESCOBEDO

Department of Chemical Engineering, University of Wisconsin—Madison, Madison, WI 53706-1691

CONTENTS

Advances in Chemical Physics, Volume 105, Monte Carlo Methods in Chemical Physics, edited by David M. Ferguson, J. Ilja Siepmann, and Donald G. Truhlar. Series Editors I. Prigogine and Stuart A. Rice.
ISBN 0-471-19630-4

I. INTRODUCTION

Perhaps the single most important feature that sets polymers apart from other materials is the large number of distinct conformations that chain molecules can adopt. Indeed, many of the unique properties of polymeric systems, such as flexibility, resilience, and elasticity, originate in their large conformational entropy, which is a consequence of the numerous degrees of freedom that a polymeric molecule has. Furthermore, different degrees of freedom can have markedly different length (and time) scales; from highly localized bond vibrations, to long-range molecular relaxation and translation. The great challenge to the simulator is to formulate a methodology capable of sampling effectively all degrees of freedom of an arbitrarily large multichain system, over all length (and time) scales of interest. Designing an effective *recipe* to conduct a Monte Carlo (MC) simulation of a simple fluid is, in most situations, a fairly straightforward task. Unfortunately, as the number of internal degrees of freedom of a molecule increases, so does the complexity of the required MC recipe. This chapter is intended to provide an overview of several MC methods that have been devised to surmount some of the difficulties associated with the simulation of polymeric fluids.

Several monographs are available that cover both the methodological aspects of simulation and relevant applications. Most of them, however, have been concerned with lattice models [1,2]. A few recent reviews have specifically addressed some of the intricacies that arise when simulating polymers in a continuum [3–7]. The field of polymer simulation is evolving so fast, however, that several newer and important techniques were not included in earlier reports. In this chapter we focus on the *methodological aspects* of simulations of continuum–space polymeric systems; no attempt has been made to give a proper and systematic account of all the important *applications* reported in the literature. Polymer simulation methods are described in a broad and systematic context, giving special attention to those recent advances that appear to be most promising for the years to come.

II. STATIC AND DYNAMIC MONTE CARLO

The ultimate goal of a MC simulation is that of obtaining accurate averages of the quantities of interest (e.g., thermophysical properties such as pressure and density). To accomplish this, the MC recipe must be able to generate important configurations of the system, that is, the quantities to be measured from these configurations should make a significant contribution to the final average properties. These representative configurations can be generated following two basic strategies, namely a static or a dynamic MC methodology.

A. Static MC

In a static Monte Carlo simulation, every new configuration of the system is generated from scratch. In the past, this approach has been employed mainly for simulation of isolated molecules, such as by building a polymer chain of specified length randomly in space. Most random configurations will possess a large interaction energy and contribute negligibly to the calculation of an average property (e.g., the end-to-end distance); many chains must therefore be constructed. Note that in this approach, individual measures must be weighted by their corresponding Boltzmann factor. This completely random approach is of little use for dense, multichain systems. Recently, however, more sophisticated static MC methods have been proposed to study dense polymeric fluids.

1. *Recursive Sampling*

Garel and Orland [9] and Grassberger and Hegger [10] have extended the recursive sampling scheme of Wall and Erpenbeck [8] to generate a population of trial conformations of homopolymer chains of length N (e.g., starting from a monomer, successive generations of dimers, trimers, ..., N-mers are produced following preset rules). This approach permits the calculation of the excess chemical potential of the N-mer system from the average number of N-mers that have survived the growth process. Garel and Orland's implementation [9] (called the "ensemble growth" method) searches chain populations in breath first, thus requiring simultaneous simulation of a larger number of ensembles. For long chains, the depth-first algorithm of Grassberger and Hegger is faster and requires less storage space, but it requires previous knowledge of $N + 1$ weights that control the growth rate of the process; these weights can be obtained from preliminary simulations.

Recursive sampling has the virtue of generating unbiased populations of chain conformations. While it has been shown to be highly efficient, some of its effectiveness is lost at low temperatures and for systems that exhibit long-range interactions [10]. Its application to complicated molecular architectures may face, however, some practical challenges if efficiency and algorithmic simplicity are to be maintained.

Recently, recursive sampling has been combined with configurational bias methods (see Section III.D). In the "Pruned-enriched Rosenbluth" method of Grassberger [11], the number of copies of a growing chain is either reduced (pruning) or increased (enrichment) based on the chain's

Rosenbluth weight; at high enough temperatures, this method can effectively simulate the free energy of long polymers ($\sim 10^4$ sites). Recursive sampling has also been used in a method for calculation of the static properties of long alkane molecules [12] (e.g., free energies). It appears that the approach has also been extended to evaluate free energies of mixing for polymer blends [12]. Along a similar vein, in their *recoil-growth* method Alexandrowicz and Wilding [13] have incorporated recursive sampling into a dynamic MC scheme.

B. Dynamic MC

Dynamic MC methods are much more common than static methods. The key idea is that starting from an initial, arbitrary configuration of the system, successive random moves are attempted to achieve a new configuration. Trial moves are accepted or rejected according to criteria that ensure that relevant configurations are sampled.

The most widely used acceptance/rejection rule was proposed by Metropolis et al. [14] almost a half a century ago. It is a rejection scheme based on the principle of reversibility between successive states of an equilibrium system. Consider proposing some arbitrary transition from configuration r'^N to configuration r^N (where r represents the coordinates of the N particles in the system). The Metropolis criterion prescribes that a trial move be accepted with probability

$$A(r^N \mid r'^N) = \min\left(1, \frac{T(r'^N \mid r^N)p(r^N)}{T(r^N \mid r'^N)p(r'^N)}\right) \tag{2.1}$$

where T denotes the probability of proposing a transition and p is the probability of finding the system in a given configuration.

An efficient dynamic MC algorithm should lead to successive configurations as uncorrelated from each other as possible, so that meaningful averages can be determined rapidly and reliably. The remainder of this chapter describes different dynamic MC methods to generate configurations of a polymeric system (e.g. the specifics of the change $r'^N \rightarrow r^N$) that, in our opinion, offer attractive convergence characteristics.

III. SIMULATION METHODS AND MOLECULAR MODELS OF POLYMERS

Simulation studies of equilibrium properties of polymeric materials can be broadly classified according to the type of information that is sought. If, on one hand, the purpose is to generate or complement thermophysical property data for specific materials, a chemically detailed model is required; that is, as many interactions between atoms as possible (or suitable groups of atoms) should be included in the representation of the system. For multichain systems, such detailed simulations are computationally demanding because of the wide spectrum of length (and time) scales that prevail in polymers. If, on the other hand, the purpose is to explore the behavior of a new class of materials, or to search for universal features in model systems, a coarse-grained model is often more useful. There are different ways of integrating out "local" or "fast' degrees of freedom. There are also different degrees of coarsening. Coarsening is usually based on smearing out the local structure of a chain backbone so as to map a group of atoms onto an *effective site*; neighboring effective sites are then connected by rigid *effective bonds* or by entropic springs (e.g., the bead-spring chain model).

A more drastic degree of coarsening can be achieved by working with chains on a lattice. From a computational point of view, simulations of simple lattice systems are significantly less demanding than those of continuum space models; consequently, they have been widely used in the past [1,2] and, more importantly, they have provided valuable insights about many aspects of polymer physics. Furthermore, in some of the more recent lattice models of polymers, such as the bond-fluctuation model [15], a much finer and flexible discretization of configuration space provides a qualitatively satisfactory description of many local properties related to the packing of polymer systems. In fact, recent studies have shown that a fairly realistic representation of a polymer (e.g., polyethylene) can be mapped onto a bond-fluctuation model, and that the latter is capable of yielding some equilibrium and dynamic properties in fair agreement with experiment [16].

Simulation techniques for on- or off-lattice systems are, in general, applicable to systems with different degrees of coarsening. Their effectiveness, however, is often model-specific. In the following sections we describe different types of moves; some of them can be applied to almost any type of molecular model (e.g., reptation, configurational-bias), and some others are applicable only to certain types of models (e.g., extended configurational bias or concerted-rotations).

A. Displacement Moves

A molecular displacement move consists of selecting a molecule of the system at random, and displacing it by a random amount in a random direction. These moves are effective for rigid, spherically symmetric molecules; however, they are of limited usefulness for dense systems of large multisite chains. More importantly, they do not sample intramolecular degrees of freedom.

In contrast, random displacements of individual sites of a chain (or a few neighboring sites), when feasible, can be a valuable tool (see Fig. 1). For this approach to be applicable, the chain backbone cannot have rigid constraints (e.g., rigid bonds and bond angles). It is particularly effective for coarse-grained models that allow wide fluctuations of individual sites around their bonded neighbors. Relevant examples are the bond-fluctuation model and certain bead-spring models such as that employed by Binder

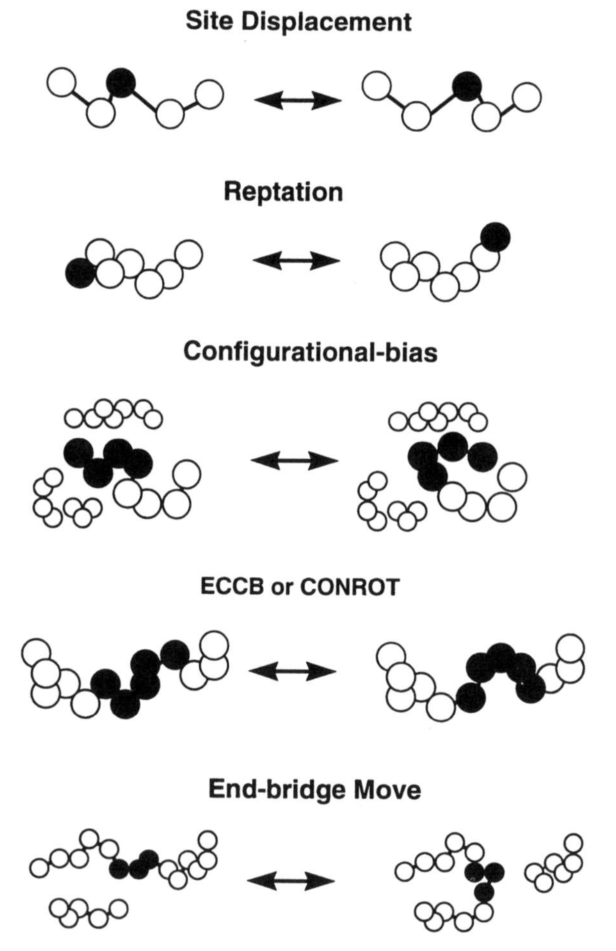

Figure 1. Illustration of the principle of operation of various elementary Monte Carlo moves; from top to bottom: single-site displacement moves, reptation moves, Continuum configurational-bias move, Extended continuum configurational-bias (ECCB) moves and also concerted-rotation moves (CONROT), and end-bridge moves.

and co-workers [17,18]. For these molecular models, single-site displacements have an acceptable success rate, even at high densities; they can move the system through configuration space reasonably well without resorting to more complicated moves and, if used alone, have the advantage of capturing certain features of the dynamic evolution of the system [16,18,19].

A translation move is accepted with probability

$$p_{\mathrm{acc}} = \min(1, \exp(-\beta\, \Delta U_{\mathrm{trans}})) \tag{3.1}$$

where $\Delta U_{\mathrm{trans}}$ denotes the energy change created by translating the relevant sites and $\beta = (k_B T)^{-1}$ (where k_B is Boltzmann's constant and T is the temperature). In Eq. (3.1) it is assumed that the probability of proposing the (forward) move is identical to that of proposing the hypothetical reverse move [e.g., the T probabilities cancel from Eq. (2.1)].

B. Reptation Moves

In a "reptation" move, a segment of the molecule is cut off from one end of the chain and appended at the other end of the chain [20] (see Fig. 1). To satisfy microscopic reversibility, it is important to choose either end of the chain with an arbitrary but fixed probability; if both ends are selected at random, the probabilities of making a "head-to-tail" or a "tail-to-head" transition are equal.

Unfortunately, simple reptation moves cannot be used to simulate branched polymers or block copolymers, because the geometry of the molecule would be altered in the process. Furthermore, for a long molecule a significant amount of time is spent moving the chain ends back and forth, while leaving its center portions unchanged. For this reason, reptation moves are usually supplemented by other moves.

There are several extensions of the basic reptation algorithm described above. For example, it can be applied to block copolymers provided that the move is accompanied by a suitable change of identity of the end site being repositioned, and of those sites at the A–B bonding point [21]. Another extension is the "general reptation" algorithm [22], in which a randomly chosen monomer on a chain is first removed and then reattached at another, random position along the chain. This move can be valid only if connectivity constraints are properly maintained, which is easier to accomplish on a lattice and in some coarse-grained continuum models [6].

C. Smart Sampling Techniques

Some of the earliest attempts to devise efficient algorithms for MC simulations of fluids in a continuum are due to Rossky et al. [27] and Pangali et al. [26], who proposed the so-called "smart Monte Carlo" and "force-bias Monte Carlo" methods, respectively. In a force-bias Monte Carlo simulation, the interaction sites of a molecule are displaced preferentially in the direction of the forces acting on them. In a smart Monte Carlo simulation, individual-site displacements are also proposed in the direction of the forces; in this case, however, a small stochastic contribution is also added to the displacements dictated by the forces. In both algorithms, the acceptance criteria for trial moves are modified to take into account the fact that displacements are not proposed at random but in the direction of intersite forces.

Many attempts to improve the efficiency of Monte Carlo simulations of fluids have also relied on the concept of "umbrella sampling" [23,24,46]. In umbrella sampling methods, a modified probability distribution function is introduced with the purpose of facilitating sampling of regions of phase space that would otherwise be difficult to explore. For a canonical-ensemble simulation, configurations are generated according to

$$\rho(r^N) = \frac{f(r^N)\exp(-\beta U(r^N))}{\int f(r^N)\exp(-\beta U(r^N))\,dr^N} \tag{3.2}$$

where $f(r^N)$ is an arbitrary function designed to facilitate efficient sampling of configuration space. The canonical average value of an arbitrary property M is then calculated according to

$$\langle M \rangle = \frac{\langle M/f \rangle}{\langle 1/f \rangle} \tag{3.3}$$

where the brackets denote an ensemble average over all of the configurations generated according to Eq. (3.2). Umbrella sampling can be used for simulations of polymers in different contexts. It can, for example, be used to facilitate conformational transitions; in that case the energy barriers associated with the relevant torsional potential-energy function would be artificially lowered (through use of function $f(r^N)$ in Eq. (3.2)) [25]. It could also be used to sample more effectively the interactions between two polymer molecules in dilute solution by introducing an artificial force that would keep them from traveling too far apart.

D. Biased Sampling Techniques

Part of the difficulty of simulating polymers resides in the fact that both intra- and intermolecular interactions play a role on the behavior of a molecule. Some of the methods mentioned above have circumvented this problem by proposing local moves that produce a minimal disruption of the overall configuration of a polymeric molecule (e.g., reptation).

More drastic configurational changes can be produced by cutting large pieces of a chain (as opposed to just one end segment), and regrowing these into completely new configurations. The expected decrease of the acceptance probability can, to some extent, be offset by biasing the regrowth of the chain into favorable positions. Configurational bias moves can be used to achieve that goal.

Frenkel [3,7] has recently given a review of configurational bias methods for simulation of polymers. An account of these methods and their application to phase equilibria calculations is also described in the first chapter in this volume (by I. Siepmann). Here we merely outline the main ideas behind them (see also Fig. 1). In a configurational bias Monte Carlo move [34,43,44], a number of segments from one end of the chain are deleted. These segments are subsequently regrown, in a sequential manner, by "scouting" the positions available to every segment and selecting with a higher probability those with a lower energy. Every time that a segment is appended to the growing chain, a number N_{samp} of trial positions is explored by calculating the energy corresponding to that particular configuration. One of these positions is "selected" with probability

$$p_i = \frac{\exp(-\beta(U_i))}{\sum_j^{N_{\text{samp}}} \exp(-\beta(U_j))} \tag{3.4}$$

where U_i is the energy of the relevant segment in position i. Once the chain has been fully regrown to its original length, a Rosenbluth weight is calculated according to

$$R_W^{\text{new}} = \prod_l \left(\frac{1}{N_{\text{samp}}} \sum_j^{N_{\text{samp}}} \exp(-\beta U_j) \right)_l \tag{3.5}$$

where the product is conducted over all of the regrown segments l of the chain under consideration. In order to satisfy microscopic reversibility, it is necessary to reconstruct the process that would have been followed to arrive at the current configuration by a configurational-bias algorithm. The relevant chain is therefore regrown again by sampling several positions for each segment, but arbitrarily selecting its original orientation. A Rosenbluth weight can then be calculated for the current (original) configuration of the chain according to

$$R_W^{\text{old}} = \prod_l \left(\frac{1}{N_{\text{samp}}} \sum_j^{N_{\text{samp}}} \exp(-\beta U_j) \right)_l \tag{3.6}$$

The overall Monte Carlo move for such a chain is accepted with probability

$$p_{\text{acc}} = \min\left(1, \frac{R_W^{\text{new}}}{R_W^{\text{old}}}\right) \tag{3.7}$$

thereby removing the bias that was introduced by generating configurations by means of a "scouting" algorithm.

A configurational bias algorithm is very general in that it can be applied to chain molecules of arbitrary complexity. It has been used successfully to simulate the thermodynamic properties, including phase equilibria, of relatively long alkane molecules [34,35,37,38,40]. However, it also suffers from important shortcomings. Perhaps the most important of these is the fact that, as the length of a chain increases, the probability of successfully regrowing a large portion of the molecule decreases drastically, thereby limiting the applicability of the method to chains of moderate length [5,50,58,71]. Fortunately, newer alternatives have now been devised to overcome some of the limitations of the original configurational bias formulation.

E. Rearrangement of Inner Sites

In the original formulation of a "crankshaft" move (sometimes called "kink" or "flip" move), one or more inner sites of a lattice chain are chosen at random and subsequently rotated around the axes formed by the two sites of the chain to which the "flipping" sites are attached [47,48]; the move is accepted according to the Metropolis criterion. The one-site crankshaft move (later referred to simply as "crankshaft") can be easily extended to off-lattice simulations of flexible chains [48]. A crankshaft algorithm must be supplemented by other types of moves to alter the configuration of chain ends (e.g., reptation or pivot). Crankshaft moves are highly localized and can therefore be effective even at high densities. Their performance can be further improved by means of biasing techniques (e.g., force bias [49] or configurational bias [50]).

The rearrangement of more than one inner site in a flexible or semiflexible chain molecule can be conveniently performed if the geometric constraints that guarantee chain closure are taken into account every time a site is repositioned [50] (see Fig. 1). Performance can be enhanced by favoring low-energy trial positions for each "growing" site (extended continuum configurational bias, or ECCB method). Since this method can be applied to inner sites of arbitrary functionality, it has been used to study linear, branched, and crosslinked polymers [50–52].

In the ECCB method, chain closure is viewed as a geometric problem whose solution entails the use of geometric rules. An alternative view has also been recently proposed in which chain closure is treated as a probabilistic problem [53–55]. The idea is that the orientation of the growing bonds is no longer sampled from a uniform distribution (on a unit sphere centered at the former neighboring site), but from a distribution that satisfies chain closure automatically. Vendruscolo [53] and Frenkel and co-workers [7,54] have discussed schemes in which analytical expressions for the statistics of ideal, non-self-avoiding (Gaussian) chains between two fixed points are used to generate trial bond orientations. The acceptance criteria for such moves must include suitable factors that take into account the nonuniform sampling introduced. Yong et al. [55], for example, employed displaced Gaussian distributions to generate trial site positions; such distributions were the natural choice for the particular polymer model employed by these authors because chains were connected by Gaussian springs. Just as in the geometric method, in the probabilistic approaches mentioned above, a bias-multisampling scheme is employed to favor low-energy configurations. Unfortunately, a direct comparison between these different methods has not been reported. In fact, only the geometric approach has been applied to (off-lattice) polymer melts [50,51]; other reports have been restricted to isolated molecules. One could speculate that a geometric approach would lose efficiency faster than a probabilistic method as the growing chain section becomes longer. For dense systems, however, only a few inner sites can be usually moved; excluded volume effects have a more significant effect on the rate of successful reconstructions than the details of a particular chain-closure scheme. Under those circumstances, all ECCB variants are likely to be equally effective. The geometric approach has the advantage of being more easily extended to complex architectures, such as asymmetric sites and bonds, arbitrary bonding potential-energy models, and multifunctional sites [51,52].

The main limitation of ECCB-type methods is that they become inefficient as the strength of intramolecular interactions increases. They cannot be efficiently applied, for example, to stiff nematogenic chains and to chain molecules with rigid bond angles (a model commonly adopted for alkane-like molecules). In those instances, different methods are necessary. The next section describes one such method.

F. Concerted Rotations

Concerted rotation moves (CONROT) [57] constitute the method of choice for rearranging inner sites of alkane-like polymer models. The CONROT method is based on the early work of Go and Scheraga for solution of the geometric problem associated with the closure of ring structures in proteins [56]. The relevant geometric constraint equations are solved numerically (these equations can have multiple solutions or none); several versions of CONROT moves have been implemented, depending on the proximity of the inner sites (to be repositioned) to a chain end [57] and on the particular scheme used to solve the equations [60]. This method is particularly useful for simulation of polymer melts and glasses. Combinations of CONROT and configurational-bias moves have been proposed to simulate linear chains [5,58] and, more recently, to both linear and cyclic peptides [59]; such combinations appear to lead to an increase in the rate of relaxation of intramolecular configurations over that obtained with either of the two methods alone.

The ideas behind concerted rotation moves have recently been developed further by Theodorou [60] and co-workers. These authors have proposed a variable-connectivity algorithm (akin to the "pseudo-kinetic" moves of Mansfield [61] and Madden [62]), in which two chain molecules A and B *react* to form two new chains C and D. In a single Monte Carlo move, one end section of chain A *attacks* the middle section of chain B. By restoring to a concerted-rotation type algorithm (end-bridging move), chain B is disconnected by deleting a trimer section, which is, in turn, used to bridge one of the open sections of chain B with the end section of chain A (see Fig. 1). The result is the formation of two new molecules C and D, whose end-to-end distances are likely to be significantly different from those of the original chains A and B.

To achieve a reasonable acceptance rate, the molecular weight of the polymers must be allowed to fluctuate. The variable-connectivity scheme therefore results in a polydisperse system. Pant and Theodorou [60] have shown how different limiting molecular-weight distributions can be easily targeted by resorting to a semigrand ensemble formulation (see Section V.B and chapter by D. Kofke, this volume); in that ensemble the temperature, pressure, spectrum of reduced chemical potentials (with respect to those of two reference species), total number of monomers, and total number of chains are fixed throughout the simulation. For linear homopolymers, the variable-connectivity method offers unprecedented rates of relaxation of the end-to-end distance. Figure 2 shows the decay of the end-to-end vector autocorrelation function for a C-78 polyethylene melt at $T = 450$ K and $P = 1$ atm [63]. The system has a uniform chain length distribution with polydispersity index of 1.08. The performance of conventional molecular dynamics simulation (NVE MD) and three MC simulations have been compared in Figure 2. One half of the moves are common for all three MC runs (which is a combination of concerted rotations, reptations, site rotations, and volume moves); the remaining half correspond to either configurational bias, end bridge, or a combination of the two. Clearly, including end-bridge moves leads to relaxation speeds that are at least one order of magnitude

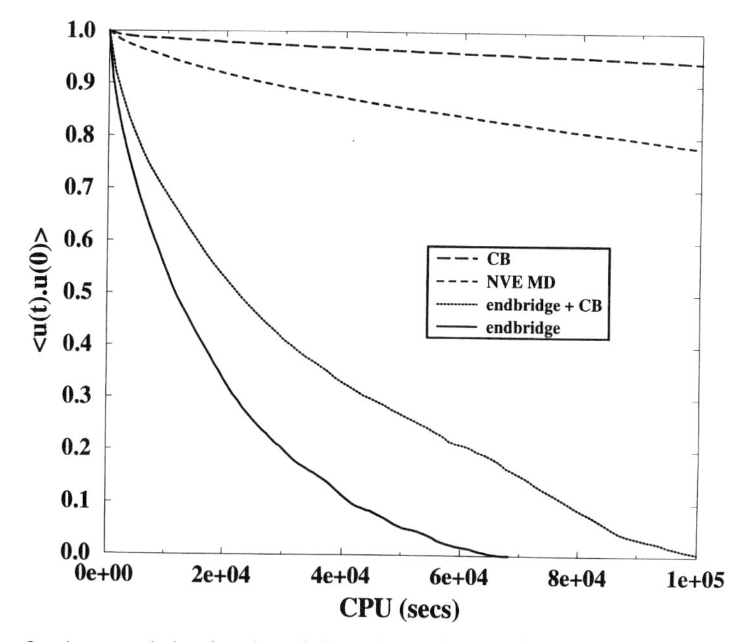

Figure 2. Autocorrelation function of the end-to-end vector for a C-78 polymethylene melt ($T = 450$ K, $P = 1$ atm) from different algorithms. The system has a uniform length distribution with polydispersity index of 1.08. For the MC runs, a common base combination of moves was used that consists of reptations (6%), rotations (6%), flips (6%), concerted rotations (32%), and volume move (0.5%). The rest of the moves were configurational-bias "CB" (49.5%, long dashed line), end bridge (49.5%, full line), and half CB (25%), half end bridge (24.5%, dotted line). The short dashed line corresponds to a NVM molecular dynamics simulation of the same system. The CPU time refers to an IBM/RS600 3CT machine [63].

faster. In their present form, however, end-bridge moves are limited to homonuclear, linear polymers.

G. Other Localized Moves

For short molecules (say, with less than 10 sites), moves that attempt a small rearrangement of all sites of the molecule at once have also been proposed [64,65]. In the translate–jiggle move of Dickman and Hall [64], a given chain is subjected to a random translation followed by a small displacement of each segment ("jiggling"); the new position of each segment is found by renormalizing the bond lengths to the preset values. Chang and Sandler [65] implemented a variation of this move in which the position of an individual site is changed by rotations of bond vectors, as opposed to translations.

Rigid molecules do not have internal degrees of freedom and, consequently, can be relaxed only by means of translations and rotations; for a description of algorithms for conducting rigid-body rotations, see Ref. 46. Orientational-bias moves (analogous to configurational bias moves) have also been proposed for rigid molecules; these, however, have been typically applied to small polar molecules [66–68].

In a *pivot* algorithm [69], a randomly chosen site of a molecule is used as a pivot joint; the shorter portion of the chain is then rotated through a random angle and, if the rotated portion does not overlap other sites in the system, the move is accepted. Although this technique can generate dramatic configurational changes, it becomes impractical in dense systems because of the high probability of encountering overlapping configurations (even if biased sampling is employed over several trial rotation angles). For this reason, pivot moves have been mostly used for single-chain simulations.

H. Collective Moves

Most of the Monte Carlo moves discussed up to this point involve the localized rearrangement of one or at most all sites of a single molecule. In contrast to such moves, collective methods involve simultaneous rearrangement of several or all molecules in the system.

1. *Cluster Moves*

It is sometimes convenient to perform rearrangements that involve several molecules or several molecular sections at once. This is often the case when working with parallel computers. This situation also arises, for instance, when strong local attractions tend to keep a group of particles together (e.g., in ionic systems). A relevant example is the formation of micelles in surfactant systems [70]. Simple micelles are typically spherically symmetric aggregates of surfactant molecules that encapsulate a component immiscible with the solvent molecules at the outside. Once the micelles self-assemble, it is convenient to introduce moves that can displace (or rotate) micelles as a whole, so that relevant conformations of the micellar state are properly sampled. Moves intended to disassemble micelles should still be conducted, although the energetic penalty involved would render most such attempts fruitless. The criterion to select which particles belong to a given cluster is to a large extent arbitrary; a typical choice is to define some suitable cutoff radius for inter-site interactions [70]. Several variants of cluster moves have been recently reviewed elsewhere [7].

A somewhat related concept is “preferential sampling” [46], in which molecules in the vicinity of a given point are sampled more often than the rest of the system. The reference point might, for example, be centered at a solute site; conformations resulting from important solute-solvent interactions are then preferentially explored. A variation of this idea has been used to simulate chain molecules dissolved in a solvent of small spherical molecules (the *hole* method [50]).

2. *Hybrid Methods*

In a hybrid method, molecules are displaced in time according to conventional molecular dynamics (MD) algorithms, specifically, by integrating Newton's equations of motion for the system of interest. Once the initial coordinates and momenta of the particles are specified, motion is deterministic (i.e., one can determine with machine precision where the system will be in the near future). In the context of Eq. (2.1), the probability of proposing a transition from a state 0 to a state 1 is determined by the probability with which the initial velocities of the particles are assigned; from that point on, motion is deterministic (it occurs with probability one). If initial velocities are sampled at random from a Maxwellian distribution at the temperature of interest, then the transition probability function required by Eq. (2.1) is given by

$$T(0 \mid 1) = \exp\left(-\beta \sum_i^N \frac{(\dot{r}_i^0)^2}{2m_i} \right) \tag{3.8}$$

where $\dot{r}_i^0$ is the initial velocity of particle i and m_i is its mass. After a series of N_{MD} molecular dynamics steps, each corresponding to a time step δt, the velocity of particle i evolves to $\dot{r}_i^1$. The distribution of velocities at time $\tau = N_{\text{MD}}\,\delta t$ is given by

$$T(1 \mid 0) = \exp\left(-\beta \sum_i^N \frac{(\dot{r}_i^1)^2}{2m_i} \right) \tag{3.9}$$

As usual, the internal potential energy $U(r^N)$ of the system is calculated from a specified intersite potential-energy function, generally by assuming pairwise additive interactions. The kinetic energy of the system is given by

$$U_{\text{kin}} = \sum_i^N \frac{\dot{r}_i^2}{2m_i}$$

The Hamiltonian of the system is given by $H = U + U_{\text{kin}}$. If δH is used to denote the change in the Hamiltonian that arises in a hybrid move then, by substituting Eqs. (3.8) and (3.9) into (2.1), the criterion for accepting that move is found to be

$$p_{\text{acc}} = \min(1, \exp(-\beta\,\delta H))$$

The MD component of a hybrid Monte Carlo method is generally carried out at constant energy; conservation of energy requires that H be constant. In practice, however, roundoff errors give rise to small changes of H that ultimately determine whether a move is accepted. Such errors become more pronounced as the size of the time step δt is increased, thereby leading to a smaller acceptance rate. Hybrid methods were originally proposed in the context of lattice-gauge calculations. Their application to simple fluids was first discussed by Mehlig et al. [28]; extensions to polymeric systems and other complex fluids have been discussed later [5,29–32].

Molecular dynamics methods play a dominant role in commercial simulation software by virtue of their relative simplicity and wide applicability, regardless of molecular architecture. Unfortunately, they are much less flexible and efficient than MC techniques for simulation in different ensembles. Hybrid methods provide a convenient means for merging MD codes with MC algorithms, thereby offering the possibility of getting some of the benefits of both approaches. Note, however, that the efficiency of such methods decreases significantly with system size.

IV. CONFORMATIONAL SAMPLING AND SIMULATION OF FREE ENERGIES

The establishment of chemical potential equilibrium (with respect to either a setpoint or phase coexistence) is the central component of most Monte Carlo schemes for simulation of the phase behavior and stability of molecular systems. Simulation of the chemical potential (or chemical potential equilibration) in a polymeric system requires more effort than the corresponding calculation for a simple fluid. The reason is that efficient conformational sampling of the polymer is implicitly required for a free-energy calculation and, in fact, the ergodicity problems described in earlier sections are often exacerbated.

In addition to standard thermodynamic integration, perhaps the most widely used idea for simulation of the chemical potential of a simple fluid is embeded in Widom's method, which gives residual contributions to the chemical potential as [46]

$$\beta\mu^{\mathrm{res}} = -\ln\langle \exp - (\beta U^{\mathrm{ghost}})\rangle_N \tag{4.1}$$

The energy appearing in Eq. (4.1), U^{ghost}, corresponds to that of a "ghost" or "test" particle that interacts directly with the system. The brackets denote an ensemble average over a system containing N real particles. The system is not influenced by the ghost molecule, whose only purpose is that of sampling the energy that an additional $(N + 1)$th particle would experience if it were to be introduced at a random point in the system.

For simple fluids at low to moderate densities, Eq. (4.1) provides one of the most reliable means for estimating the chemical potential. For long, articulated molecules, however, random insertions of a ghost particle lead to frequent overlaps with the host system that hamper collection of reliable statistics for Eq. (4.1). For molecules of intermediate size, this sampling problem can be alleviated by resorting to configurational bias ideas [36,39,42]. Instead of inserting the ghost molecule into the host system at once, it is inserted sequentially, scouting the space available to each segment at every step of a growth process; if R_W^{ghost} is computed for the ghost molecule according to Eq. (3.5), then the chemical potential is given by

$$\beta\mu^{\text{res}} = -\ln\langle R_W^{\text{ghost}}\rangle_N \tag{4.2}$$

Configurational bias techniques have been extended to other ensembles, in particular, to the Gibbs ensemble [38,41], a technique that allows direct simulation of phase equilibrium [45] (see introductory chapter by I. Siepmann).

At liquid-like densities, the configurational-bias ghost particle approach is reliable only for chain molecules of intermediate length [71]. This limitation can be partially overcome by applying other techniques, such as the method of expanded ensembles.

A. Expanded Ensemble Methods

A free-energy change between two points can be obtained from a single molecular simulation by sampling the two states of interest simultaneously. Note, however, that if the free-energy barrier that separates such states is large, the required sampling can be difficult. An illustrative example is provided by the calculation of the chemical potential for a polymeric system by a Widom-type approach; the two states of interest correspond to an N-molecule system and an $(N + 1)$-molecule system, respectively.

The difference in the free energy of a system between two thermodynamic states is related to the ratio of their partition functions. Since the partition function is proportional to the probability with which the system "visits" the relevant state, such a ratio can be obtained in a single, "expanded" simulation that samples both states. If the transition between them is difficult, additional intermediate conditions can be simulated. This principle was exploited by Lyubartsev et al. [72] to simulate changes of free energy with temperature for simple systems. Similar ideas have been implemented by others to address a variety of problems, including the simulation of chemical potentials for polymeric systems [73–76].

1. *Canonical Ensemble*

Formally, an expanded canonical ensemble can be defined through the partition function:

$$\Omega = \sum_{y=1}^{M} Q_y \exp(w_y) \tag{4.3}$$

where Q_y denotes the canonical partition function corresponding to a particular state of the system characterized by a parameter y, which assumes values from 1 (initial state) to M (end state). Each of these M states is assigned a preweighting factor w_y, whose value is selected so as to give each state a similar chance of being visited [in other words, these factors ensure that all the canonical partition functions appearing in Eq. (4.3) contribute in a commensurate manner to Ω].

Each state of the expanded ensemble is equilibrated as usual (e.g., through molecular rearrangements). However, a new type of trial move is now introduced: trial transitions between neighboring y states. Such trial moves are subject to a Metropolis-like acceptance criterion:

$$p_{\text{acc}}(y \to x) = \min\left\{1, \frac{\exp[w_x - \beta U_x]}{\exp[w_y - \beta U_y]}\right\}$$

These transitions give rise to a random walk through states (an additional weight has to be included to account for the asymmetry of end states). The probability of encountering a particular configuration of the system is

$$p_y = \frac{1}{\Omega} \exp(w_y) \exp(-\beta U_y) \tag{4.4}$$

where U_y is the total energy of the system. The ensemble average of p_y is

$$\langle p_y \rangle = \frac{Q_y}{\Omega} \exp(w_y) \tag{4.5}$$

For any two states y and x it follows that

$$\frac{Q_y}{Q_x} = \frac{\langle p_y \rangle}{\langle p_x \rangle} \exp(w_x - w_y) \tag{4.6}$$

Once a histogram of visits to each state has been constructed, the ratio of probabilities appearing in Eq. (4.6) can be calculated, thereby permitting evaluation of the ratio of the partition functions for any two x and y:

$$\beta A_y - \beta A_x = -\log\left[\frac{Q_y}{Q_x}\right] = -\log\left[\frac{\langle p_y \rangle}{\langle p_x \rangle}\right] + w_y - w_x \tag{4.7}$$

For the particular case of $x = 1$ and $y = M$, Eq. (4.7) gives the desired total free-energy change. To ensure that end states are sampled regularly, all intermediate states must be sampled with about the same frequency; in the case for which $\langle p_x \rangle = \langle p_y \rangle$, it follows from Eq. (4.7) that $\beta A_y = w_y$ (within an additive constant). This suggests an iterative scheme to progressively refine initial estimates of the preweighting factors [75].

The parameter y that defines the various states of an expanded ensemble should be chosen so as to facilitate simulation of the relevant free-energy change. For the particular case of a chemical potential evaluation, y must gradually couple and decouple a "tagged" molecule from the system. This coupling can be achieved "isotropically," e.g. by varying the strength of the intermolecular potential-energy parameters of *all* sites in the molecule simultaneously [73–75], or it can be implemented "anisotropically," that is, by changing the number of sites, n_y, of the tagged molecule (e.g., the chain length, for the case of a linear molecule) [76]. The principle of operation of this latter approach is illustrated in Figure 3. The advantage of an anisotropic coupling is that preweighting factors are related to segmental chemical potential, and, therefore, for an n-mer homopolymer we could write

$$w_y \approx \frac{n_y}{n} (\beta A_M - \beta A_1)$$

which reduces the preliminary work required to estimate preweighting factors to that of obtaining an estimate of the desired free-energy change.

The efficiency of expanded ensemble simulations can be further increased by implementing a configurational-bias scheme for the insertion and deletion processes associated with the segmental coupling–decoupling of the tagged chain [76].

2. *Other Ensembles*

The main advantage of the expanded ensemble approach is that it provides a natural framework for inserting and removing chain molecules from a system. By facilitating such insertion–deletion moves, chemical potential equilibration can be effectively attained in grand canonical and Gibbs

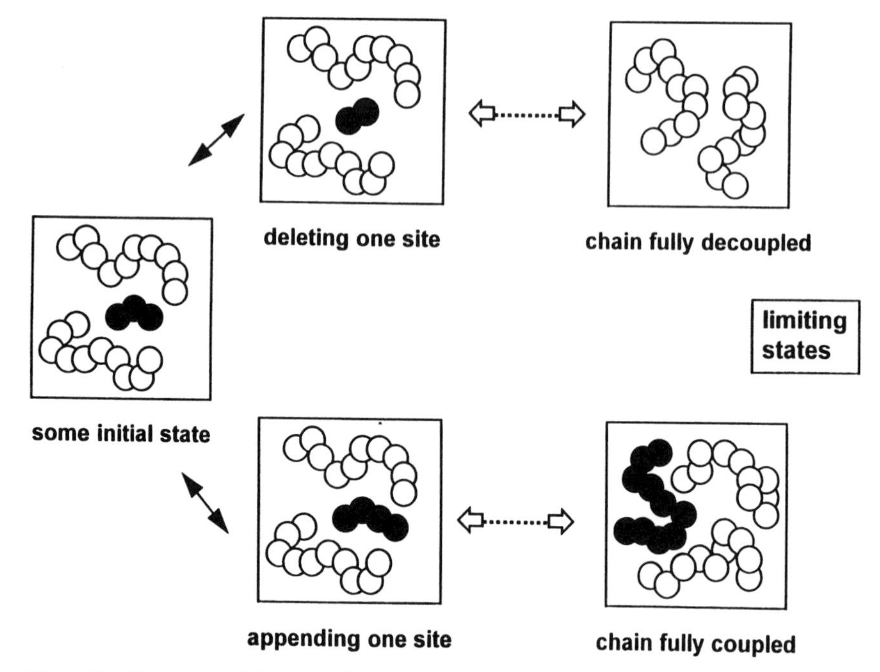

Figure 3. Illustration of the transitions between states in a Expanded Ensemble simulation of the chemical potential. The tagged chain of variable length (anisotropic coupling) is depicted in black (solid circles).

ensemble simulations [77]. In a Gibbs-type of ensemble, a simultaneous gradual coupling–decoupling occurs for a tagged molecule that is split into two sections, each residing in one of the two simulation boxes. For multicomponent systems, one *split* tagged chain is required for every species in the system. Note that the conventional Gibbs ensemble formulation can be recovered by specifying only two states of the tagged chain, namely, the fully decoupled and fully coupled states. For applications involving flexible homonuclear polymers, expanded Gibbs ensemble simulations have the advantage that the preweighting factors essentially cancel out from the acceptance criterion and, consequently, there is no need for preliminary simulations [77].

Applications of the expanded-ensemble approach have been reported for the partitioning of polymers between a slit pore and the bulk [52,77,79], and for fluid–fluid equilibria [78–80]. Figure 4 shows results of expanded Gibbs ensemble simulations of phase coexistence in a binary system of Lennard-Jones chains dissolved in their own monomer [80] (cross-inter-

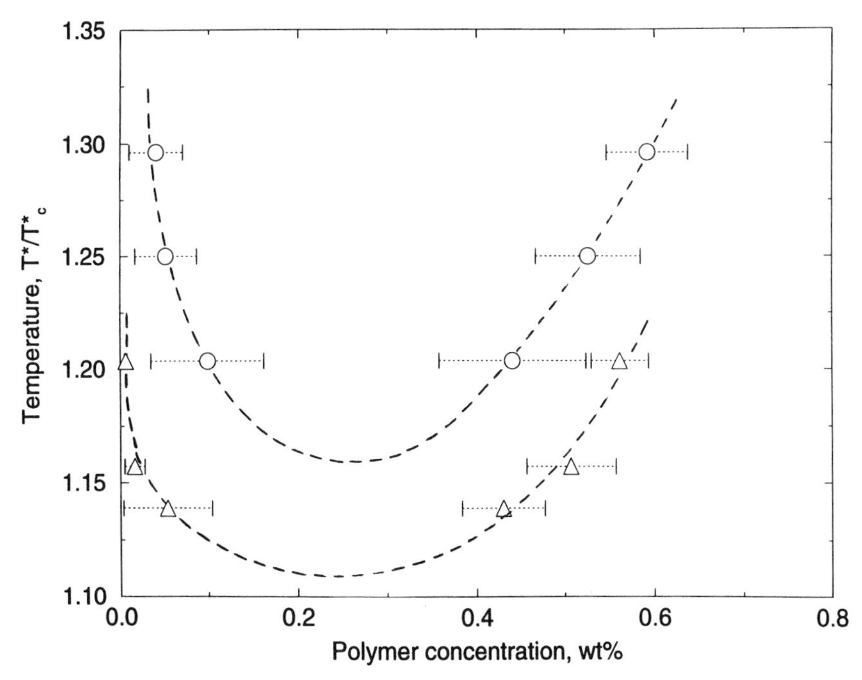

Figure 4. Demixing temperatures for Lennard-Jones polymer in Lennard-Jones solvent at $P_r = 2$ as a function of polymer concentration for 16-mers (circles) and 64-mers (triangles). Temperature and pressure units are relative to the critical solvent properties. The Lennard-Jones interaction potential energy is cut at 2.5σ and shifted [80].

action parameters are identical to like interaction parameters). The reduced pressure is fixed at $P^* = 2$, and all reduced temperatures reported are $T^* > 1$; these units are reduced with respect to the critical values for the pure solvent. At these supercritical conditions, the system exhibits a lower critical solution temperature, which decreases as the chain length increases [80].

The expanded-ensemble method has been shown to be capable of handling molecules with $O(10^2)$ sites. The obvious disadvantages of the method are that preliminary, iterative simulations are sometimes needed and that, as the number of sites per molecule increases, the number of intermediate states must increase accordingly.

B. Multistage Free-Energy Perturbation Methods

Widom's method [Eq. (4.2)] represents a single-stage free-energy perturbation method to measure the chemical potential of chain molecules (the ghost particle insertions correspond to the "perturbations"). Given the low insertion probability of long chains in dense media, the insertion could be performed in, say, N_s stages; for instance, the chain is gradually coupled in going from 1 to N_s, and at each stage a ghost incremental coupling of the chain is attempted [82]. This process implies N_s different simulations, each yielding a partial chemical potential $\Delta\mu_k$; the total chemical potential would then be given by $\mu = \Sigma_{k=1}^{N_s} \mu_k$. Multistage free-energy perturbation methods are reminiscent of thermodynamic integration methods in that a series of simulations is required; in fact, the latter method can be recast as an approximation of the former. Recently, Kofke and Cummings have critically reviewed different methods for simulation of chemical potentials for a hard sphere fluid, and identified staged-insertion methods as one of the most efficient and precise [83]. Such a conclusion could hold, to some extent, for polymeric fluids, provided that the coupling stages (which are in principle arbitrary) were suitably chosen.

The chain-increment method of Kumar and co-workers [81] can be viewed as an approximate multistage insertion method that requires a single simulation. In this method, the chemical potential associated with "appending" one ghost end segment to a chain is computed through a Widom-type sampling scheme. Since the chemical potential of a chain molecule is given by the sum of incremental chemical potentials of its segments, a true multi-stage free-energy perturbation method would entail a repeated application of the ghost appending process for chains of increasing length (from 1 to N). Although such an application would be computer intensive for long polymers, the real strength of the method lies in the fact that, for simple homopolymer systems, the segmental chemical potential becomes chain-length-independent for chains longer than a characteristic length N_c. Typically $N_c \sim O(10^1)$ and, if $N \gg N_c$, the calculation of the end-segment chemical potential $\Delta\mu_N$ could be employed to estimate the total chemical potential as $\mu_{\text{chain}} \approx N \times \Delta\mu_N$. Although such an approximation can yield valuable information, the accuracy of μ_{chain} determined by this approach is often insufficient for reliable phase equilibrium calculations.

A more prudent implementation of the chain-increment method entails the use of a two-stage insertion process. In the first stage, configurational bias can be used to obtain the chemical potential associated with the insertion of an N_c-mer chain into the system; then $\mu_{\text{chain}} \approx \mu_{N_c} + (N - N_c) \times \Delta\mu_N$. Note that μ_{N_c} and $\Delta\mu_N$ can be obtained in a single simulation run. This two-stage approach has been successfully used to estimate the chemical potential of relatively long chains (e.g., with 100 segments) required for vapor–liquid [84] and liquid–liquid [85] equilibria calculations. Simulation of polymers with complex architectures, such as branched polymers and copolymers, however, will require multiple stages.

V. IMPROVING SAMPLING THROUGH ENSEMBLE SELECTION

Conformational sampling and free energy sampling can sometimes be enhanced by working in ensembles other than the canonical, which happens to be the first choice for most studies. For example, in an open (grand canonical-type) ensemble, the allowed fluctuations in the number of molecules (and density) can actually accelerate the relaxation of the system to equilibrium by promoting drastic changes in the configurations of the system. A grand-canonical ensemble also provides a convenient approach to establish the relationship between chemical potential and other thermodynamic properties of the system.

A. Grand Canonical Ensemble

In a grand-canonical ensemble simulation of a pure fluid, the temperature, volume, and chemical potential are specified. The number of particles in the system, however, is allowed to fluctuate by means of insertions and deletions [46]. This approach shares some of the shortcomings of particle insertion methods (or Widom-type methods) for calculation of the chemical potential; for polymers, just as for direct simulation of the chemical potential (e.g., in a canonical ensemble), configurational-bias ideas [71,86] and the expanded ensemble formalism [77] can be advantageously adapted to perform grand-canonical simulations. Such implementations are straightforward extensions of the ideas discussed in previous sections and need not be given here again.

Grand-canonical simulations are somewhat more cumbersome to use than canonical or isothermal-isobaric ensemble simulations. Their implementation often requires exploratory work, because it is generally easier to anticipate or specify the density, pressure or composition of a system (one usually has some reference or an intuitive feeling about such properties), rather than its chemical potentials.

B. Semigrand Ensemble

A technique that can be particularly useful for simulation of mixtures is the so-called semigrand ensemble of Kofke and Glandt [93]. A thorough review of this technique and its applications is given in the chapter by D. Kofke (this volume). In that ensemble, ratios of the chemical potential of each component and that of a reference component are specified, whereas the composition of the system is allowed to fluctuate. The equilibrium composition is attained by trial changes of the identity of each species, at constant number of particles (such changes are accepted according to the imposed ratios of chemical potentials). Identity-exchange trials have a reasonably high acceptance rate when the components of the mixture are relatively similar (in both size and shape). When viable, these moves lead to drastic changes of the instantaneous configuration of the system, thereby leading to efficient sampling.

The semigrand ensemble method can be implemented in different forms for calculation of equilibrium properties, and phase equilibria for inert or reacting mixtures. Recently, it has been applied to simulate phase coexistence for binary polymer blends [85], where advantage was taken of the fact that identity exchanges are employed in lieu of insertions or deletions of full molecules. The semigrand ensemble also provides a convenient framework to treat polydisperse systems (see Section III.F).

C. Pseudo-Ensemble Methods

Mehta and Kofke [91] have recently proposed pseudo-ensemble simulations, in which calculations in a nominal ensemble are "mimicked" by simulation in a different ensemble or, more precisely, in a pseudo-ensemble. Mehta and Kofke implemented pseudo grand-canonical ensemble simulations in which particle deletion and insertion attempts (pertaining to the constant-μ_{GC} constraint in the grand-canonical ensemble) were replaced by volume changes. It has been shown [78] that the approach can be restated as one of attempting to transform a μVT ensemble into a NPT ensemble; such a transformation is possible if the equilibrium pressure P_{eq} corresponding to μ_{GC} can be determined; that link can be established by the thermodynamic relationship

$$dP = \rho \, d\mu, \qquad T = \text{constant} \tag{5.1}$$

If the system is in some initial state (e.g., ρ_0, P_0, and μ_0), and we seek conditions consistent with some specified chemical potential, say, μ_{GC}, numerical integration of Eq. (5.1) (a two-point boundary value problem) provides P_{eq}. Such an integration can be performed during a single simulation in several ways; for example, by employing an iterative approach based on a first-order expansion of Eq. (5.1):

$$P_{eq}^1 = P_0 + \rho_0(\mu_{GC} - \mu_0) \tag{5.2}$$

The value of P_{eq}^1 may initially be a poor estimate of P_{eq}; nevertheless, it should point in the right direction and successive, periodic refinements could then be implemented (e.g., by updating the initial point to the current conditions), thereby leading to convergence to P_{eq} and the corresponding equilibrium density.

The advantage of performing an NPT- in lieu of a μVT-ensemble simulation lies in the fact that the allegedly slow convergence associated with particle insertions can be avoided. However, use of Eq. (5.2) requires that some means be provided to evaluate the chemical potential during the simulation; fortunately, such evaluations can often be performed. An illustrative application of these ideas has been presented by Mehta and Kofke [91] for a system of dense hard spheres. Camp and Allen [92] have recently extended the pseudo grand-canonical ensemble method to the Gibbs ensemble [45]. In the latter case, particle transfer moves between coexisting phases are not attempted; instead, special volume moves are devised so as to mimic the effect of such transfers.

Mechanical equilibrium between two phases or between a system and an external reservoir can be achieved by allowing the volume of the system to fluctuate. Conventional volume moves [46], although expensive for molecular systems, are easy to implement. Specialized variants have been proposed to speed up the equilibration in systems of flexible polymers, including branched and crosslinked polymers [51,52].

If a μVT ensemble simulation can be *turned* into a ("quasi") NPT ensemble-type simulation (e.g., a pseudo-μVT ensemble), the inverse transformation (a pseudo-NPT ensemble) is also possible. The key relationship for a pseudo-NPT ensemble technique is Eq. (5.1) [78]. Such a reverse strategy can be practical only if molecular insertion and deletion moves can be performed efficiently for the system under study (e.g., by expanded ensemble moves for polymeric fluids). Replacing volume moves by particle insertions can be advantageous for polymeric and other materials that require simulation of a large system (due to the sluggishness of volume moves for mechanical equilibration of the system); such an advantage has been clearly demonstrated for a test system of dense, athermal chains [78].

Along a similar vein, the principles behind a pseudo-NPT ensemble simulation can be extended to the Gibbs ensemble formalism [78]. In one possible implementation of a pseudo Gibbs simulation for a single component system (fixed temperature), volume transfers between the two simulation boxes are never attempted; instead, two simultaneous grand canonical-type simulations are conducted (one for each box). The imposed chemical potentials for both phases are gradually and concertedly refined, starting from stable uniphase states, towards coexistence conditions of equal pressure and chemical potential. Figure 5 illustrates the operation of this algorithm.

In contrast to the standard Gibbs ensemble, this variant of the pseudo-Gibbs ensemble does not require à-priori knowledge of the location of the two-phase region; it is therefore particularly appealing for systems in which the two-phase region is very narrow. This is the case for the (narrow)

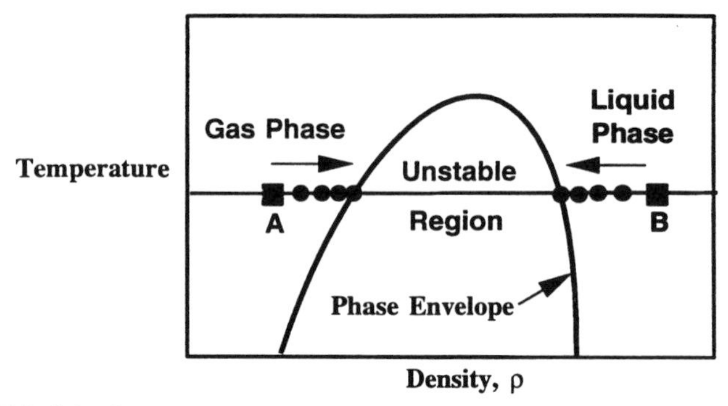

Figure 5. Principle of operation of a variant of pseudo-Gibbs ensemble. A coexistence curve (full line) is sketched on the T–ρ plane; the initial states are shown by the squares, and the sequence of points generated through pseudo-Gibbs moves are shown by circles.

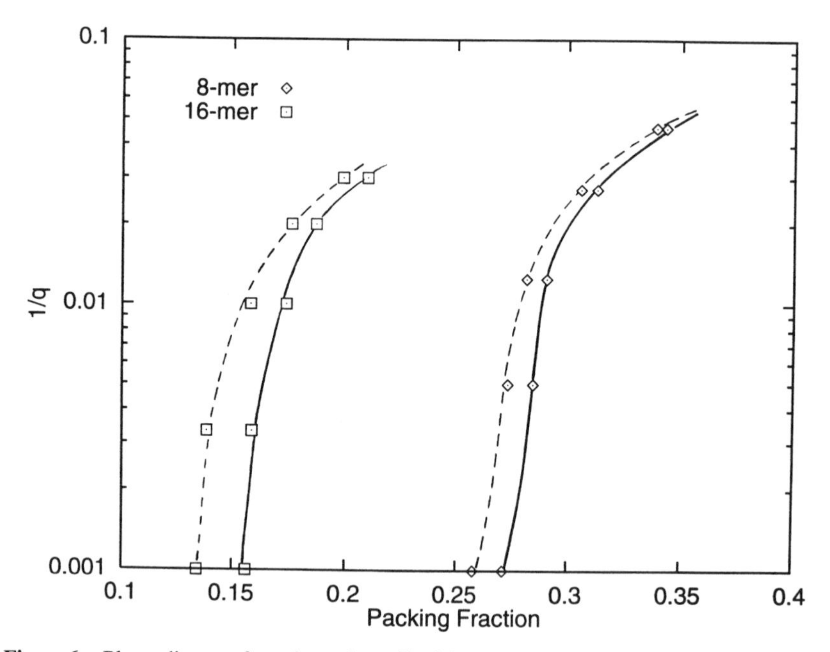

Figure 6. Phase diagram for athermal semiflexible chains of 8 (diamonds) and 16 (squares) segments generated by pseudo-Gibbs ensemble simulations (Fig. 5). The y axis shows the reciprocal persistence length (q), which provides a measure of chain flexibility; data shown at $1/q = 0.001$ actually correspond to the infinitely stiff limit ($1/q \to 0$). Chain flexibility increases as $1/q$ increases. (Adapted from Escobedo and de Pablo [79].)

isotropic–nematic phase diagrams exhibited by liquid crystals; Figure 6 shows representative coexistence curves for pure mesogenic polymers determined via a pseudo-Gibbs method (modeled as athermal, semiflexible chains) [79].

VI. HISTOGRAM REWEIGHTING

In a grand-canonical ensemble, the density and the energy of the system both fluctuate throughout the course of a simulation. It is therefore possible to construct density and energy histograms, which in essence provide the relative probability of finding the system at a given density or energy. Similarly, in a NPT ensemble it is possible to construct energy and volume histograms. As originally suggested by Ferrenberg and Swendsen [94], histograms generated under one set of thermodynamic conditions (e.g., μ, V and T) can be used to estimate the average properties of a system at some other, nearby conditions (μ', V', and T'). Furthermore, histograms generated at different conditions can be combined to generate a full thermodynamic "map" of the system, which can then be used to calculate thermodynamic properties, including phase equilibria. These ideas have been exploited in recent work to determine the phase behavior of pure polymers and blends on a lattice [95], and of pure off-lattice systems [78,96]. Unfortunately, at this point only one application of these ideas has been reported for polymer chains in a continuum [97]; most continuum studies have been limited to Lennard-Jones fluids, Stockmayer fluids, and water [96,98]. Applications to mixtures have been limited.

VII. CONCLUSION

In this review, particular attention has been given to the ability of different methods to sample the conformational space relevant for calculation of equilibrium properties, including free energies, of polymeric systems. For completeness, we have chosen to provide a broad (rather than a comprehensive) overview of methods having different degrees of difficulty. Indeed, the field of molecular simulation of polymers is rapidly evolving and numerous novel and often powerful methods emerge every year. In fact, some of the techniques discussed here are relatively recent and a lack of experience makes their outlook and future somewhat uncertain.

As should be apparent from our discussion, the efficiency of a given method is highly dependent on the details of the system under study. Size, chain length, degree of coarsening of the molecular model, presence of aggregates or inhomogeneities, topological complexity, and properties of interest are just a few of the factors that have to be considered before selecting a particular algorithm. In a sense, the specification of the details of the molecular model and the type of ensemble to be adopted is more critical than the selection of the simulation method itself; often both aspects, model setup and simulation method, are closely interconnected and the choice of one of them can determine the choice of the other.

Just as for continuum mechanics, where analytic solution of relevant equations has been gradually surpassed by numerical solution under more complicated circumstances, we expect that in the years to come numerical statistical mechanics will be the technique of choice for many new, challenging problems in many fields of science and engineering. It is likely, however, that most users will not be expert simulators. If a rational selection of methods is to be made, it is essential for such users to have knowledge of the advantages and disadvantages of a broad spectrum of techniques. This review has been written with such a need in mind; we have described a diverse selection of methods, and we have discussed in greater detail those techniques that, while being relatively simple and robust, exhibit a wide range of applicability.

ACKNOWLEDGMENTS

The authors are grateful to Professors D. Theodorou and D. Kofke for sending us preprints of their work prior to publication.

REFERENCES

1. K. Binder, ed., *Applications of the Monte Carlo Method in Statistical Physics*, Springer, New York, 1984.
2. K. Binder and D. W. Heermann, *Monte Carlo Simulation in Statistical Physics*, 2nd ed., Springer, New York, 1992.
3. D. Frenkel, "Numerical Techniques to Study Complex Liquids," in *Observation, Predictions and Simulation of Phase Transitions in Complex Fluids*, M. Baus et al., eds., Kluwer, Netherlands, 1995, pp. 357–419.
4. A. Baumgartner, "Simulations of Macromolecules," in *The Monte Carlo Method in Condensed Matter Physics*, K. Binder, ed., 2nd ed., Springer, Berlin, 1995, pp. 285–316.
5. E. Leontidis, B. M. Forrest, A. H. Widman, and U. W. Suter, *J. Chem. Soc. Faraday Trans.* **91**, 2355–2368 (1995).
6. K. Kremer, "Computer Simulation Methods for Polymer Physics," *Monte Carlo and Molecular Dynamics of Condensed Matter Systems; Conference Proceedings*, Vol. 49, K. Binder and G. Ciccotti, eds., 1996, pp. 600–650.
7. D. Frenkel and B. Smit, *Understanding Molecular Simulation*, Academic Press, 1996.
8. F. T. Wall and J. J. Erpenbeck, *J. Chem. Phys.* **30**, 634–637 (1959).
9. T. Garel and H. Orland, *J. Phys. A: Math. Gen.* **23**, L621 (1990).

10. R. Hegger and P. Grassberger, *J. Phys. A: Math. Gen.* **27**, 4069–4081 (1994). P. Grassberger and R. Hegger, *J. Phys.: Condens. Matter* **7**, 3089–3097 (1995).
11. P. Grassberger, *Phys. Rev. E* **56**, 3682–3693 (1997).
12. J. Sadanobu and W. A. Goddard III, *J. Chem. Phys.* **106**, 6722–6729 (1997).
13. Z. Alexandrowicz and N. B. Wilding, *Recoil Growth: An Efficient Simulation Method for Multi Polymer Systems*, preprint.
14. N. Metropolis, A. W. Rosenbluth, M. N. Rosenbluth, A. H. Teller, and E. Teller, *J. Chem. Phys.* **21**, 1087–1092 (1953).
15. I. Carmesin and K. Kremer, *Macromol.* **21**, 2819 (1988); H.-P. Deutsch and K. Binder, *J. Chem. Phys.* **94**, 2294 (1991); W. Paul, K. Binder, D. W. Hermann, and K. Kremer, *J. Chem. Phys.* **95**, 7726 (1991).
16. V. Tries, W. Paul, J. Baschnagel, and K. Binder, *J. Chem. Phys.* **106**, 738–748 (1997).
17. I. Gerroff, A. Milchev, K. Binder, and W. Paul, *J. Chem. Phys.* **98**, 6526 (1993).
18. K. Binder and W. Paul, *J. Polym. Sci. B: Polym. Phys.* **35**, 1–31 (1997).
19. W. Paul and N. Pistoor, *Macromolecules* **27**, 1249 (1994).
20. F. T. Wall and F. Mandel, *J. Chem. Phys.* **63**, 4592–4595 (1975); M. Bishop, D. Ceperley, H. L. Frisch, and M. H. Kalos, *J. Chem. Phys.* **72**, 3228–3235 (1980).
21. R. G. Larson, *J. Chem. Phys.* **83**, 2411–2420 (1984); ibid. **87**, 1642–1650 (1988).
22. M. Murat and T. A. Witten, *Macromolecules* **23**, 520 (1990).
23. D. Chandler, *Introduction to Modern Statistical Mechanics*, Oxford Univ. Press, New York, 1987.
24. G. M. Torrie and J. P. Valleau, *J. Comp. Phys.* **23**, 187–199 (1977).
25. W. L. Jorgensen, J. D. Madura, and C. J. Swenson, *J. Am. Chem. Soc.* **106**, 6638–6646 (1984).
26. C. Pangali, M. Rao, and B. J. Berne, *Chem. Phys. Lett.* **55**, 413–417 (1978).
27. P. Rossky, P. Doll, and H. L. Friedman, *J. Chem. Phys.* **69**, 4628–4633 (1978).
28. B. Mehlig, D. W. Heermann, and B. M. Forrest, *Phys. Rev. B* **45**, 679–685 (1992).
29. B. M. Forrest and U. W. Suter, *J. Chem. Phys.* **101**, 2616–2629 (1994).
30. B. M. Forrest and U. W. Wuter. *Mol. Phys.* **82**, 393–410 (1994).
31. D. W. Heermann and L. Yixue, *Macromol. Chem. Theory Simul.* **2**, 229 (1993).
32. D. G. Gromov and J. J. de Pablo, *J. Chem. Phys.* **103**, 8247–8256 (1995).
33. F. A. Brotz and J. J. de Pablo, *Chem. Eng. Sci.* **49**, 3015–3031 (1994).
34. J. J. de Pablo, M. Laso, and U. W. Suter, *J. Chem. Phys.* **96**, 2395–2403 (1992).
35. J. J. de Pablo, M. Laso, and U. W. Suter, *Macromolecules* **26**, 6180–6183 (1993).
36. J. J. de Pablo, M. Laso, and U. W. Suter, *J. Chem. Phys.* **96**, 6157–6162 (1992).
37. J. J. de Pablo, M. Laso, I. J. Siepman, and U. W. Suter, *Mol. Phys.* **80**, 55–63 (1993).
38. M. Laso, J. J. de Pablo, and U. W. Suter, *J. Chem. Phys.* **97**, 2817–2819 (1992).
39. J. I. Siepmann, *Mol. Phys.* **70**, 1145–1158 (1990).
40. I. J. Siepmann, S. Karaborni, and B. Smit, *Nature* **365**, 330–332 (1993).
41. G. C. A. M. Mooij, D. Frenkel, and B. Smit, *J. Phys. Condens. Matter* **4**, L255–L259 (1992).
42. D. Frenkel, G. C. A. M. Mooij, and B. Smit, *J. Phys. Condens. Matter* **4**, 3053 (1992).
43. J. Harris and S. A. Rice, *J. Chem. Phys.* **88**, 1298–1306 (1988).
44. J. I. Siepmann and D. Frenkel, *Mol. Phys.* **75**, 59 (1992).

45. A. Z. Panagiotopoulos, *Mol. Phys.* **61**, 813–826 (1987); A. Z. Panagiotopoulos, N. Quirke, M. Stapleton, and D. J. Tildesley, *Mol. Phys.* **63**, 527–545 (1988).
46. M. P. Allen and D. J. Tildesley, *Computer Simulation of Liquids*, Clarendon, Oxford, 1987.
47. P. H. Verdier and W. H. Stockmayer, *J. Chem. Phys.* **36**, 227–232 (1962).
48. A. Baumgartner and K. Binder, *J. Chem. Phys.* **71**, 2541–2545 (1979).
49. X.-J. Li and Y. C. Chiew, *J. Chem. Phys.* **101**, 2522–2531 (1994).
50. F. A. Escobedo and J. J. de Pablo, *J. Chem. Phys.* **102**, 2636–2652 (1995).
51. F. A. Escobedo and J. J. de Pablo, *J. Chem. Phys.* **104**, 4788–4801 (1996).
52. F. A. Escobedo and J. J. de Pablo, *J. Chem. Phys.* **106**, 793–810 (1997).
53. M. Vendruscolo, *J. Chem. Phys.* **106**, 2970–2976 (1997).
54. M. Dijkstra, D. Frenkel, and J.-P. Hansen, *J. Chem. Phys.* **101**, 3179–3189 (1994).
55. C. W. Yong, J. H. R. Clarke, J. J. Freire, and M. Bishop, *J. Chem. Phys.* **105**, 9666–9673 (1996).
56. N. Go and H. A. Sheraga, *Macromolecules* **3**, 178 (1970).
57. L. R. Dodd, T. D. Boone, and D. N. Theodorou, *Mol. Phys.* **78**, 961–996 (1993).
58. E. Leontidis, J. J. de Pablo, M. Laso, and U. W. Suter, *Adv. Polym. Sci.* **116**, 283–318 (1994).
59. M. W. Deem and J. S. Bader, *Mol. Phys.* **87**, 1245–1260 (1996).
60. P. V. K. Pant and D. N. Theodorou, *Macromolecules* **28**, 7224–7234 (1995).
61. M. L. Mansfield, *J. Chem. Phys.* **77**, 1554 (1982).
62. W. G. Madden, *J. Chem. Phys.* **87**, 1405 (1987).
63. V. G. Mavrantzas, D. N. Theodorou, T. D. Boone, and P. V. K. Pant, preprint (1997).
64. R. Dickman and C. K. Hall, *J. Chem. Phys.* **89**, 3168–3174 (1988).
65. J. Chang and S. Sandler, *Chem. Eng. Sci.* **49**, 2777–2791 (1994).
66. R. F. Cracknell, D. Nicholson, N. G. Parsonage, and H. Evans, *Mol. Phys.* **71**, 931–943 (1990).
67. J. C. Shelley and G. Patey, *J. Chem. Phys.* **100**, 8265–8273 (1994).
68. D. M. Tsangaris and J. J. de Pablo, *J. Chem. Phys.* **101**, 1477–1489 (1994).
69. N. Madras and A. D. Sokal, *J. Stat. Phys.* **50**, 109 (1988).
70. D. Wu, D. Chandler, and B. Smit, *J. Phys. Chem.* **96**, 4077–4083 (1992).
71. F. A. Escobedo and J. J. de Pablo, *Molec. Phys.* **89**, 1733–1755 (1996).
72. A. P. Lyubartsev, A. A. Martsinovski, S. V. Shevkunov, and P. N. Vorontsov-Velyaminov, *J. Chem. Phys.* **96**, 1776–1783 (1992).
73. P. Attard, *J. Chem. Phys.* **98**, 2225–2231 (1993).
74. R. D. Kaminsky, *J. Chem. Phys.* **101**, 4986–4994 (1994).
75. N. B. Wilding and M. Müller, *J. Chem. Phys.* **101**, 4324–4330 (1994).
76. F. A. Escobedo and J. J. de Pablo, *J. Chem. Phys.* **103**, 2703–2710 (1995).
77. F. A. Escobedo and J. J. de Pablo, *J. Chem. Phys.* **105**, 4391–4394 (1996).
78. F. A. Escobedo and J. J. de Pablo, *J. Chem. Phys.* **106**, 2911–2923 (1997).
79. F. A. Escobedo and J. J. de Pablo, *J. Chem. Phys.* **106**, 9858–9868 (1997).
80. D. G. Gromov, J. J. de Pablo, G. Luna-Barcenas, I. C. Sanchez, and K. P. Johnston, *J. Chem. Phys.* **108**, 4647–4654 (1998).
81. S. K. Kumar, I. Szleifer, and A. Z. Panagiotopoulos, *Phys. Rev. Lett.* **66**, 2935–2938 (1991).

82. K. K. Mon and R. B. Griffiths, *Phys. Rev. A* **31**, 956–959 (1985).
83. D. A. Kofke and P. T. Cummings, *Molecular Phys.* **92**, 973–996 (1997).
84. Y. Sheng, A. Z. Panagiotopoulos, S. K. Kumar, and I. Szleifer, *Macromol.* **27**, 400–406 (1994).
85. S. K. Kumar, *Phys. Rev. Lett.* **77**, 1512–1515 (1996).
86. M. Dijkstra and D. Frenkel, *Phys. Rev. E* **50**, 349–357 (1994).
87. M. Mehta and D. A. Kofke, *Chem. Eng. Sci.* **49**, 2633–2645 (1994).
88. R. Agrawal and D. A. Kofke, *Mol. Phys.* **85**, 43–59 (1995).
89. R. Agrawal and D. A. Kofke, *Mol. Phys.* **85**, 23–42 (1995).
90. M. Dijkstra and D. Frenkel, *Phys. Rev. E* **51**, 5891–5898 (1995).
91. M. Mehta and D. A. Kofke, *Mol. Phys.* **86**, 139–147 (1995).
92. P. J. Camp and M. P. Allen, *Molec. Phys.* **88**, 1459–1469 (1996).
93. D. A. Kofke, and E. D. Glandt, *Mol. Phys.* **64**, 1105–1131 (1998).
94. A. M. Ferrenberg and R. H. Swendsen, *Phys. Rev. Lett.* **63**, 1195–1198 (1989).
95. N. B. Wilding, M. Müller, and K. Binder, *J. Chem. Phys.* **105**, 802–810 (1996); M. Müller and N. B. Wilding, *Phys. Rev. E* **51**, 2079–2089 (1995).
96. N. B. Wilding, *Phys. Rev. E* **52**, 602–611 (1995).
97. G. S. Grest, M.-D. Lacasse, K. Kremer, and A. M. Gupta, *J. Chem. Phys.* **105**, 10583–10594 (1996).
98. K. Kiyohara, K. E. Gubbins, and A. Z. Panagiotopoulos, *J. Chem. Phys.* **106**, 3338–3347 (1997).

THERMODYNAMIC-SCALING METHODS IN MONTE CARLO AND THEIR APPLICATION TO PHASE EQUILIBRIA

JOHN P. VALLEAU

Chemical Physics Theory Group, Department of Chemistry. University of Toronto, Toronto, Canada M5S 3H6

CONTENTS

I. INTRODUCTION: PHASE TRANSITIONS AND MONTE CARLO

"Thermodynamic-scaling," in the context of Monte Carlo (MC) computations in statistical mechanics, refers to the technique of estimating the properties of a model system over a substantial *range* of thermodynamic states, simultaneously, in a *single* MC run. It is thus a particular application of what is known as "umbrella sampling" [1]. An umbrella sample is designed to span a substantial range of different physical situations in a single MC run, sampling on a distribution quite unlike an ordinary ensemble distribution. This can be useful in many contexts. For example one can explore potentials of mean force between elements of a system, and the activation barriers for (even slow) processes, or compare the thermodynamic stabilities of alternative conformations of a large molecule (between which transitions would not occur in a conventional simulation), and so on

Advances in Chemical Physics, Volume 105, Monte Carlo Methods in Chemical Physics, edited by David M. Ferguson, J. Ilja Siepmann, and Donald G. Truhlar. Series Editors I. Prigogine and Stuart A. Rice.
ISBN 0-471-19630-4

(for some examples, see Ref. 2); these are all in fact just determinations of relative free energies of systems variously constrained. When the constraints of interest are only those defining thermodynamic states we have *thermodynamic-scaling Monte Carlo* (ThScMC); in this case likewise we obtain automatically the relative free energies of the states throughout the range of states covered by the run. This is particularly useful in the MC study of phase transitions, and we focus on that application in this chapter.

We first discuss briefly that there are special problems associated with the MC exploration of phase transitions. The difficulties are associated with the fact that in practical computations only rather small systems, of, say, a few hundred particles, can be studied. Consider, as an example, a simple liquid–gas condensation. For a *macroscopic* (e.g., experimental) system there will be (at a subcritical temperature) a density range for which the two phases coexist at equilibrium. In a *small* system, however, the free-energy expense of this phase separation (described by the surface tension) will usually be too great to allow its occurrence. The characteristic effect in MC calculations is that both phases have much-extended ranges of metastability: one to unnaturally high density; the other to low density (see Fig. 1 for a famous example). This means that, in a MC simulation of finite length, within the coexistence density range, one is likely to persist in either one of these metastable phases without ever finding the other. This would imply that the two structures belong to distinct regions of configuration space, regions that are essentially disjoint from the standpoint of conventional MC sampling. This is an extreme example of the "quasi-ergodic" problem in Monte Carlo: the trapping of the Markov sampling chain in a subregion of the relevant part of the space during long periods. This is characteristic of the phase-separation region, and, even where it is less extreme than described above, will render the results unreliable in this region. This is disastrous, in the sense that it will mean that, using conventional Boltzmann-sampling MC, it will not even be possible to locate the coexisting phases. In particular, it is impossible to relate the free energies of those phases, for example by "thermodynamic integration" of the pressures, in the absence of reliable data joining them.

Now a *complete* evaluation of the configuration integral under the particular thermodynamic constraints (e.g., temperature and density) in the coexistence region would incorporate *all* the relevant subregions of configuration space, each with its correct weight. The expected result would, of course, be a continuous variation of free energy (and pressure), leading to a smooth van der Waals loop (in a finite system), and in principle allowing determination of the free energies and specification of the coexistence properties. But conventional MC runs cannot in practice adequately achieve the complete averaging required.

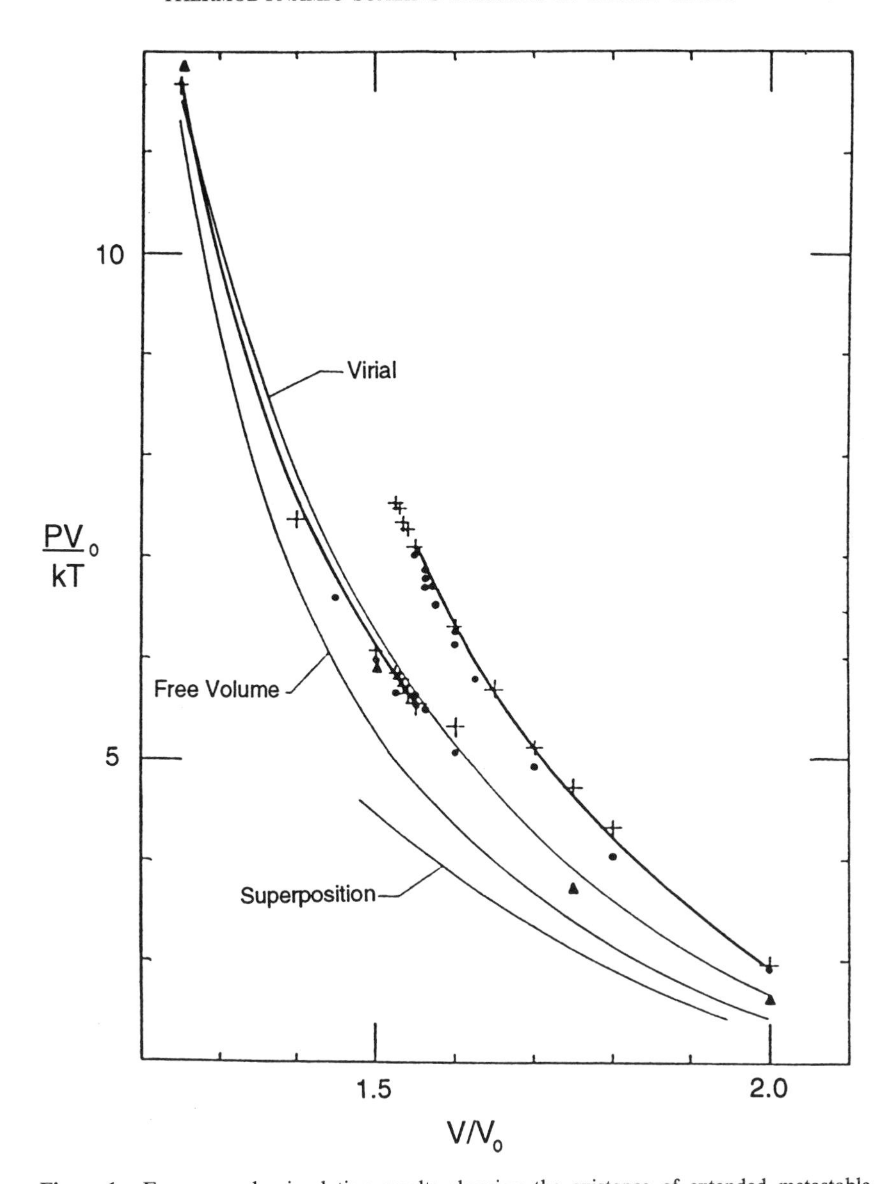

Figure 1. Famous early simulation results showing the existence of extended metastable behavior of two phases in a finite system of hard spheres. The heavy curves and plus signs (+) represent 108 and 32 particle molecular dynamics data of Alder and Wainwright; the circles are MC data due to Wood and Jacobson (V_0 is the close-packed volume). [Reprinted with permission from W. W. Wood and J. D. Jacobson, *J. Chem. Phys.* **27**, 1207 (1957).]

There are thus two requirements for a more effective MC approach. On one hand, there is a need for quite precise estimation of relative free energies, itself a classic MC problem. At the same time this will only be possible if the "quasi-ergodic" problem, characteristic of the phase-transition region, can be overcome somehow. We shall see that ThScMC addresses precisely these two aspects in which conventional MC methods are inadequate for the study of phase transitions—it samples in such a way as to visit appropriately all relevant parts of configuration space, and it automatically gives direct and rather precise determinations of relative free energy throughout the thermodynamic range studied.

Before describing this technique, one should point out that some other MC methods of addressing phase transition problems have recently been introduced. Of particular importance is the *Gibbs-ensemble Monte Carlo* (GEMC) method, due primarily to Panagiotopoulos [3]. This very pretty method eliminates the "surface tension" problem, described above, by dividing the simulation volume into two separate boxes, each containing one of the phases (and each with its own periodic boundary conditions), so that no surface is generated between the two phases. Panagiotopoulos showed what "steps" and transition probabilities were required to obtain equilibrium averages in such an ensemble, so that it leads to direct simulation of the coexisting phases. This has been successfully applied to many systems. It should be seen as a method complementary to ThScMC. It is easy to apply, and at present possibly more convenient for certain problems, such as the study of coexistent solutions. On the other hand, it cannot lead to comparable precision (especially where one of the phases is of high density, when particle transfers between the boxes become rare), and it has systematic difficulties [4] that may render it misleading close to the critical point. Both approaches are useful, therefore, the choice depending in large part on the information sought and the level of detail one wishes (in particular about the critical behaviour).

A still more recent method [5] concentrates on the critical region for fluids. It seeks to match the order-parameter distribution at the critical point to that of the Ising-system (presumed to belong to the same universality class), thereby allowing evaluation of some nonuniversal critical parameters. This is an exciting idea. Its application requires the adjustment of several fitting parameters; one hopes that it will prove to be robust and precise.

In principle it is possible to estimate the chemical potential for a given fluid state by the classic MC methods of the grand-canonical ensemble or the "Widom particle-insertion method," and these can be used to locate the coexistence curve. This has proved useful for some systems. However, these methods share a technical problem that means they can often not be used with confidence: at high densities it becomes rare to find a site for which "insertion" of a particle is allowed, or which is relevant for the Widom chemical potential estimation, so the results become unreliable. Various rather complicated attempts to overcome this problem have been only partially successful. (It can, in fact, be overcome, in the case of the "Widom" method, by an alternative application of umbrella sampling [6], but this seems usually to be less useful than applying ThScMC.) These chemical potential methods are not ideal for obtaining very precise coexistence data nor for detailed examination of the critical region itself.

The methods of thermodynamic-scaling Monte Carlo have been refined only slowly. In the original "umbrella-sampling" methodological paper [1], it was applied to cover a range of temperature for a fluid at fixed density [i.e., *temperature-scaling Monte Carlo* (TSMC)] and also to states differing instead in the Hamiltonians of the systems, as represented by their potential-energy functions [*Hamiltonian-scaling Monte Carlo* (HSMC)]. These one-dimensional ThScMC exercises are very easy to carry out, and already quite powerful for many purposes; they have been widely used (and some examples are mentioned in Section III). More useful for the study of first-order phase transitions is to "scale" with respect to the order parameter of the transition, but this is characteristically more challenging. In the case of fluids this means *density-scaling Monte Carlo* (DSMC) [7]; this was also applied to some transitions [8,9]. For several reasons, however, the real power of the method comes when "scaling" in more than one dimension, and this is not difficult to do. In particular, for the case of fluids, simultaneous *temperature-and-density-scaling Monte Carlo* (TDSMC) has shown itself extremely effective for several systems [10–12] exhibiting phase transitions. For this reason this chapter will focus especially on TDSMC, and in particular will present the detailed discussion of the ThScMC technique and its application in the context of its TDSMC variant.

The plan of the article is as follows. In the next section the ThScMC method is described, beginning from basic facts about statistical mechanics and the Markov chain Monte Carlo method, and cast in a form appropriate to two-dimensional scaling and in particular to TDSMC. Remarks about strategies for choosing an appropriate sampling distribution are included. In Section III we return to considering the simpler, one-dimensional versions of ThScMC, with respect to their technical simplicity and utility (although with no attempt to review systematically their applications). We also mention closely related work, such as "simulated tempering" [13], "multiple Markov chains" [14], and "multicanonical sampling" [15,16]; these can all be regarded as versions of one-dimensional ThScMC using particular choices of the sampling distribution and sampling technique. In Section IV we return to TDSMC by briefly reviewing its application to one test system, namely Lennard-Jonesium [11], and we conclude (Section V) with some further remarks on applying ThScMC.

II. THERMODYNAMIC-SCALING METHOD: TDSMC

To begin, focus on the MC estimation of the Helmholtz free energy A of a system of N classical particles. For simplicity of presentation we restrict ourselves for now to a one-component system of isotropic particles of mass m; its configurations can be described by a set of coordinates $\mathbf{q}^N \equiv (\mathbf{q}_1, \mathbf{q}_2, \ldots, \mathbf{q}_N)$, say. Then we have, in the canonical ensemble,

$$\beta A = -\log\left(\frac{Z}{N!\,\Lambda^{3N}}\right) \tag{2.1}$$

where $\beta \equiv (k_B T)^{-1}$ at temperature T (where k_B is Boltzmann's constant), the thermal wavelength Λ is $h(\beta/2\pi m)^{1/2}$ where h is Planck's constant, and Z is the configuration integral

$$Z \equiv \int_{(V)} d\mathbf{q}^N \exp(-\beta U(\mathbf{q}^N)) \tag{2.2}$$

where $U(\mathbf{q}^N)$ is the potential energy of the configuration $\mathbf{q}^N$ and the integral is over the volume V for each $\mathbf{q}_i$. Thus complete thermodynamic information would follow from evaluation of Z as a function of temperature and density.

Monte Carlo computations in statistical mechanics usually take the form, however, of attempts to estimate *ratios* of integrals over the configuration space, by using random-number sampling on some chosen sampling distribution, say $\pi(\mathbf{q}^N)$. The choice of this $\pi(\mathbf{q}^N)$ is crucial. Most integrals of statistical–mechanical interest have integrands that vary wildly in $\mathbf{q}^N$ space, and have substantial values only in very limited parts of the space. When this is so the sampling distribution $\pi(\mathbf{q}^N)$ must evidently be chosen to sample such regions preferentially; this is called "importance sampling."

It is important to recall what we mean by "sampling on some chosen sampling distribution, say $\pi(\mathbf{q}^N)$," since this will point up the difficulty of A estimation, and also how to overcome it. We may be able to choose an appropriate *shape* for our distribution, but ordinarily there will be no hope of *normalizing* that distribution to find an actual probability density. This limitation has important consequences. For one thing, it means that we cannot take *independent* samples in $\mathbf{q}^N$. The best we can do is to generate states $\mathbf{q}^N$ with the correct *relative* probabilities; this can be done, for an arbitrary unnormalized $\pi(\mathbf{q}^N)$, by generating a Markov chain of samples, as described in the first chapter of this volume (by Siepmann).

A second consequence is that we *cannot* estimate the value of any *individual* integral in configuration space [such as Z, in Eq. (2.2)]; for example, for an integrand $R(\mathbf{q})^N$

$$\int_{(V)} d\mathbf{q}^N R(\mathbf{q}^N) \equiv \frac{\int_{(V)} d\mathbf{q}^N \left(\frac{R}{\pi}\right)\pi}{\int_{(V)} d\mathbf{q}^N \pi} \int_{(V)} d\mathbf{q}^N \pi$$
$$\equiv \left\langle \frac{R}{\pi} \right\rangle_\pi \int_{(V)} d\mathbf{q}^N \pi(\mathbf{q}^N) \tag{2.3}$$

where $\langle\ \rangle_\pi$ signifies an average on the distribution $\pi(\mathbf{q}^N)$. It is straightforward to estimate a π average, such as $\langle R/\pi \rangle_\pi$, but the normalization integral remains unknown. On the other hand, in any *ratio* of such integrals the normalizing integrals will cancel, so we *can*, in principle, estimate such a ratio:

$$\frac{\int_{(V)} d\mathbf{q}^N R(\mathbf{q}^N)}{\int_{(V)} d\mathbf{q}^N S(\mathbf{q}^N)} \equiv \frac{\langle R(\mathbf{q}^N)/\pi(\mathbf{q}^N) \rangle_\pi}{\langle S(\mathbf{q}^N)/\pi(\mathbf{q}^N) \rangle_\pi} \tag{2.4}$$

This is the central result we use

A particular example of Eq. (2.4) would be the estimation of the canonical ensemble average of any "mechanical" quantity $M(\mathbf{q}^N)$ in a particular state (T, V), according to

$$\langle M(\mathbf{q}^N) \rangle_{\text{can}} \equiv \frac{\int_{(V)} d\mathbf{q}^N M(\mathbf{q}^N)\, e^{-\beta U(\mathbf{q}^N)}}{\int_{(V)} d\mathbf{q}^N\, e^{-\beta U(\mathbf{q}^N)}}$$
$$\equiv \frac{\langle M\, e^{-\beta U}/\pi \rangle_\pi}{\langle e^{-\beta U}/\pi \rangle_\pi} \tag{2.5}$$

"Conventional" MC work, as pioneered by Metropolis et al., is directed at such estimates; it ordinarily uses the convenient (although not optimal) choice of π as proportional to the Boltzmann distribution $\exp(-\beta U)$, when Eq. (2.5) becomes

$$\langle M(\mathbf{q}^N) \rangle_{\text{can}} = \langle M(\mathbf{q}^N) \rangle_\pi \tag{2.6}$$

In view of the above considerations it appears that we cannot estimate the free energy, since according to Eq. (2.1), that means estimating the configuration integral Z itself. This seems discouraging. It is important to note, however, that only *differences* of free energy between thermodynamic states are physically meaningful. Thus in reality we could not do better than to estimate quantities like

$$\Delta(\beta A) = \beta_j A_j - \beta_i A_i = \frac{3N}{2} \log \frac{T_i}{T_j} + \log \frac{Z_i}{Z_j} \tag{2.7}$$

where the subscripts i, j refer to two different thermodynamic states. We see that this depends only on the *ratio* Z_i/Z_j, which is exactly the sort of quantity we *can*, in principle, estimate according to Eq. (2.4).

In thermodynamic-scaling Monte Carlo (ThScMC) one estimates relative free energies like Eq. (2.7), not, however, simply between *two* states but among arbitrarily *many* states over a *substantial region* of the thermodynamic state space, within a *single* MC sampling run.

In this introductory discussion we focus especially on fluid behavior, with thermodynamic states described by (T, ρ). However, the same techniques can be (and have been) applied more generally. For example, some "order parameter" other than the volume or density might be relevant (e.g., the magnetisation of a lattice of spins). Also the "states" may differ not (or not only) in their thermodynamic constraints, but by having different potential energy functions $U_i(\mathbf{q}^N)$, and hence different Hamiltonians; this is known as Hamiltonian-scaling (HSMC). We will mention such variants below in (Section III). ThScMC is at its most powerful when "scaling" with respect to more than one variable, as in TDSMC; we stress TDSMC in what follows.

To allow TDSMC (or any density-scaling), a minor technical improvement is required. The preceding discussion, including Eqs. (2.5) and (2.7) as written, assumed that all integrals were in the *same* configuration space $\mathbf{q}^N$, of volume V^N. However, if two N-particle states i, j correspond to different densities ρ_i, ρ_j, then the two $\mathbf{q}^N$ configuration spaces will be different, having volumes V_i^N and V_j^N (where $\rho_i = N/V_i$). However, we can easily describe them in the *same* space by expressing the configurations in terms of *reduced* coordinates $\mathbf{r}$, given by

$$\mathbf{r} \equiv \frac{\mathbf{q}}{L_k} \qquad \text{with} \qquad L_k \equiv V_k^{1/3} \tag{2.8}$$

The space of $\mathbf{r}^N$ has unit volume. With *each* (reduced) configuration $\mathbf{r}^N$ there is then associated a configuration $\mathbf{q}^N$ for *every* density $\rho_k = N/V_k$, and a corresponding potential energy

$$U(\mathbf{q}^N) = U(\mathbf{r}^N, L_k) \tag{2.9}$$

the value of which depends on ρ_k through L_k [see Eq. (4.1) for an explicit example]. Using these reduced coordinates, we may rewrite the configuration integral Z [Eq. (2.2)] at a particular (T_i, ρ_i) as

$$\begin{aligned} Z(N, T_i, \rho_i) &= V_i^N \int_{(1)} d\mathbf{r}^N \exp(-\beta_i U(\mathbf{r}^N, L_i)) \\ &= V_i^N Z^{\text{ex}}(N, T_i, \rho_i) \end{aligned} \tag{2.10}$$

which defines an "excess" configuration integral Z^{ex}; the integral is over unit volume for each $\mathbf{r}$, regardless of the densities considered. It is easy to relate Z^{ex} directly to the thermodynamics, since (in view of the definition of ideality) we can rewrite Eq. (2.7), using (2.10), as an equation for the relative *excess* free energy:

$$\begin{aligned} \Delta(\beta A^{\text{ex}}) &\equiv \beta_j A_j^{\text{ex}} - \beta_i A_i^{\text{ex}} \\ &= (\beta_j A_j - \beta_i A_i) - (\beta_j A_j^{\text{id}} - \beta_i A_i^{\text{id}}) \\ &= (\beta_j A_j - \beta_i A_i) - \left(N \log \frac{V_j}{V_i} + \frac{3}{2} N \log \frac{T_i}{T_j} \right) \\ &= -\log \frac{Z_j^{\text{ex}}}{Z_i^{\text{ex}}} \end{aligned} \tag{2.11}$$

Thus the ratio of the *excess* configuration integrals gives directly the relative reduced *excess* free energies; this ratio [see Eq. (2.4)]

$$\frac{Z_j^{\text{ex}}}{Z_i^{\text{ex}}} = \frac{\langle \exp(-\beta_j U(\mathbf{r}^N, L_j))/\pi(\mathbf{r}^N) \rangle_\pi}{\langle \exp(-\beta_i U(\mathbf{r}^N, L_i))/\pi(\mathbf{r}^N) \rangle_\pi} \tag{2.12}$$

is that of simple averages taken during MC sampling on the distribution $\pi(\mathbf{r}^N)$, and is asymptotically unbiassed. In a typical TDSMC run one would collect the averages appearing in Eq. (2.12) for a quite large number of thermodynamic states in a target region of (T, ρ); then the relative free energy of every pair of those states is available from (2.12). [If the Hamiltonians were also to be different in the two "states," this would correspond to different potential-energy functions U_j, U_i in the numerator and denominator of Eq. (2.12).]

Of course, the canonical average of any configurational quantity $M(\mathbf{q}^N) = M(\mathbf{r}^N, L_i)$, at any particular thermodynamic state (T_i, ρ_i), can also be estimated by the analog of Eq. (2.5), expressed in the reduced space:

$$\langle M(\mathbf{q}^N)\rangle_{\text{can}}^{T_i,\rho_i} \equiv \frac{\int_{(1)} d\mathbf{r}^N\, M(\mathbf{r}^N, L_i)\, e^{-\beta_i U(\mathbf{r}^N, L_i)}}{\int_{(1)} d\mathbf{r}^N\, e^{-\beta_i U(\mathbf{r}^N, L_i)}}$$
$$= \frac{\langle M(\mathbf{r}^N, L_i)\, e^{-\beta_i U(\mathbf{r}^N, L_i)}/\pi(\mathbf{r}^N)\rangle_\pi}{\langle e^{-\beta_i U(\mathbf{r}^N, L_i)}/\pi(\mathbf{r}^N)\rangle_\pi} \tag{2.13}$$

This allows the estimation of structural quantities such as the pair correlation function. Thermodynamic quantities such as the internal energy and pressure can also be estimated using Eq. (2.13). However, all thermodynamic properties are, in principle, also available by using derivatives of the free energy A with respect to the temperature and/or density. Since in TDSMC the relative free energy can easily be estimated on an arbitrarily fine regular grid of (T, ρ) states, numerical differentiation is very convenient. Numerical differentiation of inherently noisy data is seldom an attractive prospect, but it turns out that in the present case it leads to marginally more precise estimates than direct estimation using Eq. (2.13); this is so even for second-order quantities such as the heat capacity and compressibility (which require of course estimates of fluctuation quantities in the direct computation). This is illustrated in Section IV. Since this use of thermodynamic estimates obtained from A differentiation [rather than by evaluating Eq. (2.13)] also avoids a lot of complicated and time-consuming data collection during the run, it is clearly to be preferred; as far as thermodynamics is concerned, therefore, one need collect only a *single* average at each state (T_i, ρ_i)—specifically, $\langle \exp(-\beta_i U(\mathbf{r}^N, L_i)/\pi\rangle_\pi$, as shown in Eq. (2.12). This is one of the TDSMC advantages of scaling *both* temperature and density simultaneously.

Among the thermodynamic functions easily calculated is of course the chemical potential. One might remark that the difficulty with the traditional MC estimation techniques for chemical potentials, namely the grand-canonical and Widom particle-insertion methods, is that one must attempt insertion or deletion of at least one whole discrete particle. This corresponds to a major disruption of the system, and is accordingly infrequently allowed (or carries negligible weight), at any substantial density, which leads to unreliability in the estimates. In this context the ThScMC method has the advantage of working instead with a fixed number of particles, and examining density variation only in terms of that of the *volume*, a continuous variable that we can study on an arbitrarily fine scale.

All of the above is formally straightforward. Evidently the key to success is in being able to choose an appropriate sampling distribution $\pi(\mathbf{r}^N)$, namely one that adequately samples the regions of $\mathbf{r}^N$ important to *every* state in the target (T, ρ) region. A MC sampling distribution that "covers" the regions of configuration space relevant to more than one physical state is known as an "umbrella sample" [1]. Since the form of $\pi(\mathbf{r}^N)$ is not known *a priori*, choosing it looks at first sight like a daunting challenge. However generating a useful umbrella distribution has turned out not to be difficult: see "Section II.A, where one approach is described.

The second basic difficulty facing conventional MC work near phase transitions was noted above: the common occurrence of quasi-ergodicity. Specifically, the sampling of a particular thermodynamic state will often become trapped for long periods in a subregion of the part of phase-space appropriate to the state; indeed, the relevant region of phase space will often contain subregions which are effectively disjoint under conventional sampling on a distribution proportional to the state's own probability density. One may hope for dramatic relief from this problem under TDSMC. The reason is that if $\pi(\mathbf{r}^N)$ is designed to cover a wide range of states then the configurations sampled in the Markov chain will in any case repeatedly travel entirely away from the region relevant to some particular state; when the chain finally returns to that region, it will do so from some new direction, having "forgotten" its previous visits. One can expect, after many such excursions, to sample all the subregions of $\mathbf{r}^N$ appropriate to the state, and with the correct relative probability. This escape from quasi-ergodicity seems to be borne out in practice.

Naturally sampling on a broad distribution, from which the results for many different physical states may be obtained, requires longer runs than conventional single-state sampling. On the other hand, *each* sampled configuration $\mathbf{r}^N$ contributes in principle to the required average $\langle \cdots \rangle_\pi$ at *every* (T, ρ) state. In practice a given configuration will contribute *substantially* only to a cluster of nearby states, of course. Nevertheless, if, as is usual, one wishes to extract data for a fairly dense grid of states in the target region, then the fact that each sampled configuration contributes to a range of these states is an important factor in the efficiency of the method. This fact also has a negative consequence, however—the estimated averages for adjacent pairs of states are not independent, but correlated. This means, in turn, that in order to make correct estimates of the uncertainty in $Z_j^{\text{ex}}/Z_i^{\text{ex}}$ [Eq. (2.12)] and hence of the relative free energies of two states i, j [Eq. (2.11)], one needs to know not only the variances of both averages in Eq. (2.12) but also their *covariance*. If, as is usually the case, the errors are estimated by using block averages (of sufficiently long blocks of the samples making up the chain), it is straightforward to estimate the covariances for each pair of states i, j at the same time as the variances for each state. It is only necessary to store the block-averages $\langle \exp(-\beta U)/\pi \rangle_{\pi}$ for every targeted state.

Usually one will want also to extract estimates of a variety of thermodynamic quantities other than the free energy itself, and this will require numerical differentiation of A. For this purpose it is most convenient to extract and record the block averages for a *fairly dense regular grid* of states—probably several hundred of them. Then the required derivatives, along with the variances and covariances that figure in estimating their errors, can easily be obtained directly using familiar formulas [17]. It is convenient to write an analysis program that derives the free energy and all the other desired thermodynamic quantities simultaneously from the block averages, along with their statistical uncertainties.

In principle, one could target an arbitrarily large range of states (T, ρ), thus carrying out any desired investigation in a *single* TDSMC run. In practice, that is not always the most efficient strategy, however. That is because the MC Markov chain constitutes a random walk in phase space. As is well known, to explore a wide space a random walk requires eventually a "time" that is quadratic in the span; this argues for dividing the investigation into overlapping runs each investigating only part of the desired range of states. On the other hand, if this is done, an additional analysis must be carried out to estimate the free energy match at the "seams" between the runs. Not only is this a nuisance, but it introduces an additional error (that of the matching) into the results for distant states. So a trade off is indicated: a guess must be made of the number of constituent runs that might optimize numerical efficiency (perhaps in some combination with human convenience); see Section IV for an example.

A. Selection of a Sampling Distribution

Although the preceding description is formally correct for an arbitrary sampling distribution $\pi(\mathbf{r}^N)$, in practice the selection of π is, of course, critical if the ThScMC method is to be workable and reasonably efficient. Since an appropriate form for π is not known a priori, finding one sounds a rather formidable task, and this may have frightened off some potential ThScMC users. The efficient generation of suitable sampling distributions is somehow the central issue in applying ThScMC methods. Happily some rather effective practical answers have been developed in this regard, and we mention some here.

What is required of $\pi(\mathbf{r}^N)$ is clear enough:

1. It should sample the subregions of $\mathbf{r}^N$ important to each of the family of targeted thermodynamic states, and with roughly comparable weights.
2. A second consideration is equally important: $\pi(\mathbf{r}^N)$ needs to be fairly uniform over a substantial region of $\mathbf{r}^N$ embracing those sub-regions, to ensure adequate mobility among the subregions.

The second condition may sometimes be satisfied automatically by the first. This will occur if the targeted states are chosen sufficiently dense in thermodynamic state space that there is strong overlap of the $\mathbf{r}^N$ subregions corresponding to adjacent states. However, the relative breadths of the relevant distributions become narrower as the system size N increases, and the second condition becomes more demanding.

It is clear from the start that in selecting $\pi(\mathbf{r}^N)$ some sort of trial-and-error procedure is required, but this is possible at various levels of sophistication. As we point out in Section III, rather crude approaches are perfectly adequate for the case of simple temperature scaling. However, if one wants to cover substantial ranges of more than one thermodynamic variable, as in TDSMC, a more elaborate strategy helps. Many strategies are possible, and the most convenient choice may depend on the nature of the model under study; there is no need to adopt a universal answer. What we do here is to point out one very practical approach (and eventually we suggest how one might improve on it); this discussion also gives some general insight about choosing π.

Consider expressing $\pi(\mathbf{r}^N)$ as a weighted sum of the Boltzmann distributions for a grid of states k in the target (T, ρ) region:

$$\pi(\mathbf{r}^N) = \sum_k \omega_k \exp(-\beta_k U(\mathbf{r}^N, L_k)) \tag{2.14}$$

In the case of TDSMC, the states might form a rectangular grid $k = (T_k, \rho_k)$ distributed uniformly in the target (T, ρ) region. If such a grid is made sufficiently dense, the overlap of the distributions of adjacent states will ensure that condition 2 is satisfied. [The set of states k appearing in Eq. (2.14) may or may not coincide with the set of states i, j for which data are being collected to evaluate the averages in Eq. (2.12) or (2.13).] With this choice [Eq. (2.14)] for $\pi(\mathbf{r}^N)$, one will at least not sample in irrelevant parts of $\mathbf{r}^N$!

It remains to choose the weights ω_k to satisfy condition (i). But it is clear that each state k will contribute an *equal* weight to $\pi(\mathbf{r}^N)$ if Eq. (2.14) is actually proportional to the sum of distributions *normalized* in $\mathbf{r}^N$. That is [as can be seen from Eq. (2.11)]

$$\omega_k \propto \exp(\beta_k A^{\text{ex}}(T_k, \rho_k)) \tag{2.15}$$

Of course, this requires the excess free energy, which is exactly what we are seeking to estimate! So this emphasizes the trial-and-error nature of determining a suitable $\pi(\mathbf{r}^N)$, and may seem a little discouraging. But in fact the result in Eq. (2.15) is most useful, as we now explain:

1. For a given system size N, it is straightforward to refine our initial guess for ω_k and thus for $\pi(\mathbf{r}^N)$. That is, a short run with even a poor choice of $\pi(\mathbf{r}^N)$ will lead by Eqs. (2.11) and (2.12) to a new estimate of βA^{ex} that is essentially unbiassed, and that will therefore give immediately a new and better $\pi(\mathbf{r}^N)$. With one or two such refinement steps one will ordinarily have an excellent distribution suitable for a long "production" run—the trial A^{ex} values in Eq. (2.15) will already be fairly close to the final estimate to be obtained in the long run, which is all that is required. The approximate self-consistency is indeed one useful test of the adequacy of the sampling distribution.

2. Having results for some system of N particles, it is easy to proceed to a larger system, say, of N'. That is because the excess free energy is an extensive property, proportional to the size N of the system (except for a small genuine N dependence in finite systems). So the N-particle results for βA^{ex} immediately give a good approximation for the ω_k of the larger system. Often this approximation will be perfectly satisfactory; if not (due to the actual N dependence of A^{ex}) a cycle of refinement as discussed above can easily be carried out. One can therefore very easily "bootstrap" the computations beginning from a small-N system. The problem of generating appropriate sampling distributions for any N is evidently solved by Eqs. (2.14) and (2.15) if only we can find a barely adequate guess for $\pi(\mathbf{r}^N)$ for a *small-N* system.

3. But for sufficiently small N even the roughest of guesses will be sufficient to get started. [This is implicit in the preceding prescription, where the fact that A^{ex} is extensive implies a low sensitivity of ω_k, Eq. (2.15), to errors in A^{ex} for small N; in addition, the distributions will overlap strongly in this case.] If a theoretical estimate exists, however crude, it will be adequate as a start; indeed any educated hunch will do for very small N. If one were completely at sea, an easy beginning might be made by doing an ordinary canonical MC run for a state near the center of the target (T, ρ) region, roughly estimating the energy and pressure; these give the local temperature and density dependence of the free energy: for a small-N system a linear approximation for the variation of A^{ex}, based on these values, will almost certainly be adequate to give an initial set of trial ω_k.

It is not difficult to monitor the sampling to ensure that $\pi(\mathbf{r}^N)$ is indeed covering the required ranges, and that it is doing so in a reasonably smooth and uniform way (see Section IV).

The scheme just described (2.14), (2.15) for generating an effective sampling distribution π has some attractive features. For one, an improved $\pi(\mathbf{r}^N)$ corresponds merely to improved estimates of βA^{ex}, a quantity that is to be estimated in any case; thus no extra computations are required while refining π. And the fact that one knows that βA^{ex} is (very nearly) proportional to N means that there is no need to start from scratch when seeking π for a larger or smaller system size N. The scheme has proved simple and effective in practice, and was used unmodified to study Lennard-Jonesium [10] and the square-well fluid [11] by TDSMC.

Nevertheless it is perhaps not quite ideal as described. That is because one may want many terms in Eq. (2.14), so that the computing time needed for repeated calculations of π may not be negligible. Furthermore this problem will worsen as one moves to larger N. [This is because the relative widths of the single-state distributions will of course narrow, so that eventually a finer grid in Eq. (2.14) will become necessary to ensure reasonable smoothness of π.] One can easily think of ways of using more-quickly-computed (and automatically smooth) functions for π, rather than Eqs. (2.14) and (2.15), while still incorporating the insight into N dependence that we gained above. (For example, one might express it as an analytic function involving a small number of parameters that could be refined on the basis of preliminary short runs—see Section III for an example in one-dimensional scaling—and which could be scaled appropriately with N.)

What we describe here is instead a very minor modification of Eqs. (2.14) and (2.15) for the purpose of economising the computation time for π. Consider some configuration $\mathbf{r}^N$, in which the separation of the closest pair of particles is r_{nn}. It is easy to see that the only terms of Eq. (2.14) that will contribute substantially to $\pi(\mathbf{r}^N)$ are those for which $\tilde{\sigma}_k$ is in the vicinity of r_{nn}, where $\tilde{\sigma}_k$ is some sort of (reduced) repulsive diameter for the density ρ_k: otherwise the energy $U(\mathbf{r}^N, L_k)$ is bound to be high. Thus it would be equally satisfactory to replace Eq. (2.14) by a sum over only a small "window" of states k with densities in that vicinity; the window will move around as r_{nn} changes. That will reduce the computing time for $\pi(\mathbf{r}^N)$ markedly. Furthermore the density spacing of the states in the window can be decreased if required at large N, and the width of the window proportionately decreased too (if necessary interpolating the A^{ex} values to give the ω_k values within the window)—so the number of terms need not now increase rapidly as N increases. A dramatic application was to the restricted primitive model (RPM) of Coulombic systems, where tests showed [12] that it is fully adequate to include terms corresponding to only a *single* moving density determined by the condition $\tilde{\sigma}_k \equiv \sigma/L_k = r_{nn}$, where σ is the hard-core diameter; the adequacy of this very simple form may well depend, however, on the hardness of the repulsive core in the RPM, coupled with the fact that the forces in the RPM are otherwise long-ranged.

III. SOME ONE-DIMENSIONAL APPLICATIONS

The earliest applications of ThScMC were unambitious: they attempted only one-dimensional "scaling," and that over cautiously small ranges. They were concerned with variations either of the potential-energy function $U(\mathbf{q}^N)$ (and hence of the system's Hamiltonian, whence HSMC) [1,19,20] or of the temperature alone (TSMC) [1,20]; for technical reasons (discussed below) these applications are extremely easy to carry out.

HSMC allows one, for example, to obtain the free energies of model systems, characterized by a parametrized set of Hamiltonians, by relating them to that of a reference model for which the free energy has already been investigated [1] (e.g., Lennard-Jones particles by reference to "soft spheres" [1,18], dipolar hard spheres by reference to plain hard spheres [19], and so on). Another example would be to find the thermodynamics of mixing, by comparing the free energy of a mixture to that of a pure fluid [20,21]. Such applications readily form part of a phase-transition study (this section). In a nice application, Frenkel and Ladd [22] studied the relative stabilities of different crystal forms by relating their free energies to the corresponding Einstein crystals. Lee and Scott [23] measured surface tension as the free energy of creating a surface; here the Hamiltonian alteration concerned only the boundary conditions!

In applying TSMC to a fluid, one covers a substantial range of temperature, typically at fixed density. This already offers a way to explore many phase transitions in some detail, by creating a map of the free energy at a discrete set of temperature-concentration states in the region of the transition. For example, such a map could be constructed [1,20,24,25] by first obtaining the free energy along some isotherm (supercritical, to avoid quasiergodic problems) by using HSMC, by thermodynamic integration, by grand-canonical MC or in some such way; TSMC can then be used to obtain the free energy relative to that on the reference isotherm as a function of temperature at several densities or concentrations, resulting in a two-dimensional chart of free energy (see Fig. 2). (It may be appropriate to take the reference isotherm at infinite temperature [25], especially for systems with a hard-core repulsion, since the hard-sphere free energy is already known to good accuracy). The map of A can then be used to locate the coexistence curve and find the thermodynamics of the transition; this procedure has been applied to both liquid–vapor [24,25] and coexistent-solution transitions [20]. (For the former, at least, it will be less precise and less economical than a TDSMC study, however.)

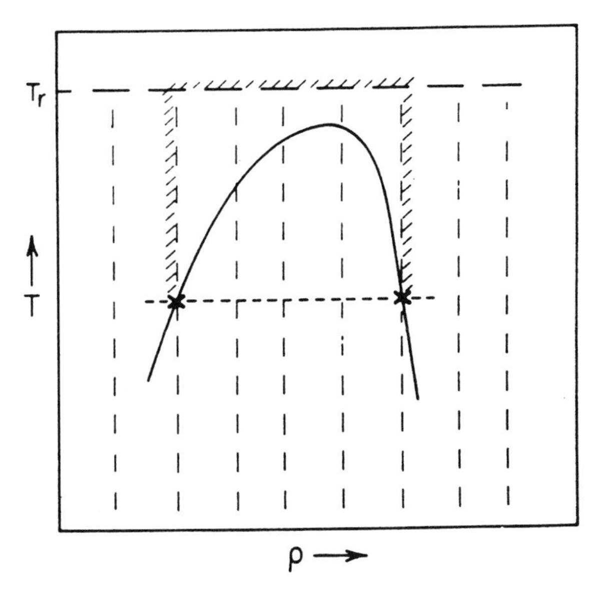

Figure 2. Schematic representation of an approach to studying a first-order transition using one-dimensional scaling: T_r marks a supercritical reference isotherm along which the free energy is obtained, by any of several methods. Then TSMC, carried out at various ρ values, relates the free energies along the isochores to those on the reference isotherm, thus creating a (T, ρ) chart of the free energy. The hatched path draws attention to the computational path by which the relative free energies of coexisting phases are actually obtained. The dashed line corresponds to the shorter path of DSMC, instead. (While in TDSMC the whole region might be covered in a single run.) (Reprinted with permission from Valleau [7], courtesy of *Journal of Computational Physics.*)

TSMC and HSMC are very undemanding with respect to the choice of a useful sampling distribution π. The effect of changing temperature is only to shift the corresponding energy distribution in a simple way. As a result to cover a range of temperatures only requires that π be a function of the energy, $\pi(U)$, such that an appropriately broad range of energy is sampled in a mildly uniform way. This is very easy to do. It can be done entirely numerically [1], recording π as a histogram; starting with some arbitrary trial π, one can progressively improve it—for example, simple by multiplying it by the inverse of the resulting nonuniform energy distributions in a series of short runs (with suitable safeguards against explosion of π in the wings of the distribution)—until an adequately smooth function is obtained over the required range. A more elegant alternative is to introduce an analytic function $\pi(U)$ with a small number of parameters that can be adjusted by looking at the resulting energy distribution output of a few successive short runs, in order to find a $\pi(U)$ of the desired ranged and smoothness. (Such a procedure is described below.) It is easy to write a small programme that will analyze the energy output and propose improved parameters. Of course, one could use more elaborate formulations of π, such as that proposed in Eqs. (2.14) and (2.15) (but now summing only over a set of *temperatures*, of course), but for TSMC the simpler approaches seem perfectly adequate and easy, and may save computer time. Similar remarks apply to the choice of π for HSMC, where it is only required that the sampling cover an adequate range of ΔU, the energy difference between the reference system and those with a modified Hamiltonian; this can be done along the lines discussed for TSMC [1].

To study transitions in the Ising lattice and similar problems, a different use of TSMC is attractive. One can simply carry out TSMC at zero field, say, without constraining the order parameter (such as the Ising magnetization). One will seek to cover a wide temperature range, including, of course, the critical temperature. Then one can carry out a complete investigation of the model in a *single* TSMC run, estimating the magnetization, heat capacity, susceptibility and free energy, for example, both above and below the critical temperature, using analogs of Eqs. (2.5) and (2.7), at as many temperatures as desired. This has been tested [26] for the Ising model and for a realization of the "random-field Ising model." It is only necessary to choose the sampling probability $\pi(\mathbf{S})$ for a configuration $\mathbf{S}$ to give an appropriately broad and smooth energy distribution. A simple and convenient form turned out to be an exponential of a polynomial of the energy:

$$\pi(\mathbf{S}) = \pi(U) \propto \exp(-\beta_0 U + a(U - U_0)^2 + b(U - U_0)^4)$$

for the required range $U_{\min} < U < U_{\max}$, where β_0 was a temperature in midrange and U_0 the corresponding mean energy; this form clearly broadens the sampled energy distribution with respect to that at β_0 alone, as required. The parameters a,b were adjusted independently in the two ranges $U < U_0$ and $U > U_0$, using short preliminary runs; a simple computer algorithm suggested improved values of a, b on the basis of output energy distributions, making this painless. The required energy range $[U_{\min}, U_{\max}]$ also emerges naturally during the preliminary runs. Figures 3 and 4 show some sample results.

This is simple and straightforward, and very efficient (since each sampled configuration is relevant to a substantial temperature range). Two points are worth making: (1) At low temperatures two directions of polarization are relatively stable. To sample both depends on repeated migrations between configurations appropriate to those low temperatures and to

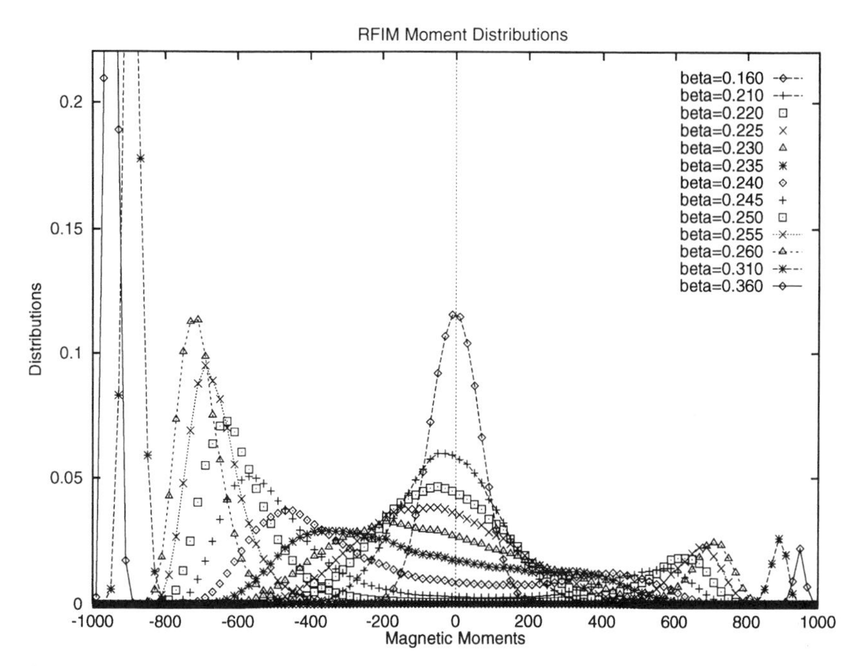

Figure 3. Distributions of magnetic moments in one realization of the random-field Ising model, at a selection of the temperatures covered by a single TSMC run. (The curves are merely a visual aid).

(rapidly mixing) supercritical temperatures; successive low-temperature excursions will explore both directions, giving eventually correct averaging. [For a truly symmetrical model, such as the zero-field Ising model itself, or with mild asymmetry, convergence can be improved by including MC steps that attempt to reverse the directions of all the spins simultaneously; this was not attempted for the (asymmetric) random-field case reported in the figures.] (2) The problem of "critical slowing down" near the critical point appears to be very greatly relieved with this TSMC sampling, but not totally eliminated. The relief presumably results from the repeated reentry into the large-fluctuation critical region, coming sometimes from more random regimes, sometimes more ordered: the sampling is forced to pass repeatedly through the critical region, with new starting points each time.

Viewed in another light, such a TSMC experiment may be seen as a way of seeking out and exploring the low-energy regions in configuration space. This is a classic task in the study of spin glasses and other fields, where one may be seeking the ground state(s) or the low-lying states in a complicated landscape. TSMC can then be viewed as an improvement on the standard

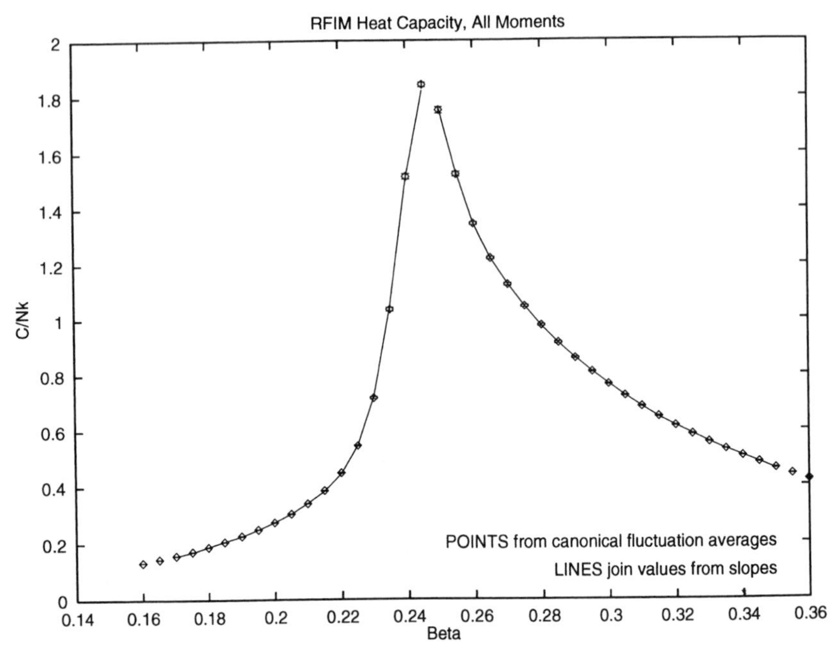

Figure 4. The heat capacity as a function of temperature, obtained in the same run as in Figure 3. The points are from canonical averages of the energy fluctuations at each temperature [using Eq. (2.13)]; values obtained by differentiation are connected by straight-line segments.

"simulated annealing" approach to such problems—it will not become trapped in any one energy minimum, but continually reverts to high-energy configurations and seeks anew.

That viewpoint explains the name "simulated tempering" for a method introduced by Marinari and Parisi [13]. In this one chooses a set of M temperatures T_p and associated weights ω_p. The system is regarded, at any step of the chain, as being at one of those temperatures, and most MC steps correspond to those of a conventional canonical MC run at that temperature. These steps are interspersed, however, with "jump" steps, in which an attempt is made to jump to an adjacent temperature, such moves being accepted or rejected in such a way that time spent in T_p is biased by the weight ω_p to ensure that comparable time is spent in each. The "jump" moves will occur only if adjacent temperatures are sufficiently close that their energy distributions overlap substantially. The temperature gaps and the ω_p must be adjusted by trial and error (like π). One can see, in fact, that this is actually just a particular form of TSMC, carried out on a weighted sum of Boltzmann distributions [see Eq. (2.14)]; it has been implemented (unnecessarily) by treating T_p as an auxiliary variable to be sampled for. This is, however, somewhat awkward, because to make full use of the computing, one must then reconstruct the thermodynamics and any other properties by reweighting appropriate histograms from the separate T_p, which may be complicated and will certainly degrade precision. An equivalent method, almost identical to "simulated tempering," was introduced by Lyubartsev and co-workers [27], and called the "expanded ensemble" method; it was applied to several systems. Like the more direct form of TSMC, these two approaches can be used either to explore phase separations or to seek energy minima.

A related method was introduced by Geyer [14]. In this case *parallel* runs are carried on simultaneously for M systems, say, at M different temperatures. Periodically one attempts to "switch" the systems at two adjacent temperatures, say T_p and T_q, accepting the switch with probability min(1, $\exp(-(\beta_p - \beta_q)(U_q - U_p))$. The switch will (again) occur frequently only if the energy distributions at adjacent temperatures overlap substantially, so the temperature gaps must be chosen with care (but no weights ω_p need now be assigned to the M temperatures). The overall sampling distribution is simply the product of Boltzmann distributions at all the temperatures; the results for a particular temperature, which come from the succession of different systems that have visited there, correspond to the canonical distribution. The point is that as the individual systems migrate to high temperatures and back, they should eventually average over all the various configurations appropriate to each temperature, again overcoming the quasi-ergodic problem. (This is the advantage over the old "multistage sampling" method (MSS) [27], which is reliable only in the absence of quasi-ergodic difficulties.) In this "multiple Markov chain" method the properties at temperatures other than the set T_p (and even at those temperatures if full use is to be made of the data) must be obtained by histogram reweighting, after first using energy-distribution overlaps to estimate a common normalization, as for MSS [27]; in particular, free energies are no longer available as the ratio of simple run averages, but must be estimated either by thermodynamic integration based on the set of temperatures, or by histogram reweighting of the kind just described. This sets the method off somewhat from TSMC, although it shares the aims of covering a substantial temperature range and avoiding quasi-ergodic restrictions (and the limitations of any one-dimensional "scaling"). It has an advantage, in that one need optimize only the temperature gaps, not π, and also disadvantages, not only in complication and loss of precision in extracting estimates but also because the required computing is now proportional to the number of temperatures required, M, and is therefore likely to increase more rapidly than it need do so in TSMC, as N increases. Both the "multiple Markov chain" and the "simulated tempering" algorithms have been applied particularly to lattice models.

The investigation of a first-order phase transition by using TSMC, as discussed above, is clearly inelegant. The coexisting phases that we need to relate are actually at the *same* temperature, but *different densities*. It is somewhat perverse, therefore, to relate their free energies by such a long and artificial path (e.g., the hatched path in Fig. 2). This was the motive for introducing density-scaling, DSMC [7], and applying it to some systems [8,9]. This development is not trivial, because whereas a temperature change only alters the breadth of the corresponding energy distribution, even a small density change in a fluid means that the nature of the configurations changes in a more serious way [7]. Correspondingly, the appropriate sampling distribution is not likely any longer to be simply a function of the energy [7]. In practice, it is in any case as easy, and much more powerful, to proceed directly to TDSMC computations, as described in Section II.

Density is the order parameter describing fluid condensation; one may instead be interested in the order parameters for other transitions: spin alignment on lattices and the like. So DSMC may be seen as a special case of general *order-parameter scaling Monte Carlo* (OSMC), for instance (and TDSMC of TOSMC, although examples of the latter, except for TDSMC, do not seem to exist so far). A particular set of choices for the sampling and its distribution π to be used in OSMC was proposed by Berg and Neuhaus [15]; this proposal and its offshoots are known as *multicanonical sampling*. The basic idea is to generate a sampling distribution that is approximately *flat* with respect to some mechanical variable; in the initial OSMC applications the variable was usually essentially the energy (the distribution to be nearly flat over the range of energies relevant to the whole range of order parameters joining the coexistent phases). The argument for the particular choice of energy, for OSMC in a many-body system, is perhaps not very obvious; its efficiency should at least depend strongly on the nature of the model under study. On the other hand the distribution can instead be chosen to be roughly flat with respect to a quantity more closely connected with the order parameter (e.g., the magnetization of a lattice or the volume in a fluid condensation) [30], and this can be a useful approach. The required sampling distributions must be generated as usual by trial and error, using techniques basically like those described above for TSMC; rather sophisticated consideration has gone into the techniques for this recursive generation of the flat multicanonical distributions on the basis of direct numerical analysis of the output of repeated MC runs [16].

[A different development of the multicanonical idea [31] recreates TSMC. Evidently finding a flat energy distribution is equivalent to studying the energy density of states, e.g., $n(E)$, for the particular model; if this is done for some range of energy, then canonical averages of any function $f(E)$ can be estimated for a corresponding temperature range, so that thermal properties such as free energy, energy, entropy, and heat capacity are available by reweighting the histogram of $n(E)$. To estimate averages of nonthermal properties requires collection of more complicated additional histograms expressing their correlations with the energy. The overall result is then that described above for the TSMC study of lattice systems. In fact, the density-of-states viewpoint is just that originally used to introduce TSMC in Ref. 1; in our applications we find it less complicated and more convenient to apply Eqs. (2.7) and (2.5) directly, without explicit evaluation of $n(E)$, and retaining flexibility regarding the precise form of π.]

We see that there remains a wide field of important applications of thermodynamic scaling carried out in just one dimension; see Section V in this regard, too. In one flamboyant development, Panagiotopoulos and coworkers [32] have even combined TSMC with Gibbs ensemble MC, in order to observe the coexisting phases directly, over a range of temperatures, in a single run! Nevertheless we feel that for simple transitions such as fluid condensation the efficiency and effectiveness of two-dimensional scaling is decisive. We return to exhibiting that by describing an example application of TDSMC.

IV. AN EXAMPLE: LENNARD-JONESIUM

This section describes the application of TDSMC to a test problem involving a phase transition. The test case is one very familiar in statistical mechanics: the Lennard-Jones (LJ) fluid and its condensation. We mention the highlights of a TDSMC study over a region of (T, ρ) embracing the critical point; a complete report will appear elsewhere [10]. (A DSMC study of one subcritical LJ isotherm has been described previously [9].)

For any state (T_k, ρ_k) the reduced potential energy U of the LJ system, when in a (reduced) configuration $\mathbf{r}^N$, is described by parameters ε, σ according to

$$\beta_k U(\mathbf{r}^N, L_k) \equiv 4\varepsilon\beta_k \sum_{i<j}^{N} \left[\left(\frac{\tilde{\sigma}_k}{r_{ij}}\right)^{12} - \left(\frac{\tilde{\sigma}_k}{r_{ij}}\right)^{6} \right] \tag{4.1}$$

where r_{ij} is the separation of particles i and j, while

$$\tilde{\sigma}_k \equiv \frac{\sigma}{L_k} \tag{4.2}$$

is the reduced LJ distance parameter (and $\beta_k \equiv 1/k_B T_k$, as before).

Systems of $N = 72{,}256$ and 500 particles were studied, within a periodic cubic box of unit volume. The system was studied for reduced densities $\rho_k^* \equiv \rho_k \sigma^3 \equiv N\tilde{\sigma}_k^3$ over the range $0.08 \leq \rho_k^* \leq 0.60$, and for reduced temperatures $T_k^* \equiv k_B T/\varepsilon \equiv 1/\varepsilon\beta_k$ in the range $1.20 \leq T_k^* \leq 1.36$ (1.20–1.40 for $N = 72$). To approximate the infinite-system energy the pair potentials were as usual truncated spherically, and the resulting thermodynamic functions corrected approximately by adding tail corrections [9]; we used a cutoff equal to half the box length. In view of the large overall density range it was decided to subdivide it, covering the ranges 0.08–0.20, 0.20–0.40, and 0.40–0.60 in three TDSMC runs, overlapping at the "seams" at 0.20 and 0.40. (As a test, the system was studied also, for $N = 72$ and 256, by using instead a single TDSMC run spanning the whole density and temperature region. As expected, this was straightforward and gave the same results, but with somewhat greater statistical errors for a given computing time.)

One could collect data at as many thermodynamic states as wished. In this case data was collected at regular intervals of 0.01 in both T^* and ρ^*, forming a rectangular grid of states—this made in all 901 thermodynamic states (1113 for $N = 72$). It is very helpful to collect the data on a *regular* grid to simplify its subsequent numerical analysis.

The sampling distribution was that given by Eq. (2.14), with the weights approximated as in Eq. (2.15). The sum in Eq. (2.14) for $\pi(\mathbf{r}^N)$ was over the same rectangular grid of thermodynamic states (T_k, ρ_k) as that described for data collection (although it need not have been, of course); this grid proved to be sufficiently dense to give an adequately smooth sampling distribution. [It was actually enhanced by extrapolation to include a surrounding border of states to ensure good sampling at the edges of the (T_k, ρ_k) region, although this was probably unnecessary.]

Even the roughest guess for the ω_k will be sufficient to start the bootstrap refinement for a quite small system (and for other systems where no prior information was available we have begun with $N = 32$ or even $N = 16$ particles, to exploit this fact). But for the LJ fluid useful information was available. We used the equation of state $P_{\text{EOS}}(T, \rho)$ of Nicolas et al. [33], which is based on many results of molecular dynamics, MC and virial series computations. From its 33 parameters it was possible to evaluate the corresponding (relative) excess free energy $A_{\text{EOS}}^{\text{ex}}(T, \rho)$, and this was used for the initial trial input to (15) with $N = 72$. Some refinement of this (as described in Section II) gave a satisfactory $\pi(\mathbf{r}^N)$. For the larger systems the estimated A^{ex} of a smaller system were scaled up, and often a slight further refinement was carried out.

The adequacy of $\pi(\mathbf{r}^N)$ can be monitored by recording various distributions observed during the run. As an example, Figure 5 shows the overall sampled distribution of the sum $\sum_{i<j} r_{ij}^{-12}$, and also (coarse) histograms of

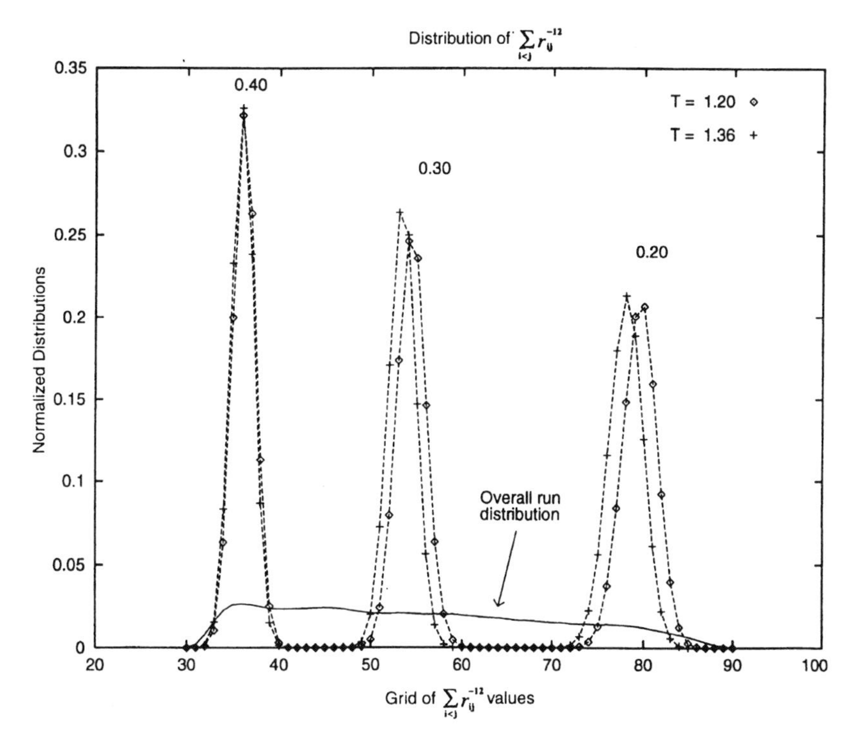

Figure 5. Confirmation of the adequacy of the sampling distribution, for a TDSMC run on the Lennard-Jones fluid. This shows that the distribution of $\sum_{i<j} r_{ij}^{-12}$, under π, is broad enough to include all the ranges appropriate to the states covered, and is reasonably smooth. Similar checks can be made for other quantities.

this quantity as extracted for several states extending over the whole density range and at the highest and lowest temperatures. The sampling under $\pi(\mathbf{r}^N)$ covers the required range in a smooth way. Similar checks were carried out for $\sum_{i<j} r_{ij}^{-6}$ and for energy distributions.

The MC runs themselves were carried out with local moves in the familiar way. At each step a trial move is attempted for a randomly-chosen single particle within a small cubic box (the size chosen to give a moderate acceptance rate). A trial move from state m to state n is accepted as usual with probability $a_{mn} = \min(1, \pi_n/\pi_m)$ (using an obvious notation). Thus the run proceeds just as in conventional MC computations, except with the nonphysical sampling distribution $\pi(\mathbf{r}^N)$. The only extra computation time, compared to conventional MC runs at a single state, comes from having this more complicated probability function π to compute, at each trial move, and from the time necessary to file the data contributing to averages for a large number of thermodynamic states. So the computation proceeds at a speed not much less than that of a conventional MC computation. (This is true unless structural data, such as correlation functions, are required for many states; if they are, the mere collection of the multidimensional data can easily eat up computing time as great as that needed to generate the Markov chain. In the example discussed here, no structural data were collected.)

To estimate relative excess free energies (and the other thermodynamic quantities), it is only necessary to collect during the run averages of the kind that appear in Eq. (2.12) for each state (T_j, ρ_j). It is convenient to tabulate the resulting reduced free energy per particle $\beta_j A_j 1/N$ relative to that of one of the states taken as a reference, $\beta_{\text{ref}} A_{\text{ref}}/N$. Thus the tabulated free energies of the three runs are based on different arbitrary reference states. However, it is easy to put them on a single scale, relative to a common reference state, by shifting all the several hundred values (in each of two of the runs), by a single constant for each run. The two required constants were chosen to give zero average discrepancy in $\beta_k A_k/N$ for the set of overlapping states along the two "seams" at $\rho^* = 0.2$ and 0.4. The r.m.s. discrepancy of the 17 pairs of values at each seem were 1.3×10^{-4} and 5×10^{-5} (for $N = 500$; slightly less for the smaller systems). These are small (note that the variation of $\beta A^{\text{ex}}/N$ over the temperature range is of order unity), and are an indication of the remarkable precision with which free energies can be estimated in ThScMC. (The corresponding variances of the reference-state shifts of the run's free energies must be included in those of the relative free energies for pairs of states covered by different runs, of course.)

As suggested above, the remarkable feature of the estimated free energies is their small statistical uncertainty, estimated by the variances and covariances of block averages during the runs. For example, Figure 6 displays the standard deviations of the differences of reduced excess free energy $(\beta A^{\text{ex}}/N)$ from that of a reference state $(T^* = 1.20, \rho^* = 0.2)$ for various densities and temperatures within a single TDSMC run. These are astonishingly small; the reduced excess free energy itself varies by 0.7 in the region. In fact, the standard deviation for the most distant states of the target region of the study [(1.20, 0.08) to (1.36, 0.60)] is under 0.1% of the difference in total reduced free energy $\beta A/N$. This precision opens the way to detailed study of phase transitions and of size-dependences.

Figure 7 shows the resulting reduced free energy along five (of the 17) isotherms studied; these are plotted as straight-line segments joining the raw free energy estimates along each isotherms; because of the density of

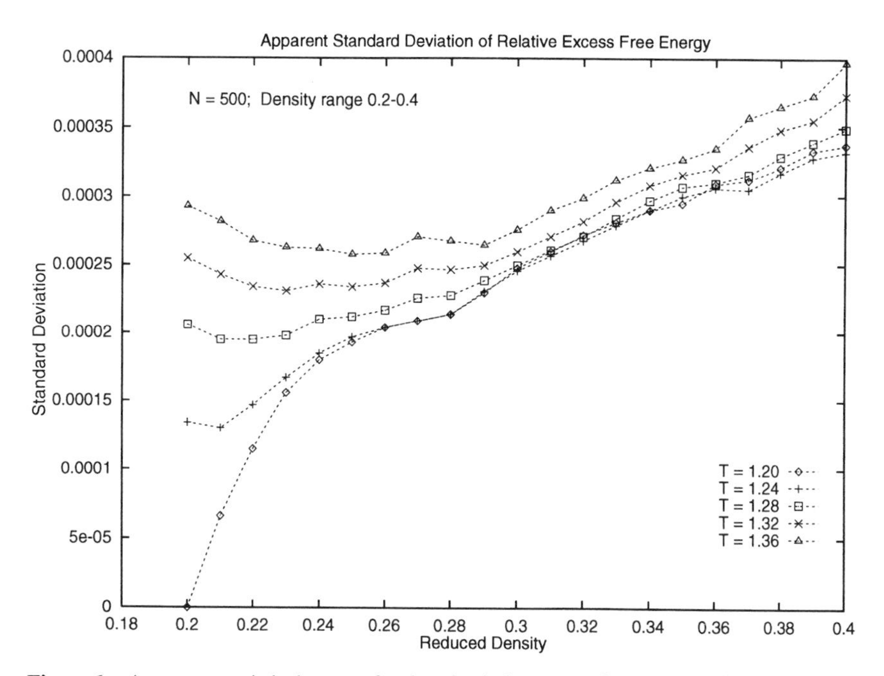

Figure 6. Apparent statistical error of reduced relative excess free energy with respect to a reference state ($T^* = 1.20$, $\rho^* = 0.2$) in a TDSMC run; the reduced excess free-energy variation over the region is of order unity, so the precision is extraordinary.

the thermodynamic states and the precision of the data they look like smooth curves! Obviously the isotherms at the lower temperatures show concave regions corresponding to the phase separation; we return to this later.

Other thermodynamic quantities can be estimated according to Eq. (2.13) by collecting the required averages during the run. Alternatively, however, all thermodynamic quantities can be generated from the Helmholtz free energy and its temperature and density derivatives. Numerical differentiation of noisy data is rarely an attractive prospect, but the precision noted above is encouraging. In the present case this was explored by calculating in *both* ways the energy, pressure, heat capacity, compressibility, and internal pressure. The two sets of results agree everywhere within their statistical uncertainties. The results derived from differentiating the free energy are actually slightly preferable, behaving more smoothly, and having slightly smaller uncertainties except at the borders and "seams" of the data region. As an example, in Figure 8 we show pressure data along some isotherms, obtained in both ways; the inset shows data on the statistical errors

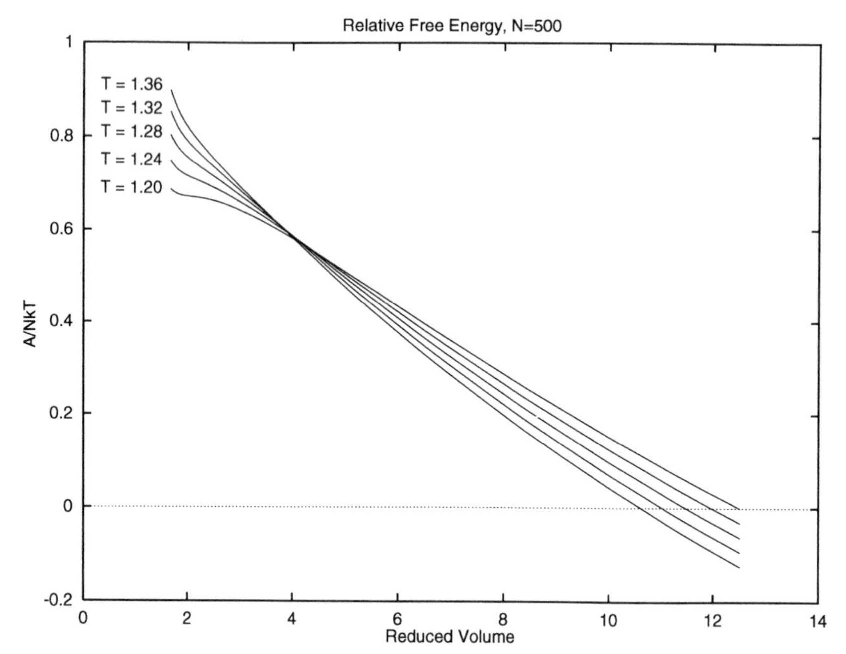

Figure 7. Reduced relative free energy over the whole density range studied ($0.08 \leq \rho^* \leq 0.60$), at five temperatures (of the 17 studied). The graphs consists of straight-line segments joining the raw data values at adjacent densities along the isotherms. They resemble smooth curves only because of the precision of the results, and the number of densities (53) for which data were recorded. At the lower temperatures one sees concave regions, indicating instability with respect to coexisting liquid and solid phases.

along one isotherm. [The numerical differentiation was performed using the original block averages $\langle e^{-\beta U}/\pi \rangle_\pi$ during the run, and the statistical errors of the derivatives were estimated using the variances and covariances of those block averages. Three-point and five-point estimates agreed well, within their uncertainties.] Even second-order quantities, requiring second derivatives of the free-energy data, could be estimated successfully, although, of course, they were noisier (but not more so than direct estimates, by Eq. (2.13) of the required fluctuation quantity). Figure 9 shows some heat capacities, as an example; the maximum near the critical density reflects the expected energy-fluctuation behavior there, and its strong size dependence for small systems.

Evidently, then, there is no point in collecting the required averages to make "direct" estimates [Eq. (2.13)] of thermodynamic quantities in TDSMC investigation; we require only, averages of the type appearing in

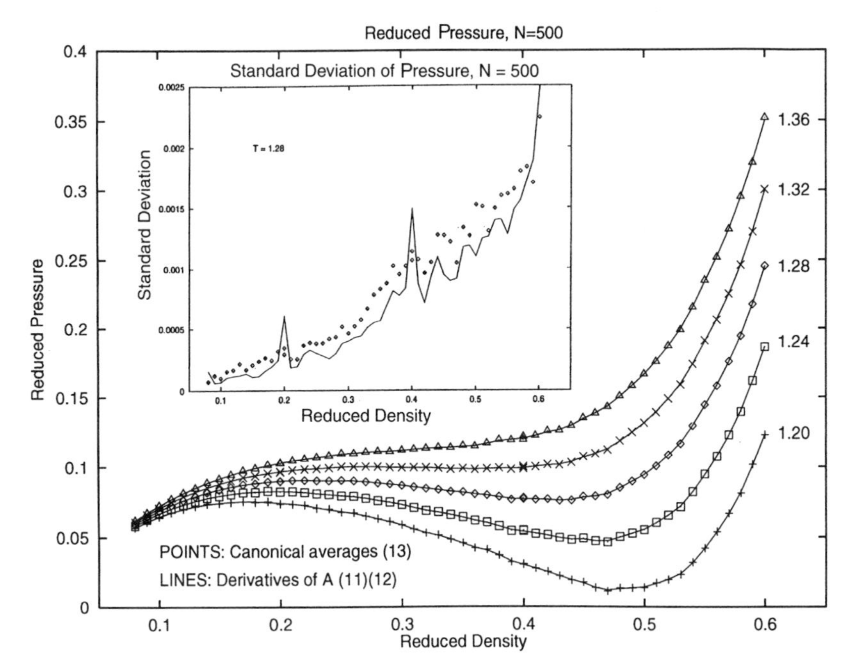

Figure 8. Five (of the 17) pressure isotherms for $N = 500$. The points come from a canonical average of the virial [using Eq. (2.13)]; line segments join values obtained by numerical differentiation of the free energy. The agreement is excellent. The inset shows the standard deviations of the former (points) and latter (joined by line segments) for one temperature. The differentiation leads to slightly smaller uncertainty except at the borders and seams.

Eq. (2.12), which we need anyway to estimate the free energies. Not only is this convenient, but it is also a saving because one can avoid the (nonnegligible) overhead computing time of filing special data for several properties at many thermodynamic states during the run.

To investigate liquid–vapor coexistence, we can use the "double-tangent" construction on the subcritical isotherms. Figure 10 shows an example: the tangent points give the coexisting densities, and the tangent's slope gives the vapor pressure. The double tangent can be sought numerically, after generating local interpolations of the free energy near the coexistence densities. In view of the small uncertainties in the free energies, the vapor pressure is obtained with extraordinary precision. The resulting pairs of coexistence densities are shown in Figure 11. An alternative method of analysis would be to fit all the free-energy data to a global equation of state, from which the coexistence properties (and all the other thermodynamic properties) could then be generated. This too has been done for the present system.

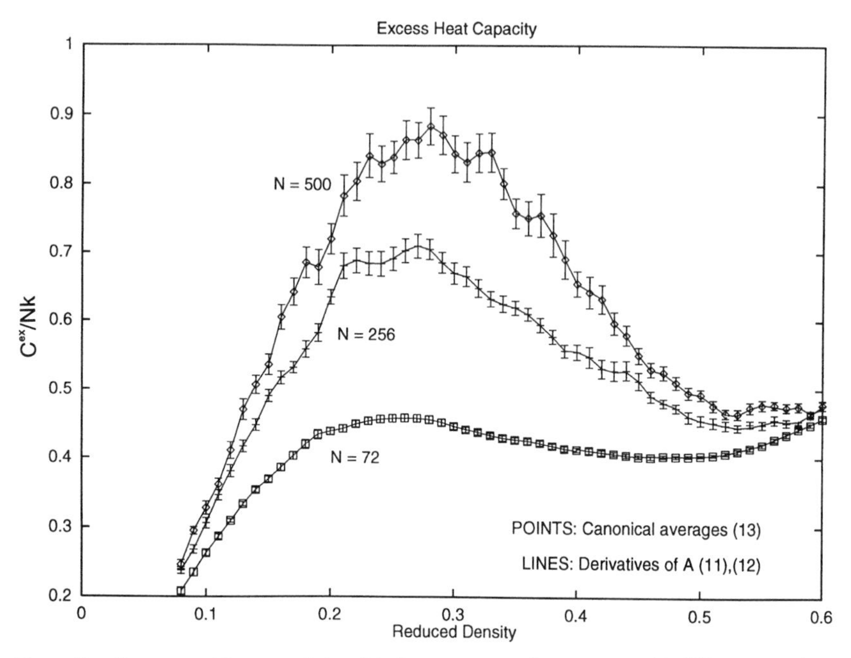

Figure 9. Heat capacities near their critical temperatures for systems with different numbers of particles N. Once again the points come from canonical averaging using Eq. (2.13), the line segments join values obtained by finding numerically the second temperature derivative of the free energy. The data exhibit the expected peak near the critical density, corresponding to the enhanced fluctuations in that region, and they show the expected strong size dependence.

Figure 11 shows, along with the points obtained "directly" as described above, the coexistence curves that follow from a double-Padé fit involving 50 parameters. The agreement is very satisfactory.

These data are replotted in a different form in Figure 12, on the assumption that the order parameter (the coexistence density gap) for the LJ system should behave in an Ising-like manner. This is reflected in the nearly straight-line behavior of much of the data; very close to the critical points the data deviate from linearity, becoming mean-field-like because of the limitation on fluctuations in a finite system. The precision of the results puts us in position to study this "finite-size crossover" and also other nonuniversal properties of the critical behavior of fluid phase transitions.

One can evidently do a "finite-size scaling" extrapolation to estimate the critical point of the bulk thermodynamic system, and can hope to extract other critical scaling information—such work will be reported elsewhere (10).

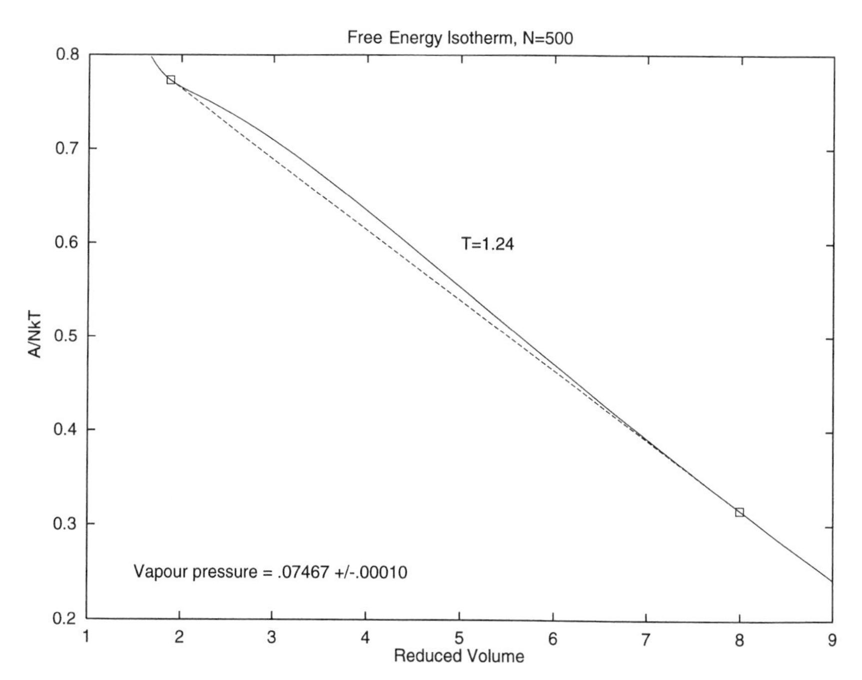

Figure 10. A portion of one subcritical isotherm. The dashed line shows the double tangent, obtained numerically after local interpolation in the vicinity of the coexisting densities; the squares mark the resulting liquid and gas phases. The slope of the double tangent gives the vapor pressure; in view of the high precision of the free energies, the vapor pressure is obtained very precisely.

A couple of technical remarks might be appropriate here:

1. We have seen that there is no point in collecting data for direct canonical averaging of higher-order thermodynamic quantities. We can as well (in fact, better) derive them from the free energy, for which the data collection expense is minor. The TDSMC runs would therefore be about as fast as conventional MC were it not for the time required repeatedly to recalculate $\pi(\mathbf{r}^N)$ as given by Eq. (2.14); since the sum in (2.14) may include many states spanning the target region, this time may not be negligible. Furthermore, as one moves to larger N, the relative widths of the Boltzmann distributions narrow, so that an ever-finer grid will be required. However, it was pointed out in Section II that this can easily be solved without loss of accuracy, by including only a "window" of densities for each configuration. (This improvement was not actually used for the reported Lennard–Jones study, but has been tested elsewhere.)

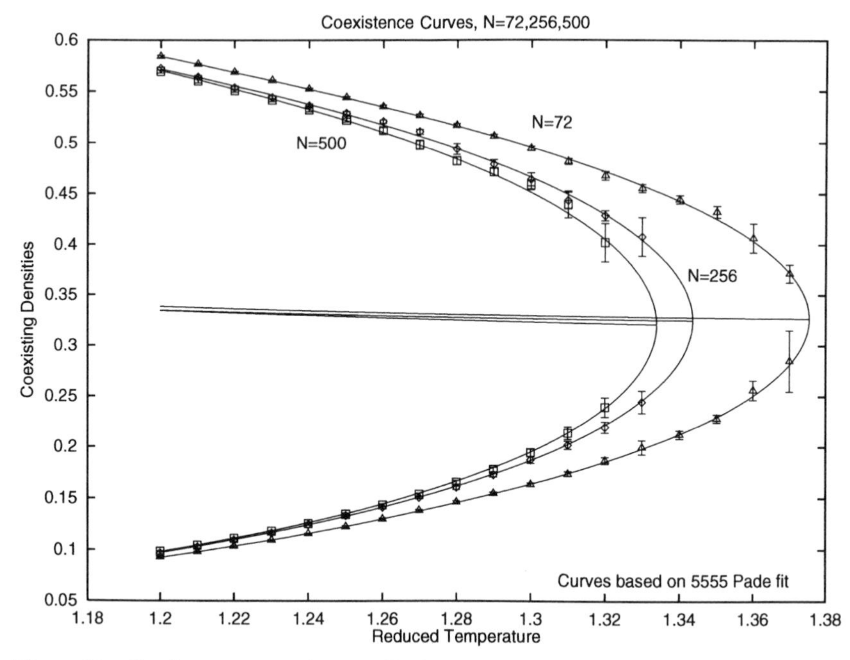

Figure 11. Coexistence curves for $N = 72$, 256, and 500 particles. The points come from direct analysis of the free-energy data on the isotherms for which they were collected; the curves derive from a global fit of a many-parameter function to all the free-energy data in the target (T, ρ) region.

2. The weighting used in $\pi(\mathbf{r}^N)$ can be rather dramatic; for instance, the ω_k for states within a single run differ by factors of up to about 10^{180}. One must keep one's nerve! But, also, one must write the programme algorithm with care to avoid losing or distorting data through overflows and such things.

Rather complete studies of phase transitions by TDSMC have recently been carried out for the square-well fluid (with the collaboration of Nikolai Brilliantov) [11] and the restricted primitive model (with Glenn Torrie) [12].

V. CONCLUSIONS

The thermodynamic-scaling approach continues its gradual development. It has been designed and tested until now mostly with the phase changes of

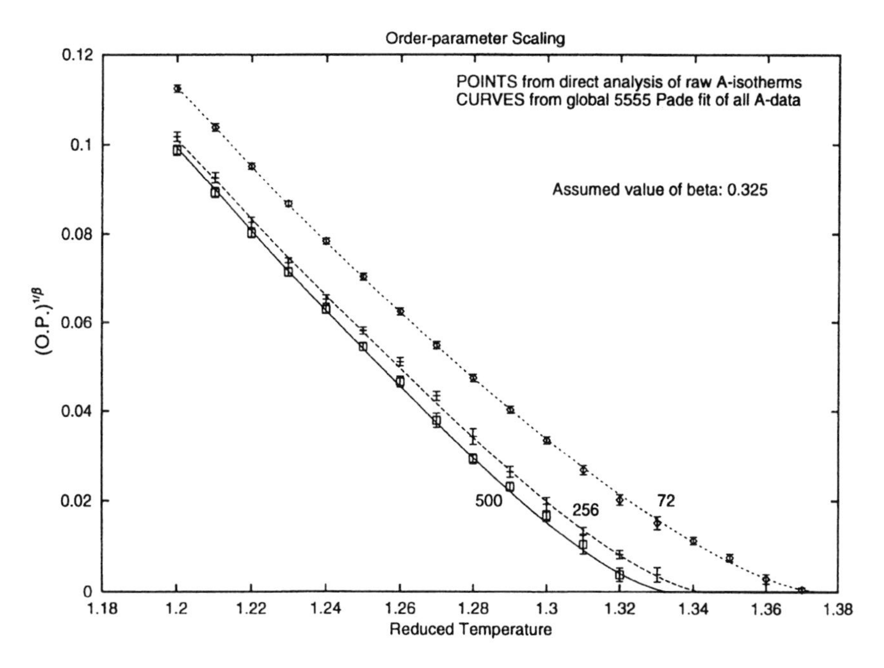

Figure 12. Order-parameter scaling plot, assuming Ising-like universality. In the absence of corrections-to-scaling terms and in a thermodynamic system, these should be straight lines. The sharp curvatures at low values arise from finite-size effects due to limitations of the density fluctuations in the small finite systems.

simple one-component systems in view, especially in fluids. We have seen that TDSMC is very successful in such cases, as it appears that the quasi-ergodic difficulties characteristic of previous MC study of such transitions can be eliminated, and it is possible to make economical estimates of free energies with remarkable precision. As a result, we can hope to study coexistence and critical properties with new confidence. One should perhaps mention that the generation of appropriate sampling distributions to carry out such investigations no longer presents a serious challenge. Such two-dimensional sampling can certainly be extended easily to analogous problems, such as the various lattice models; on the other hand, such models can also be studied very successfully using simple TSMC (as described in Section III).

It is natural to contemplate the next extensions to less-simple models. Of course, a fluid of spherical particles carrying point multipoles presents no new difficulties (and such TDSMC investigations are beginning). More care is required, however, for particles that are geometrically nonspherical, such as molecules. Like the "diameter" σ of the Lennard-Jones system [Eq. (4.2)], every physical distance λ in the model particle becomes, in the reduced coordinates convenient for density-scaling, a function of density according to $\tilde{\lambda} = \lambda/L_k$. The calculations are then, in principle, straightforward. However, they will not always proceed so rapidly as for spherical particles, because the energy calculations will be more time-consuming. Consider Lennard-Jonesium, for example; in view of the form of $U(\mathbf{r}^N, L)$, in Eq. (4.1), it is necessary to update only the two sums

$$\sum_{i<j} r_{ij}^{-6} \qquad \text{and} \qquad \sum_{i<j} r_{ij}^{-12}$$

in order to calculate the energy at *every* density. For a nonspherical particle, on the other hand, each density will require a different geometric calculation. Thus the economies of scaling will not be as dramatic. (If one is interested in data at, say, n_T temperatures and n_ρ densities, then the geometric calculation required for the $n_T n_\rho$ states is, for the LJium, just what it would have been for a single state, whereas for a nonspherical case, it will correspond to n_ρ such calculations.)

Another area of interest is the study of phase behavior in liquid solutions. We have discussed how this problem has been approached by using TSMC at various concentrations, in conjunction with a single isotherm of free energy as a function of concentration, obtained by HSMC or some more traditional method. This approach is always available and will often be the most convenient to adopt. It is natural to ask whether there might be a two-dimensional scaling, analogous to TDSMC in the case of liquid–vapor condensation, that gave comparable advantages. The order parameter will now be the concentration-difference of the coexisting phases, so the analogous experiment is *temperature-and-concentration-scaling Monte Carlo* (TCSMC). In TDSMC each configuration $\mathbf{r}^N$ is reinterpreted to find its contribution at every density; in TCSMC the analogous reinterpretation involves making different assignments of identity (as one of the solution components) to each particle in the system! This is amusing. It will clearly work efficiently, in fact, if the components are pretty similar to each other; it will probably be onerous otherwise. (Of course, it is also true that alternative approaches, such as the Gibbs ensemble, will not work very smoothly where the components are very different in size and shape. This can be relieved by devices such as "configuration-bias"; it is not clear to what extent such devices might easily be combined with TCSMC.) We have assumed that the interest was only in solutions of two unique and dissimilar components; for theoretical purposes one might be interested instead in a whole parametrized set of components, ranging from identical to each other to quite different. In that case it would be natural to consider improving on the use of TSMC alone by using a THSMC, in which one would study in a single run the ranges of interest of both the temperatures and the Hamiltonians (probably at only one relative concentration at a time); for the purposes of such a large systematic study, this approach would probably be very efficient.

We have mentioned as an effective technique the study of lattice models and similar situations by doing TSMC at fixed field, while leaving the order parameter unconstrained. It may be possible to try other things here as well. For example, one can surely improve on the convergence by scaling as well with respect to the order parameter—TOSMC. Meanwhile it might be of interest instead to cover a range of applied fields simultaneously: *temperature-and-field-scaling Monte Carlo* (TFSMC).

In fact, there are endless interesting possibilities that will be fun to explore. The important thing is not to be timid!

ACKNOWLEDGMENTS

The author appreciates the ongoing support of this research by the Natural Sciences and Engineering Research Council of Canada. He has also enjoyed the helpful advice of many colleagues, and especially Glenn Torrie, who shared in much of the early development of umbrella-sampling techniques, and Stu Whittington, always a source of insight and enthusiasm.

REFERENCES

1. G. M. Torrie and J. P. Valleau, *J. Comp. Phys.* **23**, 187 (1977).
2. J. P. Valleau, in *Proceedings of International Symposium on Ludwig Boltzmann*, G. Battimelli, M. G. Ianello, and O. Kresten, eds., Austrian Academy of Sciences, Vienna, 1993, p. 229.
3. A. Z. Panagiotopoulos, *Mol. Phys.* **61**, 813 (1987); A. Z. Panagiotopoulos, N. Quirke, M. Stapleton, and D. J. Tildesley, ibid. **63**, 527 (1988).
4. See, for example, B. Smit, Ph. de Smedt, and D. Frenkel, *Mol. Phys.* **68**, 931 (1989). See also some remarks in J. P. Valleau, *J. Chem. Phys.* **108**, 2962 (1998); a more thorough analysis is in preparation.
5. N. B. Wilding and A. D. Bruce, *J. Phys. Condensed Matter* **2**, SA 121 (1990); N. B. Wilding, *Phys. Rev. E* **52**, 602 (1995).
6. K. Ding and J. P. Valleau, *J. Chem. Phys.* **98**, 3306 (1993).
7. J. P. Valleau, *J. Comp. Phys.* **96**, 193 (1991).
8. J. P. Valleau, *J. Chem. Phys.* **95**, 584 (1991).
9. J. P. Valleau, *J. Chem. Phys.* **99**, 4718 (1993).
10. J. P. Valleau, manuscript in preparation.

11. N. Brilliantov and J. P. Valleau, *J. Chem. Phys.* **108**, 1115, 1123 (1998).
12. J. P. Valleau and G. M. Torrie, *J. Chem. Phys.* **108**, 5169 (1998).
13. E. Marinari and G. Parisi, *Europhys. Lett.* **19**, 451 (1992).
14. C. J. Geyer, "Markov Chain Monte Carlo Maximum Likelihood," *Computing Science and Statistics: Proceedings of 23rd Symposium on the Interface*, E. M. Kerimidas, ed. (Interface Foundation, Fairfax Station, 1991), p. 156.
15. B. A. Berg and T. Neuhaus, *Phys. Lett. B* **267**, 249 (1991).
16. J. A. Berg, *J. Stat. Phys.* **82**, 323 (1995); G. R. Smith and A. D. Bruce, *J. Phys. A* **28**, 6623 (1995).
17. See, for example, L. Fox and D. F. Mayers, *Computing Methods for Scientists and Engineers*, Oxford Univ. Press, 1968, Chapter 8.
18. G. N. Patey and J. P. Valleau, *Chem. Phys. Lett.* **21**, 297 (1973).
19. G. M. Torrie and J. P. Valleau, *Chem. Phys. Lett.* **28**, 578 (1974).
20. G. M. Torrie and J. P. Valleau, *J. Chem. Phys.* **66**, 1402 (1977).
21. K. Nakanishi, S. Okazaki, K. Ikari, T. Higuchi, and H. Tanaka, *J. Chem. Phys.* **76**, 629 (1982).
22. D. Frenkel and A. J. C. Ladd, *J. Chem. Phys.* **81**, 3188 (1984); D. Frenkel, *Phys. Rev. Lett.* **56**, 858 (1986).
23. C. Y. Lee and H. L. Scott, *J. Chem. Phys.* **73**, 4591 (1980).
24. I. S. Graham and J. P. Valleau, *J. Chem. Phys.* **94**, 7894 (1990).
25. K.-C. Ng, J. P. Valleau, G. M. Torrie, and G. N. Patey, *Mol. Phys.* **38**, 781 (1979), [the nature of the phase separation reported is unclear in view of later investigations—see, among other works, J.-M. Caillol, *J. Chem. Phys.* **98**, 9835 (1993).]
26. J. P. Valleau, unpublished.
27. A. P. Lyubartsev, A. A. Martsinovskii, S. V. Shevnukov, and P. N. Vorontsov-Vel'yaminov, *J. Chem. Phys.* **96**, 1776 (1992); A. P. Lyubartsev, A. Laaksonen, and P. N. Vorontsov-Vel'yaminov, *Mol. Phys.* **82**, 455 (1994).
28. J. P. Valleau and D. N. Card, *J. Chem. Phys.* **57**, 5457 (1972).
29. M. C. Tesi, E. J. Janse van Rensburg, E. Orlandini, and S. G. Whittington, *J. Stat. Phys.* **82**, 155 (1996); E. Orlandini, in *Numerical Methods for Polymeric Systems*, S. G. Whittington, ed., *IMA Volumes in Mathematics and Its Applications*, Vol. 102. Springer-Verlag, 1998 in press.
30. G. R. Smith and A. D. Bruce, *Phys. Rev. E* **53**, 6530 (1996); N. B. Wilding, *Phys. Rev. E* **52**, 602 (1995).
31. B. A. Berg and T. Celik, *Phys. Rev. Lett.* **69**, 2292 (1992); U. H. E. Hansmann and Y. Okamoto, *Phys. Rev. E* **54**, 5863 (1996); see also Ref. 16.
32. K. Kiyohara, T. Spyrouni, K. E. Gubbins, and A. Z. Panagiotopoulos, *Mol. Phys.* **89**, 965 (1996).
33. J. J. Nicolas, K. E. Gubbins, W. B. Streett, and D. J. Tildesley, *Mol. Phys.* **37**, 1429 (1979).

SEMIGRAND CANONICAL MONTE CARLO SIMULATION; INTEGRATION ALONG COEXISTENCE LINES

DAVID A. KOFKE

Department of Chemical Engineering State University of New York at Buffalo, Buffalo, NY 14260

CONTENTS

Advances in Chemical Physics, Volume 105, Monte Carlo Methods in Chemical Physics, edited by David M. Ferguson, J. Ilja Siepmann, and Donald G. Truhlar. Series Editors I. Prigogine and Stuart A. Rice.
ISBN 0-471-19630-4

I. INTRODUCTION

This chapter reviews two topics: the simulation of mixtures in a semigrand ensemble, and the evaluation of phase coexistence lines by the Gibbs–Duhem integration method. The notion of a semigrand ensemble has its roots in the work of Griffiths and Wheeler [1], who developed and applied it for the study of critical phenomena in mixtures (one might even argue that the concept goes back to the Ising model). The idea was rekindled in the work of Briano and Glandt [2], which formed the foundation for the first (off-lattice) Monte Carlo simulations in the ensemble, conducted by Kofke and Glandt [3–5]. At about the same time Sariban and Binder [6] applied semigrand Monte Carlo simulation to study a lattice model of polymer mixtures. Since then the ensemble has surfaced in a variety of places, some of which will be described in the review to follow. The Gibbs–Duhem integration method was inspired by the Gibbs ensemble methodology of Panagiotopoulos [7]. The approach was conceived by the author in 1992 during a one-month visit with the research group of Daan Frenkel in the Netherlands. The method has found a niche in applications to phase equilibria involving solids and ordered phases, although it is equally applicable to other phase-coexistence phenomena.

These two topics that form the subject of this review are disconnected in general, but there are important instances in which they may be used together to good effect. Most notably, Gibbs–Duhem integrations that trace coexistence lines for mixtures rely on the semigrand formalism, so it is appropriate that we review the latter first. We begin each review by discussing the basic idea underlying the simulation methodology, followed by a presentation of the formalism needed for its application. We review details of the methods before going on to summarize the applications each has seen to date. We finish the chapter with a detailed example of an application of the Gibbs–Duhem integration methodology.

II. SIMULATION IN SEMIGRAND ENSEMBLES

A. Concept

An ensemble is characterized by its independent variables, specifically, those thermodynamic quantities that the investigator uses to define the state of the system [8]. A familiar example is the choice between isochoric (fixed-volume) and isobaric (fixed-pressure) ensembles. The choice is usually made as a matter of convenience—data may be more easily analyzed in one or the other formalism, or an experiment may be more easily constructed with one or the other quantity held fixed. Implicit in the choice of an independent variable is the selection of its conjugate dependent variable [9]. Thus in taking the pressure as a chosen quantity, one simultaneously assigns the volume (or density) to a dependent role; it is given as an ensemble average, and it must be *measured* (or calculated) for its value to be known.

The notion of a semigrand ensemble arises when this decision is made in the context of mixture composition variables. Experience tells us that mole numbers or mole fractions are the most natural way to specify a mixture composition, and indeed anyone unfamiliar with the formalism of thermodynamics may not conceive that there exists an alternative. Of course, the component chemical potentials provide just this alternative, and their selection as independent thermodynamic variables in lieu of mole numbers is no less valid than the substitution of the pressure for the volume. In fact there are very familiar physical manifestations of semigrand ensembles in Nature. For example, in osmotic systems the amount of one component is not fixed, but takes a value to satisfy equality of chemical potential with a solvent bath. Another example is seen in systems undergoing chemical reaction, where the amounts of the various components are subject to chemical equilibrium.

A marvelous feature of molecular simulation (and thermodynamics in general) is that one need not work with the formalism (ensemble) that is most easily realized physically. Thus one may invoke a semigrand ensemble outside of applications involving osmotic or chemical equilibria. Often this option proves very convenient and desirable. The main point to keep in mind is that one *fixes* the chemical potential of some components of the mixture and one *measures* (or calculates) the composition. To proceed, one must surmount the minor conceptual hurdle of working with a system of variable composition absent any plausible reaction mechanism (or even stoichiometry). One can then play many variations on this theme.

B. Formalism

1. *Ensembles*

A natural starting point is the fundamental equation in a canonical ensemble of c components, for which the independent variables are the temperature T, the volume V, and the number of molecules of each species, N_i, $i = 1, \ldots, c$; the potential is the Helmholtz free energy A

$$A \equiv U - TS = -PV + \sum_{j=1}^{c} N_j \mu_j \tag{2.1a}$$

$$d(\beta A) = U\, d\beta - \beta P\, dV + \sum_{j=1}^{c} \beta\mu_j\, dN_j \tag{2.1b}$$

where U is the internal energy, S is the entropy, P is the pressure, and μ_i is the chemical potential of species i; the reciprocal temperature is $\beta = 1/k_B T$ where k_B is Boltzmann's constant. The latter equality in Eq. (2.1a) is a statement of Euler's relation. Legendre transformation [9] of *all* the chemical terms yields the grand-canonical ensemble, for which the independent variables are T, V, and μ_i, $i = 1, \ldots, c$.

$$L \equiv A - \sum_{j=1}^{c} N_j \mu_j = -PV \tag{2.2a}$$

$$d(\beta L) = U\, d\beta - \beta P\, dV - \sum_{j=1}^{c} N_j\, d(\beta\mu_j) \tag{2.2b}$$

Semigrand ensembles—as the term implies—lie between the canonical and the full grand-canonical ensembles. They are formulated by Legendre transformation of some, rather than all, of the chemical terms. Thus we derive an osmotic semigrand ensemble by the most straightforward variant; we transform some terms ($j = 1 \cdots m$), and leave others ($j = m + 1, \ldots, c$):

$$Y_o \equiv G - \sum_{j=1}^{m} N_j \mu_j = \sum_{j=m+1}^{c} N_j \mu_j \tag{2.3a}$$

$$d(\beta Y_o) = H\, d\beta + \beta V\, dP + \sum_{j=m+1}^{c} \beta\mu_j\, dN_j - \sum_{j=1}^{m} N_j\, d(\beta\mu_j) \tag{2.3b}$$

We prefer to continue the development using isobaric ensembles, which can be done with semigrand ensembles but not with the fully grand ensemble. We have exercised this option in the Eq. (2.3). Here $G = A - PV$ is the Gibbs free energy and $H = U + PV$ is the enthalpy.

Often it is advantageous to group some of the chemical terms together before applying the transformation. For example, addition and subtraction of $\sum_{j=1}^{c} \beta\mu_1\, dN_j$ leaves Eq. (2.1) looking different although it is, of course, inherently unchanged (except we now present its isobaric form):

$$d(\beta G) = H\, d\beta + \beta V\, dP + \beta\mu_1\, dN + \sum_{j=2}^{m} \beta\, \Delta\mu_j\, dN_j \tag{2.4}$$

where $\Delta\mu_i = \mu_i - \mu_1$ and we have exploited $dN = \sum dN_j$. In writing this form we have singled out a component (labeled "1") from the others. The selection of this "reference" component is arbitrary, although in specific applications a choice may present itself naturally. Legendre transformation can be applied to this reformulation to yield a new ensemble, for which the set of chemical potential differences $\Delta\mu_i$ is among the independent variables:

$$Y_N \equiv G - \sum_{j=2}^{m} N_j \,\Delta\mu_j = N\mu_1 \tag{2.5a}$$

$$d(\beta Y_N) = H\, d\beta + \beta V\, dP + \beta\mu_1\, dN - \sum_{j=2}^{m} N_j\, d(\beta\,\Delta\mu_j) \tag{2.5b}$$

A unique feature of this ensemble is that the total number of molecules N is an independent quantity, even though no species mole numbers are specified. This outcome is appealing because it means that a simulation can be conducted in this "isomolar semigrand" ensemble without requiring trials in which molecules are inserted in or deleted from the simulation volume. Such steps are often problematic because they can be difficult to accomplish without causing a very large fluctuation in the internal energy; in a Monte Carlo simulation this means that such trials are rarely accepted. Also, in simulating solids it is desirable to work at constant N to avoid problems with lattice defects, and even more severe problems associated with the use of periodic boundaries.

The isomolar semigrand ensemble is a specific instance of the more general case of semigrand ensembles that are formed while meeting certain constraints, such as fixing the total number of segments in a polymer mixture [10], or fixing the surface area in a model colloid. In most cases of practical interest the constraints depend linearly on the mole numbers. If there are r such constraints, they can be written

$$\sum_{j=1}^{c} g_j^{(k)} N_j = G_k, \qquad k = 1, \ldots, r \tag{2.6}$$

where $g_j^{(k)}$ is the multiplier for jth component in the kth constraint, and G_k quantifies the constraint. For example, if the kth constraint fixes the total number of segments in a polymer mixture, then $g_j^{(k)}$ represents the number of segments in a species-j chain, and G_k is the total number of segments in the system. Of course, we require $r \le c$. We choose r of the c components arbitrarily as reference species, and we can solve the set of linear equations for their molecule numbers in terms of the G_k. The general form of the solution defines the coefficients $h_k^{(j)}$

$$N_j = \sum_{k=1}^{r} h_k^{(j)} G_k + \sum_{i=r+1}^{c} h_i^{(j)} N_i, \qquad j = 1, \ldots, r \tag{2.7}$$

where for convenience of notation we have taken the first r components as the reference species. When this result is substituted into the fundamental equation, we Legendre-transform with respect to the non-reference-species molecule numbers to obtain the "constrained semigrand" ensemble

$$Y_c \equiv G - \sum_{j=r+1}^{c} N_j \mu_j^* = \sum_{k=1}^{r} \beta\mu^{(k)} G_k \tag{2.8a}$$

$$d(\beta Y_c) = H\, d\beta + \beta V\, dP + \sum_{k=1}^{r} \beta\mu^{(k)}\, dG_k - \sum_{j=r+1}^{c} N_j\, d(\beta\mu_j^*) \tag{2.8b}$$

where

$$\mu_j^* \equiv \mu_j + \sum_{i=1}^{r} h_j^{(i)} \mu_i, \qquad \mu^{(k)} \equiv \sum_{j=1}^{r} h_k^{(j)} \mu_j \tag{2.8c}$$

Our outline for formulating semigrand ensembles with constraints follows naturally from Krishna Pant and Theodorou's [10] development for polymer mixtures.

Another way to regroup the chemical terms may be suggested by the occurrence of reaction equilibria in the system of interest [11,12]. The resulting formulation is a special case of the constrained semigrand ensemble just described, but it has simplifying features and established notation that make its separate description worthwhile. If the chemical reaction is written in the general form

$$\sum \nu_i M_i = 0 \tag{2.9}$$

where ν_i is the stoichiometric coefficient (positive for product, negative for reactant) of component M_i, then the number of moles of a particular species will be

$$N_i = N_i^{\circ} + \nu_i \xi \tag{2.10}$$

where N_i° is the initial (original) number of moles of species i, and ξ is the extent of reaction. The dN_i in the fundamental equation are then written in terms of the N_i^o and ξ, thus (assuming no inert components)

$$d(\beta G) = H\,d\beta + \beta V\,dP + \sum \beta\mu_i\,dN_i^o + \left(\sum \nu_i \beta\mu_i\right) d\xi \tag{2.11}$$

Legendre transformation with respect to the last term yields a "reactive semigrand" ensemble

$$\begin{aligned} Y_r &= G - \left(\sum \nu_i \beta\mu_i\right)\xi \\ d(\beta Y_r) &= H\,d\beta + \beta V\,dP + \sum \mu_i\,dN_i^o - \xi\,d\left(\sum \nu_i \beta\mu_i\right) \end{aligned} \tag{2.12}$$

Setting the new independent variable $\sum \nu_i \beta\mu_i$ to zero puts the ensemble in correspondence with the actual chemical equilibrium that inspires its construction. Any other choice would correspond to a completely fictitious field that artificially drives the reaction one way or the other. We will not consider this variant in any detail as ensembles for reacting systems are presented more fully in the chapter by Johnson.

2. *Fugacity and Fugacity Fraction*

At times it is convenient to work with the fugacity in lieu of the chemical potential. The fugacity f_i of a species in a mixture is defined [13]

$$\beta\mu_i = \beta\mu_i^o(\beta) + \ln f_i \tag{2.13}$$

where μ_i^o is the chemical potential of species i in the ideal-gas state at unit pressure and unit mole fraction, and (as indicated) is a function of temperature alone. Consequently differential changes in the chemical potential can be written to group this temperature-explicit contribution with the enthalpic term, using

$$d(\beta\mu_i) = h_i^o\,d\beta + d\ln f_i \tag{2.14}$$

where h_i^o is the ideal-gas enthalpy per molecule for component i. Thus, for example, Eqs. (2.3) can be written

$$x_1 d\ln f_1 = h_r\,d\beta + \beta v\,dP - \sum_{j=2}^{c} x_j\,d\ln f_j \tag{2.15}$$

where $h_r = h - \sum_j x_j h_j^o$ is the residual enthalpy per molecule, v is the volume per molecule, and x_j is the mole fraction of species j. We have applied a few additional manipulations to cast this result in the form of a Gibbs–Duhem equation, which will be of use to the discussion in Section III.

The final variation of this formalism we consider employs the fugacity fraction as an independent variable [14]. The fugacity fraction of component i ξ_i is defined

$$\xi_i = \frac{f_i}{\sum_{j=1}^{c} f_j} \tag{2.16}$$

The fugacity fraction is a convenient quantity because it is bounded between zero and unity, and it has a qualitative correspondence with the mole fraction. The set of fugacity fractions are not independent, as they must sum to unity. Instead the complete set of independent chemical variables is specified by $m - 1$ fugacity fractions together with the sum $\sum_{j=1}^{c} f_j$. In terms of fugacity fractions the isomolar semigrand ensemble of Eq. (2.5) is expressed

$$d \ln\left[\sum_{n=1}^{c} f_n\right] = h_r\, d\beta + \beta v\, dP - \sum_{n=1}^{c} \frac{x_n}{\xi_n}\, d\xi_n \tag{2.17}$$

The equation is written this way because it has an appealing symmetry, but it should be remembered that the $d\xi_n$ terms are not independent. So, for example, with $c = 2$, $d\xi_1 = -d\xi_2$, and the equation is

$$d \ln[f_1 + f_2] = h_r\, d\beta + \beta v\, dP - \frac{x_2 - \xi_2}{\xi_2(1 - \xi_2)}\, d\xi_2 \tag{2.18}$$

3. *Partition Functions*

It is necessary to know the form of the partition functions to construct transition probabilities that properly sample the ensemble. For instructional purposes we record here the (semiclassical) partition functions that correspond to the osmotic and isomolar semigrand ensembles described above. In both ensembles one must average over moles of the Legendre-transformed species. Thus the osmotic semigrand ensemble partition function is

$$Y_o = \left[\prod_{j=m+1}^{c} \left(\frac{q_j}{\Lambda_j^3}\right)^{N_j}\right] \sum_{N_1=0}^{\infty} (\beta f_1)^{N_1} \sum_{N_2=0}^{\infty} (\beta f_2)^{N_2} \cdots \sum_{N_m=0}^{\infty} (\beta f_m)^{N_m} \times \left[\prod_{j=1}^{c} N_j!\right]^{-1} \int dV \int d\mathbf{r}^{(N)} \exp[-\beta(U + PV)] \tag{2.19}$$

where q_j is the molecular partition function for species j and Λ_j is its corresponding de Broglie wavelength; conveniently, for the osmotic species these contributions are canceled when the fugacity is inserted in place of the chemical potential. The isomolar semigrand partition function is written in a form similar to that above, but each sum over mole numbers must recognize the upper limit of N total molecules. This is cumbersome, and it is preferable to apply an analytic trick that converts the c coupled-limit sums over N_j into N uncoupled sums—one for each molecule—over species identities I_i [2]. The result is particularly clean when written in terms of fugacity fractions, which absorb all the molecular and momentum partition functions

$$Y_N = \frac{1}{N!} \sum_{I_1=1}^{c} \sum_{I_2=1}^{c} \cdots \sum_{I_N=1}^{c} \left[\prod_{j=1}^{c} \xi_j^{N_j} \right] \int dV \int d\mathbf{r}^{(N)} \times \exp[-\beta(U + PV)] \tag{2.20}$$

This transformation highlights the view in which the species identity forms another dimension—in addition to the spatial x, y, and z coordinates—that is sampled by each molecule. The fugacity fraction then plays the role of an external field that influences the sampling in this "direction."

C. Simulation Algorithm

Simulations in a semigrand ensemble require exploration of compositions—the average composition is a result of a semigrand simulation. In an osmotic ensemble composition sampling is conducted exactly as it is performed in a grand-canonical ensemble [15], except that insertions and deletions are not attempted for all species. Composition sampling in an isomolar semigrand ensemble is accomplished by permitting each molecule to adopt varying species identity, exactly as they adopt varying spatial coordinates to sample configuration space. There are some subtleties in the implementation. For example, the acceptance criteria differ depending on whether a molecule is chosen first, regardless of its species identity, or whether a species is first chosen, and then a molecule of that species is selected for an identity-change trial [5]. It is particularly important when working with a new ensemble [such as an instance of the constrained semigrand ensemble of Eq. (2.8)] to write down the partition function (i.e., the limiting distribution), carefully examine the trial-transition probabilities, and construct the acceptance probabilities that ensure microscopic reversibility [15]. Examples of the process have been presented several times elsewhere [5,10,16,17], and will not be repeated here. *When in doubt, it is useful to verify a proposed algorithm by performing a simulation of the ideal-gas mixture obtained when all intermolecular interactions are set to zero.* The average composition taken from such a simulation can then be compared to an exact result. As a final caveat we can add that one should be careful to include any changes in the long-range correction to the potential energy when deciding acceptance of a trial identity change [18].

D. Summary of Applications

Glandt and co-workers [2–4] originally devised the (isomolar) semigrand ensemble as a means to study polydisperse mixtures. Polydisperse mixtures in principle have an infinite number of components, so their treatment by canonical-ensemble molecular simulation can only be approximate. Isomolar semigrand Monte Carlo is particularly well suited for these systems because species-identity changes can be accomplished in a continuous fashion—this, coupled with the absence of particle insertion/deletion trials, permits the simulations to converge very well. For polydisperse mixtures, the free energy and the partition function become functionals of the chemical potential difference function $\Delta\mu(I)$. A general form for the dependence in the species identity I can be used:

$$\begin{aligned}\beta\,\Delta\mu(I) &\equiv \beta\mu(I) - \beta\mu(I_0)\\ &= c_0 \ln\left(\frac{I}{I_0}\right) + c_1(I - I_0) + c_2(I - I_0)^2 + \cdots \end{aligned} \tag{2.21}$$

Kofke and Glandt [3,4] originally simulated systems for which $1/c_2 \to 0$, $c_j = 0$ $(j \neq 2)$ to study the "nearly monodisperse" case. Subsequently it was observed [19–21] that the "infinitely polydisperse" system for which $c_j = 0$ $(j \neq 0)$ exhibited very interesting and unusual scaling properties. During this period Stapleton et al. [18] presented semigrand simulations of polydisperse mixtures following a Schultz distribution for $\Delta\mu(I)$ (nonzero c_0 and c_1). Their study is weakened by the lack of a well-defined formalism, but it is correctly implemented and was more careful than Kofke and Glandt [3,4] in accounting for the long-range correction to the energy during the Markov sampling process.

Subsequent applications of semigrand methods have been numerous, as species-identity changes have become a standard practice when simulating mixtures. We would fail in an attempt to mention all such uses, so instead we will sample some of the more interesting applications and extensions. Hautman and Klein [22] examined, by molecular dynamics a "breathing" Lennard-Jones fluid of fluctuating particle diameter; the breathing modes are introduced to better model molecules that are treated as LJ atoms. Liu and Berne [23] proposed a similar treatment as a means to accelerate the equilibration of some systems. Kofke [16] used the formalism to yield the freezing diagram of three binary mixtures of hard spheres (this was before the development of the Gibbs–Duhem integration technique). Adsorption and wetting of a Lennard–Jones mixture was examined by Fan et al. [24]. The semigrand ensemble has been used often in conjunction with Gibbs ensemble simulations of phase equilibria [7,25]. Stapleton et al. [26] performed the first simulations of phase coexistence for polydisperse fluids using a semigrand formulation of the Gibbs ensemble. Interesting also are symmetric systems, in which the like intermolecular interactions are the same for both components but differ from their cross-interactions. In a semigrand formulation it is necessary to simulate only one of the two phases, as shown by application to a square-well model [27], a Lennard-Jones mixture [24], nonadditive hard spheres [28], and the Widom–Rowlinson model [29]. It is important that such simulations be performed isobarically; otherwise an unnatural suppression of fluctuations results, exacerbating finite-size effects on approach to the critical point [30]. Simulation in a semigrand ensemble can greatly enhance sampling in polymer mixtures. Representative examples include simulations on the lattice [6,31, 32], of confined thin films [33], and off-lattice [10,34]. The work of Krishna Pant and Theodoro [10] was noted above, as they used a semigrand ensemble in which the total number of monomer segments and the total number of chains are fixed, but the chains lengths occupy a distribution obtained by moving segments from one chain to another. Finally we note again that the simulation of chemically reacting systems is naturally accomplished in a semigrand ensemble. The first efforts in this direction are due to Coker and Watts [35,36] and Kofke and Glandt [5]; both studies apply only to mole-conserving reactions. This constraint was first released by Shaw [37], whose work was clarified and applied by Smith and Triska [12] and independently by Johnson et al. [11]. We refer the reader to the chapter by Johnson elsewhere in this volume for additional information on semigrand simulations of reacting mixtures.

III GIBBS-DUHEM INTEGRATION: TRACING COEXISTENCE LINES

A. Concept

Prior to the introduction of the Gibbs ensemble, the evaluation of thermodynamic phase coexistence was a tedious affair, often requiring many simulations in an attempt to locate the state point at which the temperature, pressure, and all species fugacities were equal between two phases. The identification of this state point is often troublesome even when working with an empirical equation of state. It is all the more difficult with simulation because evaluating the thermodynamic properties at a trial solution is very expensive computationally; moreover, the evaluation of the fugacity is problematic. Two complementary approaches to fugacity evaluation are often considered [15,38,39]. Widom test-particle insertion is most convenient, as it can be conducted in a single simulation without requiring any clever biasing of configurations. Unfortunately the method is of limited reliability, and it is known to fail (more to the point, it becomes very inefficient) at high densities [40]. Complementing this approach is thermodynamic integration, which yields the fugacity through a series of simulations that connects the state of interest to some reference state of known properties. This method is very reliable, but inherently inefficient as it requires a number of simulations at "uninteresting" state points. Some efficiency can be recovered by conducting the integration in conjunction with the search for the coexistence state, but the whole procedure is still unappealing.

The key idea of the Gibbs ensemble [7,25] (GE) is that the process of searching for the coexistence point can be elegantly bundled up with the process of measuring the pressure and fugacity if the two coexisting phases are simulated simultaneously. The chemical potential measurement is done in a manner analogous to Widom insertion. The Gibbs–Duhem integration (GDI) method complements the Gibbs ensemble, in the same sense that test-particle insertion and thermodynamic integration are complementary methods for fugacity evaluation. The idea of GDI is to perform the integration *along the coexistence line*, which itself is being determined as the integration proceeds. As with the Gibbs ensemble, this procedure entails simultaneous simulation of the two (or more) coexisting phases. A very useful feature of the GDI method is that the coexistence line need not be of the traditional variety, and that phase equilibria may be evaluated along a multitude of path types, for example, one in which the intermolecular potential mutates from one form to another.

B. Formalism

1. Clapeyron Equations

The starting point for the formalism is the Gibbs–Duhem equation, which we write first for a pure fluid [13]:

$$d\mu = h\, d\beta + \beta v\, dP \tag{3.1}$$

The Clapeyron equation is derived [13] by writing this formula twice, once for each of two coexisting phases α and β. Given a point on the coexistence line, one is interested in how β $(1/k_B T)$ and P must change in concert so that the chemical potentials of the two phases change by equal amounts: $d\mu^\alpha = d\mu^\beta$. This requirement is used to eliminate $d\mu$ between the equations, and the result takes the form of a first-order differential equation

$$\left(\frac{dP}{d\beta}\right)_\sigma = -\frac{\Delta h}{\beta \, \Delta v} \tag{3.2}$$

where the σ subscript reminds us that the derivative is taken along the saturation line. The Δs here indicate a difference between the α and β phases. The right-hand side is a complicated function of pressure and temperature, but it can be evaluated by molecular simulation without much difficulty. The essence of the method is the integration of this equation using standard methods for the numerical treatment of differential equations.

The approach has been extended in two fundamental ways: (1) one may consider variations in "field" variables [1] other than temperature and pressure; and (2) one may consider additional phases, so that (for example) three-phase coexistence lines are produced. Both extensions start with a more general form of the Gibbs–Duhem equation, which we write as follows:

$$\rho_0 \, d\phi_0 = \rho_1 \, d\phi_1 + \rho_2 \, d\phi_2 + \rho_3 \, d\phi_3 + \cdots \tag{3.3}$$

In this relation, ϕ_i is a "field" variable and ρ_i is its conjugate "density" [1]. Field variables are those that must be equal among coexisting phases; examples include the temperature, pressure, fugacity fraction, or (in our interpretation) a parameter defining the intermolecular potential (such as a Lennard-Jones diameter). Examples of the corresponding conjugate density would be the molar enthalpy, molar volume, or mixture mole fraction. The "0" subscript designates the "hidden" field variable (a thermodynamic potential, or free energy) that the integration procedure is designed to keep equal between coexisting phases. With this notation the generalized Clapeyron equation for two-phase coexistence is

$$\left(\frac{d\phi_2}{d\phi_1}\right)_\sigma = -\frac{\Delta(\rho_1/\rho_0)}{\Delta(\rho_2/\rho_0)} \tag{3.4}$$

So, for example, the isothermal variation of pressure with species 2 fugacity when integrating along coexistence in an osmotic ensemble can be read directly from Eq. (2.15) (for $c = 2$)

$$\left(\frac{dP}{df_2}\right)_{\sigma, T} = \frac{\Delta(x_2/x_1)}{\Delta(\beta v/x_1)} \tag{3.5}$$

Extension to multiple-phase equilibria is straightforward, although the algebraic expressions become correspondingly more complex [13]. In the general case involving n coexisting phases, with ϕ_1 selected as the independent variable, the desired $(n-1)$ derivatives are obtained from the solution of the following system of $(n-1)$ linear equations:

$$\Delta_2^{(j)} \frac{d\phi_2}{d\phi_1} + \Delta_3^{(j)} \frac{d\phi_3}{d\phi_1} + \cdots \Delta_n^{(j)} \frac{d\phi_n}{d\phi_1} = \Delta_1^{(j)}, \qquad j = 1 \ldots n-1 \tag{3.6}$$

where $\Delta_k^{(j)}$ indicates the difference in ρ_k/ρ_0 between phase j and phase n (selected arbitrarily). For three-phase coexistence we have

$$\begin{aligned}
\frac{d\phi_2}{d\phi_1} &= \frac{1}{D}\left[\Delta_3^{(2)}\Delta_1^{(1)} - \Delta_3^{(1)}\Delta_1^{(2)}\right] \\
\frac{d\phi_3}{d\phi_1} &= \frac{1}{D}\left[-\Delta_2^{(2)}\Delta_1^{(1)} + \Delta_2^{(1)}\Delta_1^{(2)}\right] \\
D &\equiv \Delta_2^{(1)}\Delta_3^{(2)} - \Delta_2^{(2)}\Delta_3^{(1)}
\end{aligned} \tag{3.7}$$

Coexistence among more than three phases is easily handled by solving the system of linear equations (3.6) using standard numerical methods.

2. *Intermolecular-Potential-Based Field Parameters*

We digress briefly to elaborate on the use of intermolecular-potential parameters as field variables in a Gibbs–Duhem integration. Integration along such a direction may be of interest because it provides a direct, quantitative picture of how some feature of the intermolecular potential affects the phase diagram; alternatively, such a path may be taken simply to establish a coexistence point on the phase diagram—to provide a starting point for a separate, more conventional integration study.

a. General Case. The conjugate density ρ of a given field variable ϕ is defined as the derivative of an appropriate free energy (designated F) with respect to ϕ. If the corresponding partition function is designated Q such that $\beta F = -\ln Q$, then (suppressing nonsalient terms of the partition function)

$$N\rho \equiv \frac{\partial(\beta F)}{\partial \phi} \tag{3.8a}$$

$$= -\frac{1}{Q}\frac{\partial Q}{\partial \phi} \tag{3.8b}$$

$$= -\frac{1}{Q}\frac{\partial}{\partial \phi}\int \exp[-\beta U(\phi)] \tag{3.8c}$$

$$= -\frac{1}{Q}\int \frac{\partial(-\beta U)}{\partial \phi}\exp[-\beta U] \tag{3.8d}$$

$$= \left\langle \frac{\partial(\beta U)}{\partial \phi} \right\rangle \tag{3.8e}$$

which shows that the conjugate density is simply the ensemble average of the ϕ derivative of the intermolecular potential.

As an example, we can point to the system of soft spheres studied by Agrawal and Kofke [42–43]. This model is defined by an inverse-power pair potential:

$$u(r; n) = \varepsilon\left(\frac{\sigma}{r}\right)^n \tag{3.9}$$

where r is the intermolecular separation, σ and ε are size and energy parameters, and n is a parameter that characterizes the hardness of the potential: $n \to \infty$ corresponds to the hard-sphere model, and as n decreases the potential becomes softer and longer ranged. This model exhibits no vapor–liquid transition, but it does freeze. Agrawal and Kofke studied the freezing pressure as a function of the softness $s = 1/n$, from hard spheres ($s = 0$) to the limits of the model ($s = \frac{1}{3}$) (a unit s can be attained if a neutralizing background is introduced to keep the model well defined, but the study did not go this far). Thus in this application s takes the role of a field variable, and elementary calculus finds that the conjugate density is $<\partial U/\partial s> = -\beta\varepsilon n^2\langle(\sigma/r)^n \ln(\sigma/r)\rangle$.

b. Hard Potentials. Except at very small s the ensemble-averaged derivative described in the soft-sphere example is easy to evaluate in a standard simulation, and indeed it is no less straightforward than the calculation of the average intermolecular energy. This circumstance is typical of many applications. However, at the hard-sphere limit, problems arise because significant contributions to the average are made by a very narrow range of separations. This difficulty indicates a general problem that surfaces when the parameter ϕ relates to the size or shape of a discontinuous potential. In such circumstances special measures are required to evaluate the derivative. Usually such potentials can be written in the form

$$\exp[-\beta u(r;\,\phi)] = e_0(r) + e_\phi H(r - \phi) \tag{3.10}$$

where H is the Heaviside step function and r is some suitable measure of separation of the molecules. For example, a square-well potential of well depth ε, r is the center-to-center distance and we have

$$\exp[-\beta u_{\rm SW}(r;\,\phi)] = e^{\beta\varepsilon}H(r - \sigma) + (1 - e^{\beta\varepsilon})H(r - \lambda) \tag{3.11}$$

for which ϕ may be taken as the hard-core diameter σ or the well extent λ.

Two approaches exist to compute the desired derivative. A finite-difference approximation to the right-hand side of Eq. (3.8c) shows that the derivative is related to the fraction of configurations that exhibit overlap when the parameter ϕ is varied by a small amount, thus

$$N\rho = -\lim_{\Delta\phi \to 0} \frac{1}{\Delta\phi} \langle \exp[-\beta U(\phi + \Delta\phi)] \rangle_{\Delta\phi = 0} \tag{3.12}$$

where $\langle \exp[-\beta U(\phi + \Delta\phi)] \rangle_{\Delta\phi = 0}$ is the overlap fraction, measured in the ensemble for which $\Delta\phi = 0$. The right-hand side of Eq. (3.12) is easily measured for a range of $\Delta\phi$ in a single simulation, and the dependence can be fit to render a good limiting value. This interpretation was taken by the group of Mike Allen when they applied Gibbs–Duhem integration to evaluate the isotropic–nematic coexistence line for a system of hard ellipsoids, as a function of the ellipsoid aspect ratio [44]. The alternative approach is based on Eq. (3.8d), and is best for application to spherically symmetric pairwise additive potentials. In this case the configuration integral on the right-hand side of Eq. (3.8d) can be written in terms of the cavity–cavity correlation function $y(r)$, which remains smooth regardless of discontinuities in the potential [8]

$$\rho = 4\pi\phi^2 e_\phi\, y(\phi) \tag{3.13}$$

where we have applied the notation of Eq. (3.10). Thus the derivative is easily computed if the correlation function is tabulated. We employ this approach in a detailed example below of the effect of the well width on freezing in the square-well model system.

3. *Free Energies and the Emergence of New Phases*

With little additional effort it is possible to obtain values of the thermodynamic potential ϕ_0 along the integration path. This information is needed to establish the stability of the coexisting phases relative to another phase that might be adopted by the system. The construction is depicted in Figure 1. In panel (*a*) the value of ϕ_0 in the third (γ) phase is computed along the ϕ_2–ϕ_1 path that is traced by the coexistence of the α and β phases. At the intersection of the ϕ_0–ϕ_1 lines all three phases are equally stable, and a triple-point heralds the emergence of the γ phase. Subsequent integration in the ϕ_2–ϕ_1 plane must be done in a way that traces coexistence of the γ phase with either the α or β phase. Either may be taken, but one should be careful to continue ϕ_1 in the direction that finds the third phase unstable with respect to the two newly coexisting phases.

In Figure 1*b* a different construction is described. In this situation two coexistence lines are being traced by the Gibbs–Duhem method, each line having one phase (α) in common. The point where the β–α coexistence coincides with the γ–α coexistence again forms a triple point. Subsequent integration then follows coexistence between the γ and β phases. It might be

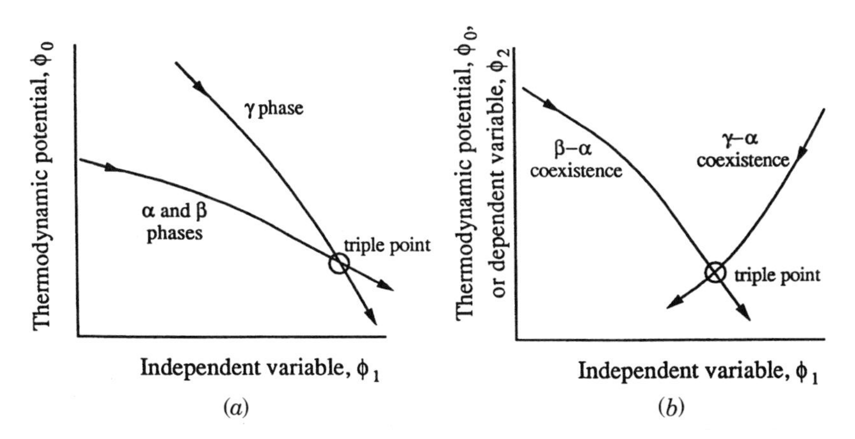

Figure 1. Schematic of construction used to detect emerging phases. (*a*) Integration is conducted along coexistence line of α and β phases, and free energy of γ phase is computed along the same path in the ϕ_2–ϕ_1 plane; thus both lines have same value of ϕ_2 at any ϕ_1. To the left of the triple point the γ phase is unstable with respect to the others; to the right it is the stable phase, so at the triple point a different coexistence should be followed (not shown). (*b*) Separate integration series are conducted along β–α and γ–α coexistence lines, respectively. Crossing of these lines in either the ϕ_0–ϕ_1 or the ϕ_2–ϕ_1 planes should occur at the same value of ϕ_1.

noted that the free energy ϕ_0 is not needed to identify three-phase coexistence in this instance, because the coexistence lines will also coincide in the ϕ_2–ϕ_1 plane. However, examination of the free energy provides a valuable check on the calculations. Examples are presented in detail below.

The best approach to evaluation of ϕ_0 along the integration path is (no surprise) thermodynamic integration. The relevant formula is easily derived from Eq. (3.3). A one-dimensional quadrature can be constructed using the derivatives that guide the integration procedure [see Eq. (3.4)]; thus

$$\phi_0^b = \phi_0^a + \int_{\phi_1^a}^{\phi_1^b} \left[\frac{\rho_1}{\rho_0} + \left(\frac{\partial \phi_2}{\partial \phi_1} \right)_\sigma \frac{\rho_2}{\rho_0} \right] d\phi_1 \tag{3.14}$$

The extension to three- and higher-phase integrations is obvious.

C. Practical Matters

1. *Choosing a Path*

The integration path is largely set by the definition of the problem; one knows beforehand the plane in which the coexistence line is desired. However application of simple transformations to the field variables can be beneficial if they reshape the coexistence line into a simpler form. The advantages can be improved accuracy, precision, and stability of the integration. A very familiar example is the conversion from pressure P to $\ln(P)$ in the characterization of vapor–liquid coexistence. The Clapeyron equation in the latter instance is

$$\left(\frac{\partial \ln P}{\partial \beta} \right)_\sigma = - \frac{\Delta h}{\beta P \, \Delta v} \tag{3.15}$$

The numerator (Δh) is not affected by the transformation, but it is already insensitive to P and it has a simple dependence on β. Away from the critical point $v^{\text{vap}} \gg v^{\text{liq}}$, so the denominator is dominated by the vapor volume. To a first approximation, the ideal-gas law applies and the denominator is unity. Thus the right-hand side does not vary much (or in a complex way) as the integration proceeds. We emphasize that the integration procedure does not make any of the approximations that we describe; rather, we consider them only to identify forms of the field variables that minimize any numerical inaccuracies inherent in the method. In the ideal situation the right-hand side varies linearly with the independent variable.

A less familiar example is provided by the work of Agrawal and Kofke to establish the complete solid–liquid coexistence line for the Lennard–Jones (6–12) fluid [45]. At sufficiently high temperatures this model behaves as a system of inverse-power (r^{-12}) soft spheres. The freezing transition of that system was characterized separately [42,43], so it is sensible to begin the Gibbs–Duhem integration of the Lennard-Jones system from this point ($\beta = 0$) and integrate toward lower temperatures (increasing β). Analysis shows that the limiting behavior of the freezing line is

$$\ln(P\beta^{5/4}) \sim \beta^{1/2}, \qquad \beta \to 0 \tag{3.16}$$

This result suggests appropriate choices for the integration path, and some quantitative renditions are presented in Figure 2. Obviously, one should work with $P\beta^{5/4}$ in lieu of the pressure P (Fig. 2*a*), and—less obvious but equally important—it is necessary to work with $\beta^{1/2}$ rather than β because the latter finds an infinite slope at the outset of the integration (Fig. 2*b*). The use of the logarithm is not required, and in fact the complete freezing line takes a simpler form if it is not applied. If one is interested only in the triple-point temperature, and not the pressure, a good value can be obtained by integrating $\beta^{1/2}$ versus $P\beta^{5/4}$ (as dependent and independent variables, respectively). Taking $\beta^{1/2}$ as the independent variable raises the concern of stepping past the triple point and causing the pressure to go negative. In contrast, because the temperature is low $P\beta^{5/4}$ can be taken to zero without too much concern for evaporation of either phase, even though they are then metastable with respect to the vapor. If the triple-point pressure is desired as well, it may be necessary to work with $\ln(\beta P^{5/4})$ to resolve the low pressure region. Some adjustment in the integration

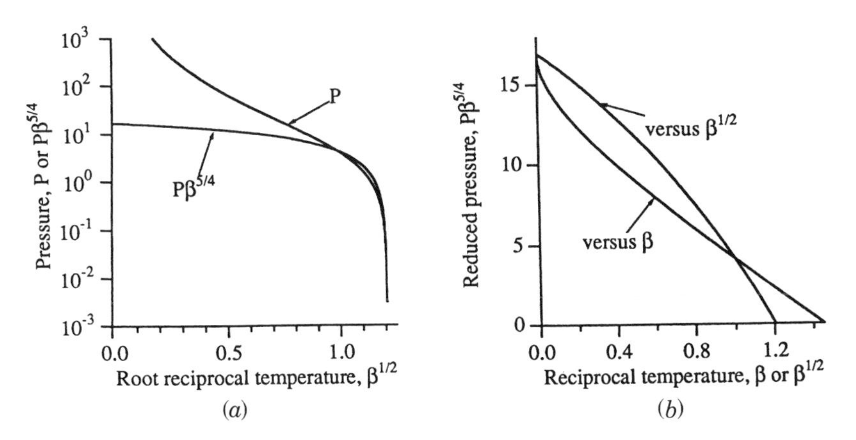

Figure 2. Solid–liquid coexistence line for the Lennard-Jones model, represented for several transformations of the pressure and temperature variables.

(change in step size, or choice of dependent vs. independent variables) is warranted as the coexistence line transitions from a shallow to a steep slope (Fig. 2*a*). The use of a variable-step algorithm (suggested below) can be particularly helpful in such cases. All things considered, the best approach for the present circumstance would be to locate the triple-point temperature in the straightforward manner just described, and then obtain the triple-point pressure at that temperature from the vapor-liquid coexistence line, which presumably is being established separately.

Clearly, a bit of planning can pay off well when contemplating an unfamiliar Gibbs–Duhem integration process.

2. *Getting It Started*

A key prerequisite for the conduct of a Gibbs–Duhem integration is a state point known to lie on the coexistence line. This state may be obtained by any established means, including the traditional free-energy matching procedure, a Gibbs ensemble simulation, or a separate Gibbs–Duhem integration. Once this point is located it is necessary (or at least very helpful) to evaluate the slope of the coexistence line at this initial state. Often this calculation is completely straightforward; for example, with an integration in the pressure-temperature plane it requires only knowledge of the enthalpies and volumes of the coexisting phases [see Eq. (3.15)]. However, in some instances the slope can be determined only via a limiting procedure, and the quantities that require measurement to evaluate it are less than obvious. A common example is presented by integrations in the composition (mole fraction) plane for mixtures. If, for example, this is done using the isomolar semigrand formalism the governing Clapeyron formula is [see Eqs. (2.18) and (3.4)]

$$\left(\frac{\partial P}{\partial \xi_2}\right)_{\sigma,\beta} = \frac{\Delta x_2/(\xi_1 \xi_2)}{\beta\,\Delta v} \tag{3.17}$$

A natural choice for the initial state of the integration is one of the species as a pure component. In this limit the numerator becomes the ratio of numbers approaching zero. Assuming that Raoult's limiting law applies, the component fugacities become proportional to the corresponding mole fractions:

$$f_1 = f_1^{\circ} x_1; \qquad f_2 = H_2 x_2, \qquad x_2 \to 0 \tag{3.18}$$

where f_1° is the fugacity of pure species 1 and H_2 is the Henry's constant of species 2 in pure component 1. With these relations it can be seen that the fugacity fraction vanishes as $x_2 H_2/f_1^\circ$ as $x_2 \to 0$, and the limiting slope is

$$\left(\frac{\partial P}{\partial \xi_2}\right)_{\sigma,\beta} = \frac{\Delta(f_1^\circ/H_2)}{\beta\,\Delta v} \tag{3.19}$$

Thus the ratio of the Henry's constant to the solvent fugacity must be evaluated in each coexisting phase to start the integration. This quantity can be measured in a simulation of pure component 1 by performing test identity changes of individual molecules from species 1 to species 2. Details of this calculation, as well as the limiting formulas appropriate to osmotic–ensemble integrations, have been presented elsewhere [17].

It is difficult to generalize the analysis of the limiting slope. Instead we will refer to the applications summarized below and in Table I. These examples may be instructive when considering other systems.

3. *Conducting the Integration*

Predictor–corrector methods [47,48] are appropriate for the integration because they require only one evaluation of the slope for each integration step. Molecular simulation is unusual in the context of the numerical treatment of differential equations, because an approximation to the slope is available before the simulation is complete. This information can be used to update the state point as the simulation proceeds. An increment in a typical GDI series entails the following steps, which for concreteness we describe for an integration in the P–β plane

1. Increment the temperature, and apply a predictor to estimate the new value of the pressure. Initiate a NPT simulation at the new temperature and the predicted pressure.
2. Collect averages for the enthalpies and volumes needed to evaluate the slope at the new state condition.
3. Apply a corrector to update the estimate of the pressure, using the slope determined from the current running averages from the full simulation.
4. Repeat from step 2 until convergence.

One should be careful with this procedure, as in principle it renders a Monte Carlo simulation a non-Markov process. The effect is likely to be benign, but the safest way to proceed is to take the corrector updates of the pressure only during the equilibration phase of the simulation (i.e., those cycles normally granted to allow the system to equilibrate to the new state conditions). In our experience the corrector iteration usually converges very quickly, well before the end of the equilibration period. As a check one can

TABLE I
Summary of Applications of Gibbs–Duhem Integration Method[a]

Model	Coexistence Phenomenon	Integration Plane	Starting Point	Ref
Lennard-Jones (LJ) (pure)	V–L	P–T	Gibbs ensemble	50,79,80
LJ (pure)	L–S; V–S	P–T	High temperature (r^{-12} soft spheres)	45
LJ (binaries)	V–L	P–μ_1; P–ξ_1	Pure component	17
LJ + point-quadrupole binary mixture (model for O_2–N_2)	V–L	P–ξ_1	Pure component	81
LJ (binaries)	V–L–S triple point	P–T–ξ_1	Pure component	46
r^{-n} soft spheres (pure)	F–S	P–softness	Hard spheres	42,43
Polydisperse hard spheres	F–S	P–polydispersity	Pure hard spheres	62,63
Model for CsI	F–S	P–T	Experiment	82
Spherocylinder chains (model for semiflexible polymers)	I–N	P–flexibility	Gibbs ensemble of fully rigid chains	66
Hard-sphere semiflexible chains	I–N		Pseudo–Gibbs ensemble	67

C_{60}	V–S	P–T	Free-energy calculations	83
Gay–Berne	V–L	P–T	Gibbs ensemble	84
Hard-core Yukawa	L–S fcc–bcc	P–T	Various	60,64
Model for colloids dispersed in polymer solution	F–S	P–$\mu_{polymer}$	Pure colloid (hard spheres)	85
Hard ellipsoids	I–N	P–elongation	Gibbs ensemble	44
Hard spherocylinders	I–N; I–R; I–Sm	P–elongation	Various	70
Square-well octamers (pure)	V–L	μ–T	Gibbs ensemble	49
Square-well octamers (symmetric binary mixture)	V–L	μ_1–T	Gibbs ensemble	49
Lattice homopolymers (pure 8-mers and pure 128-mers)	V–L	μ–T	Gibbs ensemble/grand canonical	71
Model for polymer-induced protein precipitation	F–S	P–T	High-temperature (hard spheres)	65
LJ in a planar slit	V–L	P–T, P–pore width	Free-energy calculations	74

[a] Listed are the molecular model, the type of phases in coexistence, the plane in which the integration was conducted, the known coexistence datum used to initiate the series, and the literature citation. Abbreviations for the various phases in the "Coexistence Phenomenon" column are as follows: vapor (V), liquid (L), solid (S), fluid (F), isotropic (I), nematic (N), rotator (R) and smectic (Sm). Abbreviations for the "Integration Plane" column are as defined in the text.

monitor during the production cycles the pressure that the corrector would specify if given the current averages. If sufficiently different from the current pressure, the corrector can be applied and the production cycles reinitiated at the new pressure. If convergence of the pressure is problematic, schemes can be applied involving several equilibration periods, each with running averages that include increasingly attenuated contributions from averages recorded over the previous equilibrations periods. Fortunately, such measures are rarely necessary.

Many predictor–corrector formulas are available. The original implementation of the GDI method suggested a set of constant step-size formulas of order that increased as the integration proceeded. We agree with the assessment of Escobedo and de Pablo [49] that lower-order schemes should be preferred because they are more stable, are easier to implement with a variable step size, and do not compromise accuracy—in most cases it is the uncertainty in the simulation averages that limits the overall accuracy of the integration. Escobedo and de Pablo present two second-order corrector formulas that allow for a variable step size. Both are of the form [49]

$$y_{k+1} = Ay_{k-1} + By_k + D(C_{-1}y'_{k-1} + C_0 y'_k + C_1 y'_{k+1}) \tag{3.20}$$

where $y(x)$ is the dependent variable of the integration and y' is its derivative [i.e., that expressed in Eq. (3.4)]. The coefficients depend only on the integration step sizes $h_j \equiv x_{j+1} - x_j$ and are recorded in Table II. The table also includes a Euler corrector that is suitable for taking the first step in the integration. A simple forward Euler formula is often suitable as a predictor

$$y_{k+1} = y_k + h_k y'_k \tag{3.21}$$

Higher-order predictors such as those described in the original implementation of the GDI method [50] may be warranted if the slope y' is difficult to

TABLE II
Coefficients for Use in Variable-Step-Size Corrector Formula [Eq. (3.20)], as Reported by Escobedo and de Pablo [49][a]

	A	B	C_{-1}	C_0	C_1	D
Method 1	0	1	$-r^{-1}$	$r^{-1} + 4 + 3r$	$2 + 3r$	$\frac{h_k}{6(1+r)}$
Method 2	$\frac{1}{r^2(3+2r)}$	$1 - A$	0	$r^{-1} + 2 + r$	$1 + r$	$\frac{h_k}{3+2r}$
Euler	0	1	0	1	1	$\frac{h_k}{2}$

[a] Here $r = h_{k-1}/h_k$, where $h_j = x_{j+1} - x_j$ is the integration step size. Two variants are presented, as well as a simple Euler formula suitable for starting an integration series.

evaluate accurately (and thus requires a lengthy period of simulation before a reasonable estimate of it becomes available). In this case a good predictor value will minimize the amount of iteration needed for the procedure to converge. A good predictor value is needed also if one of the phases cannot be sustained for any period in a metastable state.

No modifications to the predictor–corrector formulas are needed to conduct integrations of coexistence involving more than two phases. In such instances all dependent variables are updated simultaneously by applying the corrector formula to each of the governing differential equations; the simulation then continues with the new values in the manner outlined above.

4. *Coupling the Phases*

On approach of the critical point a small system cannot be expected to sample only one of two coexisting phases. For obvious reasons the unilateral condensation or evaporation of one of the coexisting phases defeats the GDI technique. To forestall breakdown of the method when a critical point is approached, one can introduce an artificial coupling between the phases. This coupling is designed to mimic the transfer of volume and particles between the phases—in the spirit of a Gibbs ensemble simulation—so that they remain on either side of some chosen density. This device still cannot prevent phase changes, but if they occur, they are bilateral, so that one phase vaporizes if the other condenses. The property differences required by the method then remain available for computation if one is careful to track the occurrence of these phase-identity exchanges.

The coupling is accomplished by rewriting the densities of the two phases in terms of new variables X and Y, which can be interpreted [51] as the fraction of all particles and the fraction of the total volume, respectively, held by one of the phases. Thus

$$v^{\alpha} = \frac{Y}{X}\, v; \qquad v^{\beta} = \frac{1 - Y}{1 - X}\, v \tag{3.22}$$

The "overall" volume v may be specified arbitrarily, but should lie between the expected molar volumes of the two phases. Instead of sampling the volumes of the phases independently, sampling of X and Y is performed. This change of variables is accompanied by a Jacobian

$$J(v^{\alpha}, v^{\beta}) = -N^{\alpha}N^{\beta}v\, \frac{(v^{\beta} - v^{\alpha})^3}{(v - v^{\alpha})(v^{\beta} - v)} \tag{3.23}$$

The usual acceptance criterion for volume changes must be multiplied by the ratio of the Jacobians evaluated in the trial and original states.

We are of the opinion that the tendency of a system to display unilateral or bilateral phase changes is indicative of more serious finite-size effects. Application of the GDI method (or the Gibbs ensemble for that matter) under such circumstances can provide only a qualitative picture of the phase behavior. Histogram-reweighting methods have been developed and refined to the point where they now provide a very precise description of the critical region in the thermodynamic limit. These methods should be applied if a quantitative characterization of the critical region is desired [52–57].

5. *Error Estimation and Stability*

A GDI series is subject to error from three sources:

- Stochastic errors in the simulation averages
- Systematic errors associated with the predictor–corrector scheme of numerical integration
- Errors introduced by an inaccurate initial condition, that is, an initial state point that does not lie on the true coexistence line

The magnitude of stochastic errors in the simulation averages can be quantified using standard analyses [15] and their influence on the precision of the coexistence line gauged via standard propagation-of-error method [58].

Systematic errors arising from the predictor–corrector integration do not seem to be a problem, judging from the ability of the GDI method to traverse an integration path and yield a result that agrees—within the limits of the stochastic error—with coexistence data obtained independently. Regardless, a simple test can be applied requiring no additional simulations—eliminate half the data and reintegrate the remaining set of slopes to examine the effect on the final coexistence point (alternatively, a higher-order quadrature algorithm could be applied to the full set of data). Any concerns can be alleviated by working with a smaller integration step.

The consequences of error introduced with the initial condition is a matter of stability of the GDI procedure, because in a sense each point along the integration path provides an initial condition for those that follow. If deviations from the true coexistence line give rise to even larger deviations, the integration is unstable. Certainly the opposite situation is greatly preferred, in which deviations become attenuated as the integration proceeds. Fortunately, it is not too difficult to know which situation is in effect, and to what extent.

Returning to the notation of Eq. (3.3), error in the initial state implies that the thermodynamic potential ϕ_0 is not equal between the coexisting phases because ϕ_2 is not at the coexistence value for the initial ϕ_1. The GDI procedure is designed to keep the difference of ϕ_0 between the two phases constant, so stability can be examined by estimating how a constant error in ϕ_0 translates into deviations in ϕ_2. Assuming small errors we can write, at a given point in the integration path and thus for fixed ϕ_1

$$\phi_0(\phi_2) = \phi_0(\phi_2^{eq}) + \frac{\rho_2}{\rho_0}\,\delta\phi_2 \tag{3.24}$$

where ϕ_2^{eq} is the correct value of ϕ_2 for coexistence at ϕ_1, and $\delta\phi_2 = \phi_2 - \phi_2^{eq}$ is the error. The difference in ϕ_0 between the two phases is

$$\phi_0^{\beta}(\phi_2) - \phi_0^{\alpha}(\phi_2) = \Delta\left(\frac{\rho_2}{\rho_0}\right)\delta\phi_2 \tag{3.25}$$

Assuming that the GDI procedure succeeds in keeping this difference unchanged, the error at a subsequent point in the integration can be related to the error at the initial (origin) point (subscript "o")

$$\delta\phi_2 = (\delta\phi_2)_o \frac{[\Delta(\rho_2/\rho_0)]_o}{\Delta(\rho_2/\rho_0)} \tag{3.26}$$

The important result is that the error grows or diminishes inversely with the difference in ρ_2/ρ_0 between the phases. The analysis tells us in which direction to conduct an integration, if given a choice, and permits quantification of the effects of error in the initial state. So, for example, it is advantageous to perform GDI studies of vapor–liquid coexistence by beginning close to the critical point and integrating toward lower temperature.

Examination of the first GDI data ever recorded can illustrate the analysis. Our study [50] of Lennard-Jones vapor–liquid equilibrium first attempted to integrate from low temperature using an established Gibbs ensemble (GE) datum [59]. To the author's dismay, the integration did not proceed through other GE data established at higher temperatures. Further study showed that an integration beginning from a midrange temperature found agreement with all other GE data except the low-temperature value, for which the GE pressure was 13% lower. Moreover, series beginning from the newly established GDI low-temperature datum agreed well with other data. As illustrated in Figure 3, the stability analysis characterizes very well the propagation of error that we now recognize in the low-temperature GE datum. Shown in the figure are the correct coexistence line and the one beginning from the incorrect low-temperature point. Error bars on the correct line indicate how an initial 13% error would propagate in an integration toward high temperature, as determined from the stability analysis.

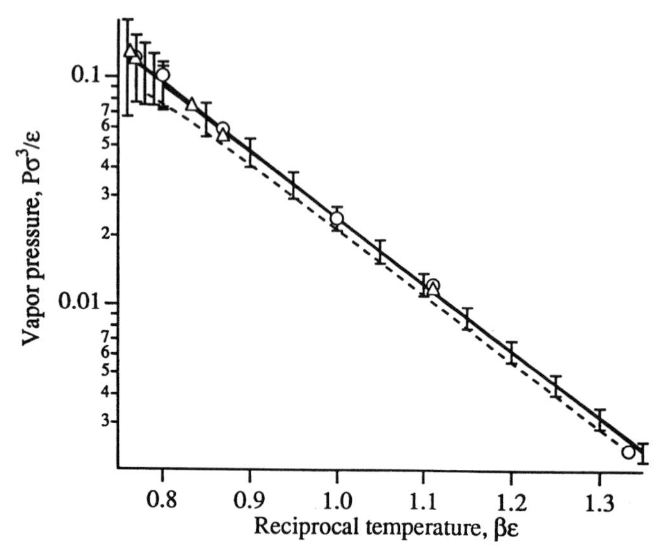

Figure 3. Vapor–liquid coexistence line for the Lennard-Jones model. Solid line is presently the best determination of the phase boundary [50]; triangles [86] and circles [59] are Gibbs ensemble data. Dashed line is obtained from Gibbs–Duhem integration beginning with the low-temperature (rightmost) Gibbs ensemble datum. "Error bars" on the true line are the stability analysis prediction of how the error in the initial datum propagates through the integration series.

The estimate exactly describes the course of the ill-fated series. This suggests another use of the stability analysis, namely, to correct a GDI series if its initial state point is subsequently found to be slightly in error. Elements of this idea are exhibited in the coexistence-line free-energy integration approach recently described by Meijer and Azhar [60].

D. Summary of Applications

Applications of the GDI method are summarized in Table I, where we list the molecular model, the type of phase behavior, the plane in which the integration was conducted, and the nature and source of the initial coexistence datum for the integration. Most of the work involves coexistence with solid or ordered phases, as this application is uniquely suited to the GDI method. However, the approach is no less effective at treating fluid–fluid equilibria, although other methods are often applicable in these cases.

The GDI method has been used to map out the entire phase diagram of the Lennard-Jones model [45,50] (exclusive of any polymorphic *fcc-hcp* transitions which may exist [61], and a few simple LJ mixtures have been studied [17] using both the isomolar and osmotic formulations of the semigrand ensemble; the triple point as a function of composition has been computed [46] for these systems. The liquid–solid coexistence line of the pure LJ system was examined against several well-established semiempirical melting rules [45]. The effect of size polydispersity on the fluid–solid phase diagram of hard spheres has been examined by Bolhuis and Kofke [62,63]; this study has application to the behavior of colloidal systems. Another model appropriate for colloids—the hard-core Yukawa potential—has been examined in two separate studies [60,64]. Using GDI and GE methods, Tavares and Sandler [65] applied a one-component potential of mean force model to study polymer-induced precipitation of amorphous and crystalline phases of proteins from solution. Two simulation studies have now been conducted to gauge the effect of rigidity on the isotropic-nematic phase behavior of polymeric systems. Dijkstra and Frenkel [66] examined a system of joined spherocylinders using the GDI method, while Escobedo and de Pablo [67] recently reported a study of hard-sphere chains; the latter does not use GDI methods but works instead with pseudo-ensemble concepts [68,69]. In a similar vein, Camp et al. [44] examined the effect of elongation on the isotropic–nematic transition of hard ellipsoids using the GDI method along a path of increasing elongation, while Bolhuis and Frenkel [70] completed a similar study for hard spherocylinders, which has a much more complex phase diagram (including smectic and rotator phases). Bolhuis and Frenkel encountered some difficulty with integration stability that they associated with a poor predictor, and they describe a slightly different integration scheme that alleviates the problem.

Escobedo and de Pablo have proposed some of the most interesting extensions of the method. They have pointed out [49] that the simulation of polymeric systems is often more troubled by the requirements of pressure equilibration than by chemical potential equilibration—that volume changes are more problematic than particle insertions if configurational-bias or expanded-ensemble methods are applied to the latter. Consequently, they turned the GDI method around and conducted constant-volume phase-coexistence simulations in the temperature–chemical potential plane, with the pressure equality satisfied by construction of an appropriate Clapeyron equation [i.e., they take the pressure as ϕ_0 of Eq. (3.3)]. They demonstrated the method [49] for vapor–liquid coexistence of square-well octamers, and have recently shown that the extension permits coexistence for lattice models to be examined in a very simple manner [71].

Another interesting variant has been described by Svensson and Woodward [72]. These workers showed how the width of a planar pore could be varied while keeping the fluid inside at a constant chemical potential, without resorting to grand-canonical methods. Instead, they work in an isotension ensemble with pressure fixed by considering how the chemical potential would otherwise change with the pore width. Their method is not an example of GDI, but it has enough features in common to warrant mention here. In particular, we see this as a very good example of ways that the GDI method can be generalized. GDI prescribes how two or more state (or potential) parameters must be adjusted simultaneously in order to keep some quantity (viz., the difference in chemical potential between two phases) unchanged. One might consider also integrations that keep other quantities fixed. Examples include fixing the composition of one phase while the pressure and temperature change (thereby permitting dew and bubble lines to be traced), or fixing the enthalpy (say) of a single-phase mixture while adjusting two or more intermolecular potential parameters (useful if fitting parameters to match experiment). We do not expect that these particular extensions would proceed as easily (or perhaps as successfully) as the GDI of phase coexistence, but they are suggestive of an even broader usefulness of the approach.

These workers have since gone on [73,74] to conduct phase-coexistence calculations in pores using these ideas. These recent studies have many notable features, including the use of free-energy perturbation formulas in lieu of thermodynamic integration to maintain the chemical potential equality, and the introduction of a restraining field to eliminate unilateral or bilateral phase changes.

E. Detailed Example: Freezing of the Square-Well Model

To close we present in some detail a study of freezing for the square-well model. This system is described by a hard core of diameter σ (which we take to be unity) that is centered within a well of depth ε extending to separations λ [see Eq. (3.11)]. Of some interest is the effect of the well extent λ on the phase diagram. A rather complete characterization was conducted by Young [75], but his results are approximate because they are based on a cell model of the solid phase. However, his study indicates that in general the square-well potential provides a very bad model for solids, as it exhibits disproportionately complex phase diagrams owing to the interplay between the sharp well cutoff and the lattice spacing of the solid. Nevertheless it remains an important model for the fluid phases, and consequently it is helpful to know the conditions under which the fluid is stable with respect to the solid. We do not attempt a complete description here, as our primary interest is to use this calculation to demonstrate various aspects of the GDI method when applied to a complex phase diagram. We will focus attention on the well extent $\lambda = 1.95$. In an earlier study [46] we applied the GDI method in the region of $\lambda = 1.50$. We have since discovered that the integration path used to reach the solid–fluid coexistence line had crossed a different first-order phase transition (from an expanded to a condensed solid), thereby invalidating the remainder of the series.

Young's cell model applied at $\lambda = 1.95$ predicts that fcc (face-centered cubic) is the stable solid at freezing, regardless of the pressure. It also indicates a polymorphic transition to the hcp (hexagonal close-packed) phase at lower temperatures. It does not find any region of stability for the bcc (body-centered cubic) solid; however for some smaller values of λ (beginning below about 1.74) it predicts that bcc is stable at the vapor–liquid–solid triple point, whereas for larger values of λ hcp or bcc may become the stable triple-point solid. This application suggests the first point to be made when attempting to compute phase diagrams by molecular simulation. All the quantitatively accurate methods (especially GDI but also GE) require that some idea of the nature and location of the phase transitions be known beforehand. Simulation is not effective at giving a quick, qualitative picture of complex phase diagrams. If the phase behavior is not otherwise obvious, one should apply an approximate treatment (such as the cell model) to construct a "roadmap" that guides the simulation-based study. Alternatively single-phase NPT simulations can be performed at various state points to get a rough picture of the phase diagram, but even in this case some rough idea of what to expect is very helpful. We will restrict our attention to coexistence between and among the two fluid phases and the fcc solid, with a cursory examination of the stability of other solids as λ is varied.

Application of the GDI method to the coexistence lines requires establishment of a coexistence datum on each. A point on the vapor–liquid line can be determined by a GE simulation. At high temperature the model behaves as a system of hard spheres, and the liquid–solid coexistence line approaches the fluid–solid transition for hard spheres, which is known [76,77]. Integration of liquid–solid coexistence from the hard-sphere transition proceeds much as described in Section III.C.1 for the Lennard-Jones example. The limiting behavior ($\beta \to 0$) finds that βP is well behaved and smoothly approaches the hard-sphere value [76,77] of 11.686 at $\beta = 0$ (unlike the LJ case, we need not work with $\beta^{1/2}$). Thus the appropriate governing equation for the GDI procedure is

$$\left(\frac{\partial \beta}{\partial \beta p}\right)_{\sigma} = -\frac{\Delta v}{\Delta u} \tag{3.27}$$

where u is the energy per molecule. The integration was conducted using eight uniform steps in βp from the hard-sphere coexistence value to $\beta p = 0$ (10,000 production cycles per simulation, with about 150 particles in each phase). The coexistence line so computed is presented in Figure 4. Removal

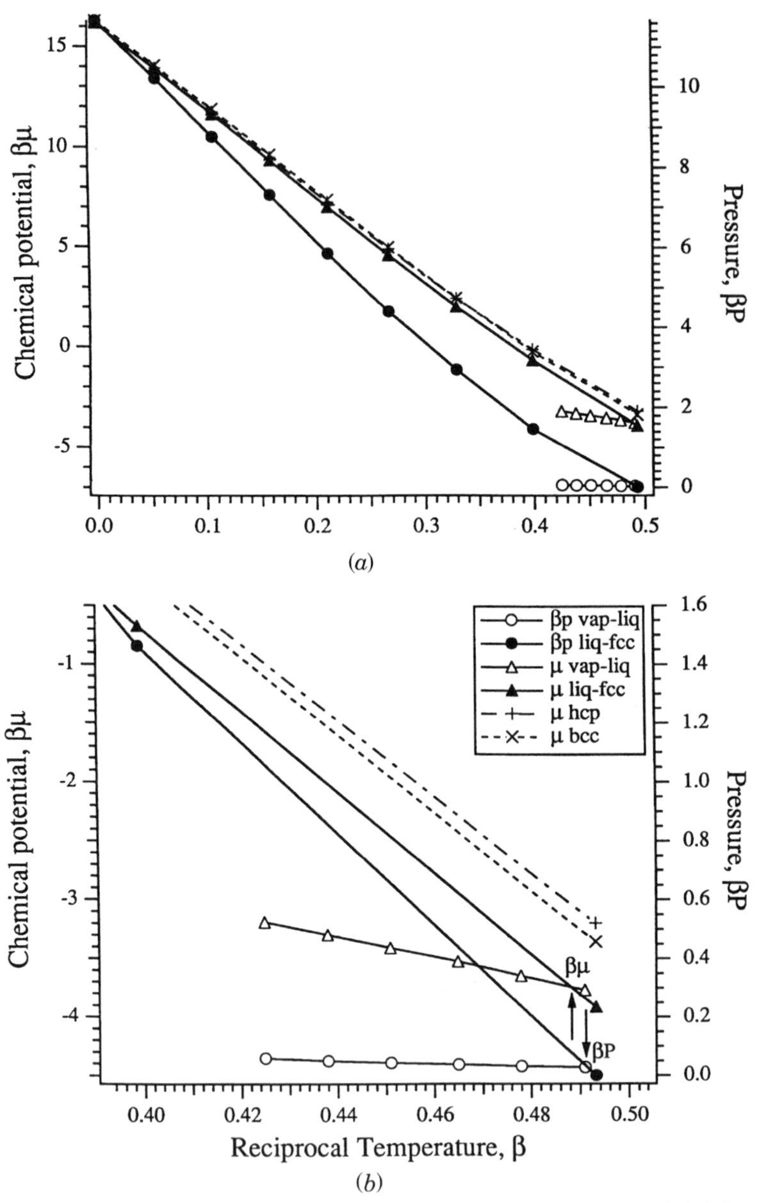

Figure 4. Coexistence for the $\lambda = 1.95$ square-well model. Lines with open symbols follow the liquid–vapor coexistence path in the pressure–temperature plane; all other lines follow the liquid–fcc coexistence path. Panel (*b*) is an enlargement of (*a*) to highlight the triple-point region. Both pressure and chemical potential (free energy) are plotted. Arrows in (*b*) point out the intersections that correspond to the triple point of vapor–liquid–fcc coexistence. A perfect integration would find these intersections at the same value of the abscissa β. Broken lines describe the chemical potentials of the hcp and bcc phases as they trace the same liquid–fcc coexistence path.

of every other integration point and application of the corrector integrator to the remaining ones (without actually redoing the simulations) alters the final value of β by 2%. This provides a conservative upper bound to the error introduced by the finite step of the integration. It indicates that the actual error due to the integrator is of the order of 0.5% or less.

A GE study of square-well vapor–liquid equilibrium was conducted by Vega et al. [78]. It includes the coexistence lines for $\lambda = 2.0$ and 1.75, which together present a good estimate of the critical point for the $\lambda = 1.95$ system. We perform the GE calculation at $\beta = 0.425$, between this estimate of the critical temperature ($\beta = 0.40$) and the cell-model estimate of the triple point ($\beta = 0.50$). The simulation yields a saturation pressure of $\beta p = 0.0575$. We initiate the GDI series from this point and integrate (β vs. $\ln \beta P$, although Eq. (3.15) would work just as well) toward lower temperatures, going just beyond the expected triple point. The resulting coexistence line is included in Figure 4.

The solid–liquid and liquid–vapor coexistence lines intersect at $\beta = 0.491$, $\beta p = 0.028$. Further verification is obtained by examining the chemical potential $\beta\mu$, which is computed along the integration path as described in Section III.B.3. Free-energy values for the initial points are available; for the solid–liquid line the hard-sphere value is known [76,77] ($\beta\mu = 16.898$), and for the liquid–vapor, it was measured during the GE simulations ($\beta\mu = -3.191$). Curves for both integrations are included in Figure 4, and at $\beta = 0.488$ they indeed intersect very close to the triple-point temperature just established.

The triple point is a key "landmark" on the phase diagram, and with the point just established, we can begin to trace out a whole line of three-phase coexistence as a function of the well extent λ. We cannot proceed very far if considering only the fcc phase, for as λ increases, the hcp and bcc phases gain in stability. The cell model calculations alerted us to the possibility of this outcome, so we performed "shadow" free-energy calculations of the hcp and bcc phases as they follow the coexistence path just established in the pressure-temperature plane. The bcc phase is so unstable at the outset of the series (hard spheres) that its structure cannot be maintained, and a single-occupancy constraint was applied [76]. The initial hcp free energy is indistinguishable from the fcc value [77], but as the solid–liquid coexistence series proceeds, it clearly rises above the free energies of the coexisting phases (Fig. 4); this behavior proves that hcp is not the stable solid at freezing. The single-occupancy bcc free energy for the hard-sphere initial condition has been established [43] ($\beta\mu = 16.32$). The bcc phase gains relative stability as it follows the fluid–fcc coexistence path, but not enough to unseat fcc as the stable solid at freezing. The shadowing process can continue as the triple-point integration proceeds in the direction of increasing

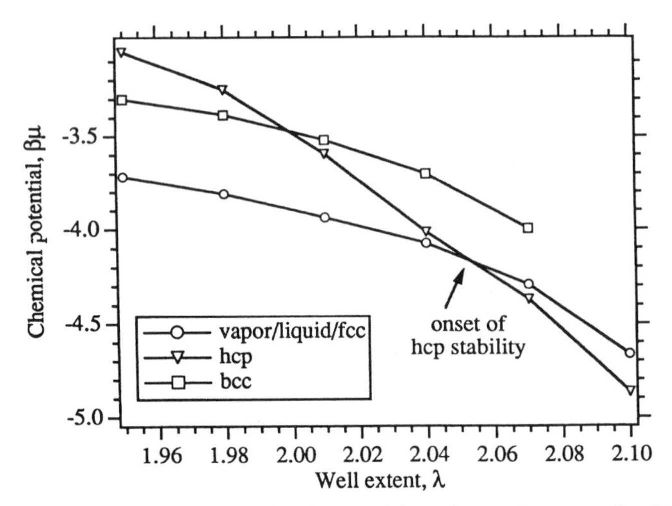

Figure 5. Chemical potential of the fcc, hcp, and bcc phases along a path of increasing square-well extent λ, with pressure and temperature set to ensure coexistence between the vapor, liquid, and fcc phases.

λ. The density conjugate to the "field" λ is measured for each phase (liquid, vapor, fcc, hcp, and bcc) in the simulation according to Eqs. (3.10), (3.11), and (3.13). Results are described in Figure 5. The figure indicates that bcc becomes increasingly unstable, but the hcp phase achieves stability at freezing for λ greater than about 2.05. The integration could continue from this point by following the liquid–vapor–hcp coexistence (or other combinations of two or three phases, if interested) as λ increases further, with the fcc and bcc phases shadowing to continue to gauge their stability with respect to hcp.

ACKNOWLEDGMENTS

This review was prepared while the author enjoyed a one-year sabbatical leave at the University of Tennessee (UT), partially supported by Oak Ridge National Laboratory (ORNL) through funding provided to UT to support the UT/ORNL Distinguished Scientist Program. ORNL is managed by Lockheed Martin Energy Research Corp. for the DOE under Contract No. DE-AC05-96OR22464. I am very grateful to Peter Cummings and Hank Cochran for their hospitality during this period. I would like to thank Fernando Escobedo and Juan de Pablo for providing manuscripts prior to publication. I am especially grateful to Ilja Siepmann for his part in coordinating this review volume and for inviting my contribution.

REFERENCES

1. R. B. Griffiths and J. C. Wheeler, *Phys. Rev. A* **2**, 1047 (1970).
2. J. G. Briano and E. D. Glandt, *J. Chem. Phys.* **80**, 3336 (1984).

3. D. A. Kofke and E. D. Glandt, *Fluid Phase Equil.* **29**, 327 (1986).
4. D. A. Kofke and E. D. Glandt, *J. Chem. Phys.* **87**, 4881 (1987).
5. D. A. Kofke and E. D. Glandt, *Mol. Phys.* **64**, 1105 (1988).
6. A. Sariban and K. Binder, *J. Chem. Phys.* **86**, 5859 (1987).
7. A. Z. Panagiotopoulos, *Mol. Phys.* **61**, 813 (1987).
8. D. A. McQuarrie, *Statistical Mechanics*, Harper & Row, New York, 1976.
9. H. B. Callen, *Thermodynamics and an Introduction to Thermostatistics*, 2nd ed., Wiley, New York, 1985.
10. P. V. Krishna Pant and D. N. Theodorou, *Macromolecules* **28**, 7224 (1995).
11. J. K. Johnson, A. Z. Panagiotopoulos and K. E. Gubbins, *Mol. Phys.* **81**, 717 (1994).
12. W. R. Smith and B. Triska, *J. Chem. Phys.* **100**, 3019 (1994).
13. K. Denbigh, *Principles of Chemical Equilibrium*, 3rd ed., Cambridge Univ. Press, Cambridge, UK, 1971.
14. S. S. Leung and R. B. Griffiths, *Phys. Rev. A* **8**, 2670 (1973).
15. M. P. Allen and D. J. Tildesley, *Computer Simulation of Liquids*, Clarendon Press, Oxford, 1987.
16. D. A. Kofke, *Mol. Simul.* **7**, 285 (1991).
17. M. Mehta and D. A. Kofke, *Chem. Eng. Sci.* **49**, 2633 (1994).
18. M. R. Stapleton, D. J. Tildesley, T. J. Sluckin, and N. Quirke, *J. Phys. Chem.* **92**, 4788 (1988).
19. D. A. Kofke and E. D. Glandt, *J. Chem. Phys.* **90**, 439 (1989).
20. D. A. Kofke and E. D. Glandt, *J. Chem. Phys.* **92**, 658 (1990).
21. D. A. Kofke and E. D. Glandt, *J. Chem. Phys.* **92**, 4417 (1990).
22. J. Hautman and M. L. Klein, *Mol. Phys.* **80**, 647 (1993).
23. Z. Liu and B. J. Berne, *J. Chem. Phys.* **99**, 6071 (1993).
24. Y. Fan, J. E. Finn, and P. A. Monson, *J. Chem. Phys.* **99**, 8238 (1993).
25. A. Z. Panagiotopoulos, in *Observation, Prediction and Simulation of Phase Transitions in Complex Fluids*, NATO ASI Series C, Vol. 460, M. Baus, ed., Kluwer Academic, New York, 1995.
26. M. R. Stapleton, D. J. Tildesley, and N. Quirke, *J. Chem. Phys.* **92**, 4456 (1990).
27. E. de Miguel, E. Martín del Rio and M. M. Telo da Gama, *J. Chem. Phys.* **103**, 6188 (1995).
28. E. Lomba, M. Alvarez, L. L. Lee, and N. G. Almarza, *J. Chem. Phys.* **104**, 4180 (1996).
29. C.-Y. Shew and A. Yethiraj, *J. Chem. Phys.* **104**, 7665 (1996).
30. D. Green, G. Jackson, E. de Miguel, and L. F. Rull, *J. Chem. Phys.* **101**, 3190 (1994).
31. H.-P. Deutsch and K. Binder, *Macromolecules* **25**, 6214 (1992).
32. H.-P. Deutsch, *J. Chem. Phys.* **99**, 4825 (1993).
33. S. K. Kumar, H. Tang, and I. Szleifer, *Mol. Phys.* **81**, 867 (1994).
34. S. K. Kumar and J. D. Weinhold, *Phys. Rev. Lett.* **77**, 1512 (1996).
35. D. F. Coker and R. O. Watts, *Mol. Phys.* **44**, 1303 (1981).
36. D. F. Coker and R. O. Watts, *Chem. Phys. Lett.* **78**, 333 (1981).
37. M. S. Shaw, *J. Chem. Phys.* **94**, 7550 (1991).
38. D. Frenkel, in *Computer Simulation in Chemical Physics*, M. P. Allen and D. J. Tildesley, eds., Kluwer Academic, New york, 1993, p. 93.

39. D. Frenkel and B. Smit, *Understanding Molecular Simulation: From Algorithms to Applications*, Academic Press, San Diego, 1996.
40. D. A. Kofke and P. T. Cummings, *Mol. Phys.* **92**, 973 (1997).
41. D. A. Kofke and P. T. Cummings, *Fluid Phase Equil.* (in press).
42. R. Agrawal and D. A. Kofke, *Phys. Rev. Lett.* **74**, 122 (1995).
43. R. Agrawal and D. Kofke, *Mol. Phys.* **85**, 23 (1995).
44. P. J. Camp, C. P. Mason, M. P. Allen, A. A. Khare, and D. A. Kofke, *J. Chem. Phys.* **105**, 2837 (1996).
45. R. Agrawal and D. Kofke, *Mol. Phys.* **85**, 43 (1995).
46. R. Agrawal, M. Mehta, and D. A. Kofke, *Int. J. Thermophys.* **15**, 1073 (1994).
47. B. A. Finlayson, *Nonlinear Analysis in Chemical Engineering*, McGraw-Hill, New York, 1980.
48. W. H. Press, S. A. Teukolsky, W. T. Vetterling, and B. P. Flannery, *Numerical Recipes: The Art of Scientific Computing*, 2nd ed., Cambridge Univ. Press, Cambridge, UK, 1992.
49. F. A. Escobedo and J. J. de Pablo, *J. Chem. Phys.* **106**, 2911 (1997).
50. D. A. Kofke, *J. Chem. Phys.* **98**, 4149 (1993).
51. B. Smit, P. De Smedt, and D. Frenkel, *Mol. Phys.* **68**, 931 (1989).
52. A. M. Ferrenberg and R. H. Swendsen, *Phys. Rev. Lett.* **61**, 2635 (1988).
53. A. M. Ferrenberg and R. H. Swendsen, *Phys. Rev. Lett.* **63**, 1195 (1989).
54. A. M. Ferrenberg, D. P. Landau, and R. H Swendsen, *Phys. Rev. E* **51**, 5092 (1995).
55. N. B. Wilding and A. D. Bruce, *J. Phys.: Condensed Matter* **4**, 3087 (1992).
56. A. D. Bruce and N. B. Wilding, *Phys. Rev. Lett.* **68**, 193 (1992).
57. K. Kiyohara, K. E. Gubbins, and A. Z. Panagiotopoulos, *J. Chem. Phys.* **106**, 3338 (1997).
58. J. R. Taylor, *An Introduction to Error Analysis*, Univ. Science Books, Mill Valley, CA, 1982.
59. A. Z. Panagiotopoulos, N. Quirke, M. Stapleton and D. J. Tildesley, *Mol. Phys.* **63**, 527 (1988).
60. E. J. Meijer and F. El Azhar, *J. Chem. Phys.* **106**, 4678 (1997).
61. Y. Choi, T. Ree, and F. H. Ree, *J. Chem. Phys.* **99**, 9917 (1993).
62. P. Bolhuis and D. A. Kofke, *Phys. Rev. E.* **54**, 634 (1996).
63. P. G. Bolhuis and D. A. Kofke, *J. Phys.: Cond. Matt.* **8**, 9627 (1996).
64. M. H. J. Hagen and D. Frenkel, *J. Chem. Phys.* **101**, 4093 (1994).
65. F. W. Tavares and S. I. Sandler, *AIChE J.* **43**, 218 (1997).
66. M. Dijkstra and D. Frenkel, *Phys. Rev. E* **51**, 5891 (1995).
67. F. A. Escobedo and J. J. de Pablo, *J. Chem. Phys.* **106**, 9858 (1997).
68. M. Mehta and D. A. Kofke, *Mol. Phys.* **86**, 139 (1995).
69. P. J. Camp and M. P. Allen, *Mol. Phys.* **88**, 1459 (1966).
70. P. Bolhuis and D. Frenkel, *J. Chem. Phys.* **106**, 667 (1997).
71. F. A. Escobedo and J. J. de Pablo, *Europhys. Lett.* **40**, 111 (1997).
72. B. Svensson and C. E. Woodward, *J. Chem. Phys.* **100**, 4575 (1994).
73. B. R. Svensson and C. E. Woodward, *J. Comp. Chem.* **17**, 1156 (1996).
74. J. Forsman and C. E. Woodward, *Mol. Phys.* **90**, 637 (1997).
75. D. A. Young, *J. Chem. Phys.* **58**, 1647 (1973).

76. W. G. Hoover and F. H. Ree, *J. Chem. Phys.* **49**, 3609 (1968).
77. D. Frenkel and A. J. C. Ladd, *J. Chem. Phys.* **81**, 3188 (1984).
78. L. Vega, E. de Miguel, L. F. Rull, G. Jackson, and I. A. McLure, *J. Chem. Phys.* **96**, 2296 (1992).
79. K. Kalospiros, N. Tzouvaras, P. Coutsikos, and D. P. Tassios, *AIChE J.* **41**, 928 (1995).
80. D. A. Kofke, *Mol. Phys.* **78**, 1331 (1993).
81. J. A. Dunne, A. L. Myers, and D. A. Kofke, *Adsorption* **2**, 41 (1996).
82. R. Boehler, M. Ross and D. B. Boercker, *Phys. Rev. B* **53**, 556 (1996).
83. M. H. J. Hagen, E. J. Meijer, G. C. A. M. Mooij, D. Frenkel, and H. N. W. Lekkerkerker, *Nature* **365**, 425 (1993).
84. E. de Miguel, E. M. del Rio, J. T. Brown, and M. P. Allen, *J. Chem. Phys.* **105**, 4234 (1996).
85. E. J. Meijer and D. Frenkel, *J. Chem. Phys.* **100**, 6873 (1994).
86. B. Smit, Ph.D. dissertation, Univ. Utrecht, The Netherlands, 1990.

MONTE CARLO METHODS FOR SIMULATING PHASE EQUILIBRIA OF COMPLEX FLUIDS

J. ILJA SIEPMANN

Department of Chemistry and Department of Chemical Engineering and Materials Science, University of Minnesota, Minneapolis, MN 55455-0431

CONTENTS

I. INTRODUCTION

Almost all molecules of industrial importance can be classified as complex molecules either because of their complex architecture (e.g., polymers and surfactants) or because of their complex interactions (e.g., water). Chain molecules are special because they can adopt many different conformations

---

Advances in Chemical Physics, Volume 105, Monte Carlo Methods in Chemical Physics, edited by David M. Ferguson, J. Ilja Siepmann, and Donald G. Truhlar. Series Editors I. Prigogine and Stuart A. Rice.
ISBN 0-471-19630-4

and relax via many different types of motions that span many decades of length (and time) scales. Complex interactions can either be very anisotropic, such as directional hydrogen bonding, or strongly dependent on the environment, like large changes in the electronic structure (polarization) caused by electric fields. The quantitative modeling of one- and multicomponent phase equilibria containing complex fluids is still in its infancy, and major challenges remain with respect to the development of efficient algorithms and transferable force fields. In this review, we will only address the algorithmic side of the problem, whereas the force field issue is entirely neglected. For a more general discussion on molecular simulation methods, the reader should always consult the excellent textbooks by Allen and Tildesley [1] and Frenkel and Smit [2].

Here it should be stressed that there is (currently) no Monte Carlo method that is *optimal* for *all* problems involving phase equilibria of complex fluids. The selection of a suitable Monte Carlo algorithm will always depend on the physical system and the detail of the model used throughout the simulations. It is generally believed that the thermodynamic scaling methods pioneered by Valleau [3,4] (see also the chapter on thermodynamic scaling in this volume) are (currently) the most efficient methods for calculations of phase equilibria for one-component monomeric fluids with spherically symmetric potentials. In contrast, when multicomponent systems containing multisite particles are considered, there is no single method which could be considered most efficient. However, at least in the subjective view of the author, the combination of the Gibbs ensemble Monte Carlo and configurational-bias Monte Carlo techniques might be considered the most general and robust of the available tools.

A fair comparison of the relative efficiencies of different algorithms for simulating phase equilibria of complex fluids cannot be easily carried out. As mentioned above, what works well for one type of problem, might not work at all for another. Furthermore, the more complicated Monte Carlo recipes have a very large number of adjustable parameters, which greatly influence the efficiency [5]. A simple example of an adjustable parameter is the maximum allowed displacement, d_{max}, for a translational move: adjusting d_{max} that, say, 99% (or only 1%) of translational moves are accepted will lead to a very inefficient sampling of phase space.

The remainder of this review is divided as follows. In Section II, methods for the calculation of free energies are discussed; the focus of Section III is on methods for the simulation of phase equilibria of chain fluids; Section IV briefly discusses algorithm for polarizable force fields; and Section V offers some perspective. Simulation methods that are suitable for continuum-space simulations and standard molecular mechanics force fields are emphasized in this review.

II. CALCULATING FREE ENERGIES OF COMPLEX FLUIDS

The position of chemical equilibria as well as the direction of all spontaneous chemical change is determined by free energies. The partition constant K of a sample molecule between two phases A and B is directly related to the Gibbs free energy of transfer [6]

$$K = \frac{\rho_A}{\rho_B} = \exp\left(\frac{-\Delta G^\circ}{RT}\right) \tag{2.1}$$

where ρ, R, and T are the number density, the molar gas constant, and the absolute temperature, respectively. Whereas the determination of mechanical properties is now routine for computer simulation, the determination of (relative and absolute) free energies and other thermal properties, which depend on the volume of phase space, remains one of the most challenging problems [1,2]. Many excellent reviews devoted to free energy calculation methods have appeared over the last 10 years [7–16]. Thus we can be very brief in our review of the traditional methods for free-energy calculations.

A. Thermodynamic Integration and Free-Energy Perturbation

Thermodynamic integration (TI) and free-energy perturbation (FEP) are the most widely used methods to calculate free-energy differences and are available in many commercial simulation packages. The TI method is based on the statistical connection between the Gibbs free energy, G, and the Hamiltonian, H, of the system

$$\Delta G = G_B - G_A = \int_0^1 \frac{\partial G_\lambda}{\partial_\lambda}\, d\lambda = \int_0^1 \left\langle \frac{\partial H_\lambda}{\partial \lambda} \right\rangle_\lambda d\lambda \tag{2.2}$$

where λ is a coupling parameter so that $H_{\lambda=0} = H_A$ and $H_{\lambda=1} = H_B$. The angular brackets, $\langle\cdots\rangle_\lambda$, denote an isobaric-isothermal ensemble average on the state defined by λ. TI of the internal energy along a line of constant density or of the pressure along an isotherm can also be used to determine the stability of phases. To use the TI expressions, it is necessary to follow a reversible thermodynamic path [1]. In the FEP method the difference between two states is treated as a perturbation (requiring that the states are not too different)

$$\Delta G = G_B - G_A = -RT \ln\left\langle \exp\left[\frac{-(H_B - H_a)}{RT}\right]\right\rangle_{P,T} \tag{2.3}$$

To alleviate the problem that most often the initial and final states are quite different, A and B can also be linked through a number of intermediate states, often called "windows." Recently, the accuracies and reliabilities of the TI and FEP methods have been investigated [16–18] by performing multiple simulations using different run lengths and numbers of windows. Kollman and co-workers [16–18] concluded that the actual errors in these types of calculation have been often underestimated and that molecular-dynamics simulations of the order of nanoseconds are required to give free energies of solvation of small molecules (methane) with an accuracy of around 10% and even longer simulations are needed to obtain reliable potentials of mean force.

B. Particle Insertion Methods

1. *Widom's Ghost Particle Insertion Method*

Widom [19,20] derived an expression that relates the chemical potential, μ, to an ensemble average using the ratio of the partition functions for systems with $N + 1$ and N particles

$$\mu = -k_B T \, ln \frac{Q_{N+1}}{Q_N} = \mu^{id} + \mu^{ex} = \mu^{id} - k_B T \ln\left\langle \exp\left(\frac{-V_{ghost}}{k_B T}\right)\right\rangle_{N,V,T} \tag{2.4}$$

where k_B and Q_N are the Boltzmann constant and the N-particle partition function of the canonical ensemble, respectively; and where V_{ghost} is the potential energy of a "ghost" particle being inserted at random to the system, but the "real" particles do not feel the presence of the ghost. Note the similarity between Eqs. (2.3) and (2.4). A great advantage of Widom's method is that the ghost particles do not disturb the system; thus, the calculation of the excess chemical potential can also be carried out on a configuration file after the simulation of the N-particle system has been completed. The main drawback of the straightforward, unbiased ghost-particle insertion method is that the statistics of insertion become very poor as the density of the fluid is increased, or as the structure of the fluid particles becomes more articulated, specifically, in a case where the insertion of a single bead is already unlikely, the insertion of a chain molecule with many beads in a random conformation will be virtually impossible. Care should be taken when inhomogeneous systems, such as interfaces, are studied where number densities and excess chemical potentials depend on position [21]. The ghost particle insertion method is not limited to the canonical ensemble and expressions similar to Eq. (2.4), but that also account for fluctuations in temperature, volume, and number of particles, have been developed for the microcanonical, isothermal-isobaric, and Gibbs ensembles [22–24].

2. *Multistep Insertion and Incremental Chemical Potential*

As noted above, Widom's original ghost-particle insertion method becomes very cumbersome at high densities or for chain molecules where the insertion probability [the Boltzmann weight in Eq. (2.4)] becomes very small. A similar problem is encountered when the interactions are very anisotropic. Since the excess chemical potential of a molecule is equal to the reversible work required for the addition of this molecule to the N-particle system, it is possible to carry out the insertion in more than one (reversible) step and to calculate the chemical potential from the sum of the individual steps. This (reversible) multistep particle insertion is, in fact, a thermodynamic integration using multiple windows. For example, for the insertion of an alkane, the typical path of the thermodynamic integration is to gradually insert methane (using many windows), then to gradually mutate methane to ethane, and so forth [15,16]. Another option would be to start with the insertion of an ideal (noninteracting) molecule and then to switch on the nonbonded interactions over multiple steps [25]. Another form of a multistep insertion is the incremental chemical potential method proposed by Kumar *et al.* [26,27], which is very convenient for the calculation of excess chemical potentials of homopolymers. In this case, the chain molecule is inserted monomer by monomer. In the thermodynamic limit and far away from an end of the polymer, the incremental chemical potential of a monomer, $\Delta\mu_{\text{monomer}}$, becomes chain-length-independent. This allows an efficient estimation of the excess chemical potential of a linear homopolymer from

$$\mu_{\text{chain}} = \mu_{\text{end}} + (M - M_{\text{end}}) \times \Delta\mu_{\text{monomer}} \tag{2.5}$$

where μ_{end} is the excess chemical potential of the chain end(s), and M and M_{end} and the total chain length (number of monomers) and the number of monomers considered as part of the chain end(s). While application of this scheme to simple homopolymer models, such as bead-spring polymer, is straightforward, care should be taken for more realistic chain models because end segments are not identical to middle segments. There has been some debate on the chain-length independence of the incremental chemical potential [27–30]. Most of the confusion might be due to finite-size effects. Considering an insertion into a finite system at constant N, V, T (canonical ensemble), the incremental chemical potential will not be chain-length-independent, since the addition of every monomer increases the overall density and hence leads to a change in the incremental chemical potential for the next monomer. This finite-size effect will be greatly reduced, when an insertion into a system at constant N, P, T is considered because the volume of the system will adjust after every monomer insertion. Finally, the (reversible) multi-step insertion approach and to a lesser extent the incremental chemical potential method suffer from the shortcoming that they require multiple simulations to measure the chemical potential and that the chemical potentials can no longer be estimated from an old configuration file.

3. *Configurational-Bias Insertion Method*

The configurational-bias insertion method [29,31–35] allows the calculation of excess chemical potentials of chain molecules with arbitrary architectures in a single simulation, that is, without the requirement for reversible multistep insertions. In the configurational-bias insertion method, the ghost chain is inserted in a biased, stepwise procedure in such a way that conformations with higher Boltzmann weights are sampled more often. This biased insertion uses a one-step scanning procedure that is based on the self-avoiding random-walk algorithm by Rosenbluth and Rosenbluth [36]. The advantage of configurational-bias insertion method is that it enables us to compute the excess chemical potential of a chain molecule as an average of many small numbers, in contrast to an unbiased insertion, which yields an average calculate from very few large numbers and an overwhelming majority of very small numbers that are essentially zero. In the configurational-bias insertion scheme the insertion of a ghost chain proceeds as follows:

1. Assuming that k_1 random trial positions for the insertion of the first monomer are generated, their Boltzmann weights, $B_k = \exp[-\beta u_1(k)]$, are calculated, and one of the k_1 possibilities, say, γ, is selected with a probability

$$p_1(\gamma) = \frac{B_\gamma}{w_1} \tag{2.6}$$

where the Rosenbluth weight, w_1, is defined as

$$w_1 = \sum_{k=1}^{k_1} B_k \tag{2.7}$$

2. For all subsequent segments $m = 2, 3, \ldots, M$, we generate k_m trial positions bonded to its preceding segment. Again one of the k_m possibilities is selected and a Rosenbluth weight is calculated analogous to Eqs. (2.6) and (2.7).

3. Once the entire chain is grown, the normalized Rosenbluth factor for this configuration Γ is calculated

$$W_\Gamma = \prod_{m=1}^{M} \frac{w_m}{k_m} \tag{2.8}$$

The number of trial sites at any given step, k_m, can be chosen freely ($k = 1$ corresponds to an unbiased insertion) and should thus be optimized to improve efficiency [38]. For chain molecules with intramolecular potentials (e.g., bond bending and dihedral potentials), it is more convenient to divide the energy into a bonded part and a nonbonded part. The bonded part can then be used to efficiently bias the selection of trial positions and only the nonbonded part is used for the selection of trial sites and the calculation of the Rosenbluth weights [34,35,37]. Using an isolated chain with intramolecular interactions as the reference state, the excess chemical potential can now be obtained from

$$\mu^{\text{ex}} = -k_B T \ln \frac{\langle W_{\text{ghost}} \rangle_{N,V,T}}{\langle W_{\text{isolated}} \rangle_T} \tag{2.9}$$

The normalized Rosenbluth factor of the isolated chain has to determined from a separate (but very fast) simulation. A related technique, the orientational-bias insertion [2,39,40] is suitable for fluids where part of the interactions is strongly orientation-dependent.

4. *Overlapping Distribution Method*

In the preceding sections, we have considered only particle insertions for the estimation of the chemical potential. However, in principle, particle removals could also be used to calculate the chemical potential according to

$$\mu = +k_B T \ln \frac{Q_N}{Q_{N+1}} = \mu^{\text{id}} + k_B T \ln \left\langle \exp\left(\frac{+V_{N+1}}{k_B T} \right) \right\rangle_{N+1,V,T} \tag{2.10}$$

where V_{N+1} is the interaction energy of particle $N+1$ with the remaining N particles of the $(N+1)$-particle system. There are some dangers with the removal of an interacting particle, and, for example, for the hard-sphere fluid an obviously incorrect result of $\mu^{\mathrm{ex}} = 0$ would be obtained. Nevertheless, sampling both particle removals and insertions can serve as a very good diagnostic tool to evaluate the reliability of chemical potential calculations [41,42]. Following Bennett's overlapping distribution method [43], it can be shown that the two functions

$$f_{\mathrm{ins}}(V_{\mathrm{ghost}}) = \ln[p_{\mathrm{ins}}(V_{\mathrm{ghost}})] - \frac{\beta V_{\mathrm{ghost}}}{2} \tag{2.11}$$

and

$$f_{\mathrm{rem}}(V_{N+1}) = \ln[p_{\mathrm{rem}}(V_{N+1})] + \frac{\beta V_{N+1}}{2} \tag{2.12}$$

where p is a probability density obtained from unweighted histograms, can be used to calculate the excess chemical potential from

$$\beta\mu^{\mathrm{ex}} = f_{\mathrm{rem}}(V_{N+1}) - f_{\mathrm{ins}}(V_{\mathrm{ghost}}) \tag{2.13}$$

The statistical accuracy of the calculated chemical potential becomes poor when the two functions do not show a significant range of overlap. A similar approach can also be used for configurational-bias insertions and removals [29,44]. More details on histogram methods can be found in the chapter by Ferguson and Garrett in this volume.

C. Finite-Size Effects

Here it should be noted that the chemical potential and other thermal properties obtained from simulation suffer from finite-size errors. Since this error can be quite large for periodic systems containing only a few hundred particles, it is always prudent to carry out some simulations for different system sizes. If an (approximate) equation of state is known for the system of interest, then the leading ($\mathcal{O}(N^{-1})$) correction can be estimated from [45]

$$\Delta\mu^{\mathrm{ex}}(N) = \frac{1}{2N}\left(\frac{\partial P}{\partial \rho}\right)\left[1 - k_{\mathrm{B}}T\left(\frac{\partial \rho}{\partial P}\right) - \rho k_{\mathrm{B}}T\,\frac{(\partial^2 P/\partial \rho^2)}{(\partial P/\partial \rho)^2}\right] \tag{2.14}$$

This expression is identical to the analytical solution for one-dimensional hard-rod systems and agrees well with numerical results for two-dimensional hard-disk systems [45]. Wilding and Müller [46] found good agreement with numerical results for three-dimensional lattice polymers.

III. SIMULATING PHASE EQUILIBRIA FOR POLYMERIC FLUIDS

There are many different routes for calculating phase equilibria that are covered in detail in other chapters of this volume: thermodynamic scaling Monte Carlo (chapter by Valleau), Gibbs–Duhem integration along coexistence lines (chapter by Kofke), and pseudo-ensemble methods (chapter by de Pablo and Escobedo). Thus these methods are not discussed here.

A. Canonical, Isobaric–Isothermal, and Grand Canonical Ensembles

A straightforward, but tedious, route to obtain information of vapor–liquid and liquid–liquid coexistence lines for polymeric fluids is to perform multiple simulations in either the canonical or the isobaric–isothermal ensemble and to measure the chemical potential of all species. The simulation volumes or external pressures (and for multicomponent systems also the compositions) are then systematically changed to find the conditions that satisfy Gibbs' phase coexistence rule. Since calculations of the chemical potentials are required, these techniques are often referred to as NVT–μ or NPT–μ methods. For the special case of polymeric fluids, these methods can be used very advantageously in combination with the incremental potential algorithm. Thus, phase equilibria can be obtained under conditions and for chain lengths where chemical potentials cannot be reliably obtained with unbiased or biased insertion methods, but can still be estimated using the incremental chemical potential ansatz [47–50].

Fluid phase equilibria can also be obtained from simulations in the grand canonical ensemble [51–53]. As proposed by Ferrenberg and Swendsen [54,55], density (or number of particles) and energy histograms can be constructed from one simulation with the thermodynamic constraints (μ, V, T). These histograms can then be used to obtain the properties for neighboring set of thermodynamic constraints (μ', V', T'). Multiple simulations spanning a range of chemical potentials and temperatures but usually all for the same volume) are performed and the coexistence curve can be estimated from the combined sets of histograms [55,56]. This approach works very well close to a critical point because here the two phases are separated by a small free-energy barrier. Thus the simulation will efficiently sample both regions of densities (numbers of particles) or compositions that correspond to the two phases. Farther away from the critical point, however, special techniques are required to allow the system to move over the free energy barrier separating the two phases [56].

B. Configurational-Bias Monte Carlo in the Gibbs Ensemble

The "ideal" ensemble for simulations of phase equilibria would be an ensemble where chemical potential, pressure, and temperature could be specified following Gibbs' condition for coexisting phases. In the late 1980s, Panagiotopoulos [57,58] developed the Gibbs ensemble, which comes very close to the ideal and is (presently) the method of the choice for the calculation of vapor–liquid and liquid–liquid phase equilibria for multicomponent systems. In the Gibbs ensemble Monte Carlo technique, two (or more) *separate* simulation boxes are utilized. Besides the conventional random displacement of particles, additional types of Monte Carlo moves are applied to ensure that the two boxes are in equilibrium, that is, have equal temperature, pressure (via volume-exchange moves) and chemical potential (via particle-exchange moves). The particular advantage of simulations in the Gibbs ensemble is that if the thermodynamic conditions are such that the system wants to phase-separate, the simulation yields, say, a vapor phase in one box and a liquid phase in the other without having to create a "real" interface. As a result for a given state point the properties of the coexisting phases, such as the partitioning of solute molecules, can be determined directly from a single simulation using a surprisingly small number of particles. The rate-determining step in a Gibbs ensemble Monte Carlo simulation is most often the equilibration of the chemical potentials between the two boxes, namely, the exchange of a molecule from one phase to the other. This particle-exchange step suffers from similar limitation as the particle insertion methods discussed in the preceding section.

To improve the sampling of insertions of flexible molecules, such as the alkanes, the configurational-bias Monte Carlo technique [29,32–35,37,59] can be used. Configurational-bias Monte Carlo replaces the conventional random insertion of entire molecules with a scheme in which the chain molecule is inserted atom by atom such that conformations with favorable energies are preferentially found. The resulting bias of the particle-swap step in the Gibbs ensemble is removed by special acceptance rules [60,61] [compare to Eq. (3.3) of first chapter in this volume]

$$\pi_{i \in V' \to i \in V} = \frac{V(N')}{V'(N+1)} \times \frac{W_{\text{ins}}}{W_{\text{rem}}} > \xi \tag{3.1}$$

where particle i moves from the box with volume V' to that with volume, and W_{ins} and W_{rem} are the Rosenbluth weights for the insertion and removal of the chain, respectively, and are calculated according to Eq. (2.8). For the

Rosenbluth removal of a chain, one of the k choices at each step must correspond to the actual position of the chain to be removed. Since a particle exchange involves simultaneous insertion and removal, acceptance of this moves requires overlapping distributions of Rosenbluth weights; thus, the dangers mentioned in the preceding section are avoided. Another advantage of the configurational-bias particle exchange step versus the configuration-bias insertion for the calculation of chemical potential is that we have more freedom to select the underlying matrix; For instance, it is more efficient to use a short potential truncation for the calculation of the Rosenbluth weights and correct for this in the acceptance rule [62]. Besides enhancing the insertion and/or removal of fluids with articulated structure in the Gibbs ensemble, the configurational-bias Monte Carlo approach also allows the efficient sampling of chain conformations.

Configurational-bias Monte Carlo in the Gibbs ensemble has been successfully applied to the calculations of single-component vapor–liquid phase equilibria of linear and branched alkanes [61,63–67], alcohols [68,69], and a fatty-acid Langmuir monolayer [70]. The extension to multicomponent mixtures introduces a case in which the smaller molecules (members of a homologous series) have a considerably higher acceptance rate in the swap move than do larger molecules. In recent simulations for alkane mixtures [71], we took advantage of this by introducing the CBMC-switch move, which combines an identity switch [72] with configurational-bias chain interconversion [73], such that molecule A is regrown as molecule B in one box, and B is regrown as A in the other box. This switch move equalizes the difference of the chemical potentials of A and B in the two boxes. The configurational-bias particle exchange is used mainly to equalize the chemical potential of the shorter alkane in the two phases. Using these techniques, the boiling-point diagram of a mixture of *n*-octane and *n*-dodecane has been calculated [71] (see Fig. 1).

Configurational-bias Monte Carlo in the Gibbs ensemble samples directly the partitioning of solute molecules between two phases and hence the partition constant can be calculated from the ratio of the number densities [see Eq. (2.1)] or, if so desired, also using the molality scale. The partition constant can then be used to determine the free energy of transfer in exactly the same way as it is done from experimental data. The partitioning of multiple solutes can be estimated from a single simulation [71]. Absolute free energies of transfer for pentane and hexane between liquid heptane and gaseous helium have been obtained with uncertainties of around 0.5 kJ/mol (relative errors smaller than 5%).

C. Finite-Size Effects and Determination of Critical Points

Due to the importance of critical points as engineering parameters, the

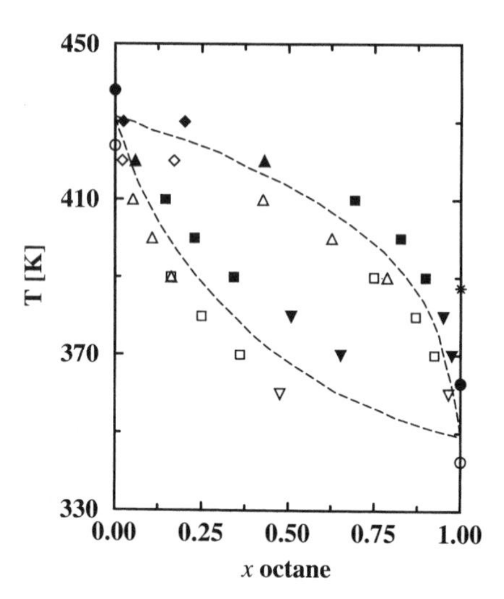

Figure 1. Boiling-point diagram for the binary system *n*-octane and *n*-dodecane at $P = 20$ kPa. Open and filled symbols are used for the TraPPE and SKS force fields, respectively. The dashed line represents the experimental data. The calculated boiling points for the pure substances are shown as circles. Simulation results for binary mixtures are depicted as diamonds, upward-pointing triangles, squares, and downward-pointing triangles for simulations containing total mole fractions of *n*-octane of 0.1, 0.25, 0.5, and 0.75, respectively.

techniques described in this section are often used to estimate critical points. This leads immediately to a discussion of finite-size effects, since simulation studies are usually carried out for fewer than 10^4 particles. The finite extent of the simulated system suppresses long-range fluctuations, while for the "real" system the fluctuations in density or composition diverge at the critical point. Thus, the finite system will exhibit nonclassical behavior only at temperatures for which all relevant fluctuations can be sampled, and a crossover to classical behavior appears as the critical temperature is approached. Smit and co-workers [74] have shown that finite-size effects are smaller in the Gibbs ensemble than in the canonical ensemble, since the former allows for fluctuations in volume and number of particles. However, care is nevertheless required in the vicinity of critical points [56,75,76]. Finite-size effects are more important, and the relative crossover temperatures are lower in two dimensions and for lattice systems (where volume fluctuations are usually not sampled) than in three dimensions and for off-lattice systems. Using Gibbs ensemble simulations, Panagiotopoulos [75] did not observe a crossover from Ising-like ($\beta \approx 0.32$) to mean-field ($\beta = 0.5$) behavior for the three-dimensional Lennard-Jones

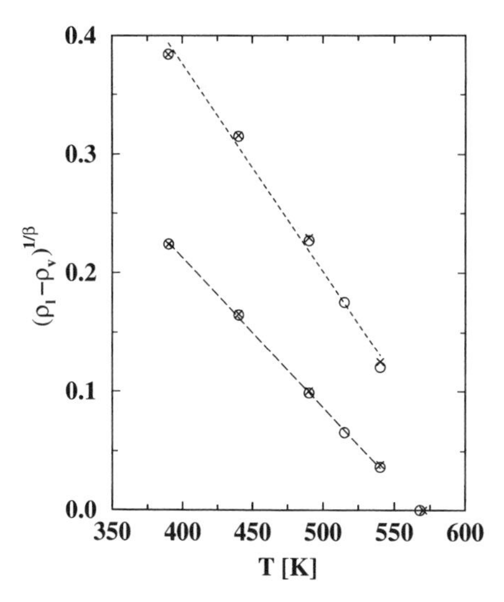

Figure 2. Finite-size effects in the vapor–liquid coexistence curve of *n*-octane (TraPPE model) obtained from simulations in the Gibbs ensemble [89]. Open circles and crosses depict results for simulations with $N = 200$ and 1600, respectively. The upper sets of points use a mean-field exponent ($\beta = 0.5$) and the lowers ones use an Ising-like exponent ($\beta = 0.32$). The estimated critical temperatures for the Ising-like exponent are also shown.

fluid. Nevertheless, the best estimate of the critical temperature obtained from the Gibbs ensemble simulations ($T_c = 1.176 \pm 0.008$) [75] is slightly lower than the obtained using finite-size scaling analysis on coexistence densities obtained from histogram reweighting in the grand-canonical ensemble ($T_c = 1.1876 \pm 0.0003$) [56]. However, for most systems of practical importance, it can be expected that the errors caused by inadequacies in the force fields will exceed the finite-size errors. Figure 2 compares Gibbs ensemble results for the vapor–liquid coexistence curve of *n*-octane obtained for systems with 200 and 1600 chains. The results for both system sizes are very similar and neither shows mean-field behavior (albeit it is hard to see the curvature for the mean-field exponent) for the temperature range that can be conveniently studied.

IV. SIMULATING PHASE EQUILIBRIA FOR POLARIZABLE FORCE FIELDS

Most standard molecular force fields model the electronic structure of atoms, ions, and molecules using partial point charges located on well-defined sites in the molecular frame. Commonly, the values of these charges remain fixed throughout the simulations; that is, they do not change depending on their environment. Thus the fixed charges are effective parameters reflecting the average environment of the particular phase for which they have been derived. This limits the applicability and transferability of fixed-charge models. For example, the failure of fixed-charge water models to deal with changes in environment has now been well documented [15,77–79]. For situations where the electronic fields are strong and heterogeneous or differ greatly between phases, polarizable models, in which the electronic structure of molecules can respond to environmental changes, should be used.

The simulation of phase equilibria for polarizable force fields is still in its infancy. To our knowledge, until now only two simulation studies for polarizable water models have been reported. Medeiros and Costas [80] carried out modified Gibbs ensemble simulations in which they vapor phase is not explicitly modeled, but assumed to obey the virial equation of state [81,82]. Kiyohara et al. [83] used grand-canonical simulations with the histogram reweighting method to calculate the coexistence curves for various polarizable water models. In polarizable models it is assumed that the system always remains in its electronic ground state (adiabatic limit). Grand-canonical and Gibbs ensemble simulations pose a special problem for polarizable models, since insertions and removals in the former and particle exchange in the latter incorporated dramatic changes in the electric environment of the molecule to be displaced (and also on its neighbors in the liquid phase). There are two principal alternatives to satisfying the adiabatic requirement. The first is to use an iterative procedure to minimize the electronic energy at every change of configuration (molecular dynamics time step or single Monte Carlo move) [80,83–85]. As has been pointed out by Frenkel and Smit [2], the iterative procedure has to be carried out to a high degree of accuracy to avoid a systematic drag force in molecular-dynamics simulations. Similarly, a high accuracy is required in Monte Carlo simulations to satisfy the detailed balance condition. Iterative schemes have been used in Monte Carlo simulations, but the fact that the electronic structure should be optimized at every move (i.e., N times per cycle) makes this method extremely computer-intensive and compares very unfavorably to molecular dynamics, where iterations are required only once for every time step. Medeiros and Costas [80] have suggested a Monte Carlo scheme that updates the electronic structure of only the molecule that is displaced. However, polarizability is a many-body effect, and Medeiros and Costas do not show that their procedure satisfies the detailed balance condition.

The more elegant solution to maintaining the adiabatic limit in molecular-dynamics simulations is the "on-the-fly" optimization pioneered by Car and Parrinello [86]. In this approach, the fluctuating charges (or the electronic density) are treated as additional dynamical variables (with fictitious mass and kinetic energy) and an extended Lagrangian formalism is used to solve the equations of motion. [86,87]. The electronic configuration fluctuates around the adiabatic value; that is, it is not exactly in its ground state. However, the fluctuations are random and can be made small by maintaining a sufficiently low (kinetic) temperature for the fluctuating charges. Recently, Martin et al. [88] have proposed a new Monte Carlo algorithm that mimicks the Car–Parrinello approach. As in the molecular-dynamics case, the fluctuating charges are treated as additional degrees of freedom and a special Monte Carlo move samples the electronic degrees of freedom. This algorithm also allows for simulations of polarizable particles in the Gibbs ensemble [90].

V. CONCLUSIONS

In this review, a broad spectrum of simulation methods for the calculation of free energies and phase equilibria is (rather briefly) discussed. Particular emphasis is given to methods that are suitable for the study of complex fluids. Dramatic progress has been made in these fields over the last 10 years, and a multitude of novel methods is added every year to our repertoire. As should be clear from our discussion, there is no single method that is the best choice for all types of physical systems. Thus, selecting the right type of algorithm (which is by no means trivial) will remain one of the most important steps in solving a given problem.

ACKNOWLEDGMENTS

We would like to thank Daan Frenkel, Berend Smit, Thanasis Panagiotopoulos, Dave Kofke, Peter Cummings, and John Valleau for many stimulating discussions on simulating phase equilibria. Financial support from the Petroleum Research Fund, administered by the American Chemical Society, and a Camille and Henry Dreyfus New Faculty Award is gratefully acknowledged. Part of the computer resources for our research were provided by the Minnesota Supercomputing Institute.

REFERENCES

1. M. P. Allen and D. J. Tildesley, *Computer Simulation of Liquids*, Oxford Univ. Press, Oxford, 1987.
2. D. Frenkel and B. Smit, *Understanding Molecular Simulation*, Academic Press, San Diego, 1996.
3. J. P. Valleau, *J. Comp. Phys.* **96**, 193–216 (1991).
4. J. P. Valleau, *J. Chem. Phys.* **99**, 4718–4728 (1993).

5. For a given physical problem the efficiency of MC^{3}S, a general Monte Carlo program written by J. I. Siepmann, M. G. Martin, and B. Chen, can be changed by many orders of magnitude by adjusting the maximum allowed displacements, the numbers of trials sites visited during CBMC growths, and the fractions of the different types of moves.
6. A. Ben-Naim, *Statistical Thermodynamics for Chemists and Biochemists*, Plenum Press, New York, 1992.
7. D. Frenkel, in *Computer Simulation in Materials Science*, M. Meyer and V. Pontikis, eds., NATO ASI Ser. E205, Kluwer: Dordrecht, 1991, pp. 85–117.
8. D. L. Beveridge and F. M. DiCapua, *Annu. Rev. Biophys. Biophys. Chem.* **18**, 431–492 (1989).
9. D. L. Beveridge and F. M. DiCapua, in *Computer Simulations of Biomolecular Systems: Theoretical and Experimental Applications*, Vol. 1, W. F. van Gunsteren, P. K. Weiner, and A. J. Wilkinson, eds., ESCOM Science, Leiden, 1989 pp. 1–26.
10. W. L. Jorgensen, in *Computer Simulations of Biomolecular Systems: Theoretical and Experimental Applications*, Vol. 1, W. F. van Gunsteren, P. K. Weiner, and A. J. Wilkinson, eds., ESCOM Science: Leiden, 1989, pp. 60–72.
11. B. M. Pettitt, in *Computer Simulations of Biomolecular Systems: Theoretical and Experimental Applications*, Vol. 1, W. F. van Gunsteren, P. K. Weiner, and A. J. Wilkinson, eds., ESCOM Science, Leiden, 1989, pp. 94–100.
12. D. A. Pearlman and P. A. Kollman, in *Computer Simulations of Biomolecular Systems: Theoretical and Experimental Applications*, Vol. 1, W. F. van Gunsteren, P. K. Weiner, and A. J. Wilkinson, eds., ESCOM Science, Leiden, 1989, pp. 101–119.
13. W. L. Jorgensen, *Acc. Chem. Res.* **22**, 184–189 (1989).
14. P. A. Kollman, *Chem. Rev.* **93**, 2395–2417 (1993).
15. P. A. Kollman, *Acc. Chem. Res.* **29**, 461–469 (1996).
16. R. J. Radmer and P. A. Kollman, *J. Comp. Chem.* **18**, 902–919 (1997).
17. C. Chipot, C. Millot, B. Maigret, and P. A. Kollman, *J. Phys. Chem.* **98**, 11362–11372 (1994).
18. C. Chipot, P. A. Kollman, and D. A. Pearlman, *J. Comp. Chem.* **17**, 112–1131 (1996).
19. B. Widom, *J. Chem. Phys.* **39**, 2808–2812 (1963).
20. B. Widom, *J. Phys. Chem.* **86**, 869–872 (1982).
21. B. Widom, *J. Stat. Phys.* **19**, 563–574 (1978).
22. D. Frenkel, in *Molecular Dynamic Simulations of Statistical Mechanics Systems*, G. Ciccotti and W. G. Hoover, eds., North-Holland, Amsterdam, 1986, pp. 151–188.
23. P. Sindzingre, G. Ciccotti, C. Massobrio, and D. Frenkel, *Chem. Phys. Lett.* **136**, 35–41 (1987).
24. B. Smit and D. Frenkel, *Mol. Phys.* **68**, 951–958 (1989).
25. M. Müller and W. Paul, *J. Chem. Phys.* **100**, 719–724 (1994).
26. S. K. Kumar, I. Szleifer, and A. Z. Panagiotopoulos, *Phys. Rev. Lett.* **66**, 2935–2938 (1991).
27. S. K. Kumar, *J. Chem. Phys.* **96**, 1490– (1992).
28. B. Smit, G. C. A. M. Mooij, and D. Frenkel, *Phys. Rev. Lett.* **68**, 3657 (1992).
29. J. J. de Pablo, M. Laso, and U. W. Suter, *J. Chem. Phys.* **96**, 6157–6162 (1992).
30. T. Spyriouni, I. G. Economou, and D. N. Theodorou, *Macromolecules*, **30**, 4744–4755 (1997).

31. J. Harris and S. A. Rice, *J. Chem. Phys.* **88**, 1298–1306 (1988).
32. J. I. Siepmann, *Mol. Phys.* **70**, 1145–1158 (1990).
33. J. I. Siepmann and D. Frenkel, *Mol. Phys.* **75**, 59–70 (1992).
34. D. Frenkel, G. C. A. M. Mooij, and B. Smit, *J. Phys.: Condensed Matter*, **4**, 3053–3076 (1992).
35. J. I. Siepmann, in *Computer Simulation of Biomolecular Systems: Theoretical and Experimental Applications*, Vol. 2, W. F. van Gunsteren, P. K. Weiner, and A. J. Wilkinson, eds., ESCOM Science: Leiden, 1993, pp. 249–264.
36. M. N. Rosenbluth and A. W. Rosenbluth, *J. Chem. Phys.* **23**, 356–359 (1955).
37. J. I. Siepmann and I. R. McDonald *Mol. Phys.* **79**, 457–473 (1993).
38. G. C. A. M. Mooij and D. Frenkel, *Mol. Simul.* **17**, 41–55 (1996).
39. R. F. Cracknell, D. Nicholson, N. G. Parsonage, and H. Evans, *Mol. Phys.* **71**, 931–943 (1990).
40. D. M. Tsangaris and J. J. de Pablo, *J. Chem. Phys.* **101**, 1477–1489 (1994).
41. K. S. Shing and K. E. Gubbins, *Mol. Phys.* **46**, 1109–1128 (1982) K. S. Shing and K. E. Gubbins, *Mol. Phys.* **49**, 1121–1138 (1983).
42. D. A. Kofke and P. T. Cummings, *Mol. Phys.* **92**, 973–996 (1997).
43. C. H. Bennett, *J. Comp. Phys.* **22**, 245–268 (1979).
44. G. C. A. M. Mooij and D. Frenkel, *J. Phys.: Condensed Matter* **6**, 3879–3888 (1994).
45. J. I. Siepmann, I. R. McDonald, and D. Frenkel, and *J. Phys.*: *Condensed Matter* **4**, 679–681. (1992).
46. N. B. Wilding and M. Müller, *J. Chem. Phys.* **101**, 4324–4330 (1994).
47. Y. J. Sheng, A. Z. Panagiotopoulos, S. K. Kumar, and I. Szleifer, *Macromolecules* **27**, 400–406 (1994).
48. Y. J. Sheng, A. Z. Panagiotopoulos, and S. K. Kumar, *Macromolecules* **29**, 4444–4446 (1996).
49. S. K. Kumar, *Phys. Rev. Lett.* **77**, 1512–1515 (1996).
50. T. Spyriouni, I. G. Economou, and D. N. Theodorou, *Macromolecules* **31**, 1430–1431 (1998).
51. G. E. Norman and V. S. Filinov, *High Temp. Res. USSR* **7**, 216–222 (1969).
52. D. J. Adams, *Mol. Phys.* **28**, 1241–1252 (1974).
53. J. Ji, T. Cagin, and B. M. Pettitt, *J. Chem. Phys.* **96**, 1333–1342 (1992).
54. A. M. Ferrenberg and R. H. Swendsen, *Phys. Rev. Lett* **61**, 2635 (1988).
55. A. M. Ferrenberg and R. H. Swendsen, *Phys. Rev. Lett* **63**, 1195 (1988).
56. N. B. Wilding, *Phys. Rev. E* **52**, 602–611 (1995).
57. A. Z. Panagiotopoulos, *Mol. Phys.* **61**, 813–826 (1987).
58. A. Z. Panagiotopoulos, N. Quirke, M. Stapleton, and D. J. Tildesley, *Mol. Phys.* **63**, 527–545 (1988).
59. J. J. de Pablo, M. Laso, and U. W. Suter, *J. Chem. Phys.* **96**, 2395–2403 (1992).
60. G. C. A. M Mooij, D. Frenkel, D., and B. Smit, *J. Phys.: Condensed Matter.* **4**, L255–L259 (1992).
61. M. Laso, J. J. de Pablo, and U. W. Suter. *J. Chem. Phys.* **97**. 2817–2819 (1992).
62. T. J. H. Vlugt, M. G. Martin, B. Smit, J. I. Siepmann, and R. Krishna, *Mol. Phys.* **94**, 727–733 (1998).

63. J. I. Siepmann, S. Karaborni, and B. Smit, *J. Am. Chem. Soc.* **115**, 6454–6455 (1993).
64. J. I. Siepmann, S. Karaborni, and B. Smit, *Nature* **365**, 330–332 (1993).
65. B. Smit, S. Karaborni, and J. I. Siepmann, *J. Chem. Phys.* **102**, 2126–2140 (1995).
66. J. I. Siepmann, M. G, Martin, C. J. Mundy, and M. L. Klein, *Mol. Phys.* **90**, 687–693 (1997).
67. N. D. Zhuravlev and J. I. Siepmann, *Fluid Phase Equil.* **134**, 55–61 (1997).
68. M. E. van Leeuwen and B. Smit, *J. Phys. Chem.* **99**, 1831–1833 (1995).
69. M. E. van Leeuwen, *Mol. Phys.* **87**, 87–101 (1996).
70. J. I., Siepmann, S. Karaborni, and M. L. Klein, *J. Phys. Chem.* **98**, 6675–6678 (1994).
71. M. G. Martin and J. I. Siepmann, *J. Am. Chem. Soc.* **119**, 8921–8924 (1997).
72. J. J. de Pablo and J. M. Prausnitz, *Fluid Phase Equil*, **53**, 177–189 (1989).
73. J. I. Siepmann and I. R. McDonald, *Mol. Phys.* **75**, 225–229 (1992).
74. B. Smit, P. de Smedt, and D. Frenkel, *Mol. Phys.* **68**, 931–950 (1989).
75. A. Z. Panagiotopoulos, *Int. J. Thermophys.* **15**, 1057–1072 (1994).
76. K. K. Mon and K. Binder, *J. Chem. Phys.* **96**, 6989–6995 (1992).
77. K. Watanabe, M. Ferrario, and M. L. Klein, *J. Phys. Chem.* **92**, 819 (1988).
78. J. C. Shelley, K. Watanabe, and M. L. Klein, *Langmuir* **9**, 916 (1993).
79. H. J. Strauch and P. T. Cummings, *J. Chem. Phys.* **96**, 864 (1992).
80. M. Medeiros and M. E. Costas, *J. Chem. Phys.* **107**, 2012 (1997).
81. D. M. Tsangaris and P. D. McMahon, *Mol. Simul.* **7**, 97 (1991).
82. D. M. Tsangaris and P. D. McMahon, *Mol. Simul.* **9**, 223 (1992).
83. K. Kiyohara, K. E. Gubbins, and A. Z. Panagiotopoulos, *Mol. Phys.*, in press.
84. A. Wallqvist and B. J. Berne, *J. Phys. Chem.* **97**, 13841 (1993).
85. J. Gao, J. J. Pavelites, and D. Habibollazabeth, *J. Phys. Chem.* **100**, 2689 (1996).
86. R. Car and M. Parrinello, *Phys. Rev. Lett* **55**, 2471 (1985).
87. M. Sprik and M. L. Klein, *J. Chem. Phys.* **89**, 7556 (1988).
88. M. G. Martin, B. Chen, and J. I. Siepmann, *J. Chem. Phys.* **108**, 3383–3385 (1998).
89. M. G. Martin and J. I. Siepmann, *J. Phys. Chem.* **102**, 2569–2577 (1998).
90. B. Chen and J. I. Siepmann, (to be submitted for publication).

REACTIVE CANONICAL MONTE CARLO

J. KARL JOHNSON

Department of Chemical and Petroleum Engineering, University of Pittsburgh, Pittsburgh, PA 15260

CONTENTS

I. INTRODUCTION

In this chapter we present a recently developed Monte Carlo method for computing the equilibrium properties of chemically reacting or associating fluids [1,2]. The method, called *reactive canonical Monte Carlo* (RCMC), efficiently samples phase space relevant to chemical reactions, and converges quickly to the equilibrium concentrations of reactants and products. In standard Metropolis Monte Carlo [3], bond breaking and bond formation are relatively rare events. This makes the standard method extremely slow to converge and subject to considerable statistical uncertainty. Reactive canonical Monte Carlo overcomes this problem by directly sampling forward and reverse reaction steps, as part of the suite of Monte Carlo moves, according to the correct stoichiometry of the reactions being sampled. The method can be applied to simultaneous chemical reactions, reactions that do not conserve the total number of molecules, and combined phase and chemical equilibria. Solvent effects on the equilibrium properties can easily be included in the simulations. The method requires

Advances in Chemical Physics, Volume 105, Monte Carlo Methods in Chemical Physics, edited by David M. Ferguson, J. Ilja Siepmann, and Donald G. Truhlar. Series Editors I. Prigogine and Stuart A. Rice.
ISBN 0-471-19630-4

only input of the intermolecular potentials and the ideal-gas partition functions for the reactants and products, along with the usual ensemble constants. No chemical potentials or chemical potential differences need to be specified for the reaction steps. RCMC simulations can be performed in the canonical, grand-canonical, isothermal–isobaric, Gibbs, and other ensembles. When RCMC is used in conjunction with an ensemble that conserves the number of particles (e.g., canonical) the number of atoms rather than the number of molecules is held fixed. When combined with the Gibbs ensemble technique [4,5], RCMC is capable of treating simultaneous chemical and phase equilibrium, such as is encountered in reactive distillation.

The motivation for developing the RCMC method stems from our previous work with associating fluids. By association we mean hydrogen-bonding and charge-transfer-type interactions, rather than the typically weaker multipole interactions. Association interactions of this type are short-ranged and highly directional in nature, somewhat like chemical bonds. Computer simulations of associating fluids were previously used to test theories of association at moderate association, or bond strength [6–9]. It is important to test the reliability of the theories at high association strengths in order to probe the limits of accuracy of a given theory. This requires that computer simulations be performed for strongly associating fluids. Standard Metropolis sampling of phase space for strongly associating fluids leads to very slow convergence and poor statistical accuracy of the equilibrium properties. This can be understood in terms of two factors:

1. Strongly associating fluids will typically have a very low fraction of monomers. With the fraction of monomers low, the probability that one monomer will encounter another monomer in the correct orientation and separation to form a bonded dimer is extremely small. Association interactions are modeled with short-range, highly anisotropic interaction potentials [8], so that a pair of monomers must be found in the correct molecular separation and orientation before bonding can occur.
2. Once a bond forms between two monomers it becomes energetically unfavorable to break the bond.

The first factor means that the approach to equilibrium will be slow, although the initial rate of bond formation in a fluid of monomers may be high. The second factor means that the fluctuations required to give an accurate statistical measure of the equilibrium composition and other properties will occur over very long simulation times. As equilibrium is approached, there is an effective slowing down of the system as bond formation and bond breaking both become rare events. The idea of RCMC is to provide a method to explicitly sample these rare events, and to do so in a way that speeds equilibration and allows for the natural fluctuations of composition around equilibrium.

The phenomenon of strong association can be viewed as a type of chemical reaction. Indeed, a method that is entirely equivalent to RCMC was developed independently by Smith and Triska [10], and based on the ideas of chemical equilibrium. Smith and Triska call their method the reaction ensemble. We shall refer to both reactive canonical Monte Carlo and the reaction ensemble methods as RCMC, since they are in fact the same. Taking the view of chemical equilibria, we can very concisely write the equations that determine the equilibrium point of a system with π phases and C components. For a system at constant temperature T and pressure p, equilibrium is reached when the total Gibbs free energy G is minimized:

$$\min G = \sum_{\alpha=1}^{\pi} \sum_{i=1}^{C} n_i^\alpha \mu_i^\alpha(T, p, \mathbf{n}^\alpha) \tag{1.1}$$

where n_i^α is the number of moles of component i in phase α, μ_i^α is the chemical potential of component i in phase α, and $\mathbf{n}^\alpha$ is the composition vector, containing the compositions, or mole numbers, of all C components in phase α. Equation (1.1) must be solved subject to material balance and nonnegativity constraints. The material balance constrains can be written in terms of the stoichiometric coefficients and the extents of reaction for each independent chemical reaction in the system.

$$n_i = \sum_{\alpha=1}^{\pi} n_{io}^\alpha + \sum_{j=1}^{R} \nu_{ij} \xi_j; \qquad i = 1, \ldots, C \tag{1.2}$$

where n_{io}^α is the initial amount of component i present in phase α, ν_{ij} is the stoichiometric coefficient of component i in reaction j, ξ_j is the molar extent of reaction for reaction j, and R is the number of independent chemical reactions. The stoichiometric coefficients are taken to be positive for products, negative for reactants, and zero for all species that do not participate in the reaction. The nonnegativity constraint requires that the number of moles of any component in any phase cannot be negative:

$$n_i^\alpha \geq 0; \qquad \alpha = 1, \ldots, \pi, \qquad i = 1, \ldots, C \tag{1.3}$$

Equation (1.1) can, in principle, be solved for any number of chemical reactions in any number of phases to give the compositions and total amounts of material in each phase. In practice, Eq. (1.1) can be very difficult to solve because the chemical potentials are nonlinear functions of the compositions, which means that Eqs. (1.1)–(1.3) form a set of nonlinear constrained equations, for which multiple solutions may exist. To make matters worse, the functional form of the chemical potentials in the liquid and dense fluid phases is not known exactly. One must therefore resort of empirical correlations for the chemical potential or activity coefficient.

Reactive canonical Monte Carlo can be used to solve Eq. (1.1) from a single computer simulation. The constraints of Eqs. (1.2) and (1.3) are automatically incorporated into the RCMC algorithm. The effects of fluid phase nonidealities (e.g., solvent effects) are rigorously accounted for, within the accuracy of the potentials used in the simulation.

II. BACKGROUND

There have been several other methods proposed for the statistical mechanical modeling of chemical reactions. We review these techniques and explain their relationship to RCMC in this section. These simulation efforts are distinct from the many quantum mechanical studies of chemical reactions. The goal of the statistical mechanical simulations is to find the equilibrium concentration of reactants and products for chemically reactive fluid systems, taking into account temperature, pressure, and solvent effects. The goals of the quantum mechanics computations are typically to find transition states, reaction barrier heights, and reaction pathways within chemical accuracy. The quantum studies are usually performed at absolute zero temperature in the gas phase. Quantum mechanical methods are confined to the study of very small systems, so are inappropriate for the assessment of solvent effects, for example.

The first method for simulating chemically reactive systems was proposed by Coker and Watts [11,12]. They presented a modified grand canonical Monte Carlo method wherein the total number of molecules is held fixed but the concentrations of the reacting species is allowed to vary. In their method a molecule is allowed to change species with a probability proportional to the exponential of the difference in chemical potentials between the two components. Thus, their method requires that the chemical potential differences be specified. Coker and Watts applied their method to the reaction

$$Br_2 + Cl_2 \rightleftharpoons 2BrCl \tag{2.1}$$

The modified grand-canonical method of Coker and Watts is applicable only to reactions that do not involve a change in the total number of molecules. Kofke and Glandt [13] corrected errors in the method Coker and Watts, and developed a method for computing chemical reaction equilibrium that they termed the *semigrand ensemble*. The semigrand ensemble method is quite versatile, and can be used to compute phase equilibria for multicomponent mixtures and the properties of polydisperse systems in addition to chemical equilibrium [13]. Simulations in the semigrand ensemble require that the ratios of the fugacities of each species in the system to a reference fugacity be specified. The reference fugacity can then be computed, for example, through the Widom particle insertion method [14] for the pure reference fluid. The semigrand method as presented by Kofke and Glandt [13] can only be used to model chemical reactions that conserve the number of molecules. The methods of Coker and Watts and Kofke and Glandt both rely on calculation of chemical potential differences. Sindzingre et al. [15,16] showed that chemical potential differences may be computed much more accurately from a simulation than the total chemical potential.

Shaw [17–20] developed the first method of simulating chemical reactions that did not require specification of chemical potentials or chemical potential differences. This was an important step forward in modeling chemical reactants because it avoided the problems associated with specifying or calculating chemical potentials, and allowed computation of chemical equilibrium in the natural isothermal-isobaric ensemble for the first time. Shaw's method is called the "$N_{\text{atoms}}pT$ ensemble," indicating that the method was derived from an atomic, rather than a molecular view point. As originally presented [17], the method was limited to reactions that conserve the total number of molecules. Shaw applied this method to computation of the equilibrium composition for the reaction

$$N_2 + O_2 \rightleftharpoons 2NO \tag{2.2}$$

at very high temperatures and pressures, consistent with conditions in shock waves. Good agreement between perturbation theory and simulations was obtained for the composition of the reacting mixture [17]. The effect of the pair potentials on the Hugoniot and the detonation velocity was later investigated [18–20]. Shaw's method is based on the idea of writing the partition function in terms of atomic coordinates and summing over all possible sets of molecular compositions that preserve the overall atomic composition. The $N_{\text{atoms}}pT$ method involves a transition probability that depends on the ideal-gas partition functions of the reactant and product molecules. Shaw later extended the $N_{\text{atoms}}pT$ ensemble method to reactions that do not conserve the number of molecules by introducing the concept of a "null particle" [19,20]. The null particle allows for changes in the number of molecules by turning on and off the interactions of any particular molecular position with the other molecules in the system. The null particle, in effect, keeps track of the extra integration variables introduced by allowing the number of molecules to change. The RCMC method is closely related to Shaw's method, and in fact, RCMC and Shaw's extension to reactions involving a change in mole number were developed independently and about at the same time [1,2,19,20]. However, RCMC and Shaw's method are not exactly the same. RCMC does not require the use of a null particle in order to model non-mole-conserving reactions, and the acceptance probability is slightly more simple in RCMC than in the $N_{\text{atoms}}pT$ ensemble. Shaw's method allows the possibility of including a bias in picking a reaction move. This bias allows one to attempt a low probability move less frequently than other moves, but accepts the move with a higher probability when it is attempted. It is this bias that actually requires the use of null particles in reactions involving a change in the number of molecules. If the bias of picking a move is removed, then Shaw's method reduces to the RCMC technique. The $N_{\text{atoms}}pT$ ensemble formulas presented in Refs. 19 and 20 are generally applicable to multiple reactions and reactions involving a change in the number of molecules. However, no $N_{\text{atoms}}pT$ simulations have yet been performed for these cases.

Three related simulation methods have recently been developed for modeling associating fluids [21–27]. These methods tackle the problem of modeling association by dividing configuration space into different regions and biasing the simulation to explore regions of configuration space important for association. The configuration-biased methods are related to simulation techniques developed for modeling sticky hard spheres [28,29]. Sticky hard spheres are a special case of the square-well fluid. The well depth is taken to infinity while the well depth goes to zero in such a way that the integral over the Boltzmann factor remains finite. Simulations of sticky hard spheres are difficult to perform for large values of the stickiness parameter. This is because the volume of configurational space that corresponds to bound configurations is much much smaller than the total volume of configurational space, and also because bound molecules then to remain to their bound state rather than breaking and forming bonds multiple times during the course of a simulation.

Shew and Mills [21–23] developed a modified Monte Carlo sampling method to simulate systems and barrier crossings. They called their new technique the subspace sampling method [21]. Equilibrium constants for a simple one-dimensional system were computed from simulations with the subspace sampling method [22]. The subspace sampling method requires specification of three parameters that control the convergence rate of the method [21,22]. These parameters need to be optimized in order to give reasonable convergence rates. The number of parameters needed can be reduced and convergence improved if the subspaces are sampled by some alternate scheme, such as smart Monte Carlo [22]. The subspace sampling method was later applied to the sticky electrolyte model, and equilibrium association constants were computed [23]. New acceptance probabilities were presented that reduced the number of free parameters for the method from three to one. Shew and Mills state that their subspace sampling method is formally equivalent to the RCMC method [23]. However, the expressions for the acceptance probabilities are quite different. Compare, for example, Eqs. (6)–(9) in Ref. 23 and Eqs. (14) and (15) of Ref. 2 for the $A + B \rightleftharpoons AB$ association reaction. The expressions for the transition probabilities in RCMC involve the ideal-gas partition functions of the molecules involved in the reaction, whereas the subspace sampling method does not contain this information, but instead relies on partitioning configuration space into the volumes accessible to the associated and dissociated species in the system. The subspace sampling method is closely related to two other configurational bias methods, the bond-bias Monte Carlo method of Tsangaris and de Pablo [24], and the association biased Monte Carlo method of Busch et al. [25].

Tsangaris and de Pablo [24] developed a method for simulating strongly associating fluids in the grand canonical or Gibbs ensembles. Their method, called *bond-biased Monte Carlo*, is also based on the idea of partitioning configuration space into bonding and nonbonding volumes. They attempt particle creations and deletions (particle transfers in the Gibbs ensemble) in a way that biases the insertion of a particle toward the subset of positions and orientations that result in the formation of a bond. Their method contains a free parameter that specifies the fraction of attempted insertions that will result in the formation of a bond. The value of this parameter affects the convergence of the simulation, but not the actual equilibrium values. Bonded dimers are still treated as monomers with respect to MC displacement and orientation moves in the bond-biased method. This leads to poor sampling of phase space for the bonded species, because large displacements or large changes in orientation of one member of a bonded dimer will certainly result in the breaking of the bond, and are not likely to be accepted. Hence, the displacements and reorientations must be made very small, resulting in poor sampling of phase space. In contrast, the RCMC method treats the bonded molecules as separate species. This allows for the reorientation and displacement of the bonded complex, rather than solely attempting reorientations and displacements of the constituent monomers. In this way, the RCMC method allows the bonded species to explore phase space more efficiently than do the bond-biased and other methods that do not allow for collective moves of the bonded clusters. The bond-biased method is restricted to calculations in open systems, namely, grand canonical or Gibbs ensembles.

Busch et al. [25–27] developed the association biased Monte Carlo (ABMC) method specifically to overcome the inefficient sampling of phase space in standard Metropolis Monte Carlo simulations of fluids with strongly associating interactions. Their specific interest is in modeling of binding in antigen–antibody systems. The ABMC method proceeds by considering the move of a single molecule within the volume of a sphere. Moves that result in bond formation are weighted according to a Boltzmann distribution of the energy. Moves that give an unbound state are assigned a weight of unity. This requires the enumeration of all possible bonded and unbonded configurations for moves within a sphere. Configuration space is subdivided into bonding and nonbonding contributions. ABMC simulations have been performed for symmetric, binary associating Lennard-Jones spheres with a single site on each particle [25], for a model antigen–antibody system consisting of hard spheres with electrostatic interactions plus two binding sites on each sphere [26], and for Lennard-Jones spheres with two binding sites on each sphere [27]. The ABMC algorithm has been extended to include moves that result in the simultaneous formation of two bonds [27]. This is an important feature for associating molecules with multiple binding sites, and for polymerization reactions. Such reactions result in chain and ring formation. The implementation of the ABMC method for chemical reactions or for complex association potentials is not straightforward. This is because ABMC requires Boltzmann weighted integrations over the regions of configuration space which lead single- or double-bond formation.

III. THEORY

The RCMC method allows for Monte Carlo moves in the reaction coordinates of a multicomponent mixture that enhances the convergence to the equilibrium composition. Reactions are sampled directly, avoiding the problem of traversing high energy states to reach the lower energy states. In this section we derive the microscopically reversible acceptance probability for RCMC.

The RCMC method is derived for the case of multiple chemical reactions occurring in a single phase. The generalization of the algorithm to multiple phases is straightforward, and involves carrying out RCMC moves within each phase, along with Gibbs ensemble moves among the phases. The grand canonical partition function for a mixture of C components can be written as

$$\Xi = \sum_{N_1=0}^{\infty} \cdots \sum_{N_C=0}^{\infty} Q(N_1, \ldots, N_C, V, T) \exp\left(\beta \sum_{i=1}^{C} N_i \mu_i\right) \tag{3.1}$$

where N_i is the number of molecules of type i, V is the total volume of the system, $\beta = 1/kT$, T is the temperature, k is the Boltzmann constant, and Q is the canonical partition function of the mixture. Equation (3.1) is valid for a reactive mixture as long as we include all possible reactants and products in the C components of the mixture. Since this is an open system, we do not have to be concerned with the material balance constraints of Eq. (1.2). The semiclassical partition function approximations [30,31] are invoked in order to factor Q into intermolecular and intramolecular parts:

$$Q = \prod_i^C \frac{q_{\text{int, i}}^{N_i}}{N_i\,!\,\Lambda_i^{3N_i}\Omega_i^{N_i}} \int \cdots \int \exp(-\beta \mathscr{U})\, d\mathbf{r}^N\, d\omega^N \tag{3.2}$$

where $q_{\text{int, i}}$ is the quantum partition function for the internal modes of an isolated molecule of type i, which includes vibrations, rotations, and similar; Λ_i is the thermal de Broglie wavelength of molecule i; and Ω_i is a normalization factor for the angular integrations. The Ω_i factor has the value 1 for spheres (no angular integrations), 4π for linear molecules, and $8\pi^2$ for nonlinear molecules. In Eq. (3.2) $\mathbf{r}^N$ denotes the center of mass vectors for all $N = \sum_i N_i$ molecules in the system. Likewise ω^N is the set of orientations for the molecules. Hence, the integration is carried out over all positions and orientations of all molecules in the system. The quantity $\mathscr{U}$ is the total configurational energy of the system. Introducing scaled coordinates, $\mathbf{s} = \mathbf{r}/V^{1/3}$, substituting Eq. (3.2) into Eq. (3.1), and rearranging gives

$$\Xi = \sum_{N_1=0}^{\infty} \cdots \sum_{N_C=0}^{\infty} \int \cdots \int \times \exp\left\{ \sum_{i=1}^{C} \left[N_i \beta \mu_i + N_i \ln\left(\frac{q_{\text{int, i}} V}{\Lambda_i^3 \Omega_i} \right) - \ln N_i\,! \right] - \beta \mathscr{U} \right\} d\mathbf{s}^N\, d\omega^N \tag{3.3}$$

The probability that the system is in a state o, with N_i molecules of type i, and so forth, is

$$\mathscr{P}_o = \frac{1}{\Xi} \exp\left\{ \sum_{i=1}^{C} \left[N_i \beta \mu_i + N_i \ln\left(\frac{q_{\text{int, i}} V}{\Lambda_i^3 \Omega_i} \right) - \ln N_i\,! \right] - \beta \mathscr{U}_o \right\} \delta^N \tag{3.4}$$

The energy $\mathscr{U}_o$ is the configurational energy of the N molecules in the specific positions and orientations $\mathbf{s}^N$ and ω^N. The term δ is a differential volume element in the scaled coordinates of the simulation box. The magnitude of δ is determined by the number of significant digits on the computer. We now consider the system undergoing a set of reactions that take the system from state o to a new state n. The composition of state n is related to the previous composition by

$$N_i' = N_i + \sum_{j=1}^{R} \nu_{ij}\xi_j; \qquad i = 1, \ldots, C \tag{3.5}$$

where N_i' is the number of molecules of type i in the new state n. The quantity ξ_j is now the molecular extent of reaction, rather than the molar extent of reaction as in Eq. (1.2). Note that ξ_j must always be an integer in Eq. (3.5). The values of ξ_j determine both the direction and the extent of the R different reactions. For example, the direction of reaction j can be reversed by replacing ξ_j with $-\xi_j$. The probability of generating state n from state o is

$$\mathscr{P}_{o,n} = \left(\prod_{i=1}^{C} N_i!\right) p_{o,n}(\xi_j) p_{o,n}^{\text{react}} p^{\text{pos}}(\overline{\nu\xi}) p_{o,n}^{\omega} \mathscr{P}_o \, p_{o,n}^{\text{acc}} \tag{3.6}$$

The factorial terms in Eq. (3.6) come from the permutation of the labels of each species. The term $p_{o,n}(\xi_j)$ is the probability of choosing the set of ξ_js that give state n. The probability of choosing a given set of reactant molecules is given by $p_{o,n}^{\text{react}}$, and $p^{\text{pos}}(\overline{\nu\xi})$ is the probability of choosing random locations for inserting "extra" product molecules. This term will be explained below. The term $p_{o,n}^{\omega}$ accounts for the probability of picking the angles of the product molecules. Probability $\mathscr{P}_o$ is given by Eq. (3.4), and $p_{o,n}^{\text{acc}}$ is the probability of the move from o to n being accepted.

We have implicitly assumed that when product molecules are created, they are placed in the locations previously occupied by the reactant molecules used to create the products. This is not a requirement of the RCMC method, but increases the simulation efficiency. Likewise, replacing reactant molecules with product molecules of similar size will enhance the efficiency of the method. If the number of reactant and product molecules is the same then no molecules need be inserted at random locations. The term $p^{\text{pos}}(\overline{\nu\xi})$ is needed for cases where the total number of molecules is not conserved. The quantity $\overline{\nu\xi}$ is the net number of molecules produced or consumed for the given set of ν_{ij}s and ξ_js,

$$\overline{\nu\xi} = \sum_{j=1}^{R} \sum_{i=1}^{C} \nu_{ij}\,\xi_j \tag{3.7}$$

If $\overline{\nu\xi} = 0$, then the number of molecules is conserved; no random insertions need be attempted. If $\overline{\nu\xi} > 0$, then there is a net increase in the number of molecules (more product than reactant molecules). In this case $\overline{\nu\xi}$ product molecules will need to be inserted at random locations in the fluid, and

$p^{\text{pos}}(\overline{\nu\xi}) = \delta^{\overline{\nu\xi}}$. If $\overline{\nu\xi} < 0$, then there is a net decrease in the number of molecules. This means that $|\overline{\nu\xi}|$ of the reactant molecule locations will be left vacant. Therefore, $p^{\text{pos}}(\chi)$ can be defined as follows:

$$p^{\text{pos}}(\chi) = \begin{cases} 1 & \text{if } \chi \leq 0 \\ \delta^{\chi} & \text{otherwise} \end{cases} \tag{3.8}$$

where χ is any integer value.

The reverse of Eq. (3.6), going from state n to state o, is generated with probability

$$\mathscr{P}_{n,o} = \left[\prod_{i=1}^{C} (N_i + \overline{\nu_i})\,!\right] p_{n,o}(-\xi_j) p_{n,o}^{\text{react}} p^{\text{pos}}(-\overline{\nu\xi}) p_{n,o}^{\omega} \mathscr{P}_n p_{n,o}^{\text{acc}} \tag{3.9}$$

where $\overline{\nu_i}$ is summed over all reactions:

$$\overline{\nu_i} = \sum_{j=1}^{R} \nu_{ij} \xi_j \tag{3.10}$$

Detailed balance requires that $\mathscr{P}_{o,n} = \mathscr{P}_{n,o}$, and hence that the reaction moves be accepted with probability

$$\mathscr{P}_{o,n}^{\text{acc}} = \min\left\{1, \frac{p_{o,n}^{\text{acc}}}{p_{n,o}^{\text{acc}}}\right\} \tag{3.11}$$

where

$$\frac{p_{o,n}^{\text{acc}}}{p_{n,o}^{\text{acc}}} = \frac{[\prod_{i=1}^{C}(N_i + \overline{\nu_i})\,!\,] p_{n,o}(-\xi_j) p_{n,o}^{\text{react}} p^{\text{pos}}(-\overline{\nu\xi}) p_{n,o}^{\omega} \mathscr{P}_n}{(\prod_{i=1}^{C} N_i\,!) p_{o,n}(\xi_j) p_{o,n}^{\text{react}} p^{\text{pos}}(\overline{\nu\xi}) p_{o,n}^{\omega} \mathscr{P}_o} \tag{3.12}$$

It can be shown from Eq. (3.8) that

$$\frac{p^{\text{pos}}(-\overline{\nu\xi})}{p^{\text{pos}}(\overline{\nu\xi})} = \delta^{-\overline{\nu\xi}} \tag{3.13}$$

The probability of choosing the reactant molecules can easily be enumerated by ordering the species such that $i = 1, \ldots, r$ are the reactants and $i = p, \ldots, C$ are the products; in other words, $\nu_i < 0$ for $i \in \{1, \ldots, r\}$, and $\nu_i > 0$ for $i \in \{p, \ldots, C\}$. For the reactions that take the system from state o to state n, we must choose $\overline{\nu_i}$ molecules of type i, where $i \in \{1, \ldots, r\}$. There are N_i ways of choosing the first, $N_i - 1$ ways of choosing the second, and $N_i + \overline{\nu_i} + 1$ ways of choosing the last molecule. Therefore, the probability of choosing the complete set of reactant molecules is

$$p_{o,n}^{\text{react}} = \left[\prod_{i=1}^{r} (N_i)(N_i - 1) \cdots (N_i + \overline{\nu_i} + 1) \right]^{-1} \tag{3.14}$$

For the reverse move from state n to state o reactant molecules must be chosen from the set $i \in \{p, \ldots, C\}$. There are $N_i + \overline{\nu_i}$ ways of choosing the first reactant molecule, $N_i + \overline{\nu_i} - 1$ ways of choosing the second, and so on. This gives a total probability of

$$p_{n,o}^{\text{react}} = \left[\prod_{i=p}^{C} (N_i + \overline{\nu_i})(N_i + \overline{\nu_i} - 1) \cdots (N_i + 1) \right]^{-1} \tag{3.15}$$

Given Eqs. (3.14) and (3.15), it can be shown that

$$\frac{[\prod_{i=1}^{C} (N_i + \overline{\nu_i})\,!\,] p_{n,o}^{\text{react}}}{(\prod_{i=1}^{C} N_i\,!\,) p_{o,n}^{\text{react}}} = 1 \tag{3.16}$$

The probability of picking the angles of the product molecules is given by

$$p_{o,n}^{\omega} = \left[\prod_{i=p}^{C} \Omega_i^{\overline{\nu_i}} \right]^{-1} \tag{3.17}$$

A similar definition for $p_{n,o}^{\omega}$ gives

$$\frac{p_{n,o}^{\omega}}{p_{o,n}^{\omega}} = \frac{\prod_{i=p}^{C} \Omega_i^{\overline{\nu_i}}}{\prod_{i=1}^{r} \Omega_i^{-\overline{\nu_i}}} = \prod_{i=1}^{C} \Omega_i^{\overline{\nu_i}} \tag{3.18}$$

The ratio of the probabilities of observing states n and o is

$$\frac{\mathscr{P}_n}{\mathscr{P}_o} = \exp\left\{ \beta \sum_{i=1}^{C} \sum_{j=1}^{R} \mu_i \nu_{ij} \xi_j + \sum_{i=1}^{C} \overline{\nu_i} \ln\left(\frac{q_{\text{int, i}} V}{\Lambda_i^3 \Omega_i} \right) + \sum_{i=1}^{C} \ln\left[\frac{N_i\,!}{(N_i + \overline{\nu_i})\,!} \right] - \beta \Delta \mathscr{U}_{o,n} \right\} \delta^{\overline{\nu\xi}} \tag{3.19}$$

where $\Delta \mathscr{U}_{o,n} = \mathscr{U}_n - \mathscr{U}_o$. The first term inside the exponent can be set to zero because at chemical equilibrium

$$\sum_{i=1}^{C} \mu_i \nu_{ij} = 0. \tag{3.20}$$

If we choose the reaction steps such that $p_{o,n}(\xi_j) = p_{n,o}(-\xi_j)$, and substitute Eqs. (3.13), (3.16), (3.18), and (3.19) into Eq. (3.11), we obtain the following expression for the probability of accepting a reaction move:

$$\mathscr{P}_{o,n}^{\mathrm{acc}} = \min\left\{1, \prod_{i=1}^{C} \frac{N_i\,!}{(N_i + \bar{v}_i)\,!} \left(\frac{q_{\mathrm{int},\,i}\, V}{\Lambda_i^3}\right)^{\bar{v}_i} \exp(-\beta \Delta \mathscr{U}_{o,n})\right\} \tag{3.21}$$

This expression, although derived in the grand canonical ensemble, is applicable to simulations in other ensembles as well.

A simulation incorporating RCMC moves can be performed by implementing reaction steps in addition to the usual MC trials in the ensemble being sampled. The RCMC part of the algorithm consists of the following steps:

1. Randomly select a single reaction, or set of reactions.
2. Randomly select a set of values for the extents of reaction, ξ_j, within predefined limits. It is important to restrict $|\xi_j|$ to relatively small values because large values of ξ_j will result in a reaction step with a very low probability of being accepted. For most situations we expect that $\xi_j = \pm 1$ will give near-optimum performance.
3. Select a set of reactant molecules at random from the system.
4. Replace the selected reactant molecules with product molecules. The replacement step must be consistent for forward and reverse moves throughout the simulation run; for example, if a type i molecule is replaced by a type j molecule in the forward move, then a type j must be replaced by a type i molecule in the reverse move.
5. Randomly insert additional product molecules or leave reactant sites vacant as dictated by the value of $\overline{v\xi}$. Efficiency is greatly enhanced if the smallest molecules are chosen for insertion.
6. Accept or reject the move with probability given by Eq. (3.21).

An alternative to step 2 is to select the values of the ξ_js at the start of the simulation and then to randomly select the sign of ξ_j in step 2 to set the direction of the reaction.

Shaw's expression for the acceptance probability for reactions that do not conserve the number of molecules is different from that given in Eq. (3.21) [19,20]. The reason for the difference is that Shaw relaxes the condition $p_{o,n}(\xi_j) = p_{n,o}(-\xi_j)$ by attempting low-probability reaction steps less frequently, but with a higher probability of being accepted when they are attempted. It is not clear whether the added complexity of Shaw's algorithm enhances the convergence and statistical sampling of the reaction steps. The necessity of introducing "null particles" into Shaw's method is a consequence of relaxing the $p_{o,n}(\xi_j) = p_{n,o}(-\xi_j)$ constraint for $\overline{v\xi} \neq 0$ reactions.

The RCMC method can be implemented within any ensemble by simply incorporating reaction steps as outlined above, along with the usual moves for that ensemble. However, there are ensembles where the RCMC moves eliminate the necessity of some trial moves. For the standard μVT ensemble simulations of mixtures, one must specify chemical potentials, and attempt particle creations and deletions for each species in the mixture. However, for the RCMC μVT ensemble one need only specify the chemical potential of one of the molecules participating in a chemical reaction. The chemical potentials for other species participating in the chemical reaction need not be specified, and no creations or deletions of those molecules need be attempted. The reaction moves take care of fixing the other chemical potentials to their correct (chemical equilibrium) values. One is free to pick which molecule to insert. Choosing the smallest molecule results in faster convergence of the grand canonical simulation, as the number of successful creations and deletions is maximized for this choice. The chemical potentials of any inert species (not participating in the reaction) must be specified, and the corresponding creation and deletion moves must be carried out. Likewise, for the Gibbs ensemble only transfers of a single reacting species need be attempted, and equality of the chemical potentials of the other reactants and products between the phases will be brought about by the RCMC moves. As in the μVT case, transfers of inert species also need to be attempted.

RCMC moves can also be applied within the hybrid Monte Carlo (HMC) framework. HMC is a method for using molecular dynamics to guide MC moves [32,33], and has been shown to be more efficient and ergodic than molecular dynamics or Monte Carlo for some classes of problems [34,35]. When implementing RCMC steps within the HMC algorithm, the velocities of the product molecules must be drawn at random from a Maxwell–Boltzmann distribution centered at the appropriate temperature.

IV. APPLICATIONS

The RCMC method has been used to model simple dimerization reactions of the form

$$2\mathrm{A} \rightleftharpoons \mathrm{B} \tag{4.1}$$

The initial motivation for our developing the RCMC method was to study the equilibrium properties of strongly associating mixtures, such as those undergoing dimerization (e.g., acetic acid) given by Eq. (4.1). We observed that conventional NVT ensemble simulations of associating fluids composed of Lennard-Jones (LJ) spheres with off-center directional square-well bonding sites do not converge when the ratio of the well depth to the temperature was greater than about 15. Figure 1 is a plot of the convergence of the fraction of monomers for a reduced temperature of $T^* = kT/\varepsilon = 1$, where k is the Boltzmann constant and ε is the LJ well depth. The square-well bonding site well depth is $\varepsilon_{\text{bond}} = 20\varepsilon$. Further details of the bonding potential can be found in Ref. 2. The traditional NVT simulation continues to show a slow drift of the fraction of monomers downward after 5×10^6 configurations. Continuing the run shows that the configurational properties for this state point do not converge, even after 5×10^7 configurations. In contrast, a simulation run of the RCMC NVT method starting from the same initial conditions converged to the correct fraction of monomers

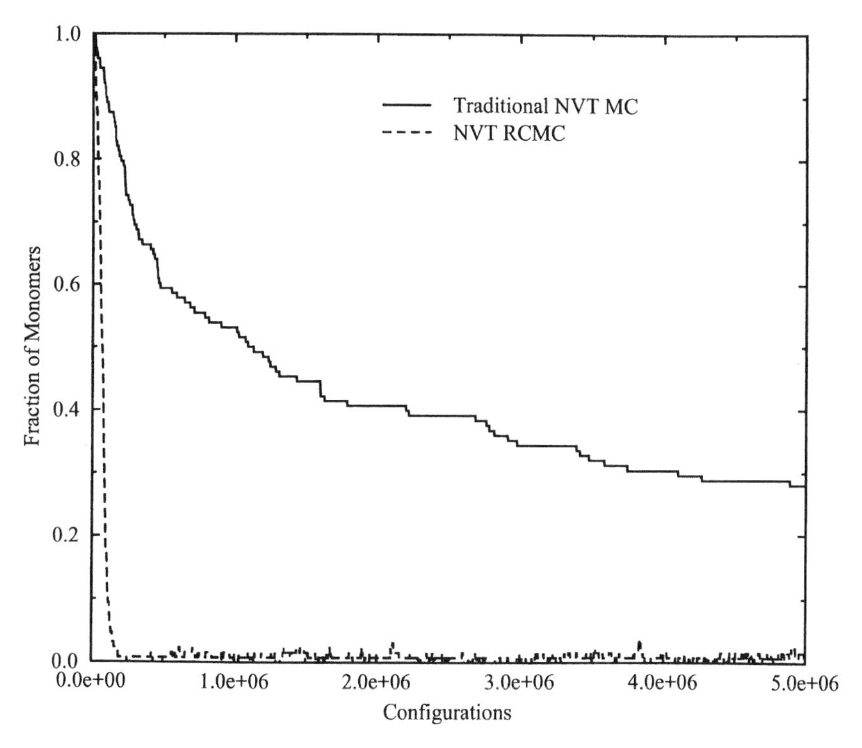

Figure 1. Simulation of an associating fluid with one square-well bonding site from traditional NVT ensemble (solid line) and RCMC NVT ensemble (dashed line) simulations. The convergence of the fraction of monomers is monitored as a function of the number of MC steps. The computational cost of an RCMC step is comparable with the cost of a traditional MC step.

within about 2×10^5 configurations. The CPU cost of RCMC and conventional NVT are roughly the same on a per move basis. In fact, for the simulations plotted in Figure 1, the RCMC method was actually faster on a per-move basis. The RCMC method drastically improves the convergence and sampling of strongly associating fluids, making simulations of strongly associating fluids possible. For systems with weaker bonding strengths, or for higher temperatures, conventional and RCMC simulations should give the same average properties. Fraction of monomers and internal energies from conventional and RCMC simulations are compared in Table I. Data at these higher temperatures ($T^* = 2.0$, 1.5) agree within the combined uncertainties of the simulations. The simulation data are plotted and compared with predictions of a theory for associating fluids in Figure 2. The theory used here is based on Wertheim's cluster resummation formalism [36–42]. Comparison of the RCMC and conventional MC simulations with theory indicates that the RCMC simulations at $T^* = 1.5$ are in slightly better agreement with the theory than are the traditional MC results. The data from the two simulation methods agree equally well with the theory at $T^* = 2.0$. This indicates that the traditional MC method begins to fail at about $T^* = 1.5$ for this bonding strength.

Lennard-Jones monomers dimerizing to form rigid-rotor diatomics according to Eq. (4.1) have been studied from standard grand canonical Monte Carlo simulations, RCMC NVT, and RCMC NpT ensemble simulations [1,2]. In the GCMC simulations the chemical potential of A was fixed and μ_B was computed from the condition of chemical equilibrium, $\mu_B = 2\mu_A$. The three methods gave average properties that agreed within

TABLE I
Comparison of Conventional NVT and RCMC Simulation Results for a Fluid with One Square-Well Bonding Site[a]

T^*	ρ_a^*	$X_M(NVT)$	$U^*(NVT)$	X_M(RCMC)	U^*(RCMC)
2.0	0.9	0.309 (15)	−10.77 (16)	0.304 (17)	−10.79 (11)
2.0	0.8	0.371 (17)	−9.93 (12)	0.374 (26)	−9.86 (15)
2.0	0.7	0.416 (15)	−9.06 (12)	0.414 (10)	−9.02 (6)
2.0	0.6	0.474 (26)	−8.02 (21)	0.470 (7)	−7.95 (8)
2.0	0.5	0.533 (13)	−6.96 (11)	0.531 (7)	−6.91 (6)
2.0	0.05	0.901 (17)	−1.17 (14)	0.902 (1)	−1.17 (1)
1.5	0.9	0.085 (5)	−13.31 (8)	0.097 (10)	−13.17 (19)
1.5	0.8	0.124 (14)	−12.5 (1)	0.102 (6)	−12.61 (5)
1.5	0.7	0.112 (15)	−12.02 (13)	0.128 (5)	−11.82 (4)

[a] With $\varepsilon_{\text{bond}} = 20\varepsilon$. For all simulations, $N_a = 256$ where N_a is the number of atoms (monomers). The numbers in parentheses are one standard deviation uncertainties in the last digits of the reported values. ρ_a^* is the reduced atomic number density.

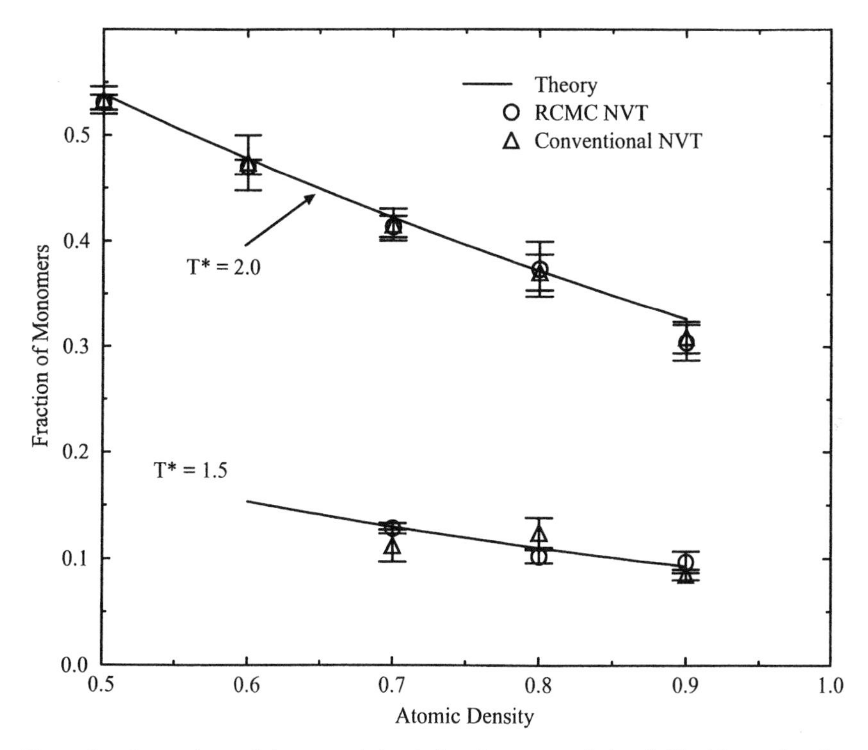

Figure 2. Comparison of theory and simulation for an associating fluid with one bonding site. Details of the theory and interaction potential are given in Ref. 2. The circles are simulation results from the RCMC method and the triangles are from conventional Monte Carlo simulations.

the uncertainties of the simulations [1,2]. The combined phase and chemical equilibrium behavior for these fluids has been studied by combining RCMC with the Gibbs ensemble. The effect of the binding strength (dissociation energy) on the phase envelope was investigated [1,2].

Smith and Triska have performed NVT and NpT RCMC simulations for several different reactions involving hard-sphere fluids [10]. They compared their results with theoretical calculations for hard-sphere mixtures. They considered an isomerization reaction of the form

$$A \rightleftharpoons B \tag{4.2}$$

where A and B are hard spheres of different diameters. They studied a dimerization reaction of the same form as Eq. (4.1), but with A and B modeled as different diameter hard spheres. The third sample reaction was

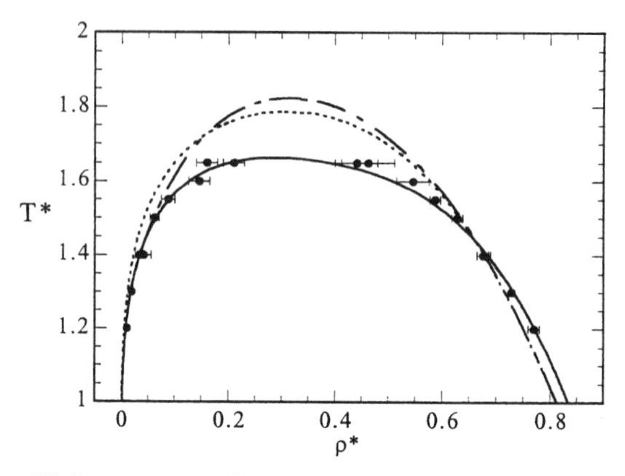

Figure 3. The equilibrium vapor and liquid densities of an associating fluid with one square-well bonding site. The circles are data from RCMC–Gibbs ensemble simulations, and the lines are calculations from three different implementations of a theory for associating fluids. The solid line uses exact values of the reference fluid radial distribution function; the dashed and long dashed-short dashed lines use the WCA and modified WCA approximations to the radial distribution function, respectively. (Reprinted with permission from Müller et al. [43]. Copyright 1995 American Institute of Physics.)

$$A + B \rightleftharpoons C \tag{4.3}$$

with A, B, and C each modeled as hard spheres with different diameters. Finally, they studied a system undergoing both reactions (4.1) and (4.2). They restricted their simulations to hard spheres in order to compare their results with theoretical predictions from the hard-sphere equation of state. Smith and Triska also repeated Shaw's simulations of nitric oxide formation given by reaction (2.2).

RCMC has been used to compute nitric oxide dimerization in the liquid and vapor phases [2]. The simulation results were found to be in good agreement with experimental data for both the liquid phase, where the association is very strong (fraction of monomers ~ 0.05), and the vapor phase, which shows a very low amount of dimerization (fraction of monomers ~ 0.99).

The phase diagrams for two different associating fluids which were computed from RCMC–Gibbs ensemble simulations and the results were compared with predictions from a theory [2]. The associating fluids were modeled as LJ spheres with one off-center square-well bonding site. Bonding site well depths of $\varepsilon_{\text{bond}} = 10\varepsilon$ and $\varepsilon_{\text{bond}} = 20\varepsilon$ were investigated. For the strongest well depth, $\varepsilon_{\text{bond}} = 20\varepsilon$, simulation results were found to be in disagreement with calculations from the theory, but the cause of the disagreement was not immediately identified. The RCMC simulations were the first to allow a test of the theory for strongly associating fluids. It was later shown that the discrepancy between the theory and simulation data was due to the use of inaccurate values of the reference fluid radial distribution function (RDF) in the theory [43]. The RDF for the reference fluid, in this case a simple LJ fluid, is needed in the theory to predict the thermodynamic properties of the associating fluid [36–42]. Comparison of RCMC simulations with data from the theory with three different approximations for the RDF is shown in Figures 3 and 4. The temperature-density projection of the phase diagram is shown in Figure 3, and Figure 4 gives the equilibrium fraction of monomers in the saturated liquid and vapor phases. The solid line in both figures was calculated by using essentially exact values for the RDF from computer simulations of the nonassociating LJ fluid. The dashed line was computed from the Weeks–Chandler–Anderson (WCA) perturbation theory approximation to the RDF [44], and the dashed-dotted line was computed from an empirical modification to the WCA RDF that improves the low-density description of the RDF [45]. It can be seen that the RCMC simulation results and the theory are in excellent agreement when accurate values for the RDF are used. This is an important example because it shows how RCMC simulations have been used to identify and correct an inaccurate approximation in a theory.

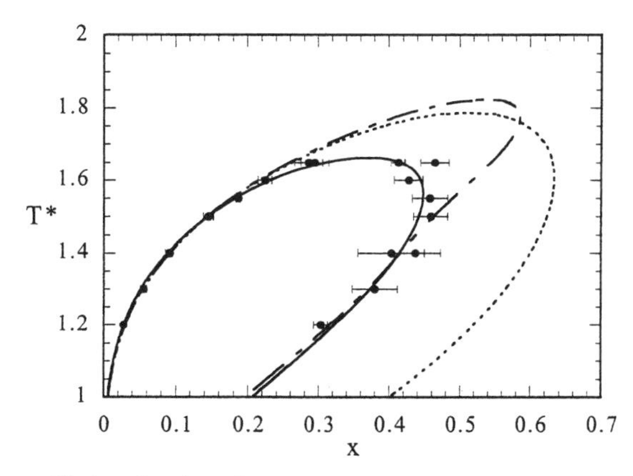

Figure 4. The equilibrium fraction of monomers in the coexisting liquid–vapor phase of an associating fluid with one square-well bonding site. The liquid-phase fractions of monomers are on the left-hand side of the figure. The circles are data from RCMC–Gibbs ensemble simulations, and the lines are calculations from three different implementations of a theory for associating fluids. The solid line uses exact values of the reference fluid radial distribution function; the dashed and long dashed-short dashed lines use the WCA and modified WCA approximations to the radial distribution function, respectively. (Reprinted with permission from Müller et al. [43]. Copyright 1995 American Institute of Physics.)

V. CONCLUSIONS

RCMC is a versatile method for modeling the equilibrium thermodynamics of chemically reacting and associating systems. The method can be applied in virtually any simulation ensemble, and can be used in conjunction with the Gibbs ensemble to compute simultaneous chemical and phase equilibrium. The RCMC method is capable of treating chemical reactions in which the total number of molecules is not conserved. This is a distinct advantage over previously proposed methods of modeling chemical reactions [11–13,17], which do not permit modeling of chemical reactions that result in a change in the number of molecules. Calculations from RCMC simulations have been compared with traditional MC methods. It was found that the RCMC method greatly accelerates convergence to the equilibrium properties of strongly associating and reacting fluids.

Although equilibrium properties can be obtained, no dynamical information is given from the RCMC method. This is because the various bonding or reaction states are sampled directly, without going through the transition states. The RCMC method is therefore not useful for probing kinetic effects.

The RCMC method may be difficult to apply to systems that involve a large number of possible reactions, such as polymerization reactions or association of molecules with two or more association sites. This is not due to a limitation of the RCMC method per se, but rather to problems associated with enumerating and bookkeeping for the many chemical reactions involved.

It should be possible to test the accuracy and transferability of pair potentials by performing RCMC simulations with realistic potentials and comparing simulation data with experimentally measured properties of reacting systems.

REFERENCES

1. J. K. Johnson, Ph.D. dissertation, Cornell Univ., Ithaca, New York, 1992.
2. J. K. Johnson, A. Z. Panagiotopoulos, and K. E. Gubbins, *Mol. Phys.* **81**, 717 (1994).
3. N. Metropolis, A. W. Rosenbluth, M. N. Rosenbluth, A. E. Teller, and E. Teller, *J. Chem. Phys.* **21**, 1087 (1953).
4. A. Z. Panagiotopoulos, *Mol. Phys.* **61**, 813 (1987).
5. A. Z. Panagiotopoulos, N. Quirke, M. Stapleton, and D. J. Tildesley, *Mol. Phys.* **63**, 527 (1988).
6. W. G. Chapman, K. E. Gubbins, C. G. Joslin, and C. G. Gray, *Fluid Phase Equil.* **29**, 337 (1986).
7. C. G. Joslin, C. G. Gray, W. G. Chapman, and K. E. Gubbins, *Mol. Phys.* **62**, 843 (1987).
8. W. G. Chapman, *J. Chem. Phys.* **93**, 4299 (1990).
9. J. K. Johnson and K. E. Gubbins, *Mol. Phys.* **77**, 1033 (1992).

10. W. R. Smith and B. Triska, *J. Chem. Phys.* **100**, 3019 (1994).
11. D. F. Coker and R. O. Watts, *Chem. Phys. Lett.* **78**, 333 (1981).
12. D. F. Coker and R. O. Watts, *Mol. Phys.* **44**, 1303 (1981).
13. D. A. Kofke and E. D. Glandt, *Mol. Phys.* **64**, 1105 (1988).
14. B. Widom, *J. Chem. Phys.* **39**, 2808 (1963).
15. P. Sindzingre, G. Ciccotti, C. Massobrio, and D. Frenkel, *Chem. Phys. Lett.* **136**, 35 (1987).
16. P. Sindzingre, C. Massobrio, G. Ciccotti, and D. Frenkel, *Chem. Phys.* **129**, 213 (1989).
17. M. S. Shaw, *J. Chem. Phys.* **94**, 7550 (1991).
18. M. S. Shaw, in *Shock Compression of Condensed Matter—1991*, S. C. Schmidt, ed., Elsevier, Amsterdam, 1992, pp. 131–134.
19. M. S. Shaw, in *High-Pressure Science and Technology—1993*, S. C. Schmidt, J. W. Shaner, G. A. Samara, and M. Ross, eds., American Institute of Physics, 1994, pp. 65–68.
20. M. S. Shaw, in *Proceedings Tenth International Detonation Symposium, July 12–16, 1993*, Office of Naval Research, Arlington, VA, 1994, pp. 401–408.
21. C.-Y. Shew and P. Mills, *J. Phys. Chem.* **97**, 13824 (1993).
22. C.-Y. Shew and P. Mills, *J. Phys. Chem.* **99**, 12980 (1995).
23. C.-Y. Shew and P. Mills, *J. Phys. Chem.* **99**, 12988 (1995).
24. D. M. Tsangaris and J. J. de Pablo, *J. Chem. Phys.* **101**, 1477 (1994).
25. N. A. Busch, M. S. Wertheim, Y. C. Chiew, and M. L. Yarmush, *J. Chem. Phys.* **101**, 3147 (1994).
26. N. A. Busch, Y. C. Chiew, M. L. Yarmush, and M. S. Wertheim, *AIChE J.* **41**, 974 (1995).
27. N. A. Busch, M. S. Wertheim, and M. L. Yarmush, *J. Chem. Phys.* **104**, 3962 (1996).
28. N. A. Seaton and E. D. Glandt, *J. Chem. Phys.* **84**, 4595 (1986).
29. W. G. T. Kranendonk and D. Frenkel, *Mol. Phys.* **64**, 403 (1988).
30. T. M. Reed and K. E. Gubbins, *Applied Statistical Mechanics*, McGraw-Hill, New York, 1973.
31. C. G. Gray and K. E. Gubbins, *Theory of Molecular Fluids*, Clarendon Press, Oxford, 1984.
32. S. Duane, A. D. Kennedy, B. J. Pendleton, and D. Roweth, *Phys. Lett. B* **195**, 216 (1987).
33. A. Brass, B. J. Pendelton, Y. Chen, and B. Robson, *Biopolymers* **33**, 1307 (1993).
34. M. E. Clamp, P. G. Baker, C. J. Stirling, and A. Brass, *J. Comp. Chem.* **15**, 838 (1994).
35. B. M. Forrest and U. W. Suter, *Mol. Phys.* **82**, 393 (1994).
36. M. S. Wertheim, *J. Stat. Phys.* **35**, 19 (1984).
37. M. S. Wertheim, *J. Stat. Phys.* **35**, 35 (1984).
38. M. S. Wertheim, *J. Stat. Phys.* **42**, 459 (1986).
39. M. S. Wertheim, *J. Stat. Phys.* **42**, 477 (1986).
40. M. S. Wertheim, *J. Chem. Phys.* **85**, 2929 (1986).
41. M. S. Wertheim, *J. Chem. Phys.* **87**, 7323 (1987).
42. M. S. Wertheim, *J. Chem. Phys.* **88**, 1145 (1988).
43. E. A. Müller, K. E. Gubbins, D. M. Tsangaris, and J. J. de Pablo, *J. Chem. Phys.* **103**, 3868 (1995).
44. J. D. Weeks, D. Chandler, and H. C. Andersen, *J. Chem. Phys.* **54**, 5237 (1971).
45. J. K. Johnson and K. E. Gubbins, *Mol. Phys.* **77**, 1033 (1992).

NEW MONTE CARLO ALGORITHMS FOR CLASSICAL SPIN SYSTEMS

G. T. BARKEMA

ITP, Utrecht University, Utrecht, The Netherlands

M. E. J. NEWMAN

Santa Fe Institute, Santa Fe, NM 87501

CONTENTS

Monte Carlo simulations of classical spin systems such as the Ising model can be performed simply and very efficiently using local update algorithms such as the Metropolis or heat-bath algorithms. However, in the vicinity of a continuous phase transition, these algorithms display dramatic critical

---

Advances in Chemical Physics, Volume 105, Monte Carlo Methods in Chemical Physics, edited by David M. Ferguson, J. Ilja Siepmann, and Donald G. Truhlar. Series Editors I. Prigogine and Stuart A. Rice.
ISBN 0-471-19630-4

slowing down, making them extremely poor tools for the study of critical phenomena. In the last 10 years or so, a number of new Monte Carlo algorithms making use of nonlocal update moves have been developed to address this problem. In this chapter we examine a number of these algorithms in some detail, and discuss their strengths and weaknesses. The outline of the chapter is as follows. In Section I, we briefly discuss the problem of critical slowing down. In Section II we describe the Wolff algorithm, which is probably the most successful of the algorithms developed so far. In Section III we discuss a number of other algorithms, such as the Swendsen–Wang algorithm, Niedermayer's algorithm, and the invaded-cluster algorithm. Each of these algorithms is introduced first in the context of the Ising model, but in Section IV we discuss how they may be generalized to systems with many valued spins such as Potts models or continuous valued spins such as the classical XY and Heisenberg models. In Section V we give our conclusions.

I. CRITICAL SLOWING DOWN

The divergence of the correlation length ξ of a classical spin system as it approaches its critical temperature means that larger and larger volumes of spins must be updated coherently in order to sample statistically independent configurations. In Monte Carlo simulations this gives rise to a correlation time τ measured in Monte Carlo steps per site, which diverges with ξ as

$$\tau \sim \xi^z \tag{1.1}$$

where z is a dynamic exponent whose value depends on the dynamics of the update method. For a system of finite dimension L close to the critical temperature, this means that the amount of CPU time required to generate an independent lattice configuration increases as L^{d+z}, where d is the dimensionality of the lattice. If we take the example of the Ising model and assume that the movement of domain walls under a dynamics employing a local update move is diffusive, then the number of Monte Carlo steps required to generate an independent spin configuration should scale as L^{d+2}, and hence $z = 2$. The scaling arguments of Bausch et al. [1] indicate that this is in fact a lower bound on the value of z for all dimensions up to 4, and numerical investigations have in general measured values slightly higher than this. The most accurate determination of z of which we are aware is that of Nightingale and Blöte [2], who measured a value of $z = 2.1665 \pm 0.0012$ for the dynamic exponent of the two-dimensional Ising model with a local update algorithm. Thus the CPU time required to perform a simulation of this model with a given degree of accuracy increases slightly faster than L^4 with system size for temperatures close to the phase transition, which severely limits the system sizes which can be studied, even with the fastest computers.

One solution to this problem is to try and find Monte Carlo algorithms which can update volumes on the order of ξ^d in one step, rather than by the slow diffusion of domain walls. Since ξ can become arbitrarily large, this implies an algorithm with a true nonlocal update move—one capable of updating an arbitrary number of spins in one time step. In the first part of this chapter we will study one of the most widely used and successful such algorithms, the Wolff algorithm, which has a measured critical exponent close to zero for the Ising model, a great improvement over the Metropolis case.

II. THE WOLFF ALGORITHM

An elegant nonlocal Monte Carlo algorithm applicable to the Ising model has been proposed by Wolff [3], based on previous work by Swendsen and Wang [4]. At each Monte Carlo step, this algorithm flips one contiguous cluster of similarly oriented spins. Clusters are built up by starting from a randomly chosen seed spin and adding further spins if they are (1) adjacent to a spin which already belongs to the cluster and (2) pointing the same direction as the spins in the cluster. Spins which satisfy these criteria are added to the cluster with some probability P_{add}. Eventually the cluster will stop growing when all possible candidates for addition have been considered and at this point the cluster is flipped over with some acceptance probability A. It is straightforward to calculate what value A should take for a given choice of P_{add}. Consider two states of the system, μ and ν, illustrated in Figure 1. They differ from one another by the flipping of a single cluster of similarly oriented spins. The crucial thing to notice is the way the spins are oriented around the edge of the cluster (which is indicated by the line in the figure). Notice that in each of the two states, some of the spins just outside the cluster are pointing the same way as the spins in the cluster. The bonds between these spins and the ones in the cluster have to be broken when the cluster is flipped to reach the other state. Inevitably, those bonds which are not broken in going from μ to ν must be broken if we make the reverse move from ν to μ.

Consider now a cluster Monte Carlo move which takes us from μ to ν. There are in fact many such moves—we could choose any of the spins in the cluster as our seed spin, and then we could add the rest of the spins to it in a variety of orders. For the moment, however, let us just consider one particular move, starting with a particular seed spin and then adding others

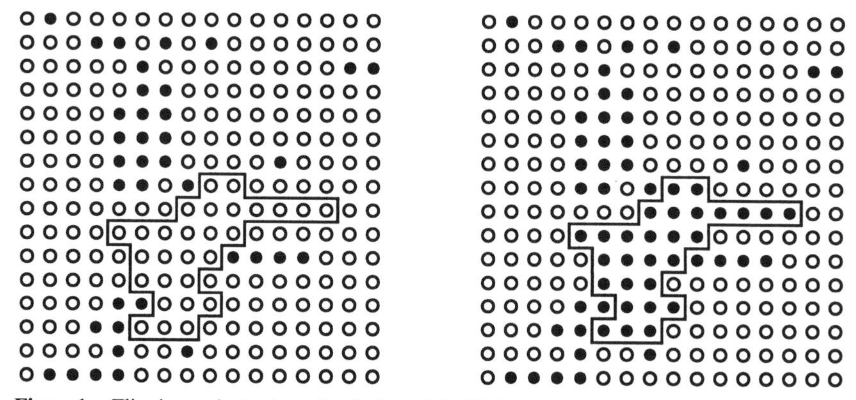

Figure 1. Flipping a cluster in a simulation of the 2D Ising model. The solid and open circles represent the up- and down-pointing spins in the model.

to it in a particular order. Consider also the reverse move, which takes us back to μ from v, starting with exactly the same seed spin, and adding others to it in exactly the same order as the forward move. The probability of choosing the seed is exactly the same in the two directions, as is the probability of adding each spin to the cluster. The only thing which changes between the two is the probability of "breaking" bonds around the edge of the cluster because the bonds which have to be broken are different in the two cases. Suppose that, for the forward move, there are m bonds which have to be broken in order to flip the cluster. These broken bonds represent correctly oriented spins which were not added to the cluster by the algorithm. The probability of not adding such a spin is $1 - P_{\text{add}}$. Thus the probability of not adding all of them, which is proportional to the selection probability $g(\mu \to v)$ for the forward move, is $(1 - P_{\text{add}})^m$. If there are n bonds which need to be broken in the reverse move, then the probability of doing it will be $(1 - P_{\text{add}})^n$, which is proportional to $g(v \to \mu)$. The condition of detailed balance then tells us that

$$\frac{g(\mu \to v)A(\mu \to v)}{g(v \to \mu)A(v \to \mu)} = (1 - P_{\text{add}})^{m-n} \frac{A(\mu \to v)}{A(v \to \mu)} = e^{-\beta(E_v - E_\mu)} \tag{2.1}$$

where $A(\mu \to v)$ and $A(v \to \mu)$ are the acceptance ratios for the moves in the two directions. The change in energy $E_v - E_\mu$ between the two states also depends on the bonds which are broken. For each of the m bonds which are broken in going from μ to v, the energy changes by $+2J$, where J is the interaction strength between neighboring spins. For each of the n bonds which are formed, the energy changes by $-2J$. Thus

$$E_\nu - E_\mu = 2J(m - n). \tag{2.2}$$

Substituting into Eq. (2.1) and rearranging, we derive the following condition on the acceptance ratios:

$$\frac{A(\mu \to \nu)}{A(\nu \to \mu)} = [e^{2\beta J}(1 - P_{\text{add}})]^{n-m}. \tag{2.3}$$

But now we notice a delightful fact; if we choose

$$P_{\text{add}} = 1 - e^{-2\beta J}, \tag{2.4}$$

then the right-hand side of Eq. (2.4) is just 1, independent of any properties of the states μ and ν, or the temperature, or anything else at all. With this choice, we can make the acceptance ratios for both forward and backward moves unity, which is the best possible value they could take. Every move we propose is accepted and the algorithm still satisfies detailed balance. The choice in Eq. (2.4) then defines the Wolff cluster algorithm for the Ising model, whose precise statement would go something like this:

1. Choose a seed spin at random from the lattice.
2. Look in turn at each of the neighbors of that spin. If they are pointing in the same direction as the seed spin, add them to the cluster with probability $P_{\text{add}} = 1 - e^{-2\beta J}$.
3. For each spin that was added in the last step, examine each of *its* neighbors to find the ones which are pointing in the same direction and add each of them to the cluster with the same probability P_{add}. (Notice that as the cluster becomes larger, we may find that some of the neighbors are already members of the cluster, in which case obviously you don't have to consider adding them again. Also, some of the spins may have been considered for addition before, as neighbors of other spins in the cluster, but rejected. In this case, they get another chance to be added to the cluster on this step.) This step is repeated as many times as necessary until there are no spins left in the cluster whose neighbors have not been considered for inclusion in the cluster.
4. Flip all the spins in the cluster.

The flipping of these clusters is a much less laborious task than in the case of the Metropolis algorithm—as we shall see, it takes a time proportional to the size of the cluster to grow it and then turn it over—so we have every hope that the algorithm will indeed have a lower dynamic exponent and less critical slowing down. In Section II.B we show that this is indeed the case. First we look briefly at how the Wolff algorithm is implemented.

The standard way to implement the Wolff algorithm is a very straightforward version of the list of steps given above. First, we choose at random one of the spins on the lattice to be our seed. We look at each of its neighbors, to see if it points in the same direction as the seed spin. If it does, then with probability P_{add}, we add that spin to the cluster and we store its coordinates on a stack. When we have exhausted the neighbors of the spin we are looking at, we pull a spin off the stack, and we start checking *its* neighbors one by one. Any new spins added to the cluster are again also added to the stack, and we go on pulling spins off the stack and looking at their neighbors, until the stack is empty. This algorithm guarantees that we consider the neighbors of every spin added to the cluster, as we should. Furthermore, the amount of time spent on each spin in the cluster is, on average, the same. Each one gets added to the stack, pulled off again at some later stage and has all of its neighbors considered as candidates for addition to the cluster. The time taken to flip over all the spins in the cluster also goes like the number of spins, so the entire time taken by one Monte Carlo move in the Wolff algorithm is proportional to the number of spins in the cluster.

A. Properties of the Wolff Algorithm

In this section we look at how the Wolff algorithm actually performs in practice. As we will see, it performs extremely well when we are near the critical temperature, but is actually a little slower than the Metropolis algorithm at very high or low temperatures.

Figure 2 shows a series of states generated by the algorithm at each of three temperatures, one above T_c, one close to T_c, and one below it. Up and down spins are represented by the black and white dots. Consider first the middle set of states, the ones near T_c. If you examine the four frames, it is not difficult to make out which cluster flipped at each step. Clearly the algorithm is doing its job, flipping large areas of spins when we are in the critical region. In the $T > T_c$ case, it is much harder to make out the changes between one frame and the next. The reason for this is that in this temperature region P_{add} is quite small [see Eq. (2.4)], and this, in turn, makes the clusters small, so it is hard to see when they flip over. Of course, this is exactly what the algorithm is supposed to do, since the correlation length here is small, and we don't expect to get large regions of spins flipping over together. In Figure 3, we have plotted the mean size of the clusters flipped by the Wolff algorithm over a range of temperatures, and, as we can see, it does indeed become small at large temperatures. When the tem-

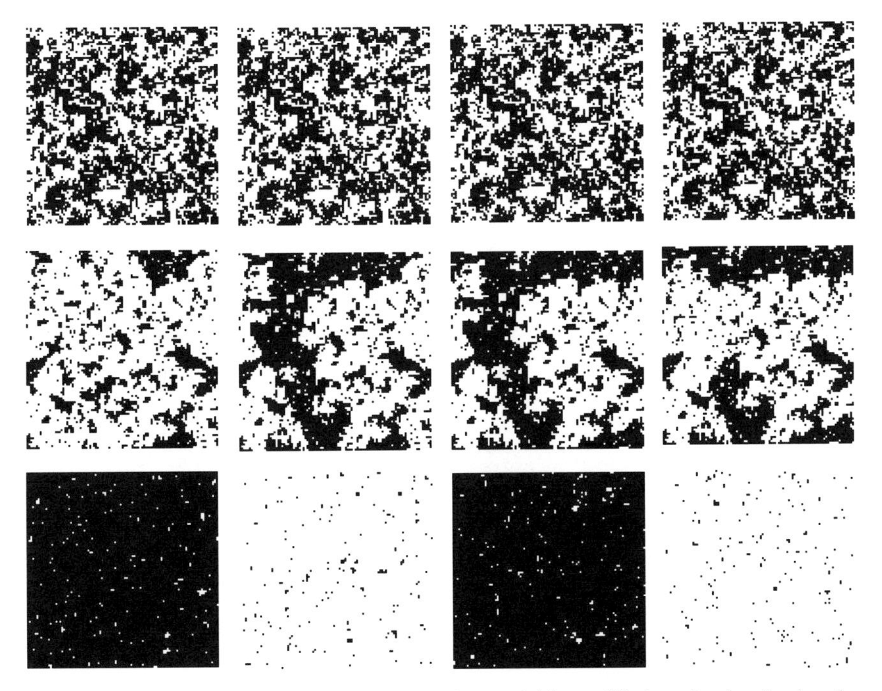

Figure 2. Consecutive states of a 100×100 Ising model in equilibrium simulated using the Wolff algorithm. The top row of four states are at a temperature $kT = 2.8\ J$, which is well above the critical temperature; the middle row is close to the critical temperature at $kT = 2.3\ J$; and the bottom row is well below the critical temperature at $kT = 1.8\ J$.

perature gets sufficiently high (around $kT = 10$), the mean size of the clusters becomes hardly greater than a single spin. In other words, the single seed spin for each cluster is being flipped over with probability one at each step, but none of its neighbors are. However, this is just exactly what the single-spin flip Metropolis algorithm does in this temperatures regime. When T is large, the Metropolis acceptance ratio is 1, or very close to it, for any transition between two states μ and ν. Thus in the limit of high temperatures, the Wolff algorithm and the Metropolis algorithm become the same thing. But notice that the Wolff algorithm will actually be the slower of the two in this case, because for each seed spin it has to go through the business of testing each of the neighbors for possible inclusion in the cluster, whereas the Metropolis algorithm only has to decide whether to flip a single spin or not, a comparatively simple computational task. Thus, even if the Wolff algorithm is a good thing near to the phase transition (which it is), there comes a point as the temperature increases where the Metropolis algorithm becomes better.

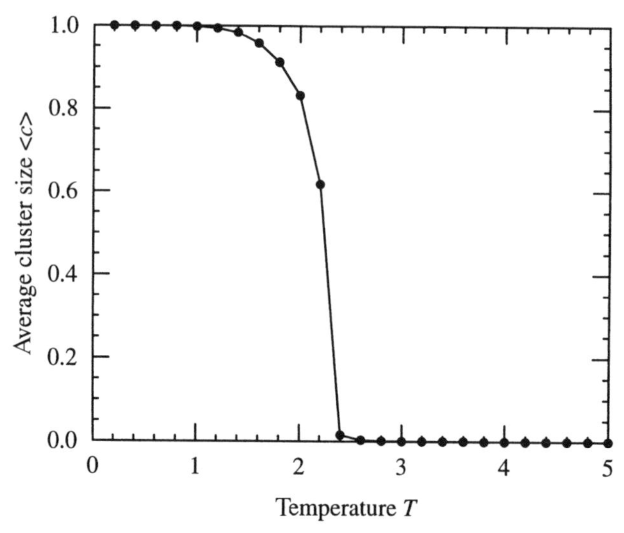

Figure 3. Mean cluster size in the Wolff algorithm as a fraction of the size of the lattice measured as function of temperature. The error bars on the measurements are not shown, because they are smaller than the points. The lines are just a guide to the eye.

Now let us turn to the simulation at low T. Looking at Figure 2, the action of the algorithm is dramatically obvious in this case—almost every spin on the lattice is being flipped at each step. The reason for this is clear. When we are well below the critical temperature, the Ising model develops a backbone of similarly oriented spins which spans the entire lattice. When we choose our seed spin for the Wolff algorithm, it is likely that we will land on one of the spins comprising this backbone. Furthermore, the probability P_{add} [Eq. (2.4)] is large when T is small, so the neighbors of the seed spin are not only likely to be aligned with it, but are also likely to be added to the growing cluster. The result is that the cluster grows (usually) to fill almost the entire backbone of spontaneously magnetized spins, and then they are all flipped over in one step.

On the face of it, this seems like a very inefficient way to generate states for the Ising model. After all, we know that what should really be happening in the Ising model at low temperature is that most of the spins should be lined up with one another, except for a few excitation spins, which are pointing the other way (see Fig. 2). Every so often, one of these excitation spins flips back over to join the majority pointing the other way, or perhaps one of the backbone spins is flipped by a particularly enthusiastic thermal excitation and becomes a new excitation spin. This of course is exactly what the Metropolis algorithm does in this regime. The Wolff algorithm, on the other hand, is removing the single-spin excitations by the seemingly extravagant measure of *flipping all the other spins in the entire lattice* to point the same way as the single lonely excitation. In fact, however, it's actually not such a stupid thing to do. First, let us point out that, since the Wolff algorithm never flips spins which are pointing in the opposite direction to the seed spin, all the excitation spins on the entire lattice end up pointing the same way as the backbone when the algorithm flips over a percolating cluster. Therefore, the Wolff algorithm gets rid of *all* the excitation spins on the lattice in a single step. Second, since we know that the Wolff algorithm generates states with the correct Boltzmann probabilities, it must presumably create some new excitation spins at the same time as it gets rid of the old ones. And indeed it does do this, as is clear from the frames in the figure. Each spin in the backbone gets a number of different chances to be added to the cluster; for most of the spins this number is just the lattice coordination number z, which is 4 in the case of the square lattice. Thus the chance that a spin will not be added to the cluster at all is $(1 - P_{\text{add}})^4$. This is a small number, but still nonzero. (It is 0.012 at the temperature $kT = 1.8\ J$ used in the figure.) Thus there will be a small number of spins in the backbone which get left out of the percolating cluster and are not flipped along with the others. These spins become the new excitations on the lattice.

The net result of all this is that after only one Wolff Monte Carlo step, all the old excitations have vanished and a new set have appeared. This behavior is clear in Figure 2. At low temperatures, the Wolff algorithm generates a complete new configuration of the model at every Monte Carlo step, although the payoff is that it has to flip very nearly every spin on every step too. In the Metropolis algorithm at low temperatures it turns out that you have to do about one Monte Carlo step per site (each one possibly flipping one spin) to generate a new independent configuration of the lattice. To see this, we need only consider again the excitation spins. The average acceptance ratio for flipping these over will be close to 1 in the Metropolis algorithm, because the energy of the system is usually lowered by flipping them. (Only on the rare occasions when several of them happen to be close together might this not be true.) Thus, it will on average just take N steps, where N is the number of sites on the lattice, before we find any particular such spin and flip it—in other words, one Monte Carlo step per site. But we know that the algorithm correctly generates states with their Boltzmann probability, so, in the same N steps that it takes to find all the excitation spins and flip them over to join the backbone, the algorithm must also choose a roughly equal number of new spins to excite out of the backbone, just as the Wolff algorithm also does. Thus, it takes one sweep of the lattice to replace all the excitation spins with a new set, generating an independent state of the lattice. (This is in fact the best possible performance that any Monte Carlo algorithm can have, since one has to allow at least one sweep of the lattice in order to generate a new configuration, so that each spin gets the chance to change its value.)

Again then, we find that the Wolff algorithm and the Metropolis algorithm are roughly comparable, this time at low temperatures. Both need to consider about N spins for flipping in order to generate an independent state of the lattice. However, again the additional complexity of the Wolff algorithm is its downfall, and the extremely simple Metropolis algorithm has the edge in speed.

So, if the Metropolis algorithm beats the Wolff algorithm (albeit only by a slim margin) at both high and low temperatures, that leaves only the intermediate regime, close to T_c in which the Wolff algorithm might be worthwhile. This, of course, is the regime in which we designed the algorithm to work well, so we have every hope that it will beat the Metropolis algorithm there, and indeed it does, very handily.

B. Correlation Time

Before we measure the correlation time of the Wolff algorithm, we need to consider exactly how it should be defined. If we are going to compare the correlation times of the Wolff algorithm and the Metropolis algorithm near to the phase transition as a way of deciding which is the better algorithm in this region, it clearly would not be fair to measure it for both algorithms in terms of number of Monte Carlo steps. A single Monte Carlo step in the Wolff algorithm is a very complicated procedure, flipping maybe hundreds of spins and potentially taking quite a lot of CPU time, whereas the Metropolis Monte Carlo step is a very simple, quick thing—we can do a million of them in a second on a good computer, although each one only flips at most one spin. In Section II we showed that the time taken to complete one step of the Wolff algorithm is proportional to the number of spins c in the cluster. Such a cluster covers a fraction c/L^d of the entire lattice, and so on average each Monte Carlo step will take an amount of CPU time comparable to $\langle c\rangle/L^d$ sweeps of the lattice using the single-spin-flip Metropolis algorithm. Thus the correct way to define the correlation time is to write $\tau \propto \tau_{\text{steps}}\langle c\rangle/L^d$, where τ_{steps} is the correlation time measured in steps (i.e., clusters flipped) in the Wolff algorithm. The conventional choice for the constant of proportionality is 1. This makes the correlation times for the Wolff and Metropolis algorithms equal in the limits of low and high temperature, for the reasons discussed in the last section. This is still not quite fair, since, as we pointed out earlier the Wolff algorithm is a little slower than the Metropolis in these regimes because of its greater complexity. However, this difference is slight compared with the enormous difference in performance between the two algorithms near the phase transition, which we will witness in a moment, so for all practical purposes we can write

$$\tau = \tau_{\text{steps}} \frac{\langle c \rangle}{L^d}. \tag{2.5}$$

In our own simulations of the 2D Ising model on a 100×100 square lattice using the two algorithms, we measure the correlation time at T_c of the Wolff algorithm to be $\tau = 2.80 \pm 0.03$ spin flips per site, by contrast with the Metropolis algorithm, which has $\tau = 2570 \pm 330$. A factor of 1000 certainly outweighs any difference in the relative complexity of the two algorithms. It is this impressive performance on the part of the Wolff algorithm, which makes it a worthwhile one to use if we are interested in the behavior of the model close to T_c.

In Figure 4 we have plotted on logarithmic scales the correlation time of the Wolff algorithm for the 2D Ising model at the critical temperature, over a range of different system sizes. The slope of the line gives us an estimate of the dynamic exponent. Our best fit, given the errors on the data points is

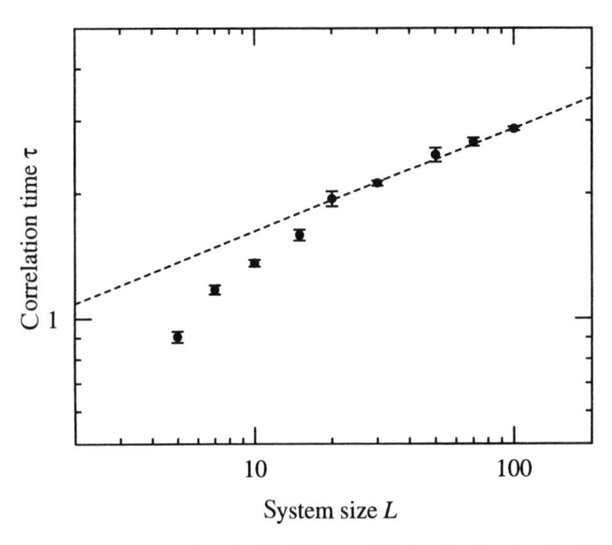

Figure 4. The correlation time τ for the 2D Ising model simulated using the Wolff algorithm. The measurements deviate from a straight line for small system sizes L, but a fit to the larger sizes, indicated by the dashed line, gives a reasonable figure of $z = 0.25 \pm 0.02$ for the dynamic exponent of the algorithm.

$z = 0.25 \pm 0.02$. This was something of a rough calculation, although our result is competitive with other more thorough ones. The best available figure at the time of the writing of this book was that of Coddington and Baillie [5], who measured $z = 0.25 \pm 0.01$. This figure is clearly much lower than the $z = 2.17$ of the Metropolis algorithm, and gives us a quantitative measure of how much better the Wolff algorithm really is.

C. The Dynamic Exponent and the Susceptibility

In fact, in studies of the Wolff algorithm for the 2D Ising model, one does not usually bother to make use of Eq. (2.5) to calculate τ. If we measure time in Monte Carlo steps (i.e., simple cluster flips), we can define the corresponding dynamic exponent z_{steps} in terms of the correlation time τ_{steps} of Eq. (2.5) thus:

$$\tau_{\text{steps}} \sim \xi^{z_{\text{steps}}}. \tag{2.6}$$

The exponent z_{steps} is related to the real dynamic exponent z for the algorithm by

$$z = z_{\text{steps}} + \frac{\gamma}{\nu} - d \tag{2.7}$$

where γ and ν are the critical exponents governing the divergences of the magnetic susceptibility and the correlation length. If we know the values of ν and γ, as we do in the case of the 2D Ising model, then we can use Eq. (2.7) to calculate z without having to measure the mean cluster size in the algorithm, which eliminates one source of error in the measurement.[1]

The first step in demonstrating Eq. (2.7) is to prove another useful result, about the magnetic susceptibility χ. It turns out that, for temperatures $T \geq T_c$, the susceptibility is related to the mean size $\langle c \rangle$ of the clusters flipped by the Wolff algorithm thus:

$$\chi = \beta \langle c \rangle. \tag{2.8}$$

In many simulations using the Wolff algorithm, the susceptibility is measured using the mean cluster size in this way.

The demonstration of Eq. (2.8) goes like this. Instead of implementing the Wolff algorithm in the way described in this chapter, imagine instead

[1] In cases such as the 3D Ising model, for which we don't know the exact values of the critical exponents, it may still be better to make use of Eq. (2.5) and measure z directly, as we did in Figure 4.

doing it a slightly different way. Imagine that at each step we look at the whole lattice and for every pair of neighboring spins which are pointing in the same direction, we make a "link" between them with the probability $P_{\text{add}} = 1 - e^{-2\beta J}$. When we are done, we will have divided the whole lattice into many different clusters of spins, as shown in Figure 5, each of which will be a correct Wolff cluster (since we have used the correct Wolff probability P_{add} to make the links). Now, we choose a single seed spin from the lattice at random, and flip the cluster to which it belongs. Then we throw away all the links we have made and start again. The only difference between this algorithm and the Wolff algorithm as we described it is that here we make the clusters first and choose the seed spin afterward, rather than the other way around. This would not be a very efficient way of implementing the Wolff algorithm, since it requires us to create clusters all over the lattice, almost all of which never get flipped, but it is a useful device for proving Eq. (2.8). Why? Well, we can write the total magnetization M of the lattice as a sum over all the clusters on the lattice thus:

$$M = \sum_i S_i c_i. \tag{2.9}$$

Here i labels the different clusters, c_i is their size (a positive integer), and $S_i = \pm 1$ depending on whether the ith cluster is pointing up or down,

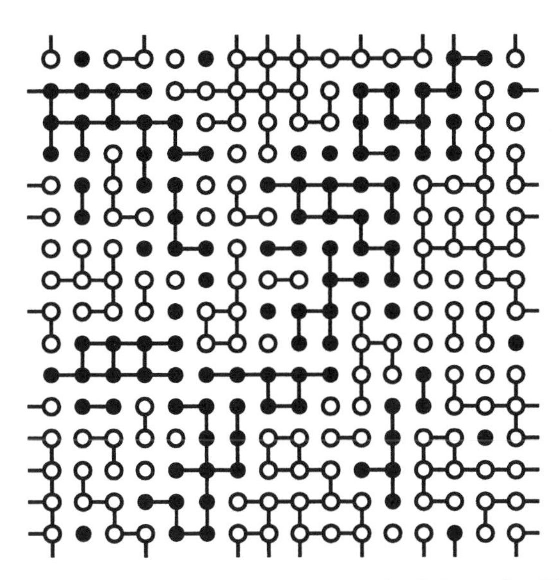

Figure 5. The whole lattice can be divided into clusters simply by putting "links" down with probability P_{add} between any two neighboring spins that are pointing in the same direction.

thereby making either a positive or a negative contribution to M. The mean-square magnetization is then

$$\begin{aligned}\langle M^2\rangle &= \left\langle \sum_i S_i c_i \sum_i S_j c_j \right\rangle \\ &= \left\langle \sum_{i\neq j} S_i S_j c_i c_j \right\rangle + \left\langle \sum_i S_i^2 c_i^2 \right\rangle.\end{aligned} \tag{2.10}$$

Now, since the average of the magnetization is zero under the Wolff algorithm (this is even true below T_c, since the algorithm flips most of the spins on the lattice at every step at low temperature), the first term in this expression is an average over a large number of quantities which are randomly either positive or negative and which will therefore tend to average out to zero. The second term, on the other hand, is an average over only positive quantities and therefore is not zero.[2] Noting that $S_i^2 = 1$ for all i, we can then write

$$\langle M^2\rangle = \sum_i \langle c_i^2\rangle, \tag{2.11}$$

or alternatively, in terms of the magnetization per spin m:

$$\langle m^2\rangle = \frac{1}{N^2}\sum_i \langle c_i^2\rangle. \tag{2.12}$$

Now consider the average $\langle c\rangle$ of the size of the clusters which get flipped in the Wolff algorithm. This is not quite the same thing as the average size $\langle c_i\rangle$ of the clusters over the entire lattice, because when we choose our seed spin for the Wolff algorithm, it is chosen at random from the entire lattice, which means that the probability p_i of it falling in a particular cluster i is proportional to the size of that cluster:

$$p_i = \frac{c_i}{N}. \tag{2.13}$$

The average cluster size in the Wolff algorithm is then given by the average over the probability of the cluster being chosen times the size of that

[2] Strictly speaking, the second term scales like the number of spin configurations in the average and the first scales like its square root, so the second term dominates for a large number of configurations.

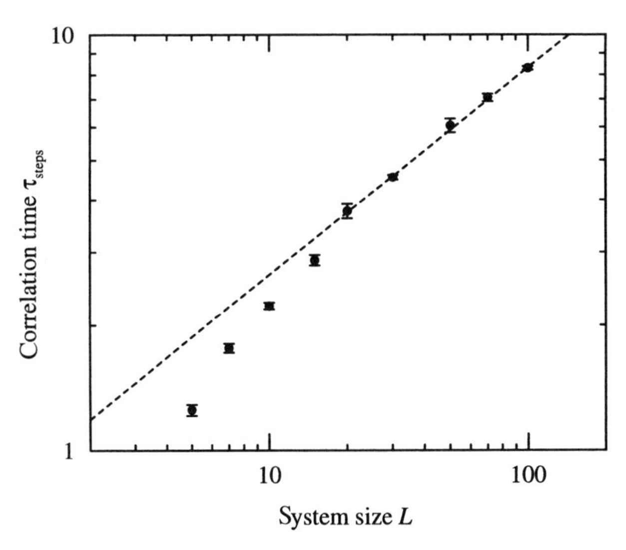

Figure 6. The correlation time τ_{steps} of the 2D Ising model simulated using the Wolff algorithm, measured in units of Monte Carlo steps (i.e., cluster flips). The fit gives us a value of $z_{\text{steps}} = 0.50 \pm 0.01$ for the corresponding dynamic exponent.

cluster:

$$\langle c \rangle = \left\langle \sum_i p_i c_i \right\rangle = \frac{1}{N} \sum_i \langle c_i^2 \rangle = N\langle m^2 \rangle. \tag{2.14}$$

Given that $\langle m \rangle = 0$ for $T \geq T_c$, we now have our result for the magnetic susceptibility as promised:

$$\chi = \beta \langle c \rangle. \tag{2.15}$$

We can now use this equation to rewrite Eq. (2.5) thus:

$$\tau = \tau_{\text{steps}} \frac{\chi}{\beta L^d} \tag{2.16}$$

which implies, in turn, that

$$\xi^{-z} \sim \xi^{-z_{\text{steps}}} \xi^{-\gamma/\nu} L^{+d}. \tag{2.17}$$

Near the critical point, this implies that

$$L^z \sim L^{z_{\text{steps}}} L^{\gamma/\nu} L^{-d}, \tag{2.18}$$

and thus

$$z = z_{\text{steps}} + \frac{\gamma}{\nu} - d \tag{2.19}$$

as we suggested earlier.

In Figure 6 we have replotted the results from our Wolff algorithm simulations of the 2D Ising model using the correlation time measured in Monte Carlo steps (i.e., cluster flips). The best fit to the data gives a figure of $z_{\text{steps}} = 0.50 \pm 0.01$. Using the accepted values $\nu = 1$ and $\gamma = \frac{7}{4}$ for the 2D Ising model [6], and setting $d = 2$, Eq. (2.19) then gives us $z = 0.25 \pm 0.01$, which is as good as the best published result for this exponent.

III. FURTHER ALGORITHMS FOR THE ISING MODEL

We have looked in detail at the Wolff algorithm for simulating the Ising model. Close to T_c this algorithm is significantly more efficient than the Metropolis one; although it is more complex than the Metropolis algorithm, the Wolff algorithm has a very small dynamic exponent, which means that the time taken to perform a simulation scales roughly like the size of the system, which is the best that we can hope for in any algorithm.

However, many other algorithms have been suggested for the simulation of the Ising model. In this section we discuss a few of these alternative algorithms.

A. The Swendsen–Wang Algorithm

After the Metropolis and Wolff algorithms, probably the most important other algorithm is the algorithm of Swendsen and Wang [4]. In fact, this algorithm is very similar to the Wolff algorithm, and Wolff took the idea for his algorithm directly from it. (Swendsen and Wang took the idea, in turn, from the work of Fortuin and Kasteleyn [7] and Sweeny [8]). Actually, we have seen the central idea behind the Swendsen–Wang algorithm already. In Section II.C we considered an alternative implementation of the Wolff algorithm in which the entire lattice of spins is divided up into clusters by making "links" with probability $P_{\text{add}} = 1 - e^{-2\beta J}$ between similarly oriented neighboring spins. We then imagined choosing a single spin from the lattice and flipping over the whole of the cluster to which it belongs. This procedure is clearly equivalent to the Wolff algorithm, although in practice it would be a very inefficient way of carrying in out.

The Swendsen-Wang algorithm divides the entire lattice into clusters in exactly the same way, with this same probability P_{add} of making a link. But then, instead of flipping just one cluster, each cluster is independently flipped with probability $\frac{1}{2}$. We notice the following facts about this algorithm.

1. The algorithm satisfies the condition of detailed balance. The proof of this fact is exactly the same as it was for the Wolff algorithm. If the number of links broken and made in performing a move are m and n, respectively (and the reverse for the reverse move), then the energy change for the move is $2J(m - n)$ (or $2J(n - m)$ for the reverse move). The selection probabilities for choosing a particular set of links differs between forward and reverse moves only at the places where bonds are made or broken, and so the ratio of the two selection probabilities is $(1 - P_{\text{add}})^{m-n}$, just as it was before. By choosing $P_{\text{add}} = 1 - e^{-2\beta J}$, we then ensure, just as before, that the acceptance probability is independent of m and n and everything else, so any choice which makes it the same in each direction, such as flipping all clusters with probability $\frac{1}{2}$ will make the algorithm correct. Notice however that many other choices would do as well. It doesn't matter how we choose to flip the clusters, although the choice made here is good because it minimizes the correlation between the direction of a cluster before and after a move. The new direction is chosen completely at random, regardless of the old one.

2. The algorithm updates the entire lattice on each move. In measuring correlation times for this algorithm, one should therefore measure them simply in numbers of Monte Carlo steps, and not steps per site as with the Metropolis algorithm. (In fact, on average, only half the spins get flipped on each move, but the number flipped scales like the size of the system, which is the important point.)

3. The Swendsen–Wang algorithm is essentially the same as the Wolff algorithm for low temperatures. Well below T_c, one of the clusters chosen by the algorithm will be a percolating cluster, and the rest will correspond to the "excitations" discussed in Section II.A, which will be very small. Ignoring these very small clusters, then, the Swendsen–Wang algorithm will tend to turn over the percolating backbone of the lattice on average every two steps (rather than every step as in the Wolff algorithm—see Fig. 2), but otherwise the two will behave almost identically. Thus, as with the Wolff algorithm, we can expect the performance of the Swendsen–Wang algorithm to be similar to that of the Metropolis algorithm at low T, although probably a little slower on average due to the complexity of the algorithm.

4. At high temperatures, the Swendsen–Wang algorithm tends to divide the lattice into very small clusters because P_{add} becomes small. As $T \to \infty$

the clusters will just be one spin each, and the algorithm will just change all the spins to new random values on each move. This is also what the Metropolis algorithm does at high temperatures in one sweep of the lattice, although again the Metropolis algorithm can be expected to be a little more efficient in this regime, since it is a simpler algorithm which takes few operations on the computer to flip each spin.

The combination of the last two points here implies that the only regime in which the Swendsen–Wang algorithm can be expected to outperform the Metropolis algorithm is the one close to the critical temperature. The best measurement of the dynamic exponent of the algorithm is that of Coddington and Baillie [6], who found $z = 0.25 \pm 0.01$ in two dimensions, which is clearly much better than the Metropolis algorithm, and is in fact exactly the same as the result for the Wolff algorithm. So the Swendsen–Wang algorithm is a pretty good algorithm for investigating the 2D Ising model close to its critical point. However, as Table I shows, for higher dimensions the Swendsen–Wang algorithm has a significantly higher dynamic exponent than the Wolff algorithm, making it slower close to T_c. The reason is that close to T_c the properties of the Ising model are dominated by the fluctuation of large clusters of spins. As the arguments of Section II.C showed, the Wolff algorithm preferentially flips larger clusters because the chance of the seed spin belonging to any particular cluster is proportional to the size of that cluster. The Swendsen–Wang algorithm, on the other hand, treats all clusters equally, regardless of their size, and therefore wastes a considerable amount of effort on small clusters which make vanishingly little contribution to the macroscopic properties of the system for large system sizes. This, coupled with the fact that the Sweden–Wang algorithm is slightly more complicated to program than the Wolff algorithm, makes the Wolff algorithm the algorithm of choice for most people. It is worth noting however, that the work of Ferrenberg et al. [9] appears to indicate that the Wolff algorithm is unusually susceptible to imperfections in the random-number

TABLE I
Comparison of Dynamic Exponent z for the Metropolis, Wolff and Swendsen–Wang Algorithms in Various Numbers of Dimensions[a]

Dimension d	Metropolis	Wolff	Swendsen–Wang
2	2.167 ± 0.001	0.25 ± 0.01	0.25 ± 0.01
3	2.02 ± 0.02	0.33 ± 0.01	0.54 ± 0.02
4		0.25 ± 0.01	0.86 ± 0.02

[a] Figures are taken from Coddington and Baillie [5], Matz et al. [10], and Nightingale and Blöte [2]. To our knowledge, the dynamic exponent of the Metropolis algorithm has not been measured in four dimensions.

generator used to implement the algorithm. Although this result is probably highly sensitive to the way in which the algorithm is coded, it may be that the Swendsen–Wang algorithm would be a more sensible choice for those who are unsure of the quality of their random numbers.

B. Niedermayer's Algorithm

Another variation on the general cluster algorithm theme was proposed by Niedermayer [12]. His suggestion is really just an extension of the ideas used in the Wolff and Swendsen–Wang algorithms. In fact, Niedermayer's methods are very general and can be applied to all sorts of models including glassy spin models. Here we will just consider their application to the ordinary Ising model.

Niedermayer pointed out that it is not necessary to constrain the "links" with which we make clusters to be only between spins which are pointing in the same direction. In general, we can define two different probabilities for putting links between sites—one for parallel spins and one for antiparallel ones. The way Niedermayer expressed it, he considered the energy contribution E_{ij} that a pair of spins i and j makes to the Hamiltonian. In the case of the Ising model, for example

$$E_{ij} = -Js_i s_j. \tag{3.1}$$

He then wrote the probability for making a link between two neighboring spins as a function of this energy $P_{\text{add}}(E_{ij})$. In the Ising model E_{ij} can take only two values $\pm J$, so the function $P_{\text{add}}(E_{ij})$ needs to be defined only at these points, but for some of the more general models Niedermayer considered it needs to be defined elsewhere as well. Clearly, if for the Ising model we make $P_{\text{add}}(-J) = 1 - e^{-2\beta J}$ and $P_{\text{add}}(J) = 0$, then we recover the Wolff algorithm or the Swendsen–Wang algorithm, depending on whether we flip only a single cluster on each move, or many clusters over the entire lattice—Niedermayer's formalism is applicable in either case. To be concrete about things, let us look at the case of the single-cluster, Wolff-type version of the algorithm.

Let us apply the condition of detailed balance to the algorithm. Consider, as we did in the case of the Wolff algorithm, two states of our system which differ by the flipping of a single cluster. (You can look again at Figure 1 if you like, but bear in mind that, since we are now allowing links between antiparallel spins, not all the spins in the cluster need be pointing in the same direction.) As before, the probability of forming the cluster itself is exactly the same in the forward and reverse directions, except for the contributions which come from the borders. At the borders, there are some pairs of spins which are parallel and some which are antiparallel. Suppose that in the forward direction there are m pairs of parallel spins at the border—these correspond to bonds which have to be broken in flipping the cluster—and n pairs which are antiparallel—bonds which will be made when we flip. By definition, no links are made between any of these border pairs, and the probability of that happening is $[1 - P_{\text{add}}(-J)]^m [1 - P_{\text{add}}(J)]^n$. In the reverse direction the corresponding probability is $[1 - P_{\text{add}}(-J)]^n[1 - P_{\text{add}}(J)]^m$. Just as in the Wolff case, the energy cost of flipping the cluster from state μ to state ν is

$$E_\nu - E_\mu = 2J(m - n). \tag{3.2}$$

Thus, the appropriate generalization of the acceptance ratio relation [Eq. (2.3)] is

$$\frac{A(\mu \to \nu)}{A(\nu \to \mu)} = \left[e^{2\beta J} \frac{1 - P_{\text{add}}(-J)}{1 - P_{\text{add}}(J)} \right]^{n-m}. \tag{3.3}$$

Any choice of acceptance ratios $A(\mu \to \nu)$ and $A(\nu \to \mu)$ which satisfies this relation will satisfy detailed balance. For the Wolff choice of P_{add}, we get acceptance ratios that are always unity, but Niedermayer pointed out that there are other ways to achieve this. In fact, all we need to do is choose P_{add} to satisfy

$$\frac{1 - P_{\text{add}}(-J)}{1 - P_{\text{add}}(J)} = e^{-2\beta J} \tag{3.4}$$

and we will get acceptance ratios which are always one. Niedermayer's solution to this equation was $P_{\text{add}}(E) = 1 - \exp[\beta(E_0 - E)]$ where E_0 is a free parameter whose value we can choose as we like. Note, however, that this quantity is supposed to be a probability for making a bond, so it is not allowed to be less than zero. Thus the best expression we can write for the probability $P_{\text{add}}^{(ij)}$ of adding a link between sites i and j is

$$P_{\text{add}}(E_{ij}) = \begin{cases} 1 - e^{\beta(E_0 - E_{ij})} & \text{if } E_{ij} > E_0 \\ 0 & \text{otherwise.} \end{cases} \tag{3.5}$$

And this defines Niedermayer's algorithm. In Figure 7, we have plotted P_{add} as a function of E for one particular choice of the constant E_0.

Notice the following things about this algorithm:

1. As long as all E_{ij} on the lattice are greater than or equal to E_0, the right-hand side of Eq. (3.3) is always unity, so the two acceptance ratios can

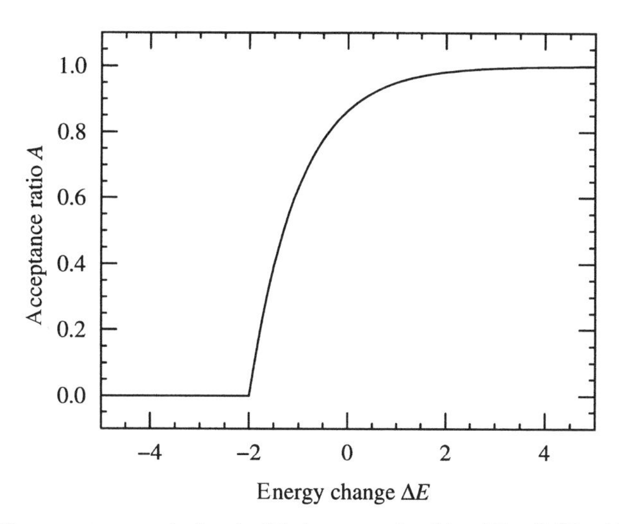

Figure 7. The acceptance ratio for the Niedermayer algorithm [Eq. (3.5)], with $\beta = 1$ and $E_0 = -2$ as a function of energy.

be chosen to be one for every move. Since we are at liberty to choose E_0 however we like, we can always choose it to satisfy this condition by making it less than or equal to the smallest value that E_{ij} can take, which is $-J$ in the case of the Ising model. This gives us a whole spectrum of Wolff-type algorithms for various values $E_0 \leq -J$, which all have acceptance ratios of one. As E_0 becomes increasingly negative, the probabilities: $P_{\text{add}}(\pm J)$ tend closer and closer to one, making the clusters formed larger and larger. This gives us a way of controlling the sizes of the clusters formed in our algorithm, all the way up to clusters which encompass (almost) every spin on the lattice at every move.

2. If we choose E_0 be greater than the smallest possible value of E_{ij} (which is $-J$ in the Ising case), then the right-hand side of Eq. (3.3) is no longer equal to one, and we can no longer choose the acceptance ratios to be unity. Instead, we have

$$\frac{A(\mu \to \nu)}{A(\nu \to \mu)} = [e^{2\beta J} e^{\beta(E_0 - J)}]^{n-m} = [e^{2\beta(E_0 + J)}]^{n-m}. \tag{3.6}$$

Just as with the Metropolis algorithm, the optimal choice of the two acceptance ratios is then to make the larger of the two equal to one, and choose the smaller to satisfy this equation. If we do this, we again achieve detailed balance, and we now have an algorithm which, as E_0 is made larger and larger, produces smaller and smaller clusters, although it does so at the expense of an exponentially decreasing acceptance ratio for cluster moves which increase the energy of the system.

3. If E_0 is chosen to be larger than the largest possible value of E_{ij}, which is $+J$, then $P_{\text{add}}^{(ij)} = 0$ for all pairs of spins i, j, so every cluster formed has only one spin in it and the acceptance ratios are given by

$$\frac{A(\mu \to \nu)}{A(\nu \to \mu)} = [e^{2\beta J}]^{n-m}. \tag{3.7}$$

Bearing in mind that m is the number of neighbors of this single spin which are pointing in the same direction as it is, and n is the number which point in the opposite direction, we can see that this is exactly the same as the acceptance ratio for the Metropolis algorithm.

Thus we see that by varying the parameter E_0, Niedermayer's cluster algorithm includes as special cases both the Metropolis algorithm and the Wolff algorithm, and interpolates smoothly from one to the other and beyond, varying the average cluster size from one spin all the way up to the entire lattice. The trouble with the algorithm is that no one really knows what value one should choose for E_0. Niedermayer himself conjectured that the Wolff choice $E_0 = -J$ might not give the optimal correlation time and that shorter ones could be achieved by making other choices. He gave preliminary evidence in his 1988 paper [11] that, in some cases at least, a larger value of E_0 than this (i.e., a value which produces smaller clusters on average, at the expense of a lowered acceptance probability) gives a shorter correlation time. However, to our knowledge no one has performed an extensive study of the dynamic exponent of the algorithm as a function of E_0, so for the moment at least, the algorithm remains an interesting extension of the Wolff idea which has yet to find use in any large-scale simulations.

C. The Limited-Cluster Algorithm

An alternative technique for varying the sizes of the clusters flipped by the Wolff algorithm has been studied by Barkema and Marko [12], who proposed a general method for placing constraints on the clusters generated in the Wolff algorithm. In their algorithm, cluster growth proceeds just as in the normal Wolff algorithm, except when the addition of a site would violate some condition that we wish to preserve. For example, we might want to place a limit on the total number of spins in the cluster, or limit the radius of the cluster around the initial seed spin. If the addition of a certain spin would violate such a condition, we simply make the probability of adding it to the cluster zero, instead of the usual P_{add}. On its own, this procedure would violate detailed balance since it alters the selection probability for the move $g(\mu \to \nu)$ away from its carefully chosen Wolff value. However, the requirement of detailed balance only places a condition on the product of the selection probability for a move and the acceptance ratio. Thus it is possible to compensate for a change in $g(\mu \to \nu)$ by making an opposing change to the acceptance ratio $A(\mu \to \nu)$. By doing this, detailed balance is restored and the desired condition is met, albeit at the expense of changing the acceptance ratio from the optimal value of 1 which it has in the Wolff algorithm.

In detail, the method works like this. Consider the two states μ and ν, and suppose that for a move which takes us from μ to ν there are m bonds which have to be broken in order to flip the cluster. These broken bonds represent correctly oriented spins which are not added to the cluster by the algorithm. For most of these broken bonds, the probability of not adding such a spin is $1 - P_{add}$ as in the usual Wolff algorithm, but there will now be some number m_0 that were rejected because of the applied constraint and thus had no chance of being added. Thus the probability of not adding all the broken bonds, which is proportional to the selection probability $g(\mu \to \nu)$ for the forward move, is $(1 - P_{add})^{m - m_0}$. If n bonds are broken in making the reverse move from ν to μ, n_0 of which had no chance of being added since they violated our constraint, then the selection probability $g(\nu \to \mu)$ will be proportional to $(1 - P_{add})^{n - n_0}$. The ratio of the forward and backward selection probabilities is thus the same as in the Wolff algorithm, except for a factor of $(1 - P_{add})^{m_0 - n_0}$. In order to restore detailed balance, we compensate for this factor by introducing an acceptance ratio of

$$A(\mu \to \nu) = \begin{cases} e^{-\beta J(m_0 - n_0)} & \text{if } m_0 - n_0 > 0 \\ 1 & \text{otherwise.} \end{cases} \tag{3.8}$$

What is the purpose of introducing constraints such as these? In the context of the normal Ising model, a constraint can be enforced to limit the cluster to a certain region. This allows us to divide the lattice into a number of regions, in each of which cluster flips can be performed independently of the others. This leads to an efficient algorithm for simulating the model on a parallel (distributed) computer which shows near-linear speedup in the number of processors employed [13]. Periodically, the division of the lattice has to be rearranged in order to obtain ergodicity.

Another application of the idea of cluster constraints can be found in the simulation of the conserved-order-parameter Ising model. Clusters of up- and down-pointing spins are grown with a hard constraint on their mass, and if both clusters reach the same mass, an exchange of the spins is proposed.

Size constraints have also proved useful in the simulation of the random-field Ising model (RFIM). The RFIM is normally simulated using a single-spin-flip Metropolis algorithm. It is also possible to apply the Wolff algorithm to the problem by growing Wolff clusters ignoring the random fields and incorporating their energy contribution into an acceptance ratio for the cluster flip [14]. However, in practice, this algorithm works poorly in the critical region of the model because the RFIM has a critical temperature lower than that of the normal Ising model. This means that the clusters grown are nearly always large percolating clusters, which have little chance of being flipped over because the random fields pin them in one direction or the other. This problem can be overcome if the size of the cluster is limited to a radius r, which is chosen at random on each step from a distribution as $1/r^2$ [15].

D. Multigrid Methods

Also worthy of mention is a class of methods developed by Kandel and others [16,17], which are referred to as *multigrid methods*. These methods are also aimed at reducing critical slowing down and accelerating simulations at or close to the critical temperature. Multigrid methods have not been used to a very great extent in large-scale Monte Carlo simulations because they are considerably more complex to program than the cluster algorithms discussed above. However, they may yet prove useful in some contexts because they appear to be faster than cluster algorithms for simulating very large systems.

The fundamental idea behind these methods is the observation that, with the divergence of the correlation length at the critical temperature, we expect to see fluctuating domains of spins of all sizes up to the size of the entire lattice. The multigrid methods therefore split the CPU time of the simulation up, spending varying amounts of time flipping blocks of spins of various sizes. This idea certainly has something in common with the ideas behind the Wolff and Swendsen–Wang algorithms, but the multigrid methods are more deliberate about flipping blocks of certain sizes, rather than allowing the sizes to be determined by the temperature and configuration of the lattice. Kandel and his co-workers gave a number of different, similar algorithms, which all fall under the umbrella of multigrid methods. Here, we describe one example, which is probably the simplest and most efficient such algorithm for the Ising model. The algorithm works by grouping the spins on the lattice into blocks and then treating the blocks as single spins and flipping them using a Metropolis algorithm. In detail, what one does is this.

In the Swendsen–Wang algorithm we made "links" between similarly oriented spins, which effectively tied those spins together into a cluster, so that they flipped as one. Any two spins which were not linked, were free to flip separately—there was not even a ferromagnetic interaction between them to encourage them to point in the same direction. In the present multigrid method, two adjacent spins can be in three different configurations: they can be linked as in the Swendsen–Wang case so that they must flip together, they can have no connection between them at all so that they can flip however they like, or they can have a normal Ising interaction between them of strength J which encourages them energetically to point in the same direction, but does not force them to as our links do. By choosing one of these three states for every pair of spins, the lattice is divided up into clusters of linked spins which either have interactions between them, or which are free to flip however they like. The algorithm is contrived so that only clusters of one or two spins are created. No clusters larger than two spins appear. The procedure for dividing up the lattice goes like this.

1. We take a spin on the lattice, and examine each of its neighbors in turn. If a neighbor is pointing in the opposite direction to the spin, then we leave it alone. In Kandel's terminology, we "delete" the bond between the two spins, so that they are free to assume the same or different directions with no energy cost. If a neighbor is pointing in the same direction as our spin, then we make a link between the two with the same probability $P_{\text{add}} = 1 - \mathrm{e}^{-2\beta J}$ as we used in the Wolff and Swendsen–Wang algorithms. Kandel calls this "freezing" the bond between the spins.
2. Since we want to create clusters of only two spins at most, we stop looking at neighbors once we have created a link to any one of them. In fact, what we do is to keep the normal Ising interactions between the spin and all the remaining neighbors that we have not yet looked at. We do this regardless of whether they are pointing in the same direction as our first spin. Kandel describes this as "leaving the bond active."
3. Now we move onto another spin and do the same thing, and in this way cover the entire lattice. Note that, if we come to a spin which is adjacent to one we have considered before, some of the spin's bonds will already have been frozen, deleted, or marked as active. In this case we leave those bonds as they are, and only go to work on the others which have not yet been considered. Notice also that if we come to a spin and it has already been linked ("frozen") to another spin, then we know immediately that we need to leave the interactions on all the remaining bonds to that spin active.

In this way, we decide the fate of all the spins on the lattice, dividing them into clusters of one or two, joined by bonds which may or may not have interactions associated with them. Then we treat those clusters as single spins, and we carry out the Metropolis algorithm on them, for a few sweeps of the lattice.

But this is not the end. Now we do the whole procedure again, treating the clusters as spins, and joining them into bigger clusters of either one or two elements each, using exactly the same rules as before. (Note that the lattice of clusters is not a regular lattice, as the original system was, but this does not stop us from carrying out the procedure just as before.) Then we do a few Metropolis sweeps of this coarser lattice too. And we keep repeating the whole thing until the size of the blocks reaches the size of the whole lattice. In this way, we get to flip blocks of spins of all sizes from single spins right up to the size of the entire system. Then we start taking the blocks apart again into the blocks that made them up, and so forth until we get back to the lattice of single spins again. In fact, Kandel and co-workers used a scheme where at each level in the blocking procedure they either went towards bigger blocks ("coarsening") or smaller ones ("uncoarsening") according to the following rule. At any particular level of the procedure we look back and see what we did the previous times we got to this level. If we coarsened the lattice the previous 2 times we got to this point, then on the third time, we uncoarsen. This choice has the effect of biasing the algorithm toward working more at the long length scales (bigger, coarser blocks).

Well, perhaps you can see why the complexity of this algorithm has put people off using it. The proof that the algorithm satisfies detailed balance is even more involved, and, since you're probably not dying to hear about it right now, we'll refer you to the original paper for the details [17]. In the same paper it is demonstrated that the dynamic exponent for the algorithm is in the region of 0.2 for the two-dimensional Ising model—a value similar to that of the Wolff algorithm. Before we dismiss the multigrid method out of hand, however, let us point out that the simulations do indicate that its performance is superior to cluster algorithms for large systems. These days, with increasing computer power, people are pushing simulations toward larger and larger lattices, and there may well come a point at which using a multigrid method could win us a significant speed advantage.

E. The Invaded-Cluster Algorithm

Finally, in our round-up of Monte Carlo algorithms for the Ising model, we come to an unusual algorithm proposed by Jon Machta and co-workers called the "invaded-cluster algorithm" [18]. The thing which sets this algorithm apart from the others we have looked at so far is that it is not a general purpose algorithm for simulating the Ising model at any temperature. In fact, the invaded-cluster algorithm can only be used to simulate the model at the critical point; it does not work at any other temperature. What's more, for a system at the critical point in the thermodynamic limit the algorithm is just the same as the Swendsen–Wang cluster algorithm of Section III.A. So what's the point of the algorithm? Well, there are two points. First, the invaded-cluster algorithm can find the critical point all on its own—we don't need to know what the critical temperature is beforehand in order to use the algorithm. Starting with a system at any temperature, the algorithm will adjust its simulation temperature until it finds the critical value T_c. This makes the algorithm very useful for actually measuring the critical temperature, something which normally requires either finite-size scaling or Monte Carlo RG. Second, the invaded cluster algorithm equilibrates extremely quickly. Although the algorithm is equivalent to the Swendsen–Wang algorithm once it reaches equilibrium at T_c, its behavior while getting there is very different, and in a direct comparison of equilibration times between the two, say for systems starting at $T = 0$, there is really no contest. For large system sizes, the invaded-cluster algorithm can reach the critical point as much as a hundred times faster than the Swendsen–Wang algorithm. (A comparison with the Wolff algorithm yields similar results—the performance of the Wolff and Swendsen–Wang algorithms is comparable.)

So, how does the invaded cluster algorithm work? Basically, it is just a variation of the Swendsen–Wang algorithm in which the temperature is continually adjusted to look for the critical point. The algorithm finds the fraction of links at which a percolating backbone of spins first forms across the lattice and uses this fraction to make successively better approximations to the critical temperature at each Monte Carlo step. In detail, here's how it goes:

1. We choose some starting configuration of the spins. It doesn't matter what choice we make, but it could, for example, be the $T = 0$ state in which all spins are aligned.
2. We add links at random between pairs of similarly oriented nearest-neighbor sites until one of the clusters formed reaches percolation.
3. Once the links are made, we flip each cluster on the lattice separately with probability $\frac{1}{2}$, just as we do in the normal Swendsen–Wang algorithm. Then the whole procedure is repeated from step 2 again.

Why does this algorithm work? Well, for any given configuration of the lattice, the maximum number of links which can be made between spins is equal to the number of pairs of similarly oriented nearest-neighbour spins. Let us call this number n. The number of links m needed to achieve the percolation of one cluster will normally be smaller than this maximum value. A given step of the algorithm is equivalent to a single step of the Swendsen–Wang algorithm with $P_{add} = m/n$. (The number of links m required for percolation is fairly constant, and most of variation in P_{add} comes from n.) Substituting this value into Eq. (2.4) and rearranging, we can then calculate the effective temperature of the Monte Carlo step:

$$kT_{eff} = -\frac{2J}{\log(1 - P_{add})} = -\frac{2J}{\log(1 - m/n)}. \tag{3.9}$$

Consider what happens if the system is below the critical temperature. In this case, the spins are more likely to be aligned with their neighbors than they are at the critical temperature, and n is therefore large. This gives us a value of T_{eff} which is higher than T_c. In other words, when the system is below the critical temperature, the algorithm automatically chooses a temperature $T_{eff} > T_c$ for its Swendsen–Wang procedure. Conversely, if the system is at a temperature above the critical point, neighboring spins are less likely to be aligned with one another than they are at T_c, giving a lower value of n and a correspondingly lower value of T_{eff}. So, for a state of the system above the critical temperature, the algorithm automatically chooses a temperature $T_{eff} < T_c$ for its Swendsen–Wang procedure.

The algorithm therefore has a kind of negative feedback built into it, which always drives the system toward the critical point, and when it finally reaches T_c, it will stay there, performing Swendsen–Wang Monte Carlo steps at the critical temperature for the rest of the simulation. Note that we do not need to know what the critical temperature is for the algorithm to work. It finds T_c all on its own, and for that reason the algorithm is a good way of measuring T_c. The same feedback mechanism also drives the algorithm toward T_c faster than simply performing a string of Swendsen–Wang Monte Carlo steps exactly at T_c. This point is discussed in more detail below.

Before we review the result of simulations using the invaded-cluster algorithm, let us briefly examine its implementation. By and large, the algorithm is straightforward to program, but some subtlety arises when we get to the part about testing for a percolating cluster. Since we need to do this every time a link is added to the lattice, it is important that we find an efficient test.

There are various criteria one can apply to test for percolation on a finite lattice. Here we use one of the simplest: we assume a cluster to be percolating when it "wraps around" the periodic boundary conditions on the lattice. In other words, the cluster percolates when there is a path from any site leading across the lattice and back to the same site which has an integrated displacement which is nonzero. This test can be efficiently implemented as follows.

All sites belonging to the same cluster are organized into a tree structure in which there is one chosen "reference site." All the other sites in the cluster are arranged in a hierarchy below this reference site. The immediate daughter sites of the reference site possess vectors which give their position relative to the reference site. Other sites are arranged as granddaughter sites, with vectors giving their position relative to one of the daughter sites, and so on until every site in the cluster is accounted for. The vector linking any site to the reference site can be calculated by simply ascending the tree from bottom to top, adding the vectors at each level until the reference site is reached.

The particular way in which the sites in a cluster are arranged on the tree is dictated by the way in which the cluster grows. At the start of each Monte Carlo step no links exist and each site is a reference site. Each time a link is added, we must either join together two separate clusters to make one larger cluster, or we join together two sites which already belong to the same cluster. Since our criterion for percolation is the appearance of a cluster which wraps around the boundary conditions, it follows that percolation can only appear on moves of the latter type. In other words, we need only check for percolation when we add a link joining two sites which are already members of the same cluster. The procedure is therefore as follows. Each time we add a link, we use the tree structures to determine the reference site of each of the two sites involved. These reference sites tell us which clusters the sites belong to. When a link is made between two sites belonging to different clusters, we have to combine the two clusters into one, which we do by making one of the reference sites the daughter of the other. When we make a link between sites in the same cluster, we don't need to combine two clusters, but we do need to check for percolation. To do this, we take the lattice vectors from each of the two sites to the reference site of the cluster and subtract them. Normally, this procedure should result in a difference vector of length one—the distance between two nearest-neighbour spins. However, if the addition of a link between these two spins causes the cluster to wrap around the boundary conditions, the difference of the two vectors will instead be a unit vector *plus* a vector of length the dimension of the lattice. In this case we declare the cluster to the percolating, and stop adding links.

The results for the invaded-cluster algorithm are impressive. Machta et al. [18] found equilibration times 20 or more times faster than the Swendsen–Wang algorithm at T_c for the two- and three-dimensional Ising systems they examined, and of course, the invaded-cluster algorithm allowed them to measure the value of T_c, which is not directly possible with the normal Swendsen–Wang algorithm. Figure 8 shows their results for the critical temperature of a set of 2D Ising systems of different sizes. Because there is always a finite chance of producing a percolating cluster on a finite lattice with any value of P_{add}, no matter how small (as long as it's not zero), the average estimate of T_c which the algorithm makes on a finite system tends to be a little high. However, we can easily extrapolate to the limit $L = \infty$ using finite-size scaling. In this case we do this by plotting the measured T_c as a function of L^{-1} and extrapolating to the $L^{-1} = 0$ axis. The result is $kT_c = 2.271 \pm 0.002$, which compares favorably with the known exact result of $kT_c = 2.269$, especially given the small amount of CPU time taken by the simulation. (The runs were 10,000 Monte Carlo steps, for each system size.)

The invaded-cluster algorithm can also be used to measure other quantities at the critical temperature—magnetization, for example, or internal energy. However, a word of caution is in order here. For lattices of finite size—which, of course includes all the lattices in our Monte Carlo simulations—the invaded-cluster algorithm does not sample the Boltzmann distribution exactly. In particular, the fluctuations in quantities measured using the algorithm are different from those you would get in the Boltzmann distribution. To see this, consider what happens once the algorithm has equilibrated to the critical temperature. At this point, as we argued

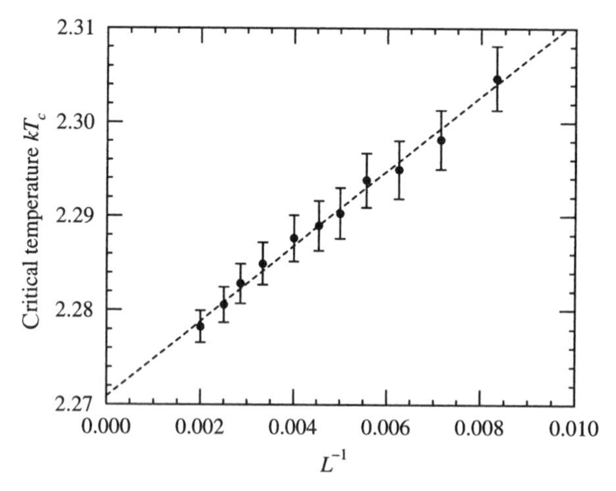

Figure 8. The critical temperature of the two-dimensional Ising model measured using the invaded cluster algorithm for systems of a variety of sizes from $L = 120$ up to $L = 500$. Here they are plotted against L^{-1}, and the extrapolation to $L = \infty$ gives an estimate of $kT_c = 2.271 \pm 0.002$ for the critical temperature in the thermodynamic limit. The data are taken from Machta et al. [18].

before, it should stop changing the temperature and just become equivalent to the Swendsen–Wang algorithm at T_c, which, as we know, certainly samples the Boltzmann distribution correctly. However, in fact, because the lattice is finite, statistical variations in the order in which we generate the links on the lattice will give rise to variations in the temperature T of successive steps in the simulation. The negative feedback effect that we described above will ensure that T always remains close to T_c, but the size of fluctuations is very sensitive to small changes in temperature near T_c and as a result the measured fluctuations are not a good approximation to those of the true Boltzmann distribution. Thus the invaded-cluster algorithm is not suitable for measuring, for instance, the magnetic susceptibility χ or the specific heat C of the Ising model at T_c, both of which are determined by measuring fluctuations. On the other hand, one could use the algorithm to determine the value of T_c, and then use the normal Wolff or Swendsen–Wang algorithm to perform a simulation at that temperature to measure χ or C.

IV. MANY-VALUED AND CONTINUOUS-SPIN MODELS

Cluster Monte Carlo algorithms are applicable not only to the Ising model. They also work for other types of spin models. In this section and the next we consider how they can be modified to work with Potts models and continuous-spin models.

A. Potts Models

Potts models suffer from critical slowing down in just the same way as the Ising model. All the algorithms discussed in this chapter can be generalized to Potts models. Here we discuss the example of the Wolff algorithm, whose appropriate generalization is as follows:

1. Choose a seed spin at random from the lattice.
2. Look in turn at each of the neighbors of that spin. If they have the same value as the seed spin, add them to the cluster with probability $P_{\text{add}} = 1 - e^{-\beta J}$. (Note that the 2 has vanished from the exponent; the two-state Potts model, for example, is equivalent to the Ising model except for a factor two in the interaction constant J.)
3. For each spin that was added in the last step, examine each of *its* neighbors to find which ones, if any, have the same value and add each of them to the cluster with the same probability P_{add}. (As with the Ising model, we notice that some of the neighbors may already be members of the cluster, in which case you don't have to consider adding them again. Also, some of the spins may have been considered

for addition before, as neighbors of other spins in the cluster, but rejected. In this case, they get another chance to be added to the cluster on this step.) This step is repeated as many times as necessary until there are no spins left in the cluster whose neighbours have not been considered for inclusion in the cluster.

4. Choose at random a new value for the spins in the cluster, different from the present value, and set all the spins to that new value.

The proof that this algorithm satisfies detailed balance is exactly the same as it was for the Ising model. If we consider two states μ and ν of the system which differ by the changing of just one cluster, then the ratio $g(\mu \to \nu)/g(\nu \to \mu)$ of the selection probabilities for the moves between these states depends only on the number of bonds broken m and the number made n around the edges of the cluster. This gives us an equation of detailed balance, which reads

$$\begin{aligned}\frac{g(\mu \to \nu)A(\mu \to \nu)}{g(\nu \to \mu)A(\nu \to \mu)} &= (1 - P_{\text{add}})^{m-n}\frac{A(\mu \to \nu)}{A(\nu \to \mu)} \\ &= e^{-\beta(E_\nu - E_\mu)}\end{aligned} \tag{4.1}$$

just as in the Ising case. The change in energy is also given by the same expression as before, except for a factor of 2:

$$E_\nu - E_\mu = J(m - n) \tag{4.2}$$

and so the ratio of the acceptance ratios $A(\mu \to \nu)$ and $A(\nu \to \mu)$ for the two moves is

$$\frac{A(\mu \to \nu)}{A(\nu \to \mu)} = [e^{\beta J}(1 - P_{\text{add}})]^{n-m}. \tag{4.3}$$

For the choice of P_{add} given above, this is just equal to one. Equation (4.3) is satisfied by making the acceptance ratios equal to 1, and with this choice the algorithm satisfies detailed balance.

As in the case of the Ising model, the Wolff algorithm gives an impressive improvement in performance near to the critical temperature. Arguments similar to those of Bausch et al. [1] indicate that the dynamic exponent of the Metropolis algorithm for a Potts model should have a lower bound of 2. By contrast, Baillie and Coddington [19] have measured a dynamic exponent of $z = 0.60 \pm 0.02$ for the Wolff algorithm in the 2D, $q = 3$ case. As before, the single-spin-flip algorithms come into their own well away from the critical point, because critical slowing down ceases to be a problem and the relative simplicity of these algorithms over the Wolff algorithm tends to give them the edge. One's choice of algorithm should therefore (as always) depend on exactly which properties of the model one wants to investigate, but the Wolff algorithm is definitely a good choice for examining critical properties.

The Swendsen–Wang algorithm can also be generalized for use with Potts models in a very simple fashion, as can all of the other algorithms described in Section III.

B. Continuous-Spin Models

Cluster algorithms can also be generalized to models with continuous spins. Let us take the XY model as our example. A version of the Wolff algorithm for this model was given by Wolff in his original paper in 1989. It also works for similar models in higher dimensions. The idea is a simple one—one chooses at random a seed spin to start a cluster and a direction vector $\hat{\mathbf{n}}$.[3] Then we treat the components $\hat{\mathbf{n}} \cdot \mathbf{s}_i$ of the spins in that direction roughly in the same way as we did the spins in the Ising model. A neighbor of the seed spin whose component in this direction has the same sign as that of the seed spin can be added to cluster by making a link between it and the seed spin with some probability P_{add}. If the components point in opposite directions then the spin is not added to the cluster. When the complete cluster has been built, it is "flipped" by reflecting all the spins in the plane perpendicular to $\hat{\mathbf{n}}$.

The only complicating factor is that, in order to satisfy detailed balance, the expression for P_{add} has to depend on the values of the spins which are joined by links thus:

$$P_{\text{add}}(\mathbf{s}_i, \mathbf{s}_j) = 1 - \exp[-2\beta(\hat{\mathbf{n}} \cdot \mathbf{s}_i)(\hat{\mathbf{n}} \cdot \mathbf{s}_j)]. \tag{4.4}$$

Readers may like to demonstrate for themselves that, with this choice, the ratio of the selection probabilities $g(\mu \to \nu)$ and $g(\nu \to \mu)$ is equal to $e^{-\beta\Delta E}$, where ΔE is the change in energy in going from a state μ to a state ν by flipping a single cluster. Thus, detailed balance is obeyed as in the Ising case by an algorithm for which the acceptance probability for the cluster flip is 1. This algorithm has been used, for example, by Gottlob and Hasenbusch [20] to perform extensive studies of the critical properties of the Heisenberg model.

[3] In two dimensions this is easy, we simply choose an angle between 0 and 2π to represent the direction of the vector. In three dimensions we need to choose both a ϕ and a θ. Angle ϕ is again uniformly distributed between 0 and 2π, and θ is given by $\theta = \cos^{-1}(1 - 2r)$, where r is random real number uniformly distributed between zero and one.

Similar generalizations to continuous spins are possible for all the algorithms discussed in Section III.

V. CONCLUSIONS

We have discussed a number of recently developed cluster algorithms for the simulation of classical spin systems, including the Wolff algorithm, which flips single clusters of spins at each update step and almost completely eliminates critical slowing down from the simulation of the Ising model; the Swendsen–Wang algorithm, which updates the entire lattice at each step and has performance similar to, though not quite as good as, the Wolff algorithm; Neidermayer's algorithm, a variation of the Wolff algorithm which allows one to tune the size of clusters flipped; the limited cluster algorithm, another technique for controlling the sizes of clusters; multigrid methods, which may be even more efficient close to criticality than the Wolff algorithm, although they achieve this at the expense of considerable programming complexity; and the invaded-cluster algorithm, which self-organizes to the critical point of the system studied and permits accurate measurements of T_c to be made with little computational effort. We have also discussed briefly how algorithms such as these can be generalized to Potts models and continuous-spin models.

ACKNOWLEDGMENTS

The authors would like to thank Eytan Domany, Jon Machta and Alan Sokal for useful discussions about cluster algorithms, and Daniel Kandel for supplying a copy of his Ph.D. thesis.

REFERENCES

1. R. Bausch, V. Dohm, H. K. Janssen, and R. K. P. Zia, *Phys. Rev. Lett.* **47**, 1837 (1981).
2. M. P. Nightingale and H. W. J. Blöte, *Phys. Rev. Lett.* **76**, 4548 (1996).
3. U. Wolff, *Phys. Rev. Lett.* **62**, 361 (1989).
4. R. H. Swendsen and J.-S. Wang, *Phys. Rev. Lett.* **58**, 86 (1987).
5. P. D. Coddington and C. F. Baillie, *Phys. Rev. Lett.* **68**, 962 (1992).
6. L. Onsager, *Phys. Rev.* **65**, 117 (1944).
7. C. M. Fortuin and P. W. Kasteleyn, *Physica* **57**, 536 (1972).
8. M. Sweeny, *Phys. Rev. B*, **27**, 4445 (1983).
9. A. M. Ferrenberg, D. P. Landau, and Y. J. Wong, *Phys. Rev. Lett.* **69**, 3382 (1992).
10. R. Matz, D. L. Hunter, and N. Jan, *J. Stat. Phys.* **74**, 903 (1994).
11. F. Niedermayer, *Phys. Rev. Lett.* **61**, 2026 (1988).
12. G. T. Barkema and J. F. Marko, *Phys. Rev. Lett.* **71**, 2070 (1993).
13. G. T. Barkema and T. MacFarland, *Phys. Rev. E* **50**, 1623 (1994).

14. Vl. S. Dotsenko, W. Selke, and A. L. Talapov, *Physica A* **170**, 278 (1991).
15. M. E. J. Newman and G. T. Barkema, *Phys. Rev. E* **53**, 393 (1996).
16. D. Kandel, Ph.D. thesis, Weizmann Institute of Science, Rehovot, Israel, 1991.
17. D. Kandel, E. Domany, and A. Brandt, *Phys. Rev. B* **40**, 330 (1989).
18. J. Machta, Y. S. Choi, A. Lucke, T. Schweizer, and L. V. Chayes, *Phys. Rev. Lett.* **75**, 2792 (1995).
19. C. F. Baillie and P. D. Coddington, *Phys. Rev. B* **43**, 10617 (1991).
20. A. P. Gottlob and M. Hasenbusch, *Physica A* **201**, 593 (1993).

AUTHOR INDEX

Numbers in parentheses are reference numbers and indicate that the author's work is referred to although his name is not mentioned in the text. Numbers in *italic* show the pages on which the complete references are listed.

SUBJECT INDEX